Purinergic Signaling in Neuroinflammation

Purinergic Signaling in Neuroinflammation

Editors

Dmitry Aminin

Peter Illes

MDPI • Basel • Beijing • Wuhan • Barcelona • Belgrade • Manchester • Tokyo • Cluj • Tianjin

Editors
Dmitry Aminin
Lab of Bioassay
G.B. Elyakov Pacific Institute
of Bioorganic Chemistry
Far Eastern Branch of the
Russian Academy of Sciences
Vladivotok
Russia

Peter Illes
Rudolf-Boehm-Institut für
Pharmakologie und
Toxikologie
Universitaet Leipzig
Leipzig
Germany

Editorial Office
MDPI
St. Alban-Anlage 66
4052 Basel, Switzerland

This is a reprint of articles from the Special Issue published online in the open access journal *International Journal of Molecular Sciences* (ISSN 1422-0067) (available at: www.mdpi.com/journal/ijms/special_issues/Purinergic_Inflammation).

For citation purposes, cite each article independently as indicated on the article page online and as indicated below:

LastName, A.A.; LastName, B.B.; LastName, C.C. Article Title. *Journal Name* **Year**, *Volume Number*, Page Range.

ISBN 978-3-0365-7687-9 (Hbk)
ISBN 978-3-0365-7686-2 (PDF)

Contents

Preface to "Purinergic Signaling in Neuroinflammation"

It is currently apparent that the extracellular ATP's physiological effect is mediated by its interaction with specific purinergic receptors. All purinergic receptors are divided into P1-purinoreceptors (the main ligand adenosine) and P2-purinoreceptors (the main ligands ATP/ADP, UTP/UDP). Each of the subtypes is divided into a number of families. For instance, P2 receptors are divided into P2X and P2Y receptors according to the mechanism by which their effect is realized: P2Y are G-protein-coupled receptors, while P2X receptors are ligand-operated ion channels (or ionotropic receptors). P2X receptors are important molecular therapeutic targets, the malfunctioning of which leads to severe complications in the physiology of humans and animals and causes dangerous diseases. The search for compounds that can modulate the function of purinergic receptors can lead to the creation of new drugs that are effective in central and peripheral nervous system and immune system disease treatment, including neuroinflammation, hypoxia/ischemia, epilepsy and neuropathic pain. In this Special Issue, we wish to offer a platform for high-quality publications on the latest advances in the identification of P2X/Y- and P1 (A1, A2A, A2B, A3)-receptor blockers, functions and regulation by them; the characterization of these receptor signaling networks and crosstalk; mechanisms underlying the role of purinoceptors in neurodegenerative illnesses as well as chronic neuronal changes following acute noxious damage and therapeutic opportunities associated with regulation of purinergic receptor activity. This issue will be of interest to researchers working on cell signaling, neurology and immunology, and also to chemical biologists interested in drug discovery as well as clinicians.

Dmitry Aminin and Peter Illes
Editors

International Journal of
Molecular Sciences

MDPI

Editorial

Purinergic Signaling in Neuroinflammation

Dmitry Aminin [1,2] and Peter Illes [3,4,*]

[1]   G.B. Elyakov Pacific Institute of Bioorganic Chemistry, Far Eastern Branch of the Russian Academy of Sciences, 690022 Vladivostok, Russia; daminin@piboc.dvo.ru

[2]   Department of Biomedical Science and Environmental Biology, Kaohsiung Medical University, Kaohsiung City 80708, Taiwan

[3]   Rudolf Boehm Institute for Pharmacology and Toxicology, University of Leipzig, 04107 Leipzig, Germany

[4]   International Collaborative Center on Big Science Plan for Purine Signaling, Chengdu University of TCM, Chengdu 610075, China

*   Correspondence: peter.illes@medizin.uni-leipzig.de

Citation: Aminin, D.; Illes, P. Purinergic Signaling in Neuroinflammation. *Int. J. Mol. Sci.* **2021**, *22*, 12895. https://doi.org/10.3390/ijms222312895

Received: 23 November 2021
Accepted: 25 November 2021
Published: 29 November 2021

Publisher's Note: MDPI stays neutral with regard to jurisdictional claims in published maps and institutional affiliations.

ATP is stored in millimolar concentrations within the intracellular medium but may be released to extracellular sites either through the damaged plasma membrane or by means of various transporters. Extracellular ATP or its enzymatic breakdown products, ADP, AMP, and adenosine, may then stimulate a range of membrane receptors (Rs). These receptors are classified as belonging to two types termed P2 or P1. P2Rs can be, in addition, subdivided into the ligand-activated P2X and the G protein-coupled P2Y types. Adenosine acts on the P1 type of receptor. A further classification identifies seven mammalian subtypes of P2X1-7 and eight mammalian subtypes of P2YRs (P2Y1, P2Y2, P2Y4, P2Y6, P2Y11, P2Y12, P2Y13, P2Y14). P1Rs are either positively (A2A, A2B) or negatively (A1, A3) coupled to adenylate cyclase via the respective G proteins. Already, such a high number of receptors suggests that purine-mediated effects at the cellular but especially whole organism level have an immense variability. Whereas P2XRs respond only the ATP, P2YRs are sensitive to ATP/ADP, UTP/UDP, or UDP–glucose. Inspection of some articles in this Special Issue will teach us that the nucleoside guanosine probably possesses a receptor of its own, that nucleotides can be gradually degraded metabolically to functionally active nucleotides/nucleosides (see above), and indirect effects by stimulating the synthesis or decomposition of purines/pyrimidines may also increase functional diversity. Eventually, P2/P1Rs may interact both with each other as well as with other neurotransmitter receptors. It is, of course, important to note that, in many cases, receptor (sub)type-preferential agonists and highly selective antagonists are available for pharmacological analysis.

The fascinating complexity of the "purinome" (the cluster of agonists, receptors, and enzymes participating in purinergic signaling) and the involvement of its components in (patho)physiological functions underline their regulatory importance. Defective P2XRs are causative factors, e.g., in male infertility (P2X1), hearing loss (P2X2), pain/cough (P2X3), neuropathic pain (P2X4), inflammatory bone loss (P2X5), and faulty immune reactions (P2X7). Purinergic signaling may regulate immune reactions through P2/P1Rs situated at immune cells in the periphery (macrophages, lymphocytes) and in the central nervous system (microglia). P2X4,7 and P2Y4,6,12, as well as A2A,3Rs, are located at microglia in the CNS. They steer microglial process motility, microglial migratory activity, microglial phagocytosis (pinocytosis), and the release of pro-inflammatory cytokines, chemokines, growth factors, proteases, reactive oxygen and nitrogen species, cannabinoids, and probably also the secretion of excitotoxic ATP and glutamate. Consequently, the "purinome" is involved in the neurodegenerative illnesses Alzheimer's disease, Parkinson's disease, Huntington's disease, multiple sclerosis, amyotrophic lateral sclerosis, neuropathic pain, post-epilepsy and post-ischemia neuronal damage, etc. These latter findings prompted the two editors of this Special Issue to invite a few experts to contribute their ideas on purinergically induced neuroinflammation. This choice of publications consists of two original articles and eight reviews, providing new insights to our present knowledge.

One of the original papers by Braune et al. [1] deals with the G protein-coupled GPR17, which was initially discovered as an orphan receptor and was found to be a target of both cysteinyl-leukotrienes and the uracil nucleotides uridine, UDP, and UDP–glucose. Montelucast, a selective antagonist of GPR17, largely facilitated the outgrowth of neuronal fiber networks from the substantia nigra/ventral tegmentum to the prefrontal cortex in an organotypic co-culture system. This effect appeared to be due to antagonism of endogenous ligands activating GPR17, because a selective agonist of this receptor, PSB-16484, reversed growth promotion by the GPR17 antagonist Montelucast. Another original article of Sophocleous et al. [2] proved the presence of P2Y2 and P2X4Rs at a canine macrophage cell line. P2Y2Rs mediated the mobilization of Ca^{2+} from its intracellular stores, while the low levels of P2X4Rs might modulate this effect. Canine cells are feasible alternatives to rodent cell systems for drug approval procedures.

The residual review articles deal with ischemia, neonatal seizures, chronic pain, retinal disorders, multiple sclerosis, and osteogenesis/adipogenesis. Coppi et al. [3] report on neuronal damage generated by cerebral ischemia (occlusion of the middle cerebral artery) in the whole animal or in vitro by oxygen–glucose deprivation in hippocampal slices. A2BRs were found to participate in the early glutamate-mediated excitotoxicity responsible for neuronal and synaptic loss in the CA1 hippocampal cells. By contrast, after the ischemic stimulus, the same receptors have protective roles in tissue damage and functional impairment. In this context, Chojnowski et al. [4] referred to the fact that guanosine, which is released under brain ischemia or trauma into the extracellular milieu, counteracted the destructive events occurring during ischemic conditions (e.g., glutamatergic excitotoxicity, reactive oxygen and nitrogen species production). Neonatal seizures are a particularly drug-resistant form of epilepsy. Menéndez Méndez et al. [5] provided data suggesting that P2X7R antagonists, previously investigated in adult epilepsy, have the most promise in neonatal seizures.

Vincenzi et al. [6] contributed extensively to the role of adenosine in pain regulation. Most of the anti-nociceptive effects of adenosine have been found to depend upon A1Rs located at peripheral, spinal, and supra-spinal sites. A2A and A2BRs have been found to be more controversial, since their activation led to both pro- and anti-nociceptive consequences. More recently, allosteric activators have been proposed to improve efficacy and limit side effects of endogenous adenosine. Trapero et al. [7] have addressed endometriosis-associated pain, which depended on inflamed endometrial tissue localized outside the uterine cavity. Altered extracellular ATP hydrolysis, due to changes in ectonucleotidase activity, leads to accumulation of ATP in the endometriosis microenvironment and activates pain-inducing P2X3Rs at sensory neurons. Pharmacological inhibition of this receptor-type appears to be an adequate therapeutic option.

Sidoryk-Wegrzynowicz and Struzyska [8] have dedicated their review to astroglial and microglial purinergic P2X7Rs as major contributors to neuroinflammation in multiple sclerosis. Hársing et al. [9] discussed findings demonstrating that energy deprivation causes in the retina an increased release of the excitotoxic ATP and glutamate, mediated by P2X7 and NMDARs, respectively. P2YR agonists facilitate the uptake of glycine by the glycine transporter 1; the resulting lower extracellular concentrations of the NMDAR co-agonistic glycine reduces neurodegenerative events in the retina. Finally, Eisenstein et al. [10] analyze the dichotomous effects of the two G protein-coupled stimulatory and the two G protein-coupled inhibitory adenosine receptors on adipogenesis and osteogenesis within the bone marrow to maintain bone health, as well as its relationship to obesity.

We are confident that, already, this limited number of papers will emphasize the great importance of purinergic signaling in neuroinflammation and the requirement for therapies directed to the involved receptors, release mechanisms, and metabolizing enzymes. It is our strong conviction that the subject deserves a subsequent Special Issue with the same or even higher number of contributions. For this issue, we intend to invite further specialists to elucidate research fields complementary to the presently discussed ones.

Author Contributions: Conceptualization, planning and design, D.A. and P.I.; original draft and improvement: D.A. and P.I. All authors have read and agreed to the published version of the manuscript.

Funding: This work was supported by the Russian Science Foundation, Grant No. 19-12-00047 and by "The Project First-Class Disciplines Development" Grant No. CZYHW1901 of the Chengdu University of TCM.

Institutional Review Board Statement: Not applicable.

Informed Consent Statement: Not applicable.

Data Availability Statement: The data presented in this study are available upon request from the corresponding authors of the respective publications.

Conflicts of Interest: The authors declare no conflict of interest.

References

1. Braune, M.; Scherf, N.; Heine, C.; Sygnecka, K.; Pillaiyar, T.; Parravicini, C.; Heimrich, B.; Abbracchio, M.P.; Müller, C.E.; Franke, H. Involvement of GPR17 in Neuronal Fibre Outgrowth. *Int. J. Mol. Sci.* **2021**, *22*, 11683. [CrossRef] [PubMed]
2. Sophocleous, R.A.; Miles, N.A.; Ooi, L.; Sluyter, R. P2Y2 and P2X4 Receptors Mediate Ca^{2+} Mobilization in DH82 Canine Macrophage Cells. *Int. J. Mol. Sci.* **2020**, *21*, 8572. [CrossRef] [PubMed]
3. Coppi, E.; Dettori, I.; Cherchi, F.; Bulli, I.; Venturini, M.; Lana, D.; Giovannini, M.G.; Pedata, F.; Pugliese, A.M. A2B Adenosine Receptors: When Outsiders May Become an Attractive Target to Treat Brain Ischemia or Demyelination. *Int. J. Mol. Sci.* **2020**, *21*, 9697. [CrossRef] [PubMed]
4. Chojnowski, K.; Opielka, M.; Nazar, W.; Kowianski, P.; Smolenski, R.T. Neuroprotective Effects of Guanosine in Ischemic Stroke-Small Steps towards Effective Therapy. *Int. J. Mol. Sci.* **2021**, *22*, 6898. [CrossRef] [PubMed]
5. Menéndez Mendez, A.; Smith, J.; Engel, T. Neonatal Seizures and Purinergic Signalling. *Int. J. Mol. Sci.* **2020**, *21*, 7832. [CrossRef] [PubMed]
6. Vincenzi, F.; Pasquini, S.; Borea, P.A.; Varani, K. Targeting Adenosine Receptors: A Potential Pharmacological Avenue for Acute and Chronic Pain. *Int. J. Mol. Sci.* **2020**, *21*, 8710. [CrossRef] [PubMed]
7. Trapero, C.; Martin-Satue, M. Purinergic Signaling in Endometriosis-Associated Pain. *Int. J. Mol. Sci.* **2020**, *21*, 8512. [CrossRef] [PubMed]
8. Sidoryk-Wegrzynowicz, M.; Struzyynska, L. Astroglial and Microglial Purinergic P2X7 Receptor as a Major Contributor to Neuroinflammation during the Course of Multiple Sclerosis. *Int. J. Mol. Sci.* **2021**, *22*, 8404. [CrossRef] [PubMed]
9. Hársing, L.G., Jr.; Szénási, G.; Zelles, T.; Köles, L. Purinergic-Glycinergic Interaction in Neurodegenerative and Neuroinflammatory Disorders of the Retina. *Int. J. Mol. Sci.* **2021**, *22*, 6209. [CrossRef] [PubMed]
10. Eisenstein, A.; Chitalia, S.V.; Ravid, K. Bone Marrow and Adipose Tissue Adenosine Receptors Effect on Osteogenesis and Adipogenesis. *Int. J. Mol. Sci.* **2020**, *21*, 7470. [CrossRef] [PubMed]

International Journal of
Molecular Sciences

Article

Gallic Acid Alleviates Visceral Pain and Depression via Inhibition of P2X7 Receptor

Lequan Wen [1,†], Lirui Tang [1,†], Mingming Zhang [2], Congrui Wang [3], Shujuan Li [3], Yuqing Wen [2], Hongcheng Tu [4], Haokun Tian [1], Jingyi Wei [4], Peiwen Liang [3], Changsen Yang [1], Guodong Li [2] and Yun Gao [2,5,*]

1   Joint Program of Nanchang University and Queen Mary University of London, Nanchang University, 461 Bayi Avenue, Nanchang 330006, China; jp4217118145@qmul.ac.uk (L.W.); jp4217118218@qmul.ac.uk (L.T.); jp4217118147@qmul.ac.uk (H.T.); jp4217119235@qmul.ac.uk (C.Y.)
2   Department of Physiology, Basic Medical College, Nanchang University, 461 Bayi Avenue, Nanchang 330006, China; zhangmm1117@126.com (M.Z.); wenyuqing0211@163.com (Y.W.); gc77li@163.com (G.L.)
3   Second Clinic Medical College, Nanchang University, 461 Bayi Avenue, Nanchang 330006, China; ncusuqyzxwcr@163.com (C.W.); zololo116@163.com (S.L.); lpwhafffe@163.com (P.L.)
4   Basic Medical College, Nanchang University, 461 Bayi Avenue, Nanchang 330006, China; 18322950360@wo.cn (H.T.); v2899001289@163.com (J.W.)
5   Jiangxi Provincial Key Laboratory of Autonomic Nervous Function and Disease, 461 Bayi Avenue, Nanchang 330006, China
*   Correspondence: gaoyun@ncu.edu.cn; Tel.: +86-791-86360586
†   These authors contributed equally to this work.

Citation: Wen, L.; Tang, L.; Zhang, M.; Wang, C.; Li, S.; Wen, Y.; Tu, H.; Tian, H.; Wei, J.; Liang, P.; et al. Gallic Acid Alleviates Visceral Pain and Depression via Inhibition of P2X7 Receptor. *Int. J. Mol. Sci.* **2022**, *23*, 6159. https://doi.org/10.3390/ijms23116159

Academic Editors: Dmitry Aminin and Natalia Gulyaeva

Received: 7 April 2022
Accepted: 30 May 2022
Published: 31 May 2022

Publisher's Note: MDPI stays neutral with regard to jurisdictional claims in published maps and institutional affiliations.

Abstract: Chronic visceral pain can occur in many disorders, the most common of which is irritable bowel syndrome (IBS). Moreover, depression is a frequent comorbidity of chronic visceral pain. The P2X7 receptor is crucial in inflammatory processes and is closely connected to developing pain and depression. Gallic acid, a phenolic acid that can be extracted from traditional Chinese medicine, has been demonstrated to be anti-inflammatory and anti-depressive. In this study, we investigated whether gallic acid could alleviate comorbid visceral pain and depression by reducing the expression of the P2X7 receptor. To this end, the pain thresholds of rats with comorbid visceral pain and depression were gauged using the abdominal withdraw reflex score, whereas the depression level of each rat was quantified using the sucrose preference test, the forced swimming test, and the open field test. The expressions of the P2X7 receptor in the hippocampus, spinal cord, and dorsal root ganglion (DRG) were assessed by Western blotting and quantitative real-time PCR. Furthermore, the distributions of the P2X7 receptor and glial fibrillary acidic protein (GFAP) in the hippocampus and DRG were investigated in immunofluorescent experiments. The expressions of p-ERK1/2 and ERK1/2 were determined using Western blotting. The enzyme-linked immunosorbent assay was utilized to measure the concentrations of IL-1β, TNF-α, and IL-10 in the serum. Our results demonstrate that gallic acid was able to alleviate both pain and depression in the rats under study. Gallic acid also reduced the expressions of the P2X7 receptor and p-ERK1/2 in the hippocampi, spinal cords, and DRGs of these rats. Moreover, gallic acid treatment decreased the serum concentrations of IL-1β and TNF-α, while raising IL-10 levels in these rats. Thus, gallic acid may be an effective novel candidate for the treatment of comorbid visceral pain and depression by inhibiting the expressions of the P2X7 receptor in the hippocampus, spinal cord, and DRG.

Keywords: gallic acid; P2X7 receptor; visceral pain; depression; hippocampus; spinal cord; dorsal root ganglion

1. Introduction

Irritable bowel syndrome (IBS) is a gastrointestinal disorder that is characterized by altered defecation habits, abdominal discomfort, and abdominal pain. It is said that IBS affects the lives of 10–15% of the global population [1]. Since IBS is the most prevalent functional gastrointestinal disorder (FGID) in chronic visceral pain [2–5], we selected it as the investigatory target in this study representing chronic visceral pain. The severity of IBS symptoms varies from person to person, from enervating to mild [6]. Moreover, in IBS patients, long-term neuroplastic changes have occurred in the brain–gut axis, which results in chronic abdominal pain [7,8]. Visceral hyperalgesia may be related not only to peripheral mechanisms within the intestinal wall but also to increased neurotransmitters released in the spinal cord and brain [9]. Aside from the nociceptive symptoms, IBS is commonly accompanied by other intestinal or non-intestinal comorbidities, 20–60% of which involve anxiety or depression [2]. Both mucosal inflammation and neuroinflammation are involved in the pathophysiology of IBS [10], which might be responsible for the comorbidity development of visceral pain and depression. In this study, we constructed the comorbidity of visceral pain and depression of IBS models for research purposes and tried to identify a common effective target for the comorbid visceral pain and depression.

To date, increasing evidence suggests that purinergic receptors are strongly related to visceral hyperalgesia; commonly studied ones are the P2X1, P2X3, P2X2/3, P2X7, P2Y1, and P2Y2 receptors [11–16]. The P2X7 receptor, which is the central factor in the process of inflammation [17], is found to have an enhanced effect in visceral hyperalgesia [15,18]. Furthermore, Antonio et al. observed that the inhibition of the P2X7 receptor expression at the nerve terminals with oxidized ATP could suppress inflammation pain [19]. Additionally, Jarvis proved that the P2X7 receptors in microglia participate in neuropathic pain [20]. It was also suggested that the P2X7 receptors could regulate the production of IL-1β, thus inducing inflammation and neuropathic pain [21,22]. Coincidentally, there is growing evidence showing that the P2X7 receptor is a crucial player in depression. Various studies have demonstrated the central role of the P2X7 receptor in the processes involved in major depressive disorders, such as damaged monoaminergic neurotransmission [23,24], enhanced glutamatergic neurotransmission [25], neuroinflammatory response [26], and repressed neuroplasticity [24,27]. Additionally, our previous study demonstrated the augmentative effect of the P2X7 receptor on comorbid diabetic neuropathic pain and depression [28]. The close association between the P2X7 receptor and inflammation and the fact that IBS patients display both mucosal and neural inflammation motivate us to assume that, by inhibiting the P2X7 receptor, comorbid chronic visceral pain and depression could be alleviated.

Gallic acid (GA), which is found in a wide variety of fruits, nuts, and plants (e.g., rhubarb, eucalyptus, Cornus), is a polyphenol organic compound that is also known as 3,4,5-trihydroxy benzoic acid [29,30]. Its anti-inflammatory effects in various diseases have been demonstrated in many studies [31], for example, diabetes mellitus [32], psoriasis [33], gouty arthritis [34], paraquat-induced renal injury [35], etc. Gallic acid may prevent the production of inflammatory factors downstream of NF-κB, such as IL-1β, TNF-α, and thioredoxin-like protein-4B [36]. Moreover, gallic acid can also mitigate pro-inflammatory responses by reducing the secretion of pro-inflammatory mediators, e.g., NO, PGE2, IL-6, etc., in a dose-dependent manner [37]. In addition to the anti-inflammatory effect of gallic acid, it was found that gallic acid could cross through the liposome membrane to react with the 1,1-diphenyl-2-picryl-hydrazyl (DPPH) free radical and had an antioxidant effect in preventing the injury of oxidative stress in neurodegenerative diseases [38]. Furthermore, it was shown to have anti-depressant properties in chronic stress mice models [39], arsenic-induced brain injury rat models [40], and post-stroke depression rat models [41]. The anti-inflammatory and anti-depressive properties of gallic acid render it a possible candidate to treat comorbid visceral pain and depression.

In this study, we aimed to study the potential beneficial effects of gallic acid on comorbid visceral pain and depression, to determine whether gallic acid can alleviate the comorbidity by affecting the P2X7 receptors in the hippocampus, spinal cord, and dorsal root ganglion (DRG) and to investigate the possible mechanism.

2. Results

2.1. Molecular Docking of Gallic Acid to P2X7 Receptors

The results of molecular docking show that gallic acid binds the P2X7 receptor at a binding pocket made up by P2X7 receptor B and C chains via hydrogen bonds. Figure 1 shows the binding patterns of gallic acid and the P2X7 receptor in different fields, where different colors represent different side chains. The results of molecular docking also show that the binding affinity of gallic acid to the P2X7 receptor is 6.4 (kcal/mol) (Table 1). The absolute value of binding affinity >6 kcal/mol being set as the standard, the binding affinity of gallic acid to the P2X7 receptor was considered good.

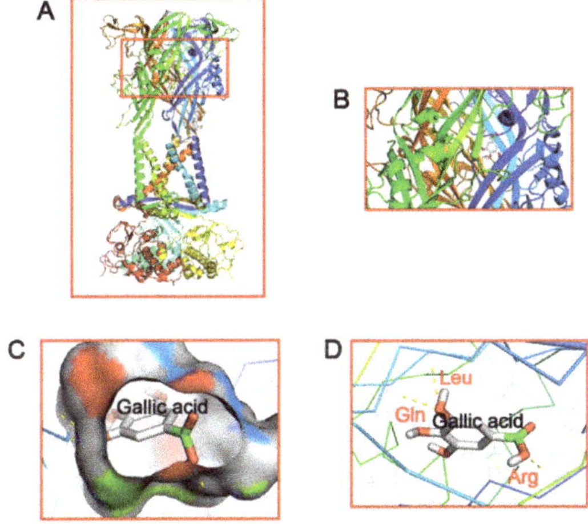

Figure 1. Molecular docking of gallic acid (GA) to the P2X7 receptor. The simulation modeling of GA docking to the P2X7 receptor was performed by a computer. The molecular docking prediction of GA to the P2X7 receptor was performed using AutoDock 4.2. The front view (**A**), top view (**B**) and enlarged views (**C,D**) indicate the perfect match for GA to interact with the P2X7 receptor.

Table 1. Molecular docking score of P2X7 receptor docking and gallic acid.

Mode	Affinity	Dist from Best Mode	
	(kcal/mol)	rmsdl.b.	rmsdu.b.
1	−6.4	0.000	0.000
2	−6.3	16.856	18.085
3	−6.1	1.228	3.775
4	−5.7	17.977	19.492
5	−5.6	10.916	13.051
6	−5.6	10.910	13.236
7	−5.6	10.866	12.913
8	−5.6	9.819	11.322
9	−5.5	27.190	27.842

The predicted binding affinity is in kcal/mol (energy). * rmsd: RMSD values were calculated relative to the best mode and only used movable heavy atoms. Two variants of RMSD metrics are provided: rmsd (RMSD lower bound: matches each atom in one conformation with itself in the other conformation, ignoring any symmetry) and rmsd/ub (RMSD upper bound: rmsd/lb [c1,c2] = max [rmsd'{c1,c2}, rmsd'{c2,c1}]; and rmsd' matches each atom in one conformation with the closest atom of the same element type in the other conformation), which differ in how the atoms are matched in the distance calculation. There was a strong reaction between ligand and protein, and the molecular docking of gallic acid to P2X7 was stable.

2.2. The Effect of Gallic Acid on Hyperalgesia Threshold of Rats with Comorbid Visceral Pain and Depression

A total of 139 male seven-day-old suckling rats were selected for CRD. After 14 days of CRD and normal feeding to adulthood (8 weeks), 51 male rats were fully consistent with visceral pain and depression in behavioral tests, and the modeling rate was about 36% (comorbidity/overall × 100%). All depressive behaviors were caused by natural visceral pain rather than through manual intervention.

The pain threshold was assessed using the AWR score. The score for each rat was the average score of two independent observers, each of whom conducted one observation every 30 min for three rounds. The scores of rats in the model group were significantly higher than those in the sham group under all pressures before treatment ($p < 0.01$), indicating that the visceral pain model was successfully established with a decreased pain threshold. The gallic acid intragastric administration (IA) protocol was performed once a day for 28 days, while the P2X7shRNA and ncRNA injection protocols were performed once a day for 7 days. The model group was intragastrically administered the same volume of solvent (DMSO + pure water) as the gallic acid preparation in the model + GA group. After 4 weeks of gallic acid IA on a daily basis or 1 week of P2X7shRNA intrathecal injection on a daily basis, the AWR score was identified as being significantly lower than that in the model group ($p < 0.01$) under the pressures of 20, 40, and 60 mm/Hg (Figure 2A–C). However, under the pressure of 80 mm/Hg, the remission effects of the gallic acid and P2X7shRNA were not evident ($p < 0.05$), as shown in Figure 2D. Under each pressure, the scores of the model and model + ncRNA groups remained significantly higher than those of the model + GA group and the model + P2X7shRNA group ($p < 0.01$), demonstrating that hyperalgesia was diminished after treatment with gallic acid or P2X7 shRNA. This indicated that both P2X7 knockdown and gallic acid could reduce the pain sensitivity of rats with visceral pain.

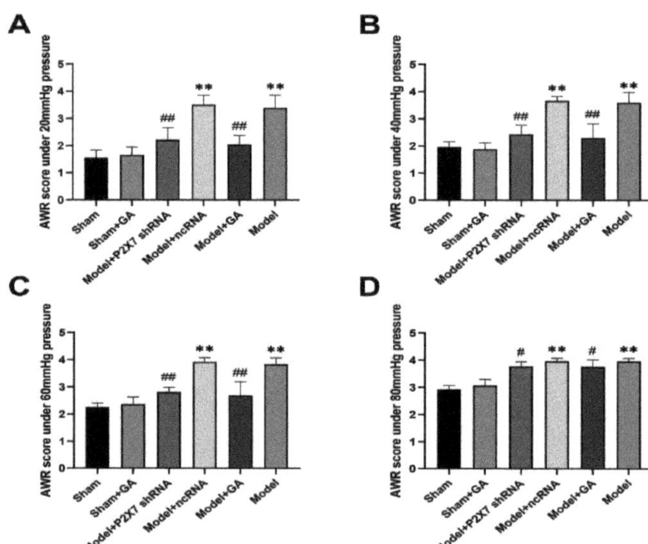

Figure 2. The chronic visceral hypersensitivity in rats was reflected in the AWR score in different groups under 20 mm/Hg ($F_{(5,53)}$ = 51.642, $p < 0.001$). (**A**) 40 mm/Hg ($F_{(5,53)}$ = 49.327, $p < 0.001$). (**B**) 60 mm/Hg ($F_{(5,53)}$ = 56.527, $p < 0.001$). (**C**) and 80 mm/Hg pressure ($F_{(5,53)}$ = 59.456, $p < 0.001$). (**D**) Values are means ± SEM. p-value was calculated by ANOVA. ** $p < 0.01$ vs. sham group; # $p < 0.05$ and ## $p < 0.01$ vs. model group.

2.3. The Effect of Gallic Acid on Depression Levels of Rats with Comorbid Visceral Pain and Depression

The weight of the selected rats was between 180 g and 250 g, and the rats were over 8 weeks old. The degree of depression was measured using three behavioral tests: the open field test (OFT), the sucrose preference test (SCPT), and the forced swimming test (FST). Before gallic acid IA, the results of the OFT demonstrated that the comorbidity model rats moved over a shorter distance (Figure 3A) ($p < 0.01$) and spent less time in the center of the field than those in the sham group (Figure 3B) ($p < 0.01$). After 4 weeks of gallic acid IA or 1 week of intrathecal injection of P2X7shRNA, the moving distances and time spent in the center of the field significantly increased for model + GA and model + P2X7shRNA rats ($p < 0.01$).

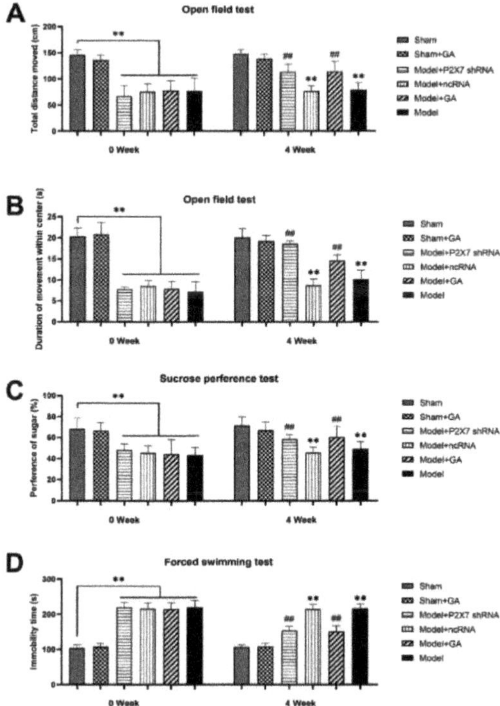

Figure 3. The depression levels of rats were reflected in the results of three independent behavioral tests (OFT, SCPT, and FST). Total moving distance (**A**) (F(5,53) = 42.329, $p < 0.001$) and duration of movement (**B**) (F(5,53) = 83.529, $p < 0.001$) within the center of the field before (0 week, 56 days age) and after (4 week) treatment in the OFT (5 min); preference of sugar (**C**) (F(5,53) = 15.273, $p < 0.001$) before and after treatment in the SCPT; IT (**D**) (F(5,53) = 140.105, $p < 0.001$) before and after treatment in the FST (5 min). The data in the first six columns of all bar charts are the data from before the treatment, and the data in the last six columns are the data from after the treatment. Every histogram bar includes the values from more than nine different samples. Values are means ± SEM. ** $p < 0.01$ vs. sham group; ## $p < 0.01$ vs. model group.

Combining the results of the SCPT and FST, the depression levels of rats were represented in the SCPT rates (sugar water consumption volume/total liquid consumption volume) and immobility time (IT). The results of the SCPT manifest that the comorbidity model rats had no preferences between sugar water and pure water; therefore, the SCPT rates were close to 50%. The comorbidity model rats presented shorter IT than sham rats, suggesting that the model group rats were more likely to be desperate in such oppressive environment ($p < 0.01$). However, rats treated by gallic acid or P2X7shRNA exhibited significantly increased SCPT rate values (Figure 3C) ($p < 0.01$) and reduced IT in the FST (Figure 3D) ($p < 0.01$) as compared with the model group.

The above results from the three behavior tests, i.e., OFT, SCPT, and FST, indicated that treatment with gallic acid or P2X7shRNA could relieve depression-like symptoms in the model rats.

2.4. Confirming Established Rat Visceral Pain Model by H&E Staining

IBS is a type of functional bowel disorder (the intestinal expression of a gut–brain interaction disorder [42]) and is one of the most common diseases to involve visceral pain. However, CRD may damage the rectal structure in the process of modeling, resulting in organ damage, such as ulceration. Therefore, H&E staining was performed to exclude this. The results of H&E staining showed that the tissue structure of the colonic wall in each group was complete and uniform; the mucosal surface was smooth; and the intestinal glands in the lamina propria were regular. There were no obvious edemas in the surrounding stroma and no infiltrations of neutrophils, monocytes, or macrophages. It was shown that the modeling method did not lead to structural damages to the intestinal tracts of the rats, which was in accordance with IBS manifestation (Figure 4). By combining the behavioral test results of the rats (AWR score, OFT, SCPT, and FST), it can be inferred that the rat model of visceral pain with depression was successfully established in this study.

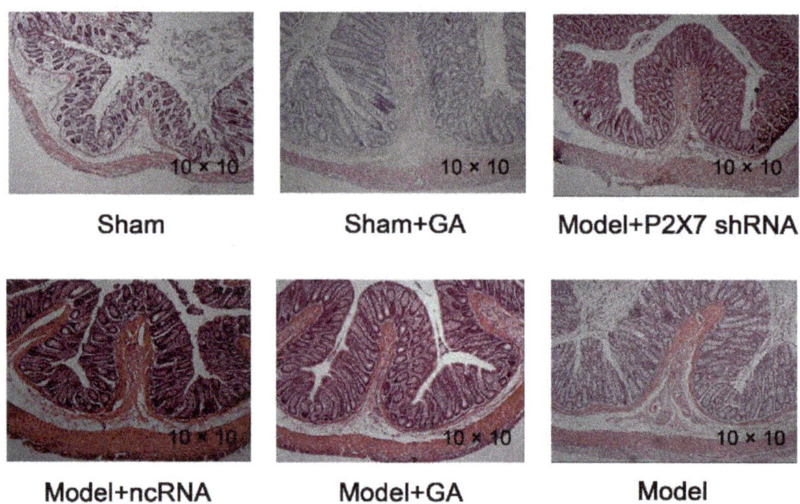

Figure 4. H&E staining of rat rectum tissues from each group under $100\times$ field of view.

2.5. Effects of Gallic Acid on P2X7 Expression in the Hippocampi, Spinal Cords, and DRGs of Rats with Comorbid Visceral Pain and Depression

The mRNA levels and protein concentrations of the P2X7 receptor in the hippocampi, spinal cords, and DRGs of rats from each group were measured by qRT-PCR (Figure 5A–C) and Western blotting (Figure 5D–F). In all three tissue types, the expression levels of the P2X7 protein and mRNA in the model group were significantly higher than those in the sham groups ($p < 0.01$). By contrast, the mRNA and protein levels of the P2X7 receptor in the model + GA and model + P2X7shRNA groups were significantly lower than those of the model group ($p < 0.01$). Nevertheless, no significant differences were observed between the model + ncRNA group and the model group. These results demonstrated that gallic acid and P2X7shRNA could significantly reduce the expression of the P2X7 receptor in these tissues.

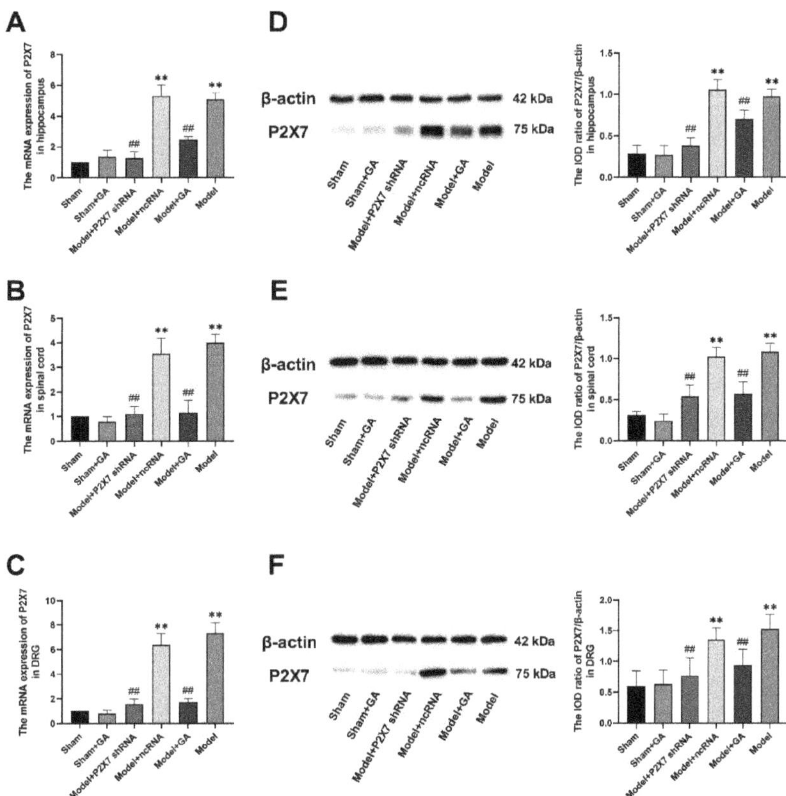

Figure 5. The expression of P2X7 receptor mRNA in the hippocampus was confirmed by qRT-PCR ($F_{(5,30)}$ = 131.475, $p < 0.001$) (**A**). The expression of P2X7 receptor mRNA in the spinal cord was confirmed by qRT-PCR ($F_{(5,30)}$ = 84.012, $p < 0.001$) (**B**). The expression of P2X7 receptor mRNA in the DRG was confirmed by qRT-PCR ($F_{(5,30)}$ = 31.043, $p < 0.001$) (**C**). β-Actin was used as the housekeeper gene in all qRT-PCRs. The relative expression of the P2X7 protein was detected by Western blotting in the hippocampus ($F_{(5,30)}$ = 68.997, $p < 0.001$) (**D**), spinal cord ($F_{(5,30)}$ = 62.397, $p < 0.001$) (**E**), and DRG ($F_{(5,30)}$ = 15.084, $p < 0.001$) (**F**). Values are means ± SEM. N = 6 per group. ** $p < 0.01$ vs. sham group; ## $p < 0.01$ vs. model group.

The alterations in the expressions of the P2X7 receptor in the hippocampus and DRG were also confirmed in the immunofluorescence results. In the immunofluorescent image of the hippocampus (Figure 6A), green represents the P2X7 receptor, and red represents GFAP. However, in the DRG (Figure 7A), the colors are the opposite to facilitate differentiation. Despite the changes in the P2X7 receptor levels in both the DRG and hippocampus, the findings also show that, in the hippocampus, the levels and number of GFAP decreased in the model with gallic acid and the P2X7shRNA administration groups as compared with the model and ncRNA addition groups ($p < 0.01$). Additionally, the gallic acid and the P2X7shRNA administration groups demonstrated a low intensity of co-expressions of GFAP and P2X7 receptors, as was expected (Figures 6B and 7B). These results suggested that gallic acid could reduce the co-expressions of GFAP and P2X7 receptors in hippocampus and DRG.

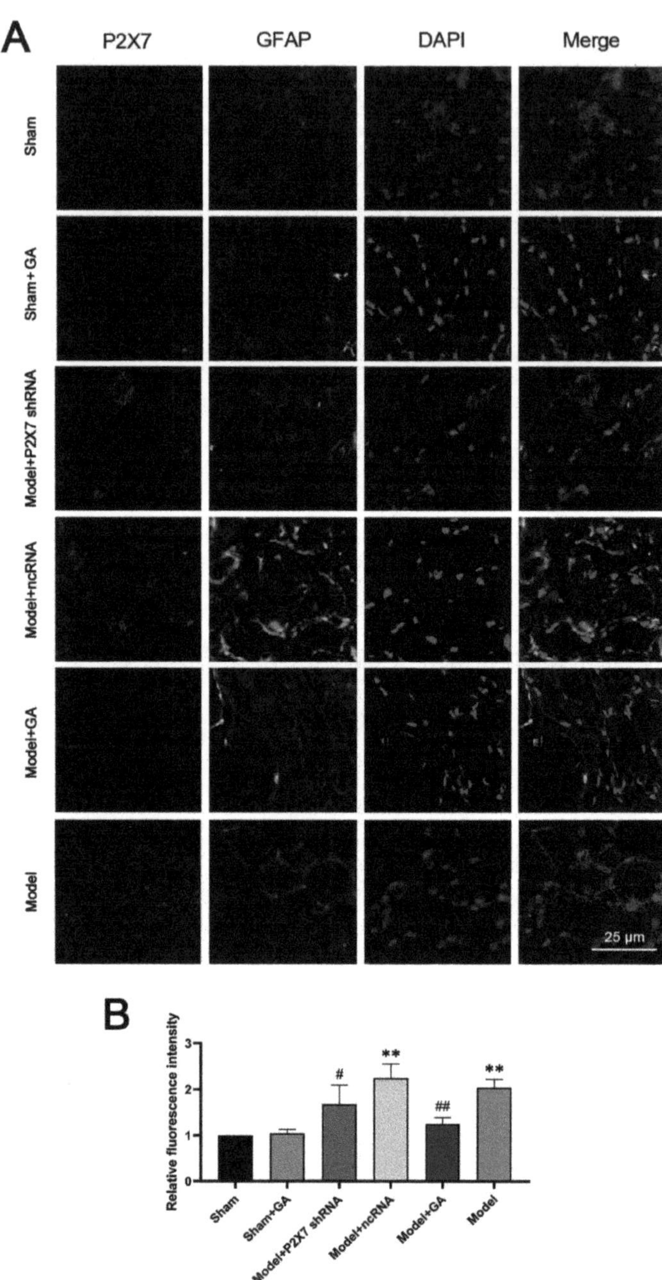

Figure 6. The effects of gallic acid on the co-expression of P2X7 and glial fibrillary acidic protein (GFAP) in the DRG. The blue signal indicates nuclei; the green signal indicates GFAP; and the red signal indicates the P2X7 receptor (**A**). The relative fluorescence intensity analysis (yellow) of the DRG (**B**) ($F_{(5,30)}$ = 31.126, $p < 0.001$). Values are means ± SEM. N = 6 per group. ** $p < 0.01$ vs. sham group; # $p < 0.05$ and ## $p < 0.01$ vs. model group.

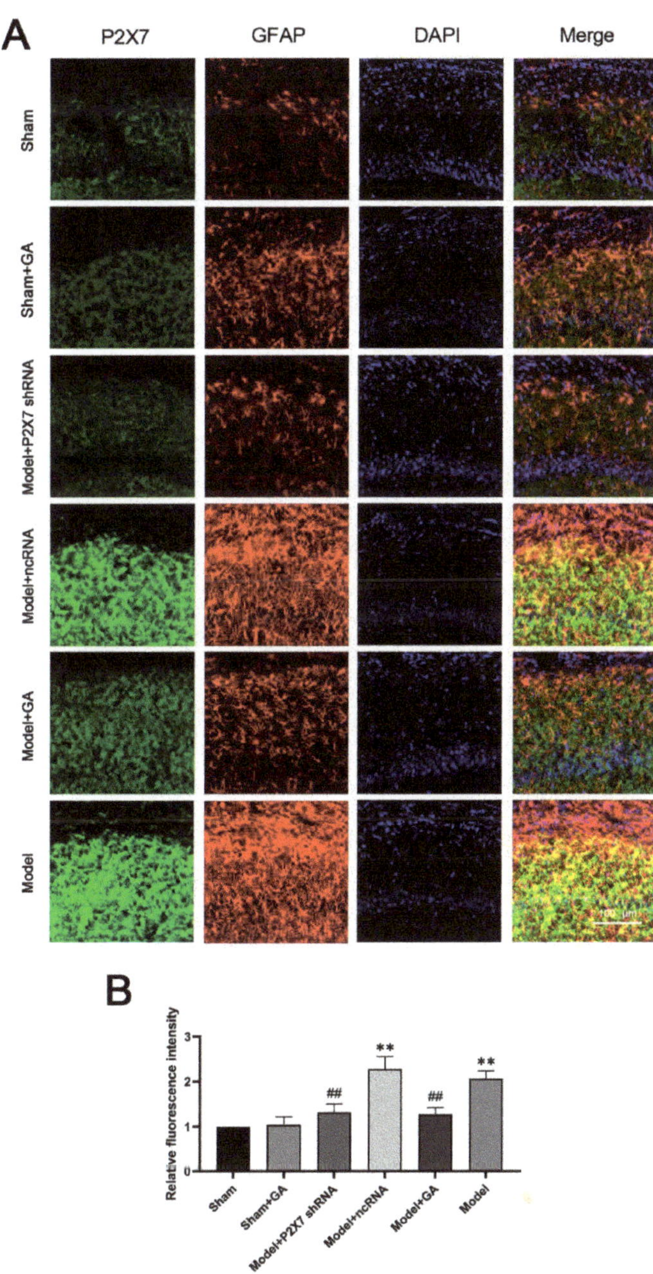

Figure 7. The effects of gallic acid on the co-expression of P2X7 and glial fibrillary acidic protein (GFAP) in the hippocampus CA1 area. The blue signal indicates nuclei; the green signal indicates the P2X7 receptor; and the red signal indicates GFAP (**A**). Relative fluorescence intensity analysis (yellow) of the hippocampus (**B**) (F(5,30) = 54.229, $p < 0.001$). Values are means $\pm$ SEM. N = 6 per group. ** $p < 0.01$ vs. sham group; ## $p < 0.01$ vs. model group.

2.6. Effects of Gallic acid on ERK1/2 Phosphorylation in the Hippocampi, Spinal Cords, and DRGs of Rats with Comorbid Visceral Pain and Depression

The levels of ERK1/2 and Phospho-ERK1/2 (p-ERK1/2) in the hippocampus, spinal cord, and DRG were measured by Western blotting. The concentration of ERK1/2 in each tissue from different groups was basically identical, but the levels of p-ERK1/2 in the model and ncRNA groups were significantly higher than those in the sham group ($p < 0.01$). This suggests that the phosphorylated (activated) state of the ERK1/2 protein performed significantly in the course of the comorbidity. Meanwhile, the expressions of p-ERK1/2 in the model + GA and model + P2X7shRNA groups were greatly lower than that in the model group. Therefore, gallic acid and P2X7shRNA treatments in rats with visceral pain and depression largely reduced p-ERK1/2 expression ($p < 0.01$) (Figure 8). There were no significant differences in the expression level of p-ERK1/2 between the model + ncRNA group and the model group.

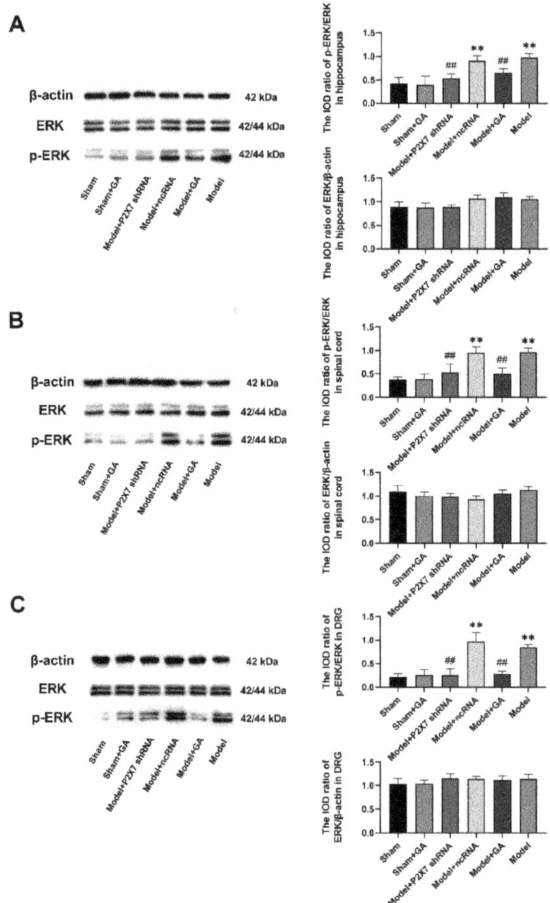

Figure 8. The effects of GA on ERK and phosphorylation of ERK1/2 in the hippocampus (**A**) (F(5,30) = 25.222, $p < 0.001$), spinal cord (**B**) (F(5,30) = 30.171, $p < 0.001$), and DRG (**C**) (F(5,30) = 55.567, $p < 0.001$) were determined by Western blotting. β-Actin was used as the housekeeper gene in all tissue types. Values are means ± SEM. N = 6 per group. ** $p < 0.01$ vs. sham group; ## $p < 0.01$ vs. model group.

2.7. Effects of Gallic Acid on Serum IL-1β, IL-10, and TNF-α in Rats with Comorbid Visceral Pain and Depression

The concentrations of IL-1β, IL-10, and TNF-α in the serum of rats in each group were detected by ELISA. Both IL-β and TNF-α are pro-inflammatory factors. The results showed that IL-1β and TNF-α in the model group were significantly higher than those in the sham groups ($p < 0.01$), whereas gallic acid and P2X7shRNA significantly decreased IL-β and TNF-α levels as compared with the model groups ($p < 0.01$) (Figure 9A,B). As an anti-inflammatory factor, the concentration of IL-10 exhibited an opposite trend; it was significantly lower in the model group than in the sham group ($p < 0.01$), while gallic acid and P2X7shRNA significantly increased serum IL-10 levels in visceral pain and depression rats ($p < 0.01$) (Figure 9C). There were no significant differences between the model group and the model + ncRNA group in all experiments. Therefore, the data demonstrated that gallic acid and P2X7 shRNA had evident anti-inflammatory effects and attenuated the comorbidity.

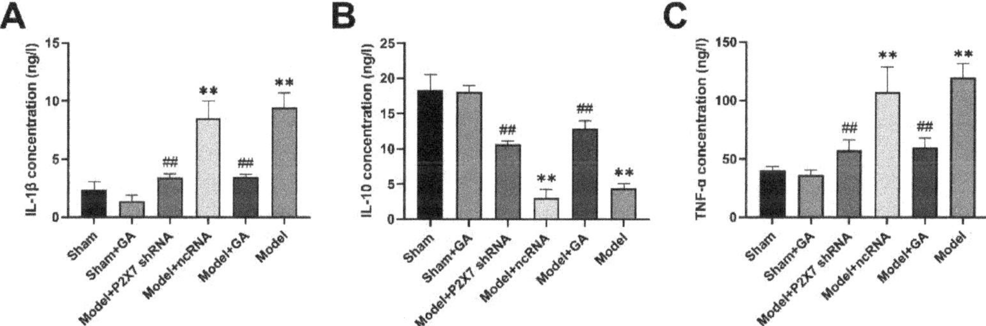

Figure 9. The effects of GA on the concentration of interleukin 1β (IL-1β) (**A**) (F(5,30) = 85.322, $p < 0.001$); interleukin 10 (IL-10) (**B**) (F(5,30) = 172.816, $p < 0.001$); and tumor necrosis factor α (TNF-α) (**C**) (F(5,30) = 56.655, $p < 0.001$) in the serum of rats in each group were assessed using ELISA. Values are means ± SEM. N = 6 per group. ** $p < 0.01$ vs. sham group; ## $p < 0.01$ vs. model group.

2.8. Effects of Gallic Acid on mRNA Levels of IL-1β, IL-10, TNF-α, and BDNF in the Hippocampus of Rats with Comorbid Visceral Pain and Depression

The expressions of IL-1β, IL-10, TNF-α, and brain-derived neurotrophic factor (BDNF) at the mRNA level in the hippocampus of rats in each group were detected by qRT-PCR. The results showed that the expressions of both IL-1β and TNF-α in the model and model + ncRNA groups were significantly higher than those in the sham group ($p < 0.01$), while gallic acid and P2X7shRNA significantly decreased their expressions (Figure 10A,B). However, the expression trend of IL-10 and BDNF was opposite to that of IL-1β and TNF-α. Their expressions in both the model group and the ncRNA group were downregulated as compared with the sham group (Figure 10C,D). By contrast, gallic acid and P2X7shRNA were able to reverse the changes in IL-10 and BDNF. The data suggested that neuroinflammation was crucial in the development of the comorbidity, which could be significantly suppressed by gallic acid or P2X7 shRNA. The concentration change in BDNF was negatively related to the depressive level. Therefore, we affirmed that the expression of the P2X7 receptor was negatively associated with BDNF concentration, and gallic acid could alleviate depression via P2X7 receptor downregulation and BDNF elevation.

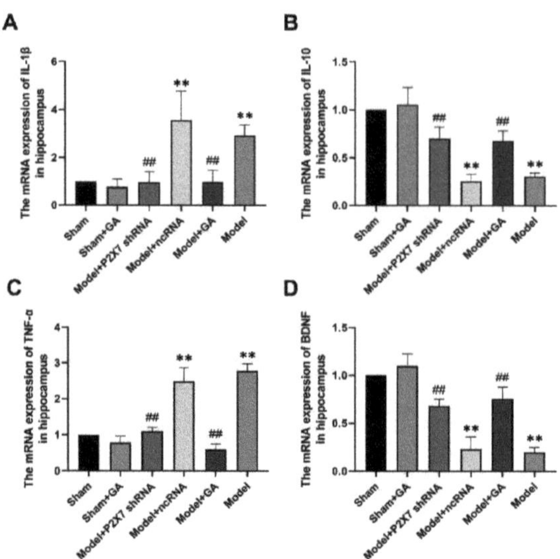

Figure 10. The effects of GA on the mRNA level of IL-1β (**A**) (F(5,30) = 24.350, *p* < 0.001); IL-10 (**B**) (F(5,30) = 62.094, *p* < 0.001), TNF-α (**C**) (F(5,30) = 123.893, *p* < 0.001); and brain-derived neurotrophic factor (BDNF) (**D**) (F(5,30) = 94.780, *p* < 0.001) in the hippocampus of rats in each group were assessed using qRT-PCR. β-Actin was used as the housekeeper gene in all qRT-PCR. Values are means ± SEM. N = 6 per group. ** *p* < 0.01 vs. sham group; ## *p* < 0.01 vs. model group.

3. Discussion

The co-occurrence of pain and depression is commonly seen in clinics. It has been reported that the depletion of both serotonin and norepinephrine seen in depression patients interferes with the pain modulatory system [43]. Moreover, many patients with chronic pain are found to suffer from major depressive disorder (MDD) [43,44]. It is well substantiated that pain at baseline, pain severity, and chronicity are statistically related to MDD [45]. Comorbid pain and depression patients show both unsatisfactory pain and depression medication responses [46,47].

The incidence of visceral pain in patients suffering depression is very high. IBS is the most common disease associated with chronic visceral pain, and it is estimated that 20–60% of IBS patients also deal with depression [2]. In this study, a rat model of comorbid visceral pain and depression was used to conduct experiments. Rats were treated with a series of colorectal balloon distension (CRD) applications as neonates resulting in visceral pain that persisted into adulthood [48]. The IBS model was verified by both abnormally high AWR scores and structurally normal rectums in the H&E staining images of IBS rats. The co-existence of depression was also testified by the apathetic performances in the SCPT, OFT, and FST. It emerged that around 84.1% (visceral pain/overall × 100%) of SD rats (male) developed visceral pain after neonatal CRD, of which 43.6% (comorbidity/visceral pain × 100%) displayed both visceral pain and depression. The results of H&E staining showed that the tissue structure of the colonic wall in each group was complete and uniform, which demonstrated that the modeling method did not lead to structural damage to the intestinal tract of the rats. Thus, the hypersensitivity of the nervous system is crucial in visceral pain development in this study [49]. It has been reported that neonatal CRD contributes to the vulnerability of hippocampal microglia, which are more susceptible to being sensitized during adult CRD and to release a plethora of cytokines. This brings about a reduction in the hippocampal glucocorticoid receptor [50], consequentially augmenting corticotropin-releasing hormone (CRH), stimulating HPA, central neuronal sensitization, and spinal sensitization [51]. Thus, in this study, many rats suffering from visceral pain naturally exhibited depression-like behavior without additional stimulation. However, only few studies have established a model of comorbid visceral pain and depression as the investigation subject. Thus, our work provides a novel approach that can be used to understand the underlying molecular mechanisms involved in comorbid visceral pain and depression.

Gallic acid is a phenolic acid that is anti-inflammatory, antioxidant, and anti-depressive [37,38]. Previous findings show that the neuroprotective effect of gallic acid has been verified under many pathological conditions, such as neurotoxicity related to glutamate [52], cobalt chloride [53], arsenic [40], and aluminum chloride [54]; neurodegeneration induced by type-2 diabetes [55] and metabolic syndrome [56]; etc. Upon aflatoxin B1 neural toxification, the application of gallic acid displays neuroprotective properties via anti-inflammatory, antioxidant, and anti-apoptosis mechanisms [57]. Our laboratory has been investigating the pharmacological effects of gallic acid and proved that it can be neuroprotective against neuropathic pain via inhibiting the P2X7 receptor-mediated NF-κB/STAT signaling pathway [58]. In our study, we showed the combination of gallic acid and the P2X7 receptor by molecular docking analysis, which was also supported in a report by Yang Runan et al. [58]. This study is aimed at investigating the effect of gallic acid on the comorbid visceral pain and depression and determining whether gallic acid could affect the expression of the P2X7 receptor.

In this study, we found that the expressions of the P2X7 receptor in the hippocampi, spinal cords, and dorsal root ganglions (DRGs) of rats with visceral pain and depression increased significantly as compared with normal rats. The results show that the P2X7 receptor could modulate IBS, which is consistent with previous reports in our laboratory [15]. As a potent mediator of inflammation, the P2X7 receptor's relationship with inflammatory markers has been well studied [59]. Resting cells originally possess the inactive precursor of casp-1-procaspase-1 (procasp-1). Once ATP binds to the P2X7 receptor, the activated P2X7 receptor conducts K^+ efflux; then, procasp-1 is proteolytically activated into casp-1 in the "IL-1β inflammasome protein complex" [60]. Thereafter, casp-1 converts pro-IL-1β (inactive) into IL-1β (active), and IL-1β is released into the pericellular space [61]. Additionally, it was also demonstrated that the P2X7 receptor could modulate the release of IL-18 from monocytes [62]. The contact of IL-18 and the αβ heterodimeric receptor causes the synthesis of other cytokines, e.g., IL-6, IL-8, TNF-α, IL-1, and IFN-γ [63]. Our results support that IBS rats with depression exhibited higher P2X7 receptor expression in the spinal cord and dorsal root ganglia—the key central factors in the lower portion of the GIT sensory system [64]. Furthermore, the P2X7 receptor is vital for NLRP3 inflammasome activation, which was found to facilitate depressive behaviors [65]. In this study, we also witnessed significantly increased P2X7 receptor expression in the hippocampus of comorbidity model rats, suggesting that elevated hippocampal P2X7 receptor expression might induce depressive behaviors. Thus, our findings substantiated that increased P2X7 receptor expression in the hippocampus, spinal cord, and DRG may promote visceral pain and depression. In this research study, we also investigated the effects of gallic acid and P2X7 shRNA treatment on visceral-pain-associated depression. Our results indicate that gallic acid or P2X7 shRNA treatment can diminish spinal cord, DRG, and hippocampus P2X7 receptor expressions, further substantiating that gallic acid has a similar down-regulating effect on the P2X7 receptor expression as P2X7 shRNA. Animal behavioral tests suggested that a lowered pain threshold and elevated depression degree in models were restored to normal via gallic acid or P2X7 shRNA administration. Therefore, the generation of comorbid visceral pain and depression is associated with the elevation in P2X7 receptor expression, which gallic acid helps restore to normal. Furthermore, serum and hippocampal IL-1β and TNF-α exhibited a similar alteration trend among model rats as the P2X7 receptor, while serum and hippocampal IL-10 and hippocampal BDNF exhibited an inverse tendency. It has become common to treat depression according to the inflammatory mechanism and pro-inflammatory cytokines (such as IL-1 β, IL-6, and TNF-α). The decrease in the IL-10 factor caused by low-grade colonic mucositis is considered to play an important role in the pathophysiology of IBS [66]. The levels of pro-inflammatory cytokines are often considered to be screening biomarkers to predict whether the anti-depression treatment is effective or not [67], as it has been widely shown that neuroinflammation promotes MDD [68]. Moreover, BDNF, a neurotrophin that is of great significance for the neurons in key brain circuits associated with cognition and emotion, has anti-depressive activities [69,70]. In this study, the changes in IL-1β, IL-10, and TNF-α ulteriorly show that gallic acid improves the inflammatory environment both peripherally and centrally via the P2X7 receptor inhibition. Treatments with gallic acid and P2X7 shRNA successfully increased the BDNF level in the hippocampus of comorbid rats, indicating alleviated depression. The interaction of gallic acid and the P2X7 receptor has been testified by whole-cell patch-clamp tests [58], and the combination of gallic acid and the P2X7 receptor was found by molecular docking analysis in our research study. Incorporating all the results mentioned, we may draw the conclusion that the P2X7 receptor may be a common target for the comorbid visceral pain and depression, and the protective effect of gallic acid in comorbid rats is related to its anti-inflammatory property and increased BDNF in the hippocampus through P2X7 receptor inhibition.

To further investigate how gallic acid and P2X7 shRNA act on the mitigation of the comorbidities, the phosphorylation of ERK1/2 was explored by Western blotting. Mitogen-activated protein kinases (MAPKs) are a group of protein kinases that include p38, ERK1/2, and c-Jun N-terminal kinase (JNK) [71], among which the ERK subfamily has been found to be closely related to visceral pain and depression [28,72]. In a previous study, we found that P2X7 receptor shRNA reduced the increased levels of p-ERK1/2 in the DRGs, spinal cords, and hippocampi of rats with diabetic neuropathic pain and depression [28]. In our visceral pain and depression rat model, higher p-ERK1/2 levels (the activated form of ERK) were detected and could be blocked by treatment with either gallic acid or P2X7shRNA. Thus, it could be postulated that the mechanism of gallic acid relieving comorbid visceral pain and depression may be related to the inhibition of the ERK1/2 pathway.

One shortcoming of this study is that we did not investigate other possible pharmacological pathways of gallic acid. Gallic acid has multiple pharmacological effects, such as anti-inflammatory, antioxidant, and anti-apoptosis effects [73]. However, whether gallic acid acts on the P2X7 receptor alone or on other pathways remain to be defined in studies. Furthermore, the exact molecular mechanism still needs further research in the future for us to better understand the role of gallic acid in comorbid visceral pain and depression. The P2X7 receptor is thought to participate in both innate and adaptive immunity due to its wide presence on nearly all immune cells [74]. Many mechanisms of autoimmune diseases, neoplastic diseases, and degenerative diseases have been found to be associated with the P2X7 receptor [75–77]. Our study indicates that gallic acid can inhibit the P2X7 receptor, thus improving inflammatory conditions, which are shared by a plethora of autoimmune, neoplastic, and degenerative diseases. In order to determine whether gallic acid has therapeutic effects on those diseases, more studies that focus on its therapeutic range and its relationship with the P2X7 receptor are required.

4. Material and Methods

4.1. Molecular Docking

The P2X7.pdb file of the P2X7 receptor protein sequences was downloaded from http://www.rcsb.org/pdb/home/home.do (Accessed on 16 June 2021), and the gallic acid.sdf file was downloaded from https://pubchem.ncbi.nlm.nih.gov/. (Accessed on 19 June 2021). After pretreating with pyMOL software (The PyMOL Molecular Graphics System, Version 2.0 Schrödinger, LLC.) to remove small molecule ligands, dehydrate, and hydrogenate, autodock tools software (La Jolla, CA, USA) was applied for molecular docking based on python [28].

4.2. Animal and Treatment

SD rats were obtained from the Department of Animal Science, Jiangxi University of Traditional Chinese Medicine. The procedures were approved by the Animal Care and Use Committee at Nanchang University Medical School (SYKX2015-0001) and were performed according to the International Association for the Study of Pain's ethical guidelines for pain research on animals. All the suckling rats along with their mother were housed in one plastic cage until day 21 after birth. At this point, the male rats were separated into other cages according to the experimental design, while the female rats were placed into other cages for mating. Female rats were excluded because the menstrual cycle would affect pain sensitivity [78].

Eight-day-old male suckling rats were selected to undergo neonatal colorectal dilation (CRD) [79]. In total, 93 suckling rats (n = 139) were used to establish the group of visceral pain combined with depression and 46 suckling rats to establish the sham operation group. During the period from the 8th day to the 21st day, neonatal CRD was regularly performed every day until the mother and baby were separated to establish the comorbidity (visceral pain + depression). The sham operation (sham) group was stroked on the anus with Vaseline at the same time.

The hyperalgesia threshold and the level of depression of the rats were determined using behavioral tests (including OFT, FST, SCPT, and AWR scores as detailed below) when the rats reached 8 weeks. During this period, the rats did not undergo any depression treatment. After CRD treatment, rats that met the depressive requirements were named model and selected to be divided into four groups. The rats in the model + GA group were administered gallic acid (20 mg/kg) (Macklin Biochemical, Shanghai, China) intragastrically for 28 days; the rats in the model group were administered the same amount of solvent (DMSO + pure water) intragastrically for 28 days. The rats in the model + P2X7shRNA group were injected with P2X7shRNA intrathecally every day for 1 week; the rats in the model + ncRNA group were injected with non-code shRNA (ncRNA) intrathecally every day for 1 week. All rats were subjected to the same behavioral tests after treatment. The rats that underwent a sham operation during neonatal age were divided into two groups. The sham group continued without any treatment, and the sham + GA group was intragastrically administered the same concentration of gallic acid as mentioned above for 28 days. The experimental design is shown in Figure 11.

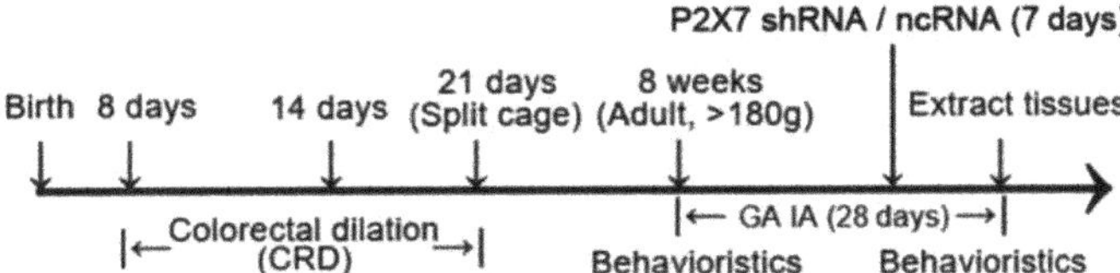

Figure 11. The flow chart showing the experimental design.

4.3. Drugs and Chemicals

The transfection complex, consisting of shRNA (P2X7shRNA or non-code shRNA) and transfection reagent at a ratio of 1:2 (µg/µL), was prepared using an Entranster™ in vivo transfection kit (Engreen Biosystem Company of Beijing), according to the manufacturer's instructions. The complex was intrathecally injected into rats of the model + P2X7shRNA and model + ncRNA groups. The P2X7shRNA sequences are as follows:

5′-CACCGTGCAGTGAATGAGTACTACGAATAGTACTCATTCACTGCAC-3′ and *3′-CACGTCACTTACTCATGATGCTTATCATGAGTAAGTGACGTGAAAA-5′.*

According to the literature [29] and the preliminary results from our laboratory, we decided to use gallic acid administrated intragastrically at a dosage of 20 mg/kg. The composition of gallic acid solution was 80 mg gallic acid + 160 µL DMSO + 9840 µL pure water (1:2:123).

4.4. Neonatal CRD

The 8-day-old male suckling rats were divided into 2 groups for different treatments. The model group was treated with neonatal CRD twice a day, from day 8 to day 21. This was performed with an angioplasty balloon and a sphygmomanometer. Firstly, the balloon smeared with Vaseline (Unilever, CT, USA) was inserted rectally into the descending colon of each rat. Then, a sphygmomanometer was applied to add 60 mmHg of pressure to the balloon for 1 min, after which the rat was released for a 30 min rest before the next round of CRD applications. In the sham group, the rats underwent 2 rounds of anal smearing with Vaseline on a daily basis from day 8 to day 21.

4.5. Adult CRD

After all the processed rats reached 8 weeks, adult CRD [79] was conducted to assess the pain threshold of each rat. Before CRD, each rat was intraperitoneally injected with 10% chloral hydrate for sedation. Then, a Vaseline-smeared balloon catheter was inserted rectally into the descending colon of each rat. After a 30 min interval for adaptation, the rat was placed on a flat table, and a sphygmomanometer was used to add 20, 40, 60, and 80 mm/Hg of pressure in a serial order. The pain threshold of each rat was determined by its behavior under each pressure level. According to the Abdominal Withdraw Reflex (AWR) score [79], head shaking represents the first degree of pain; slight abdominal muscle contraction without abdominal lifting represents the second degree of pain; abdominal lifting represents the third degree of pain; and an arching back represents the fourth and final degree of pain. By comparing the performance of the neonatal CRD group with the sham group, the rats with visceral pain that perceived more intense pain and thus had a lower pain threshold could be distinguished. These results were independently obtained and averaged by two observers without knowing the experimental details.

4.6. Sucrose Preference Test (SCPT)

After 24 h of liquid fasting, rats were kept in distinct cages containing 2 identical bottles holding 100 mL of sucrose water (10 g/L) and 100 mL of pure water. The sucrose preference (SCP) value of each rat was noted 1 h after bottle placement by measuring both pure water reduction (ΔP) and sucrose water consumption (ΔS) (SCP $= \Delta S/\Delta S + \Delta P \times 100\%$). The sucrose preference test (SCPT) is commonly used to evaluate anhedonia in animals, which refers to a reduced capacity to experience happiness [80]. By comparing the SCP values of the visceral pain group and the sham group, the rats with naturally derived comorbid visceral pain and depression were identified [28].

4.7. Open Field Test (OFT)

The open field test (OFT) imitates unsafe surroundings, evaluates animals' autonomous behavior, and reveals how tense the animals are [80]. A 30 min dark adaptation was conducted before initiating the test. Then, one rat at a time was carefully placed in the center of a 40 cm × 50 cm × 60 cm open field. Two seconds after the Canon Powershot A610 camera (Canon Co. local distributer, Tehran, Iran) detected the animal, MATLAB (MathWorks Co., Natick, MA, USA) began to record the total traveled distance and the route for 5 min. The field was sanitized with 75% ethanol solution before the initiation of the following test [28].

4.8. Forced Swimming Test (FST)

The forced swimming test (FST) rat model was used to test the desperate depressive behavior [80]. An 80 cm high glass cylinder with an inner diameter of 40 cm containing 30 cm deep water at approximately 20 °C was utilized to hold one rat a time. Every rat was forced to swim for 5 min, and their motions were recorded. Thereafter, the immobility time (IT) of each test was calculated, which accounted for the total time the rat was immotile. By incorporating the performances of individual rats in all 3 behavioral tests, the ascertained visceral pain and depression rat models were separated [28].

4.9. Tissue Extraction

After 4 weeks of intragastric administration of gallic acid, rats were anesthetized with 10% chloral hydrate. DRGs, L4-L5 spinal cords, and whole brains were removed from a small number of rats and fixed in 4% PFA at 4 °C for 2 h. Subsequently, the tissues were dehydrated with 30% sucrose solution (in 4% PFA) at 4 °C for 24 h, during which the liquid was exchanged every 8 h. The other rats were beheaded for blood collection in a 20 mL EP tube. After centrifugation (3000 rpm/min) for 15 min, the upper serum was moved and stored at −80 °C. Then, their rectums, L4-L5 spinal cords, DRGs, and hippocampi were extracted and flushed with phosphate-buffered saline (PBS). The rectums were kept in EP tubes filled with PFA. Half of the remaining tissue was stored in EP tubes filled with RNA storage solution; the other half was left in PBS. All tissues were stored at −20 °C for further use.

4.10. Western Blotting

The stored spinal cords, DRGs, and hippocampi were put into different 2 mL homogenizers for the first round of homogenization in 98% RIPA lysis buffer, which is a mixture of 50 mM Tris-Cl (pH 8.0), 150 mM NaCl, 0.1% sodium dodecyl sulfate, 1% Nonidet P-40, 0.02% sodium deoxycholate, 100 mg/mL phenylmethylsulfonyl fluoride, and 1 mg/mL aprotinin. Next, 1% protease inhibitor and 1% phosphatase inhibitor were added into the homogenizers for the second round of grinding until no solid tissues were visible. Subsequently, all the homogenizers were left on ice for 30 min sedimentation before pouring the liquid into EP tubes and centrifuging at 12,000 rpm at 4 °C for 10 min. Thereafter, the supernatants were moved and mixed with 6 × loading buffer. The mixtures were heated in boiling water before cooling. They were then stored at −80 °C.

Extracted proteins were loaded on 12% sodium dodecyl sulfate–polyacrylamide gel for electrophoresis with a Bio-Rad system. The protein samples of the hippocampus and spinal cord in each well were 5–10 μg, while the protein samples of the DRG were 15–20 μg. Then, the protein bands were transferred onto polyvinylidene fluoride membranes (PVDF membranes), which were then blocked with 5% skim milk in 1×TBST (Cwbio, Beijing, China) for 2 h. After a gentle rinse, the membrane was immersed in the primary antibodies against β-actin (ZSGB-Bio, Beijing, China), the P2X7 receptor (Alomone, Jerusalem, Israel), extracellular signal–regulated kinases 1/2 (ERK 1/2) (Cell Signaling Technology, Danvers, MA, USA), and phosphorylated (p)-ERK1/2 (Cell Signaling Technology, Danvers, MA, USA) overnight at 4 °C. Thereafter, the membrane went through 3 rounds of 10 min washing with TBST before incubation in second antibodies horseradish peroxidase–conjugated secondary goat anti-rabbit IgG (ZSGB-Bio, Beijing, China) and goat anti-mouse IgG (ZSGB-Bio, Beijing, China) in blocking buffer on ice. Following another 3 rounds of 10 min washing, chemiluminescent solution (Advansta, Menlo Park, CA, USA) was dripped onto the membrane, which was then placed in the exposure machine (BIO-RAD, Hercules, CA, USA) for visualization of the protein bands. The grey density levels of the bands were measured with ImageJ.

4.11. Enzyme-Linked Immunosorbent Assay (ELISA)

The concentrations of IL-1β, TNF-α, and IL-10 in the collected serum samples were measured using relevant ELISA kits (Boster Biological Technology, WuHan, China). In brief, 50 μL of blank control buffer and 6 diluted standard solutions (720, 360, 180, 90, 45, and 22.5 μg/L) were added into the 96-well plate in triplicates. A mixture of 10 μL of serum samples of the different groups and 40 μL of standard solution were also added into the plate in triplicate. Then, the plate was sealed using a sealing membrane and incubated in a water bath for 30 min at 37 °C. Subsequently, the plate was washed for 5 times, and color developing agents were added into each well, followed by 10 min incubation at 37 °C in the dark. Finally, 50 μL of termination agent was added, and the absorbance of each well was measured using a microplate reader at 450 nm.

4.12. Quantitative Real-Time PCR (qRT-PCR)

The relevant materials were all ribozyme-free, and the homogenizers utilized went through acid soakage, diethyl pyrocarbonate soakage, autoclaving, and drying before use. The tissues (spinal cord, DRG, and hippocampus) were homogenized in RNA lysis solution, and RNA extraction was performed using a kit, according to the manufacturer's instructions (Transgen, Beijing, China). Thereafter, 2 µL of the RNA from each sample was converted into complementary DNA using a RevertAid™ HMinus First Strand cDNA Synthesis Kit (TransGen, Beijing, China). The sequences of the primers used in the experiment were constructed using Primer Express 3.0 Software (Thermo Fisher Scientific, CA, USA). The sequences of the 6 pair primers were as follows.

β-actin forward *5′-TAAAGACCTCTATGCCAACA-3′* and reverse *3′-CACGATGGAGGG GCCGGACTCATC-5′*.

P2X7 forward *5′-GATGGATGGACCCACAAAGT-3′* and reverse *3′-GCTTCTT TCCCTTCCTCAGC-5′*.

IL-1β forward *5′-CCTATGTCTTGCCCGTGGAG-3′* and reverse *5′-CACACACTA GCAGGTCGTCA-3′*.

BDNF forward *5′-CCTCTGCTCTTTCTGCTGGA-3′* and reverse *5′-GCTGTGAC CCACTCGCTAAT-3′*.

TNF-α forward *5′-CACGTCGTAGCAAACCACCAA-3′* and reverse *3′-GTTGGTTGT CTTTGAGATCCAT-5′*.

IL-10 forward *5′-CGGGAAGACAATAACTGCACCC-3′* and reverse *5′-CGGTTAGCAG TATGTTGTCCAGC-3′*.

The concentration of cDNA from each independent sample was 1000 ng/µL–1200 ng/µL. Quantitative real-time PCR was achieved using 0.1 mL 8-strip thin-wall PCR tubes with 300 ng of cDNA and 19.5 µL of master mix per well. β-Actin was used as housekeeper gene in all qRT-PCRs. Quantitative real-time PCR was conducted using StepOne (ABI, Thermo Fisher Scientific, CA, USA). The $\Delta\Delta CT$ method was used to quantify the expression of each gene, with CT as the threshold cycle. The relative levels of target genes normalized to the individual sample with the lowest CT are presented as $2^{-\Delta\Delta CT}$. Each independent sample was tested three times to obtain the average value. The experimental results include six groups of different independent samples.

4.13. Hematoxylin–Eosin Staining (H&E Staining)

A total of 10 µm of the specimens was prepared with the rectum tissues stored in PFA. After splicing the wax blocks containing the rectums, the specimens were kept at 37 °C overnight. Then, the specimens were baked in an oven at 60 °C for 2–3 h before two rounds of dewaxing in xylene I and xylene II, respectively. Subsequently, the specimens went through a 5 min hydration in 100, 95, 75, and 50% ethanol in serial order, followed by 5 min of washing with PBS. Thereafter, the specimens were dyed with hematoxylin, which was followed by 5 min of rinsing under running water. Then, the specimens were differentiated with 1% hydrochloric acid alcohol before another a 1 h rinse under running water. Eventually, the specimens were colored for 15 s using eosin, dehydrated, and sealed using neutral resin.

4.14. Double-Label Immunofluorescence

DRGs, spinal cords, and hippocampi were sliced into specimens. First, these underwent 3 rounds of 5 min PBS washing and 4% paraformaldehyde fixation. Then, another 3 rounds of 5 min PBS washing were conducted before 1 h blocking with gout serum at 37 °C. Thereafter, the specimens were incubated in a mixed antibody solution containing both anti-P2X7 (1:100) (Alomone, Jerusalem, Israel) and anti-GFAP (1:100) (Invitrogen, Carlsbad, CA, USA) at 4 °C overnight. Subsequently, the specimens were washed with PBS for 5, 10, and 15 min before being incubated in another mixed antibody solution for 60 min. For this, 2 protocols were utilized: the first was goat anti-rabbit tetramethylrhodamine (TRITC) 1:200 (Thermo Fisher Scientific, Carlsbad, CA, USA) and goat anti-mouse fluorescein isothiocyanate (FITC) 1:200 (Thermo Fisher Scientific, Carlsbad, CA, USA); the second was goat anti-mouse TRITC and goat anti-rabbit FITC 1:200. Then, the specimens were washed 3 times in PBS for 5 min and stained with 4′,6-diamidino-2-phenylindole (DAPI) for 5 min. The final products were sealed with anti-fluorescence attenuation agent.

4.15. Statistical Analysis

Statistical analyses were performed using SPSS26 software (IBM, New York, NY, USA) and GraphPad Prism (8.0.2, GraphPad software, San Diego, CA, USA). Data were analyzed by one-way analysis of variance (ANOVA) followed by the LSD post hoc test for multiple comparisons. The results are expressed as the mean ± standard error of the mean (SEM) and were considered statistically significant at $p < 0.05$.

5. Conclusions

In summary, our study suggests that the P2X7 receptor may be an effective target for both visceral pain and depression, and gallic acid could alleviate visceral pain and depressive behavior in rats by inhibiting the expression of the P2X7 receptor in the hippocampus, spinal cord, and DRG. The possible mechanism of gallic acid may be closely related to the inhibition of ERK1/2 phosphorylation and inflammatory cytokines. This study suggests that gallic acid is an effective drug to alleviate visceral pain combined with depression and is worthy of further clinical applications in the future.

Author Contributions: Conceptualization, L.W. and Y.G.; methodology, L.T., M.Z., C.W. and Y.G. software, L.W. and H.T. (Hongcheng Tu); validation, L.W., L.T., S.L. and M.Z.; investigation, Y.W.; resources, P.L.; data curation, J.W. and C.Y.; writing—original draft preparation, L.W., L.T. and Y.G.; writing—review and editing, L.W., L.T. and Y.G.; visualization, L.W. and H.T. (Hankun Tian); supervision, G.L. and Y.G.; project administration, Y.G.; funding acquisition, Y.G. and G.L. All authors have read and agreed to the published version of the manuscript.

Funding: This study was supported by grants from the National Natural Science Foundation of China (Nos. 81760152, 81970749 and 82160161).

Institutional Review Board Statement: The study was conducted according to the guidelines of the Declaration of Helsinki and approved by the Institutional Review Board of Nanchang University, China (SYKX2015-0001).

Informed Consent Statement: Not applicable.

Data Availability Statement: The datasets generated during and/or analyzed during the current study are available from the corresponding author on reasonable request.

Acknowledgments: We thank Shangdong Liang and Guodong Li for assisting us in the experimental design. We thank Ziwei Zhang and Ting Wen for assisting in the conduction of the experiments.

Conflicts of Interest: The authors declare no conflict of interest.

References

1. Shen, J.; Zhang, B.; Chen, J.; Cheng, J.; Wang, J.; Zheng, X.; Lan, Y.; Zhang, X. SAHA Alleviates Diarrhea-Predominant Irritable Bowel Syndrome Through Regulation of the p-STAT3/SERT/5-HT Signaling Pathway. *J. Inflamm. Res.* **2022**, *15*, 1745–1756. [CrossRef] [PubMed]
2. Moloney, R.D.; Johnson, A.; Mahony, S.O.; Dinan, T.; Meerveld, B.G.-V.; Cryan, J.F. Stress and the Microbiota-Gut-Brain Axis in Visceral Pain: Relevance to Irritable Bowel Syndrome. *CNS Neurosci. Ther.* **2015**, *22*, 102–117. [CrossRef] [PubMed]
3. Louwies, T.; Ligon, C.O.; Johnson, A.C.; Meerveld, B.G. Targeting epigenetic mechanisms for chronic visceral pain: A valid approach for the development of novel therapeutics. *Neurogastroenterol. Motil.* **2018**, *31*, e13500. [CrossRef] [PubMed]
4. Saha, L. Irritable bowel syndrome: Pathogenesis, diagnosis, treatment, and evidence-based medicine. *World J. Gastroenterol.* **2014**, *20*, 6759–6773. [CrossRef]
5. Radovanovic-Dinic, B.; Tesic-Rajkovic, S.; Grgov, S.; Petrovic, G.; Zivković, V. Irritable bowel syndrome—From etiopathogenesis to therapy. *Biomed. Pap.* **2018**, *162*, 1–9. [CrossRef]
6. Enck, P.; Aziz, Q.; Barbara, G.; Farmer, A.D.; Fukudo, S.; Mayer, E.A.; Niesler, B.; Quigley, E.M.M.; Rajilic-Stojanovic, M.; Schemann, M.; et al. Irritable bowel syndrome. *Nat. Rev. Dis. Prim.* **2016**, *2*, 1–24. [CrossRef]

7. Dothel, G.; Barbaro, M.R.; Boudin, H.; Vasina, V.; Cremon, C.; Gargano, L.; Bellacosa, L.; De Giorgio, R.; Le Berre-Scoul, C.; Aubert, P.; et al. Nerve Fiber Outgrowth Is Increased in the Intestinal Mucosa of Patients with Irritable Bowel Syndrome. *Gastroenterology* **2015**, *148*, 1002–1011.e4. [CrossRef]
8. Vermeulen, W.; De Man, J.G.; Pelckmans, P.A.; De Winter, B.Y. Neuroanatomy of lower gastrointestinal pain disorders. *World J. Gastroenterol.* **2014**, *20*, 1005–1020. [CrossRef]
9. Elsenbruch, S. Abdominal pain in Irritable Bowel Syndrome: A review of putative psychological, neural and neuro-immune mechanisms. *Brain Behav. Immun.* **2011**, *25*, 386–394. [CrossRef]
10. Ng, Q.X.; Soh, A.Y.S.; Loke, W.; Lim, D.Y.; Yeo, W.-S. The role of inflammation in irritable bowel syndrome (IBS). *J. Inflamm. Res.* **2018**, *11*, 345–349. [CrossRef]
11. Ochoa-Cortes, F.; Liñán-Rico, A.; Jacobson, K.A.; Christofi, F.L. Potential for Developing Purinergic Drugs for Gastrointestinal Diseases. *Inflamm. Bowel Dis.* **2014**, *20*, 1259–1287. [CrossRef] [PubMed]
12. Weng, Z.J.; Wu, L.Y.; Zhou, C.L.; Dou, C.Z.; Shi, Y.; Liu, H.R.; Wu, H.G. Effect of electroacupuncture on P2X3 receptor regulation in the peripheral and central nervous systems of rats with visceral pain caused by irritable bowel syndrome. *Purinergic Signal.* **2015**, *11*, 321–329. [CrossRef] [PubMed]
13. Galligan, J.J. Enteric P2X receptors as potential targets for drug treatment of the irritable bowel syndrome. *J. Cereb. Blood Flow Metab.* **2004**, *141*, 1294–1302. [CrossRef] [PubMed]
14. Luo, Y.; Feng, C.; Wu, J.; Wu, Y.; Liu, D.; Wu, J.; Dai, F.; Zhang, J. P2Y1, P2Y2, and TRPV1 Receptors Are Increased in Diarrhea-Predominant Irritable Bowel Syndrome and P2Y2 Correlates with Abdominal Pain. *Am. J. Dig. Dis.* **2016**, *61*, 2878–2886. [CrossRef]
15. Liu, S.; Shi, Q.; Zhu, Q.; Zou, T.; Li, G.; Huang, A.; Wu, B.; Peng, L.; Song, M.; Wu, Q.; et al. P2X$_7$ receptor of rat dorsal root ganglia is involved in the effect of moxibustion on visceral hyperalgesia. *Purinergic Signal.* **2014**, *11*, 161–169. [CrossRef]
16. Xu, G.-Y.; Shenoy, M.; Winston, J.H.; Mittal, S.; Pasricha, P.J. P2X receptor-mediated visceral hyperalgesia in a rat model of chronic visceral hypersensitivity. *Gut* **2008**, *57*, 1230–1237. [CrossRef]
17. Adinolfi, E.; Giuliani, A.L.; De Marchi, E.; Pegoraro, A.; Orioli, E.; Di Virgilio, F. The P2X7 receptor: A main player in inflammation. *Biochem. Pharmacol.* **2018**, *151*, 234–244. [CrossRef]
18. Liu, P.-Y.; Lee, I.-H.; Tan, P.-H.; Wang, Y.-P.; Tsai, C.-F.; Lin, H.-C.; Lee, F.-Y.; Lu, C.-L. P2X7 Receptor Mediates Spinal Microglia Activation of Visceral Hyperalgesia in a Rat Model of Chronic Pancreatitis. *Cell. Mol. Gastroenterol. Hepatol.* **2015**, *1*, 710–720.e5. [CrossRef]
19. Dell'Antonio, G.; Quattrini, A.; Cin, E.D.; Fulgenzi, A.; Ferrero, M.E. Relief of inflammatory pain in rats by local use of the selective P2X7 ATP receptor inhibitor, oxidized ATP. *Arthritis Care Res.* **2002**, *46*, 3378–3385. [CrossRef]
20. Jarvis, M.F. The neural–glial purinergic receptor ensemble in chronic pain states. *Trends Neurosci.* **2010**, *33*, 48–57. [CrossRef]
21. Chessell, I.P.; Hatcher, J.P.; Bountra, C.; Michel, A.D.; Hughes, J.P.; Green, P.; Egerton, J.; Murfin, M.; Richardson, J.; Peck, W.L.; et al. Disruption of the P2X7 purinoceptor gene abolishes chronic inflammatory and neuropathic pain. *Pain* **2005**, *114*, 386–396. [CrossRef] [PubMed]
22. Clark, A.; Staniland, A.A.; Marchand, F.; Kaan, T.K.Y.; McMahon, S.; Malcangio, M. P2X7-Dependent Release of Interleukin-1 and Nociception in the Spinal Cord following Lipopolysaccharide. *J. Neurosci.* **2010**, *30*, 573–582. [CrossRef] [PubMed]
23. Gölöncsér, F.; Baranyi, M.; Balázsfi, D.; Demeter, K.; Haller, J.; Freund, T.F.F.; Zelena, D.; Sperlágh, B. Regulation of Hippocampal 5-HT Release by P2X7 Receptors in Response to Optogenetic Stimulation of Median Raphe Terminals of Mice. *Front. Mol. Neurosci.* **2017**, *10*, 325. [CrossRef] [PubMed]
24. Csölle, C.; Baranyi, M.; Zsilla, G.; Kittel, Á.; Gölöncsér, F.; Illes, P.; Papp, E.; Vizi, E.S.; Sperlágh, B. Neurochemical Changes in the Mouse Hippocampus Underlying the Antidepressant Effect of Genetic Deletion of P2X7 Receptors. *PLoS ONE* **2013**, *8*, e66547. [CrossRef]
25. Mayhew, J.; Graham, B.; Biber, K.; Nilsson, M.; Walker, F. Purinergic modulation of glutamate transmission: An expanding role in stress-linked neuropathology. *Neurosci. Biobehav. Rev.* **2018**, *93*, 26–37. [CrossRef] [PubMed]
26. Iwata, M.; Ota, K.T.; Li, X.-Y.; Sakaue, F.; Li, N.; Dutheil, S.; Banasr, M.; Duric, V.; Yamanashi, T.; Kaneko, K.; et al. Psychological Stress Activates the Inflammasome via Release of Adenosine Triphosphate and Stimulation of the Purinergic Type 2X7 Receptor. *Biol. Psychiatry* **2015**, *80*, 12–22. [CrossRef]
27. Otrokocsi, L.; Kittel, Á.; Sperlágh, B. P2X7 Receptors Drive Spine Synapse Plasticity in the Learned Helplessness Model of Depression. *Int. J. Neuropsychopharmacol.* **2017**, *20*, 813–822. [CrossRef]
28. Guan, S.; Shen, Y.; Ge, H.; Xiong, W.; He, L.; Liu, L.; Yin, C.; Wei, X.; Gao, Y. Dihydromyricetin Alleviates Diabetic Neuropathic Pain and Depression Comorbidity Symptoms by Inhibiting P2X7 Receptor. *Front. Psychiatry* **2019**, *10*, 770. [CrossRef]
29. Mansouri, M.T.; Soltani, M.; Naghizadeh, B.; Farbood, Y.; Mashak, A.; Sarkaki, A. A possible mechanism for the anxiolytic-like effect of gallic acid in the rat elevated plus maze. *Pharmacol. Biochem. Behav.* **2013**, *117*, 40–46. [CrossRef]
30. Guo, P.; Anderson, J.D.; Bozell, J.J.; Zivanovic, S. The effect of solvent composition on grafting gallic acid onto chitosan via carbodiimide. *Carbohydr. Polym.* **2016**, *140*, 171–180. [CrossRef]
31. Bai, J.; Zhang, Y.; Tang, C.; Hou, Y.; Ai, X.; Chen, X.; Zhang, Y.; Wang, X.; Meng, X. Gallic acid: Pharmacological activities and molecular mechanisms involved in inflammation-related diseases. *Biomed. Pharmacother.* **2020**, *133*, 110985. [CrossRef] [PubMed]
32. Xu, Y.; Tang, G.; Zhang, C.; Wang, N.; Feng, Y. Gallic Acid and Diabetes Mellitus: Its Association with Oxidative Stress. *Molecules* **2021**, *26*, 7115. [CrossRef] [PubMed]

33. Zamudio-Cuevas, Y.; Andonegui-Elguera, M.A.; Aparicio-Juárez, A.; Aguillón-Solís, E.; Martínez-Flores, K.; Ruvalcaba-Paredes, E.; Velasquillo-Martínez, C.; Ibarra, C.; Martínez-López, V.; Gutiérrez, M.; et al. The enzymatic poly(gallic acid) reduces pro-inflammatory cytokines in vitro, a potential application in inflammatory diseases. *Inflammation* **2020**, *44*, 174–185. [CrossRef] [PubMed]

34. Lin, Y.; Luo, T.; Weng, A.; Huang, X.; Yao, Y.; Fu, Z.; Li, Y.; Liu, A.; Li, X.; Chen, D.; et al. Gallic Acid Alleviates Gouty Arthritis by Inhibiting NLRP3 Inflammasome Activation and Pyroptosis Through Enhancing Nrf2 Signaling. *Front. Immunol.* **2020**, *11*, 580593. [CrossRef]

35. Nouri, A.; Heibati, F.; Heidarian, E. Gallic acid exerts anti-inflammatory, anti-oxidative stress, and nephroprotective effects against paraquat-induced renal injury in male rats. *Naunyn-Schmiedebergs Arch. Fur Exp. Pathol. Und Pharmakol.* **2020**, *394*, 1–9. [CrossRef]

36. Hsiang, C.-Y.; Hseu, Y.-C.; Chang, Y.-C.; Kumar, K.S.; Ho, T.-Y.; Yang, H.-L. Toona sinensis and its major bioactive compound gallic acid inhibit LPS-induced inflammation in nuclear factor-κB transgenic mice as evaluated by in vivo bioluminescence imaging. *Food Chem.* **2013**, *136*, 426–434. [CrossRef]

37. Bensaad, L.A.; Kim, K.H.; Quah, C.C.; Kim, W.R.; Shahimi, M.; Bensaad, L.A.; Kim, K.H.; Quah, C.C.; Kim, W.R.; Shahimi, M. Anti-inflammatory potential of ellagic acid, gallic acid and punicalagin A&B isolated from Punica granatum. *BMC Complement. Altern. Med.* **2017**, *17*, 47. [CrossRef]

38. Lu, Z.; Nie, G.; Belton, P.S.; Tang, H.; Zhao, B. Structure–activity relationship analysis of antioxidant ability and neuroprotective effect of gallic acid derivatives. *Neurochem. Int.* **2005**, *48*, 263–274. [CrossRef]

39. Chhillar, R.; Dhingra, D. Antidepressant-like activity of gallic acid in mice subjected to unpredictable chronic mild stress. *Fundam. Clin. Pharmacol.* **2012**, *27*, 409–418. [CrossRef]

40. Samad, N.; Jabeen, S.; Imran, I.; Zulfiqar, I.; Bilal, K. Protective effect of gallic acid against arsenic-induced anxiety−/depression-like behaviors and memory impairment in male rats. *Metab. Brain Dis.* **2019**, *34*, 1091–1102. [CrossRef]

41. Nabavi, S.; Habtemariam, S.; Di Lorenzo, A.; Sureda, A.; Khanjani, S.; Daglia, M. Post-Stroke Depression Modulation and in Vivo Antioxidant Activity of Gallic Acid and Its Synthetic Derivatives in a Murine Model System. *Nutrients* **2016**, *8*, 248. [CrossRef] [PubMed]

42. Aboubakr, A.; Cohen, M.S. Functional Bowel Disease. *Clin. Geriatr. Med.* **2020**, *37*, 119–129. [CrossRef] [PubMed]

43. Bair, M.J.; Robinson, R.L.; Katon, W.; Kroenke, K. Depression and pain comorbidity: A literature review. *Arch. Intern. Med.* **2003**, *163*, 2433–2445. [CrossRef] [PubMed]

44. Williams, L.S.; Jones, W.J.; Shen, J.; Robinson, R.L.; Weinberger, M.; Kroenke, K. Prevalence and impact of depression and pain in neurology outpatients. *J. Neurol. Neurosurg. Psychiatry* **2003**, *74*, 1587–1589. [CrossRef] [PubMed]

45. Hilderink, P.H.; Burger, H.; Deeg, D.J.; Beekman, A.T.; Voshaar, R.C.O. The Temporal Relation Between Pain and Depression: Results from the longitudinal aging study Amsterdam. *Psychosom. Med.* **2012**, *74*, 945–951. [CrossRef]

46. Dworkin, R.H.; Gitlin, M.J. Clinical Aspects of Depression in Chronic Pain Patients. *Clin. J. Pain* **1991**, *7*, 79–94. [CrossRef]

47. Bair, M.J.; Robinson, R.L.; Eckert, G.J.; Stang, P.E.; Croghan, T.W.; Kroenke, K. Impact of Pain on Depression Treatment Response in Primary Care. *Psychosom. Med.* **2004**, *66*, 17–22. [CrossRef]

48. Gil, D.W.; Wang, J.; Gu, C.; Donello, J.E.; Cabrera, S.M.; Al-Chaer, E.D. Role of sympathetic nervous system in rat model of chronic visceral pain. *Neurogastroenterol. Motil.* **2015**, *28*, 423–431. [CrossRef]

49. Li, Q.; Zhang, X.; Liu, K.; Gong, L.; Li, J.; Yao, W.; Liu, C.; Yu, S.; Li, Y.; Yao, Z.; et al. Brain-derived neurotrophic factor exerts antinociceptive effects by reducing excitability of colon-projecting dorsal root ganglion neurons in the colorectal distention-evoked visceral pain model. *J. Neurosci. Res.* **2012**, *90*, 2328–2334. [CrossRef]

50. Zhang, G.; Zhao, B.-X.; Hua, R.; Kang, J.; Shao, B.-M.; Carbonaro, T.M.; Zhang, Y.-M. Hippocampal microglial activation and glucocorticoid receptor down-regulation precipitate visceral hypersensitivity induced by colorectal distension in rats. *Neuropharmacology* **2016**, *102*, 295–303. [CrossRef]

51. Meerveld, B.G.-V.; Johnson, A.C. Stress-Induced Chronic Visceral Pain of Gastrointestinal Origin. *Front. Syst. Neurosci.* **2017**, *11*, 86. [CrossRef] [PubMed]

52. Maya, S.; Prakash, T.; Madhu, K. Assessment of neuroprotective effects of Gallic acid against glutamate-induced neurotoxicity in primary rat cortex neuronal culture. *Neurochem. Int.* **2018**, *121*, 50–58. [CrossRef] [PubMed]

53. Akinrinde, A.; Adebiyi, O. Neuroprotection by luteolin and gallic acid against cobalt chloride-induced behavioural, morphological and neurochemical alterations in Wistar rats. *NeuroToxicology* **2019**, *74*, 252–263. [CrossRef] [PubMed]

54. Maya, S.; Prakash, T.; Goli, D. Evaluation of neuroprotective effects of wedelolactone and gallic acid on aluminium-induced neurodegeneration: Relevance to sporadic amyotrophic lateral sclerosis. *Eur. J. Pharmacol.* **2018**, *835*, 41–51. [CrossRef] [PubMed]

55. Abdel-Moneim, A.; Yousef, A.I.; El-Twab, S.M.A.; Reheim, E.S.A.; Ashour, M.B. Gallic acid and p-coumaric acid attenuate type 2 diabetes-induced neurodegeneration in rats. *Metab. Brain Dis.* **2017**, *32*, 1279–1286. [CrossRef]

56. Diaz, A.; Muñoz-Arenas, G.; Caporal-Hernandez, K.; Vázquez-Roque, R.; Lopez-Lopez, G.; Kozina, A.; Espinosa, B.; Flores, G.; Treviño, S.; Guevara, J. Gallic acid improves recognition memory and decreases oxidative-inflammatory damage in the rat hippocampus with metabolic syndrome. *Synapse* **2020**, *75*, e22186. [CrossRef]

57. Adedara, I.A.; Owumi, S.E.; Oyelere, A.K.; Farombi, E.O. Neuroprotective role of gallic acid in aflatoxin B$_1$-induced behavioral abnormalities in rats. *J. Biochem. Mol. Toxicol.* **2020**, *35*, e22684. [CrossRef]

58. Yang, R.; Li, Z.; Zou, Y.; Yang, J.; Li, L.; Xu, X.; Schmalzing, G.; Nie, H.; Li, G.; Liu, S.; et al. Gallic Acid Alleviates Neuropathic Pain Behaviors in Rats by Inhibiting P2X7 Receptor-Mediated NF-κB/STAT3 Signaling Pathway. *Front. Pharmacol.* **2021**, *12*, 680139. [CrossRef]
59. Di Virgilio, F.; Dal Ben, D.; Sarti, A.C.; Giuliani, A.L.; Falzoni, S. The P2X7 Receptor in Infection and Inflammation. *Immunity* **2017**, *47*, 15–31. [CrossRef]
60. Martinon, F.; Burns, K.; Tschopp, J. The inflammasome: A molecular platform triggering activation of inflammatory caspases and processing of proIL-beta. *Mol. Cell* **2002**, *10*, 417–426. [CrossRef]
61. Ferrari, D.; Pizzirani, C.; Adinolfi, E.; Lemoli, R.M.; Curti, A.; Idzko, M.; Panther, E.; Di Virgilio, F. The P2X$_7$Receptor: A Key Player in IL-1 Processing and Release. *J. Immunol.* **2006**, *176*, 3877–3883. [CrossRef] [PubMed]
62. Mehta, V.B.; Hart, J.; Wewers, M.D. ATP-stimulated release of interleukin (IL)-1beta and IL-18 requires priming by lipopolysaccharide and is independent of caspase-1 cleavage. *J. Biol. Chem.* **2001**, *276*, 3820–3826. [CrossRef] [PubMed]
63. Dinarello, C.A. The IL-1 family and inflammatory diseases. *Clin. Exp. Rheumatol.* **2002**, *20*, S1–S13.
64. Mayer, E.A.; Raybould, H.E. Role of visceral afferent mechanisms in functional bowel disorders. *Gastroenterology* **1990**, *99*, 1688–1704. [CrossRef]
65. Xu, Y.; Sheng, H.; Bao, Q.; Wang, Y.; Lu, J.; Ni, X. NLRP3 inflammasome activation mediates estrogen deficiency-induced depression- and anxiety-like behavior and hippocampal inflammation in mice. *Brain Behav. Immun.* **2016**, *56*, 175–186. [CrossRef] [PubMed]
66. Chang, L.; Adeyemo, M.; Karagiannidis, I.; Videlock, E.J.; Bowe, C.; Shih, W.; Presson, A.P.; Yuan, P.-Q.; Cortina, G.; Gong, H.; et al. Serum and Colonic Mucosal Immune Markers in Irritable Bowel Syndrome. *Am. J. Gastroenterol.* **2012**, *107*, 262–272. [CrossRef]
67. Roman, M.; Irwin, M.R. Novel neuroimmunologic therapeutics in depression: A clinical perspective on what we know so far. *Brain Behav. Immun.* **2019**, *83*, 7–21. [CrossRef]
68. Troubat, R.; Barone, P.; Leman, S.; DeSmidt, T.; Cressant, A.; Atanasova, B.; Brizard, B.; El Hage, W.; Surget, A.; Belzung, C.; et al. Neuroinflammation and depression: A review. *Eur. J. Neurosci.* **2020**, *53*, 151–171. [CrossRef]
69. Mondal, A.C.; Fatima, M. Direct and indirect evidences of BDNF and NGF as key modulators in depression: Role of antidepressants treatment. *Int. J. Neurosci.* **2018**, *129*, 283–296. [CrossRef]
70. Phillips, C. Brain-Derived Neurotrophic Factor, Depression, and Physical Activity: Making the Neuroplastic Connection. *Neural Plast.* **2017**, *2017*, 260130. [CrossRef]
71. Johnson, G.L.; Lapadat, R. Mitogen-Activated Protein Kinase Pathways Mediated by ERK, JNK, and p38 Protein Kinases. *Science* **2002**, *298*, 1911–1912. [CrossRef] [PubMed]
72. Li, Z.-Y.; Huang, Y.; Yang, Y.-T.; Zhang, D.; Zhao, Y.; Hong, J.; Liu, J.; Wu, L.-J.; Zhang, C.-H.; Wu, H.-G.; et al. Moxibustion eases chronic inflammatory visceral pain through regulating MEK, ERK and CREB in rats. *World J. Gastroenterol.* **2017**, *23*, 6220–6230. [CrossRef] [PubMed]
73. Zhou, D.; Yang, Q.; Tian, T.; Chang, Y.; Li, Y.; Duan, L.-R.; Li, H.; Wang, S.-W. Gastroprotective effect of gallic acid against ethanol-induced gastric ulcer in rats: Involvement of the Nrf2/HO-1 signaling and anti-apoptosis role. *Biomed. Pharmacother.* **2020**, *126*, 110075. [CrossRef] [PubMed]
74. Burnstock, G.; Boeynaems, J.-M. Purinergic signalling and immune cells. *Purinergic Signal.* **2014**, *10*, 529–564. [CrossRef]
75. Lara, R.; Adinolfi, E.; Harwood, C.A.; Philpott, M.; Barden, J.A.; Di Virgilio, F.; McNulty, S. P2X7 in Cancer: From Molecular Mechanisms to Therapeutics. *Front. Pharmacol.* **2020**, *11*, 793. [CrossRef]
76. Verderio, C.; Matteoli, M. ATP mediates calcium signaling between astrocytes and microglial cells: Modulation by IFN-gamma. *J. Immunol.* **2001**, *166*, 6383–6391. [CrossRef]
77. Yi, Y.-S. Role of inflammasomes in inflammatory autoimmune rheumatic diseases. *Korean J. Physiol. Pharmacol.* **2018**, *22*, 1–15. [CrossRef]
78. Greenspan, J.D.; Craft, R.M.; LeResche, L.; Arendt-Nielsen, L.; Berkley, K.J.; Fillingim, R.; Gold, M.; Holdcroft, A.; Lautenbacher, S.; Mayer, E.A.; et al. Studying sex and gender differences in pain and analgesia: A consensus report. *Pain* **2007**, *132* (Suppl. S1), S26–S45. [CrossRef]
79. Al–Chaer, E.D.; Kawasaki, M.; Pasricha, P.J. A new model of chronic visceral hypersensitivity in adult rats induced by colon irritation during postnatal development. *Gastroenterology* **2000**, *119*, 1276–1285. [CrossRef]
80. Hao, Y.; Ge, H.; Sun, M.; Gao, Y. Selecting an Appropriate Animal Model of Depression. *Int. J. Mol. Sci.* **2019**, *20*, 4827. [CrossRef]

International Journal of
Molecular Sciences

MDPI

Review

Inherent P2X7 Receptors Regulate Macrophage Functions during Inflammatory Diseases

Wenjing Ren [1,2], Patrizia Rubini [1,2], Yong Tang [1,2], Tobias Engel [3,4] and Peter Illes [1,2,5,*]

1   International Collaborative Centre on Big Science Plan for Purinergic Signalling, Chengdu University of TCM, Chengdu 610075, China; pathlesswoods@126.com (W.R.); patrizia.rubini@gmx.de (P.R.); tangyong@cdutcm.edu.cn (Y.T.)

2   School of Acupunct3ure and Tuina, Chengdu University of TCM, Chengdu 610075, China

3   Department of Physiology and Medical Physics, Royal College of Surgeons in Ireland, University of Medicine and Health Sciences, D02 YN77 Dublin, Ireland; tengel@rcsi.ie

4   FutureNeuro, SFI Research Centre for Chronic and Rare Neurological Diseases, RCSI University of Medicine and Health Sciences, D02 YN77 Dublin, Ireland

5   Rudolf Boehm Institute for Pharmacology and Toxicology, University of Leipzig, 04107 Leipzig, Germany

*   Correspondence: peter.illes@medizin.uni-leipzig.de

Abstract: Macrophages are mononuclear phagocytes which derive either from blood-borne monocytes or reside as resident macrophages in peripheral (Kupffer cells of the liver, marginal zone macrophages of the spleen, alveolar macrophages of the lung) and central tissue (microglia). They occur as M1 (pro-inflammatory; classic) or M2 (anti-inflammatory; alternatively activated) phenotypes. Macrophages possess P2X7 receptors (Rs) which respond to high concentrations of extracellular ATP under pathological conditions by allowing the non-selective fluxes of cations (Na^+, Ca^{2+}, K^+). Activation of P2X7Rs by still higher concentrations of ATP, especially after repetitive agonist application, leads to the opening of membrane pores permeable to ~900 Da molecules. For this effect an interaction of the P2X7R with a range of other membrane channels (e.g., P2X4R, transient receptor potential A1 [TRPA1], pannexin-1 hemichannel, ANO6 chloride channel) is required. Macrophage-localized P2X7Rs have to be co-activated with the lipopolysaccharide-sensitive toll-like receptor 4 (TLR4) in order to induce the formation of the inflammasome 3 (NLRP3), which then activates the pro-interleukin-1β (pro-IL-1β)-degrading caspase-1 to lead to IL-1β release. Moreover, inflammatory diseases (e.g., rheumatoid arthritis, Crohn's disease, sepsis, etc.) are generated downstream of the P2X7R-induced upregulation of intracellular second messengers (e.g., phospholipase A2, p38 mitogen-activated kinase, and rho G proteins). In conclusion, P2X7Rs at macrophages appear to be important targets to preserve immune homeostasis with possible therapeutic consequences.

Keywords: macrophages; P2X7R; pore formation; inflammasome activation; inflammatory diseases

Citation: Ren, W.; Rubini, P.; Tang, Y.; Engel, T.; Illes, P. Inherent P2X7 Receptors Regulate Macrophage Functions during Inflammatory Diseases. *Int. J. Mol. Sci.* **2022**, 23, 232. https://doi.org/10.3390/ijms23010232

Academic Editor: Bruno Guigas

Received: 24 November 2021
Accepted: 21 December 2021
Published: 26 December 2021

Publisher's Note: MDPI stays neutral with regard to jurisdictional claims in published maps and institutional affiliations.

1. Introduction

While millimolar concentrations of ATP are stored in the cell interior, where they are used under anaerobic conditions as an energy supply, this nucleotide can also escape to the extracellular space through discontinuities generated by metabolic/mechanical damage to the cell membrane or by means of membrane transporters and channels. Extracellular ATP has been characterized as a signaling molecule coordinating cellular and, thereby, whole organism functions [1,2]. Extracellular ATP or its enzymatic breakdown products, ADP, AMP and adenosine (see below), may then stimulate a range of membrane receptors (Rs) [3]. These receptors are classified as belonging to two types termed P2 and P1 (Figure 1). In addition, P2Rs can be subdivided into the ligand-activated P2X [2,4,5], and the G protein-coupled P2Y receptor types [6,7]. Adenosine acts on the P1 receptor type, which is also G protein-coupled. P1Rs are either stimulating (A2A, A2B) or inhibiting (A1, A3) adenylate cyclase production via the mediation of G proteins [8]. A further classification identifies seven mammalian subtypes of P2XRs (P2X1-7) and eight mammalian subtypes of P2YRs (P2Y$_1$, P2Y$_2$, P2Y$_4$, P2Y$_6$, P2Y$_{11}$, P2Y$_{12}$, P2Y$_{13}$, P2Y$_{14}$). Whereas P2XRs respond only to ATP, P2YRs respond to ATP/ADP, UTP/UDP, or UDP-glucose. Signaling through P2/P1Rs is rapidly terminated by the conversion of ATP to adenosine and eventually to the inactive inosine within the extracellular space by the activity of ecto-nucleotidases [9,10]. The four major groups of ecto-nucleotidases are the ecto-nucleoside triphosphate diphosphohydrolases (NPTDases), ecto-5′-nucleotidase, ectonucleotide pyrophosphatase/phosphodiesterases, and alkaline phosphatases. Three related family members of NPTDase are expressed in the mammalian brain [11]. NTPDase1 (CD39) hydrolyses nucleoside-5′-triphosphates and -diphosphates eventually to nucleoside-5′-monophosphates equally well. Ecto-5′nucleotidase (CD73) degrades nucleoside-5′monophosphates to the respective nucleoside, e.g., AMP to adenosine.

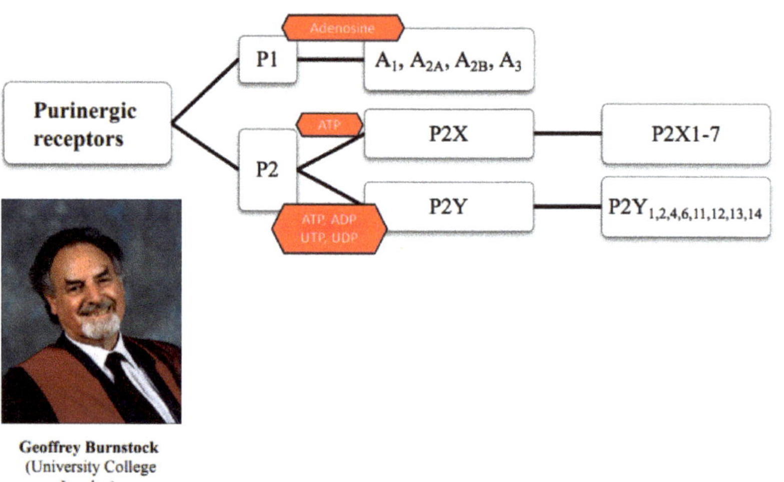

Figure 1. Classification of purinergic receptors. P1Rs consist of four subtypes and respond to the endogenous agonist adenosine. P2Rs consist of two subtypes, the ligand-gated cationic channels P2X (P2X1–7) responding to ATP only, and the G protein-coupled receptors P2Y (P2Y$_{1,2,4,6,11,12,13,14}$). The principal agonists of P2YRs are the following: ATP/ADP (P2Y$_{1,11,12,13}$), UTP/UDP (P2Y$_{2,4,6}$), and UDP-glucose (P2Y$_{14}$). Purinergic receptors were classified by the late Geoffrey Burnstock (see photo), together with his colleague Maria Pia Abbracchio [3].

ATP binds to P2X and, after degradation into its metabolites ADP and adenosine, indirectly activates P2Y and P1Rs, respectively; P2/P1Rs are located at virtually all subsets of immune cells, which has been recognized to be related to a range of biological actions in the immune system [12]. P2X7Rs are distinguished from other P2XRs by their longer C-termini and a very low affinity for ATP when compared to the other P2XRs; this is thought to be the cause of their participation in pathophysiological reactions [13,14]. The P2X7R is widely distributed and functional at the innate cells of the adaptive immune system constituting the first line of defense against invading pathogens. These cells are lymphocytes, granulocytes, macrophages, dendritic cells in peripheral tissues [15,16], and microglia, the resident macrophages of the central nervous system (CNS) [17]. An increasing number of studies show that P2X7Rs play a crucial role in the functions of macrophages by stimulating: (1) the inflammasome, leading to the production of interleukin-1β (IL-1β) and IL-18; (2) the stress-related protein kinase pathway, resulting in apoptosis; (3) the mitogen-activated protein kinase pathway, leading to generation of reactive oxygen and nitrogen species; and (4) phospholipase D, initiating phagosome/lysosome fusion [18]. This review focuses on the state of our present knowledge about P2X7Rs in macrophages during inflammatory processes, and discusses the possibility of P2X7Rs as therapeutic targets to alleviate the deleterious consequences of macrophage activation.

2. The P2X7R at Macrophages

Macrophages were first defined by Elie Metchnikoff (1845–1916) and characterized by their phagocytotic activities to maintain tissue repair and integrity [19]. They are unique innate immune cells that play a prominent role in the host defense by virtue of their ability to rapidly recognize, engulf, and kill pathogens and apoptotic cells critically required for the maintenance of homeostasis. Macrophages are remarkable mononuclear phagocytes that efficiently clear approximately 2×10^{11} erythrocytes each day and remove the worn-out cells and debris generated by tissue remodeling [20]. Macrophages are derived from monocytic precursors in the blood and bone marrow. However, growing evidence suggests that tissue-resident, or tissue-specific, macrophages are embryonically derived and self-renewing [21]. The tissue-resident macrophages are also generated from yolk sac progenitors in spleen, liver, lung, skin, and brain [22]. These macrophages are persistent and are maintained into adulthood by embryonic progenitors, without being replaced by bone marrow- and blood monocyte-derived cells [23]. Macrophages are distributed to various organs and are spread over the whole body, including Langerhans cells of the skin, Kupffer cells of the liver, marginal zone macrophages of the spleen, alveolar macrophages of the lung, and microglia of the CNS [20,24].

The ability of the immune system to recognize molecules that are broadly shared by pathogens is, in part, due to the presence of immune receptors called toll-like receptors (TLRs). In response to TLR ligands (e.g., lipopolysaccharide [LPS], a constituent of the cell membrane of gram negative bacteria) and the cytokine interferon-γ (IFN-γ), macrophages undergo activation to a pro-inflammatory, classic type termed M1 [25,26]. M1 macrophages are amoeboid in shape, are able to phagocytose pathogenic bacteria, and typically release destructive inflammatory mediators such as pro-inflammatory cytokines (interleukin-1β [IL-1β], tumor necrosis factor-α [TNF-α]), chemokines, proteases, reactive oxygen/nitrogen species, and probably also the excitotoxic ATP and glutamate by vesicular exocytosis [27,28]. Macrophages may also undergo M2 activation, which clear cellular debris through phago-cytosis and release numerous protective factors (IL-4, IL-13, nerve growth factor [NGF], fibroblast growth factor [FGF]) [25,26]. M2 macrophages, which can be further divided into M2a, M2b, M2c, and M2d, are characterized by promotion of tissue remodeling, and tumor progression that are associated with resolution of chronic inflammation [25,29,30]. These highly specialized cells often contribute to tissue homeostasis in all organs.

P2XRs show similar tertiary and quaternary architecture, further confirming the hypothesis that all P2XRs belong to the same structural and evolutionary group (Figure 2) [31]. The P2X7R is a non-selective cationic channel gated by high concentrations of ATP leading to permeation of Na^+, K^+, and Ca^{2+} [13,14]. While already somewhat lower concentrations of ATP open the cationic channel, still higher concentrations, especially on repeated application, will create a much larger aqueous pore and allows permeation to molecules of up to 900 Da [13,32] (see later). The P2X7R, similar to the other members of the P2XR family, consists of three subunits (one large extracellular loop, two transmembrane regions, and N- and C-terminal ends) forming a receptor, but each subunit has a much longer C-terminus than that of the other P2XRs [14,33]. The agonist-binding pouch is located at the contact points of two neighboring subunits. There is an increasing recognition that P2X7Rs plays a pivotal role in virtually all immune cell types, especially macrophages [12].

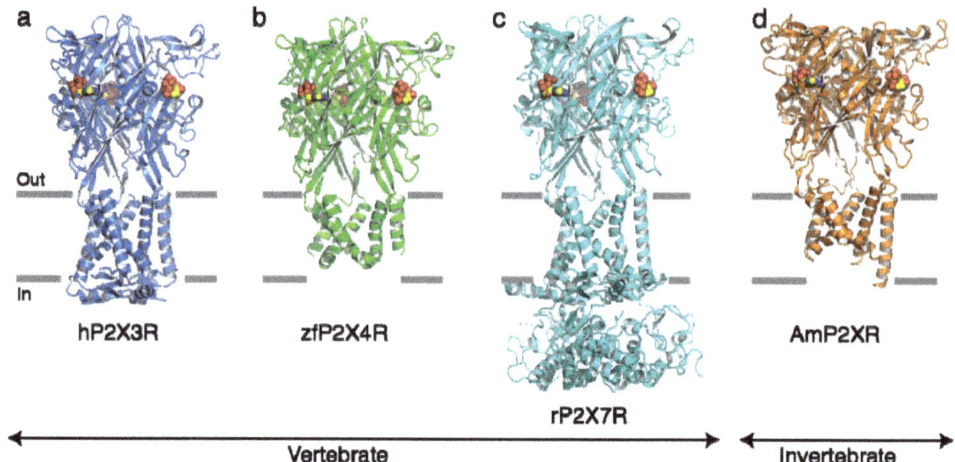

Figure 2. Structures of selected P2XRs. (**a**) Structure of the human P2X3R bound to ATP [34]. The hP2X3R is shown in blue; ATP is shown as spheres in a–d (carbon is yellow, oxygen is red, nitrogen is blue, and phosphorus is orange). Horizontal grey bars indicate the approximate location of the membrane bilayer defining the extracellular (out) and intracellular (in) milieu. (**b**) Structure of the zebrafish P2X4R bound to ATP [35]. The zf P2X4R is shown as a green. (**c**) Structure of the rat P2X7R bound to ATP [36]. The rP2X7R is shown in cyan. (**d**) Structure of the invertebrate *Amblyomma maculatum* P2XR receptor bound to ATP [37]. The AmP2XR is shown in orange. Note the structural similarity between vertebrate and invertebrate P2XRs. For structures having undergone heavy truncations, some membrane spanning helices, as well as the N- and C-termini are lacking in their intracellular sides. Reproduced from [31] with permission.

The P2X7R is widely expressed by myeloid and lymphoid immune cells, as well as by mast cells [15,31]. P2X7Rs can be detected in the majority of blood monocytes, and the expression of immunoreactive P2X7Rs is greatly enhanced, almost 10-fold, as these monocytes develop into differentiated macrophages [38]. Thus, the expression of P2X7Rs is augmented after inflammation, triggering monocytes to differentiate into macrophages [32,39]. The expression of P2X7Rs on monocytes/macrophages was four- to five-fold greater than on lymphocytes because of the larger size and surface area of the former cell type. In contrast, P2X7Rs have been shown to be weakly expressed on neutrophils and platelets [40]. The increased expression of P2X7Rs in activated macrophages was associated with the ability of these cells to differentiate into multinucleated giant cells to form syncytia [39]. The proliferation, differentiation, and apoptosis of macrophages were also mediated by P2X7R activation [41,42]. Recently, it became increasingly apparent that P2X7Rs play a potential role in the macrophage immune responses to pathogens and disorders, such as inflammation [43], cancer [44], oxidative stress [45], and virus infection [46].

3. The Regulation of Pore Formation on Macrophages via P2X7Rs

The P2X7R is abundantly expressed on macrophages and exhibits a variety of functions in innate and adaptive immune responses. Numerous studies reveal that the sequelae of P2X7R activation on macrophages includes the generation of membrane currents [47,48], membrane permeabilization, uptake of large molecules [49], massive perturbations of Na^+, K^+, and Ca^{2+} homeostasis [24,50], inflammasome activation, and interleukin processing [51–53]. Cell membrane blebbing [47,54] and spontaneous cell fusion [55] of macrophages are also mediated by P2X7Rs.

As already mentioned, the activation of P2X7Rs has an ability to open a typical ion channel for small cations, including Ca^{2+}, and subsequently a particular membrane pore for larger molecules (e.g., the fluorescent dyes YO-PRO-1 and ethidium bromide). It has been shown that intracellular Ca^{2+} in murine peritoneal macrophages [52] or human macrophages [47] is elevated after the activation of P2X7Rs by a high concentration of ATP. Although a current carried by the inward flux of Na^+ and Ca^{2+} is generated after the stimulation of all P2XRs (including P2X7Rs), the degree of current desensitization is quite diverse [56]. Noteworthy, P2X7Rs are considered slow desensitizers and, therefore, their activation leads to a continuous, long-lasting influx of Ca^{2+}.

Repetitive activation of P2X7Rs with ATP evoked a biphasic membrane current in human macrophages, which is supposed to reflect large plasma membrane pore generation for the polyatomic cationic dye YO-PRO-1 [47]. Upon stimulation with ATP, a whole-cell membrane current is activated immediately that initially desensitizes and subsequently facilitates upon prolonged channel opening; P2X7Rs promote cation-selective dye uptake while excluding anions (such as anionic calcein) [47]. The application of P2X7R antagonists, such as A-438079 and A-804598, prevents both the current responses to ATP and the ATP-induced uptake of large cationic fluorescent dyes in cultured human microglia [57]. This process is related to several other immunological functions, especially in inflammasome activation [58,59]. Some other effects, such as activation of p38 MAP kinase [60,61], activation of phospholipase D [62,63], the production of reactive oxygen/nitrogen species [56], and killing of *Mycobacterium tuberculosis* [64,65] were also involved in the opening of large membrane pores. P2X7R occupation was further associated with the release of IL-1β, interferon-γ, and reactive oxygen/nitrogen species in murine macrophages. Pore formation is essential for triggering ATP-induced interleukin processing in the immune system.

A particularly intensively discussed issue is whether the initially opened cationic channel dilates and, thereby, establishes the larger diameter pore permeable to cationic dyes or whether two different channels are involved in this effect [66]. Originally, it was suggested, based on equilibrium potential (V_{rev}) measurements with the whole-cell patch clamp technique, that the ion conducting pathway undergoes progressive dilation [67]. However, this suggestion was recently refuted, because the shift in V_{rev} in a medium in which the counter ion of intracellular K^+ was $NMDG^+$ instead of Na^+, emerged due to time-dependent alterations in the concentration of intracellular ions rather than channel dilation [68]. Moreover, during long-lasting activation of P2X7Rs, the single-channel current amplitude and the permeation characteristics remained constant [69].

The P2X7R C-terminal tail constitutes about 40% of the whole protein; it distinguishes P2X7Rs from the other P2XR types and plays a key role in P2X7R ion channel function [70]. The deletion or massive truncation of the C-terminus prevents effects mediated by receptor activation, such as dye uptake and membrane blebbing, but also alters channel kinetics [71]. In addition, there is indication that P2X7R-induced cytolytic pore formation at macrophages is regulated by its unique C-terminal domain [72].

Pannexins are a family of vertebrate proteins identified by their homology to the invertebrate innexins [73]. While innexins are responsible for forming gap junctions in invertebrates, the pannexins have been shown to predominantly exist as large transmembrane channels connecting the intracellular and extracellular space, allowing the passage of ions and small molecules between these two compartments. In contrast to pannexins, connexins are gap junction forming proteins in mammals which may also exist in the hemichannel form exerting analogous effects to pannexins. Pannexin-1 has been identified as a P2X7R-associated protein [74,75]; it is an ATP-permeable channel ubiquitously expressed in macrophages and activation of the pannexin-1 channel mediates ATP release to induce macrophage cell death [76] (Figure 3). This channel was suggested to be responsible for the activation of the large pore induced by P2X7Rs in macrophages and was identified as an upstream molecule essential for the activation of the inflammasome/caspase-1/IL-1β complex in macrophages [59]. The same authors reported that pannexin-1 co-immunoprecipitates with the P2X7R protein, and selective inhibition of pannexin-1 reverses P2X7R-mediated dye uptake without altering P2X7R protein expression [59]. Thus, pannexin-1 in the macrophage membrane has been identified to be a large pore stimulated by P2X7Rs.

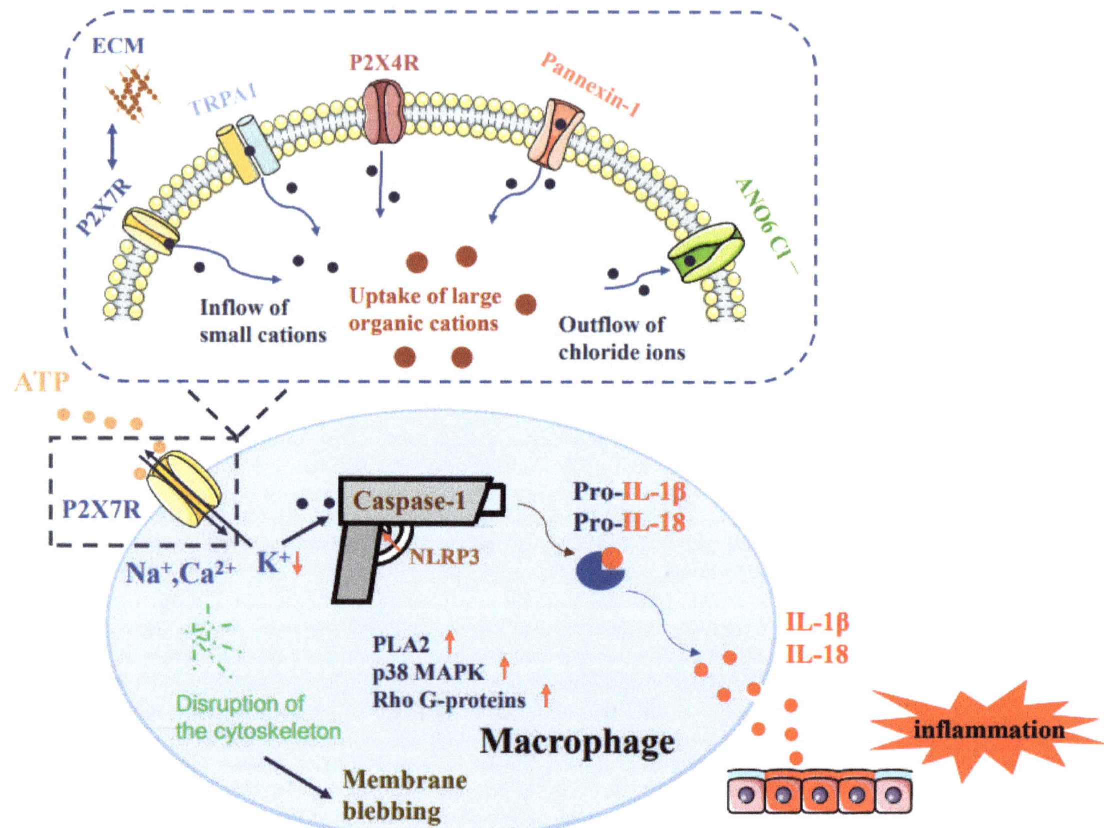

Figure 3. Role of P2X7Rs in macrophages in inducing inflammation. The activation of P2X7Rs by high concentrations of ATP drives the influx of Na⁺/Ca²⁺ and efflux of K⁺ through this plasma membrane-localized receptor channel. A decrease in the intracellular K⁺ concentration is, in co-operation with the stimulation of the toll-like receptor 4 (TLR4) by lipopolysaccharide, a stimulus for the composition and activation of the nucleotide-binding, leucine-rich repeat, pyrin domain containing 3 (NLRP3). NLRP3 then activates caspase-1, which degrades pro-interleukin-1β (pro-IL-1β) to IL-1β. IL-1β is released from macrophages either by membrane diffusion or packed in extracellular vesicles. P2X7R stimulation also leads to the activation of phospholipase A2 (PLA2), p38 mitogen-activated kinase (MAPK), and rho family G-proteins. The pro-inflammatory cytokines IL-1β and IL-18 cause inflammation. In addition, the stimulation of P2X7Rs initiate the opening of large membrane pores in co-operation with the extracellular matrix (ECM), transient receptor potential A1 (TRPA1) channels, P2X4R-channels, pannexin-1 hemichannels, and anoctamin (ANO6) Cl⁻ channels. These membrane pores allow the uptake of otherwise non-permeable cationic molecules of up to 900 Da, triggering the already-mentioned further effects of the P2X7R. Figuratively spoken, P2X7Rs provide bullets to NLRP3/caspase 1, and NLRP3 triggers the gun to release the pro-inflammatory cytokines (see the central part of the lower panel).

However, pannexin-1-induced dye uptake is also observed independent of P2X7R occupation by ATP [77]. Moreover, it was also reported (in contrast to the previously mentioned data [74]) that the pannexin-1 antagonist probenecid and interference RNA (RNAi) targeting of pannexin-1 did not affect the P2X7R macroscopic current in mouse peritoneal macrophages [75]. Similarly, it was found that a range of pannexin-1 inhibitors did not block the ATP-induced cationic dye uptake in cultured human microglia [57].

However, the activation of Cl^- channels appears to be necessary for ATP-induced dye uptake in mouse and human macrophages, and the activity of caspase-1, which is essential for interleukin release, is also blocked by Cl^- channel antagonists [57]. Anoctamin 6 (ANO6), a putative Ca^{2+}-activated Cl^- and non-selective anionic channel, contributes to P2X7R pore formation, and YO-PRO-1 uptake produced by P2X7R activation is attenuated by knockdown of ANO6 [54]. ANO6 may, therefore, constitute another downstream target of P2X7Rs in macrophages. Moreover, P2X7Rs interact with P2X4Rs via their C-termini and disruption of the P2X7/P2X4R interaction hinders responses to high concentrations of ATP in murine macrophages [78].

These rather conflicting results regarding the P2X7R pore formation suggest that more than one channel is probably responsible for dye uptake, and pannexin-1 and/or ANO6 associated with P2X7Rs participate to a certain extent in this process. Nevertheless, a possibility that pannexin-1 allows dye uptake by altering the function of plasma membrane lipids and membrane transporters cannot be ruled out either. In fact, the panda P2X7R, when purified and reconstituted into liposomes, forms a dye permeable pore in the absence of other cellular components [79]. P2X7R channel activity is facilitated by phosphatidylglycerol and sphingomyelin, but dominantly inhibited by cholesterol. Thus, the P2X7R constitutes a lipid composition-dependent dye-permeable pore, whose opening is facilitated by palmitoylated cysteines near the pore-lining helix.

Along this line of thinking, it was recently suggested that current facilitation and macropore formation involve functional complexes comprised of P2X7R and TMEM16, a family of Ca^{2+}-activated ion channel/scramblases [80]. Macropore formation entails two distinct large molecular permeation components, one of which requires functional complexes featuring the TMEM16F subtype, the other likely being direct permeation through the P2X7R pore itself. This idea perfectly complements the previous view that the P2X7R channel allows the passage of large cationic molecules immediately from its initial activation, but at a much slower pace than that of the small cations Na^+, K^+, and Ca^{2+} [16,81].

Membrane blebbing is a characteristic feature of injured cells [82]. Extensive membrane blebbing in murine and human macrophages occurs after prolonged activation of P2X7Rs [54,55,83]. This process is completely blocked by the P2X7R antagonist A-804598 [47], and is associated with reversible disruption of the cytoskeleton, activation of phospholipase A2 (PLA2), p38 MAPK, and Rho G-proteins [47,60,84]. P2X7R driven K^+ efflux and Ca^{2+} influx is also involved in macrophage membrane blebbing.

The macrophage exocytosis, proliferation and cell morphology are regulated by P2X7Rs as well. Convincing evidence proves that P2X7Rs play a crucial role in macrophage exocytosis induced by single-walled carbon nanotubes [85]. The proliferation of microglia almost completely depends on P2X7Rs, and microglia density is also decreased in P2X7R deficient embryos [86]. Additionally, IL-1β, mediated by activation of the P2X7R pore, is crucial for microglial proliferation [87]. Ionized calcium binding adaptor molecule 1(Iba1), as a cytoskeleton protein specific only for microglia and macrophages, is frequently involved in cell migration, and Iba1 silencing enhances P2X7R function [88].

4. The Role of P2X7Rs at Macrophages in Inflammatory Diseases

The P2X7R pore is believed to be an essential component of regulating inflammasome activation in macrophages [33,89]. Inflammation leads to cell damage and massive outflow of ATP into the extracellular space. Macrophages are equipped with a battery of pattern recognition receptors that stereotypically detect pathogen-associated molecular patterns (PAMPs), such as LPS, from bacterial infection or danger-associated molecular patterns (DAMPS), such as ATP [66,90,91]. Activation of macrophages stimulates the release of IL-1β in a two-steps process: the first being the stimulation of toll-like receptor 4 (TLR4) by LPS, leading to accumulation of cytoplasmic pro-IL-1β, and the second being the ATP-dependent stimulation of P2X7Rs, promoting nucleotide-binding, leucine-rich repeat, pyrin domain containing 3 (NLRP3) inflammasome-mediated caspase-1 activation and secretion of IL-1β [92–95]. Caspase-1 generates IL-1β from pro-IL-1β by enzymatic degradation. It is important to note that the decrease of intracellular K^+ after P2X7R stimulation initiates P2X7R-dependent NLRP3 inflammasome activation [15,96].

Indeed, brief stimulation of P2X7Rs is sufficient to initiate the processing of IL-1β in macrophages [97,98]. There is considerable evidence linking P2X7R function with caspase-1 and TLR4 signaling pathways, and prolonged P2X7R-ion channel stimulation is known to induce cytolytic cell death and Ca^{2+} overload as a macrophage death trigger [99]. The release of the typical pro-inflammatory cytokines IL-1β and IL-18 can be observed after about 20–30 min of P2X7R activation and leads to macrophage death [100].

P2X7R signaling also protects against bacterial infections through enhancing bacterial killing by macrophages, which is independent of the inflammasome. ATP release through connexin channels is instrumental in inhibiting inflammation and bacterial burden, and ATP protects against sepsis through activation of P2X7Rs in macrophages by enhancing intracellular bacterial killing [101].

In addition to the classic inflammatory pathways, the interaction between the P2X7R and other proteins located on macrophages is also essential for the inflammatory process. The P2X7R cross-talk with the extracellular matrix (ECM) can be exerted via paxillin. Paxillin is a multi-domain protein that localizes at the ECM; ATP triggers a P2X7R-paxillin interaction in murine and human macrophages to promote NLRP3 deubiquitination and K^+ efflux [102].

Additional inflammasome-independent but P2X7R-dependent pro-inflammatory mechanisms have also been described. (1) A recent study suggested that transient receptor potential ankyrin 1 (TRPA1) was considered to co-localize with P2X7Rs in human cultured macrophages; it mediates Ca^{2+} influx in response to BzATP, partly inhibited by pharmacological blockers of TRPA1 [103]. This process is related to ATP-induced oxidative stress and inflammation. (2) P2X7Rs mediate the pro-inflammatory function of human beta-defensin 2 (HBD2) and human beta-defensin 3 (HBD3), although none of these molecules interact directly with P2X7R but rather induce the release of ATP [104]. (3) The stimulation of P2X7Rs by ATP enhances the transport of extracellular cGAMP into macrophages and subsequently activates STING, which is crucial for the anti-tumor immune response [44].

In contrast to their role in pro-inflammation, P2X7Rs might play an anti-inflammatory role in M2 macrophages [105]. The P2X7R function on anti-inflammation depends on the release of potent anti-inflammatory proteins, such as Annexin A1 [105]. Furthermore, in intermediate M1/M2-polarized macrophages, extracellular ATP acts through its pyrophosphate chains, to inhibit IL-1β release by other stimuli through two independent mechanisms: (1) inhibition of ROS production and (2) trapping of the inflammasome complex through intracellular clustering of actin filaments [106,107].

5. Peripheral Inflammatory Diseases

Multiple studies indicate that the activation of P2X7Rs in macrophages is involved in peripheral inflammatory diseases, including rheumatoid arthritis [53], Crohn's disease [108], liver fibrosis [109,110], sepsis [101,111], renal inflammation [112,113], and pulmonary inflammation [114]. Almost all peripheral inflammatory diseases share the same, classic inflammatory pathway in macrophages with the following components: P2X7R stimulation, NLRP3 and caspase-1 activation, and IL-1β/IL-18 release. However, the fine tuning of the inflammatory pathways is different in various diseases and pathological states.

In rheumatoid arthritis patients, ATP release and P2X7R activation are significantly increased by anti-citrullinated protein antibodies (ACPAs) which are targeted against citrullinated proteins/peptides utilized as rheumatoid arthritis biomarkers [53]. ACPAs promoted IL-1β production by macrophages derived as peripheral blood mononuclear cells. ACPAs interacted with CD147 to enhance the interaction between CD147 and integrin β1 and, in turn, activated the Akt/NF-κB signaling pathway. The nuclear localization of p65 promoted the expression of NLRP3 and pro-IL-1β, resulting in priming.

The involvement of P2X7R-mediated NLRP3 inflammasome activation in IL-1β production in mouse macrophages supplemented with human hepatic stellate cells might contribute to extracellular matrix deposition and suggests that blockade of the P2X7R-NLRP3 inflammasome axis represents a potential therapeutic target for liver fibrosis [109,110]. P2X7Rs have been shown to modulate human THP-1 macrophages interacting with human hepatocytes in cell culture, thereby regulating lipid accumulation in hepatocytes [52].

However, P2X7Rs play a totally different role in sepsis. P2X7R activation on mice peritoneal macrophages suppresses sepsis-induced inflammation and augments killing of intracellular bacteria; connexin hemichannels may have a role in this process [101]. However, an opposite idea was also presented. In sepsis, a systemic blockade of P2X7Rs mitigates inflammatory responses and maintains intestinal barrier function partly by inhibiting the activation of M1 macrophages via the ERK/NF-κB pathway [111].

Cytokine storm, defined as macrophage activation syndrome (MAS), is characterized by the release of multiple pro-inflammatory cytokines (IL-1β, IL-18, IL-6, IL-2, IL-7, TNF-α) and chemokines (CCL2, CCL3) from macrophages [115]. MAS is involved in coronavirus disease-19 (COVID-19), and the blockade of pro-inflammatory cytokines by an anti-IL-6 or IL-6R antibody, such as Tocilizumab, has already been employed for the treatment of MAS and COVID-19 in patients [116,117]. The recombinant human IL-1 receptor antagonist Anakinra is also considered to be a possible therapy for COVID-19 in patients [118]. The numerous similarities between COVID-19 symptoms and those induced by P2X7R activation make it likely that low molecular weight P2X7R antagonists are probably appropriate therapeutic approaches for this disease [119].

6. Neuroinflammation

In the meantime, it is a well-known fact that P2X7Rs amplify CNS damage in neurodegenerative diseases, such as Alzheimer's Disease, Parkinson's disease, amyotrophic lateral sclerosis, multiple sclerosis, post-ischemic conditions, and neurodegeneration as a cause or consequence of epilepsy [120]. It is equally broadly accepted that the resident macrophages of the brain, the P2X7R-bearing microglial cells, mediate this effect. Extracellular β-amyloid (Aβ), a pathognomonic hallmark of Alzheimer's disease, is surrounded by microglia, and the stimulation of P2X7Rs by high local concentrations of ATP originating from damaged CNS cells results in degeneration of nearby neurons [121]. Genetic depletion or pharmacological inhibition of P2X7Rs ameliorated the symptoms observed in various Alzheimer's disease mouse models [122].

The expression of microglial P2X7Rs has been enhanced in a rat model of Parkinson's disease, induced by the intranigral injection of LPS; the application of the P2X7R antagonist Brilliant Blue G (BBG) reduced the activation of microglia and the loss of nigral dopamine neurons by decreasing the phosphorylation level of p38 MAPK [123]. Under these conditions, the deleterious effects of microglial activation are due to the release of cytokines, nitric oxide, and reactive oxygen species [124].

P2X7Rs play a two fold role in the course of amyotrophic lateral sclerosis (ALS) depending on the disease state [125]. BBG decreased microgliosis in an ALS mouse model at late pre-onset rather than at the asymptomatic or pre-symptomatic phases [125]. Consistent with the previous results, the application of the P2X7R antagonist A-804598 also alleviated disease progression in late pre-onset ALS by blocking the phagocytotic activity of SOD-1-G93A mouse microglia [126]. P2X7R antagonists modulate ALS progression by changing the polarization of microglia in SOD1-mutant mice [125,126]. At the late pre-onset phase of the disease, the overexpression of M1 microglial markers is downregulated by BBG treatment, while the anti-inflammatory M2 markers are concomitantly upregulated.

Similar changes in microglial polarization were observed at the later stage of multiple sclerosis or its rodent model, experimental autoimmune encephalomyelitis (EAP) [127]. In addition, P2X7Rs were reported to increase IL-1β, IL-6, and TNF-α release from microglia to induce neuroinflammation at a very early stage of multiple sclerosis, and the prophylactic use of P2X7R antagonists were found to delay the onset and to ameliorate disease progression in the mouse model of EAP [128].

P2X7R expression on microglia was increased during epilepsy induced by the intra-amygdala injection of kainic acid; P2X7R antagonists reduced the accompanying microgliosis and produced enduring suppression of epileptic seizures in mice injected with intra-accumbal kainic acid [129]. The typical P2X7R-dependent inflammatory pathways, including IL-1β, may participate in the ensuing neuronal damage [130].

Microglial P2X7Rs participate also in post-ischemic neurodegeneration. P2X7Rs are widely expressed on microglial cells in both hemispheres after a monolateral brain infarction; in this case, transition from the M1 to the M2 state occurs and a partial protective effect against the post-ischemic neurological impact develops in rats [131].

7. P2X7R Splice Variants and Polymorphisms

A number of P2X7R isoforms derived from alternative splicing were identified both in humans and rodents [31,132]. Some variants are expressed and functional, for example the human (h) P2X7B-R, and mouse and rat P2X7R variants "k". In addition, several non-synonymous, intronic, or missense small nucleotide polymorphisms (SNPs) have been reported for the *hP2RX7* [133]. Macrophages/microglia of various species, including humans, are endowed with gain-of-function or loss-of-function P2X7R SNPs, with important consequences for their physiology/pathophysiology.

Early investigations characterized various polymorphisms, but did not couple their identification with population genetic studies in order to define possible illnesses linked to these mutations. The rs3751143 polymorphic mutant of *P2RX7* coding for the E496A-P2X7R impaired the ATP-induced IL-1β release from human monocytes [134]. It was shown for this SNP that monocytes expressed a non-functional receptor; when transfected into HEK293 cells, at low density, the receptor was non-functional, but regained function at a high receptor density [135]. Apparently, the glutamic acid at position 496 was required for optimal assembly of the P2X7R. Another study characterized the gain-of-function SNP rs2297595 (A166G) and demonstrated that the cysteine-rich domain 1 of P2X7Rs is critical for regulating P2X7R pore function [133]. rs1718119 (A348T-P2X7R) combined with the wild-type P2X7R exhibited enhanced ATP-induced ethidium uptake [136]. The SNP rs28360457 (R307Q) owing a mutation within the ATP binding site caused a massive loss-of-function, especially when it was expressed together with rs1653624 (I568N) [137].

More recently, some of these SNPs were suggested to be involved in various illnesses. Especially, the killing of intracellular bacteria surviving in phagosomes within macrophages depended on undisturbed P2X7R function. It was found that the 1513C allele (coding for E486A; rs200141401) is a risk factor in the development of extrapulmonary and pulmonary tuberculosis in many ethnic populations [18,138,139]. Observation of low P2X7R function in subjects with symptomatic *Toxoplasma gondii*-infected human macrophages showed that the loss-of-function rs3751143 SNP (E496A) significantly reduced P2X7R-mediated parasite killing [140,141].

The risk of age-related macular degeneration was increased by the existence of a rare functional haplotype of the *P2RX4* and *P2RX7* genes in macrophages [142]. Whereas the transfection of wild-type P2X7R in HEK293 cells conferred robust phagocytosis towards latex beads, co-expression of T315C-P2X7R with the G150R-P2X7R (rs28360447) almost completely inhibited phagocytotic activity. Thus, the impairment of the normal scavenger function of macrophages and microglia impaired removal of subretinal deposits and predisposed individuals towards macular degeneration.

A particularly interesting issue is the role of *P2RX7* polymorphism in osteoclast pathophysiology [143]. Osteoclasts are bone cells that are derived from the hematopoietic lineage and are functionally/biologically closely related to macrophages [144]. Genetic linkage studies have shown a clear association between *P2RX7* SNPs in the development of bone loss, osteoporosis, and risk of fractures [145]. Bone mass reduction was found to be, e.g., associated with the loss-of-function SNPs rs1718119 (A348T) and rs3751143 (E496A) [146].

Linkage studies also suggested that SNPs of *P2RX7* are associated with diverse psychiatric and neurological illnesses. In this case, a microglia-based neuro-inflammatory reaction might be causally involved, e.g., in mood disorders. It has been suggested that the SNP rs2230912 coding for the Q460R-P2X7R indicates a predisposition for major depressive disorder (MDD) [66,147,148]. Nonetheless, in the meantime, numerous clinical data failed to confirm this assumption [149,150]. In accordance with the doubts cast on the causal relationship between Q460R-P2X7R and MDD, the ATP-induced inward current was the same through the wild-type P2X7R and the Q460R polymorphic receptor when transfected into HEK293 cells [151]. However, co-expression of the wild-type receptor with the Q460R polymorphic receptor resulted in inhibition of calcium influx and current response to ATP [152]. Similarly, conditional humanized mice co-expressing both P2X7R variants showed alterations in their sleep quality resembling signs of a prodromal MDD state [153]. In conclusion, haplotypes formed between various *P2RX7* SNPs rather than a single receptor-polymorphism was thought to be responsible for MDD predisposition [154].

MDD and bipolar disorder are aggregated in families, and epidemiological studies have found in both cases evidence for genetic susceptibility. However, in the case of bipolar disorder a different *P2RX7* polymorphism (rs1718119 coding for A348T) was reported to be involved in this disease in contrast to that responsible for MDD [148,155].

With respect to neurological illnesses the rare SNP rs28360457 (R307Q) with absent pore formation was suggested to protect against neuroinflammation in multiple sclerosis [156]. The R307Q-P2X7R responsible for this effect was found to be located at monocytes/macrophages.

8. Conclusions

P2X7Rs are highly expressed on macrophages, appear to be crucial for proliferation, differentiation, and apoptosis of this cell type, and are involved in the macrophage immune response to pathogens and peripheral/central disorders of diverse origin. The activation of P2X7Rs initiates pore formation in the plasma membrane of macrophages allowing the entry of large cations into the intracellular space; this leads through interaction with pannexin-1, TRPA1, P2X4, and ANO6 Cl^- channels to inflammasome activation, and consequently, (neuro)inflammation (Figure 3). The assembly of the P2X7R-NLRP3-caspase-1 complex plays a key role in the induction/progression of rheumatoid arthritis, Crohn's disease, liver fibrosis, sepsis, renal inflammation, pulmonary inflammation, and the amplification of neurodegenerative or mood diseases. In conclusion, P2X7Rs might be therapeutic targets to preserve immune homeostasis via their function on macrophages.

Author Contributions: Writing—original draft preparation, W.R.; Writing—review and editing: Y.T., P.R., T.E., and P.I.; Funding acquisition: Y.T., T.E., and P.I. All authors have read and agreed to the published version of the manuscript.

Funding: Our work was made possible by a generous grant ('The Project First-Class Disciplines Development'; CZYHW1901) of the Chengdu University of Traditional Chinese Medicine to Y.T. and P.I. in order to build up "The International Collaborative Center for Purinergic Signaling", and grants of the State Administration of Foreign Affairs to support the stays of P.I. and P.R. in Chengdu (G20190236012). Financial support from the Innovation Team and Talents Cultivation Program of National Administration of Traditional Chinese Medicine (ZYYCXTD-D-202003) and Sichuan Science and Technology Program (2019YFH0108, 22GJHZ0007) are also gratefully acknowledged. We thank the support by the Science Foundation Ireland (17/CDA/4708), and co-funds from the European Regional Development Fund and by FutureNeuro industry partners (16/RC/3948).

Institutional Review Board Statement: Not applicable.

Informed Consent Statement: Not applicable.

Data Availability Statement: Not applicable.

Acknowledgments: This work is dedicated to the recently deceased Geoffrey Burnstock, our ideal and teacher.

Conflicts of Interest: The authors declare no conflict of interest.

References

1. Burnstock, G. Purinergic nerves. *Pharmacol. Rev.* **1972**, *24*, 509–581.
2. Huang, Z.; Xie, N.; Illes, P.; Di Virgilio, F.; Ulrich, H.; Semyanov, A.; Verkhratsky, A.; Sperlagh, B.; Yu, S.-G.; Huang, C.; et al. From purines to purinergic signalling: Molecular functions and human diseases. *Signal Transduct. Target. Ther.* **2021**, *6*, 162. [CrossRef] [PubMed]
3. Abbracchio, M.P.; Burnstock, G. Purinoceptors: Are there families of P2X and P2Y purinoceptors? *Pharmacol. Ther.* **1994**, *64*, 445–475. [CrossRef]
4. Burnstock, G. Purinergic Signaling in the Cardiovascular System. *Circ. Res.* **2017**, *120*, 207–228. [CrossRef] [PubMed]
5. Jarvis, M.F.; Khakh, B.S. ATP-gated P2X cation-channels. *Neuropharmacology* **2009**, *56*, 208–215. [CrossRef]
6. Abbracchio, M.P.; Burnstock, G.; Boeynaems, J.M.; Barnard, E.A.; Boyer, J.L.; Kennedy, C.; Knight, G.E.; Fumagalli, M.; Gachet, C.; Jacobson, K.A.; et al. International Union of Pharmacology, L.V.III: Update on the P2Y G protein-coupled nucleotide receptors: From molecular mechanisms and pathophysiology to therapy. *Pharmacol. Rev.* **2006**, *58*, 281–341. [CrossRef]
7. Jacobson, K.A.; Delicado, E.G.; Gachet, C.; Kennedy, C.; von Kügelgen, I.; Li, B.; Miras-Portugal, M.T.; Novak, I.; Schöneberg, T.; Perez-Sen, R.; et al. Update of P2Y receptor pharmacology: IUPHAR Review 27. *Br. J. Pharmacol.* **2020**, *177*, 2413–2433. [CrossRef]
8. Fredholm, B.B.; IJzerman, A.P.; Jacobson, K.A.; Linden, J.; Müller, C.E. International Union of Basic and Clinical Pharmacology. LXXXI. Nomenclature and classification of adenosine receptors—An update. *Pharmacol. Rev.* **2011**, *63*, 1–34. [CrossRef]
9. Zimmermann, H.; Zebisch, M.; Sträter, N. Cellular function and molecular structure of ecto-nucleotidases. *Purinergic Signal.* **2012**, *8*, 437–502. [CrossRef]
10. Yegutkin, G.G. Enzymes involved in metabolism of extracellular nucleotides and nucleosides: Functional implications and measurement of activities. *Crit. Rev. Biochem. Mol. Biol.* **2014**, *49*, 473–497. [CrossRef] [PubMed]
11. Zimmermann, H.; Braun, N. Ecto-nucleotidases–molecular structures, catalytic properties, and functional roles in the nervous system. *Prog. Brain Res.* **1999**, *120*, 371–385. [PubMed]
12. Burnstock, G.; Boeynaems, J.M. Purinergic signalling and immune cells. *Purinergic Signal.* **2014**, *10*, 529–564. [CrossRef] [PubMed]
13. Surprenant, A.; Rassendren, F.; Kawashima, E.; North, R.A.; Buell, G. The cytolytic P2Z receptor for extracellular ATP identified as a P2X receptor (P2X7). *Science* **1996**, *272*, 735–738. [CrossRef]
14. North, R.A. Molecular physiology of P2X receptors. *Physiol. Rev.* **2002**, *82*, 1013–1067. [CrossRef]
15. Di Virgilio, F.; Dal Ben, D.; Sarti, A.C.; Giuliani, A.L.; Falzoni, S. The P2X7 Receptor in Infection and Inflammation. *Immunity* **2017**, *47*, 15–31. [CrossRef] [PubMed]
16. Di Virgilio, F.; Schmalzing, G.; Markwardt, F. The Elusive P2X7 Macropore. *Trends Cell Biol.* **2018**, *28*, 392–404. [CrossRef]
17. Illes, P.; Rubini, P.; Ulrich, H.; Zhao, Y.; Tang, Y. Regulation of Microglial Functions by Purinergic Mechanisms in the Healthy and Diseased CNS. *Cells* **2020**, *9*, 1108. [CrossRef]
18. Miller, C.M.; Boulter, N.R.; Fuller, S.J.; Zakrzewski, A.M.; Lees, M.P.; Saunders, B.M.; Wiley, J.S.; Smith, N.C. The role of the P2X7 receptor in infectious diseases. *PLoS Pathog.* **2011**, *7*, e1002212. [CrossRef]
19. Lavin, Y.; Merad, M. Macrophages: Gatekeepers of tissue integrity. *Cancer Immunol. Res.* **2013**, *1*, 201–209. [CrossRef]
20. Mosser, D.M.; Edwards, J.P. Exploring the full spectrum of macrophage activation. *Nat. Rev. Immunol.* **2008**, *8*, 958–969. [CrossRef]
21. Jakubzick, C.V.; Randolph, G.J.; Henson, P.M. Monocyte differentiation and antigen-presenting functions. *Nat. Rev. Immunol.* **2017**, *17*, 349–362. [CrossRef] [PubMed]
22. Wynn, T.A.; Chawla, A.; Pollard, J.W. Macrophage biology in development, homeostasis and disease. *Nature* **2013**, *496*, 445–455. [CrossRef] [PubMed]
23. Ginhoux, F.; Jung, S. Monocytes and macrophages: Developmental pathways and tissue homeostasis. *Nat. Rev. Immunol.* **2014**, *14*, 392–404. [CrossRef]
24. Layhadi, J.A.; Fountain, S.J. P2X4 Receptor-Dependent Ca^{2+} Influx in Model Human Monocytes and Macrophages. *Int. J. Mol. Sci.* **2017**, *18*, 2261. [CrossRef]
25. Sica, A.; Mantovani, A. Macrophage plasticity and polarization: In vivo veritas. *J. Clin. Investig.* **2012**, *122*, 787–795. [CrossRef] [PubMed]
26. Amici, S.A.; Dong, J.; Guerau-de-Arellano, M. Molecular Mechanisms Modulating the Phenotype of Macrophages and Microglia. *Front. Immunol.* **2017**, *8*, 1520. [CrossRef]
27. Kigerl, K.A.; Gensel, J.C.; Ankeny, D.P.; Alexander, J.K.; Donnelly, D.J.; Popovich, P.G. Identification of two distinct macrophage subsets with divergent effects causing either neurotoxicity or regeneration in the injured mouse spinal cord. *J. Neurosci.* **2009**, *29*, 13435–13444. [CrossRef]
28. Klaver, D.; Thurnher, M. Control of Macrophage Inflammation by P2Y Purinergic Receptors. *Cells* **2021**, *10*, 1098. [CrossRef] [PubMed]

29. Duluc, D.; Delneste, Y.; Tan, F.; Moles, M.P.; Grimaud, L.; Lenoir, J.; Preisser, L.; Anegon, I.; Catala, L.; Ifrah, N.; et al. Tumor-associated leukemia inhibitory factor and IL-6 skew monocyte differentiation into tumor-associated macrophage-like cells. *Blood* **2007**, *110*, 4319–4330. [CrossRef]
30. Martinez, F.O.; Sica, A.; Mantovani, A.; Locati, M. Macrophage activation and polarization. *Front. Biosci.* **2008**, *13*, 453–461. [CrossRef]
31. Illes, P.; Müller, C.E.; Jacobson, K.A.; Grutter, T.; Nicke, A.; Fountain, S.J.; Kennedy, C.; Schmalzing, G.; Jarvis, M.F.; Stojilkovic, S.S.; et al. Update of P2X receptor properties and their pharmacology: IUPHAR Review 30. *Br. J. Pharmacol.* **2021**, *178*, 489–514. [CrossRef] [PubMed]
32. Buell, G.; Chessell, I.P.; Michel, A.D.; Collo, G.; Salazzo, M.; Herren, S.; Gretener, D.; Grahames, C.; Kaur, R.; Kosco-Vilbois, M.H.; et al. Blockade of human P2X7 receptor function with a monoclonal antibody. *Blood* **1998**, *92*, 3521–3528. [CrossRef]
33. Burnstock, G.; Knight, G.E. The potential of P2X7 receptors as a therapeutic target, including inflammation and tumour progression. *Purinergic Signal.* **2018**, *14*, 1–18. [CrossRef]
34. Mansoor, S.E.; Lü, W.; Oosterheert, W.; Shekhar, M.; Tajkhorshid, E.; Gouaux, E. X-ray structures define human P2X3 receptor gating cycle and antagonist action. *Nature* **2016**, *538*, 66–71. [CrossRef]
35. Hattori, M.; Gouaux, E. Molecular mechanism of ATP binding and ion channel activation in P2X receptors. *Nature* **2012**, *485*, 207–212. [CrossRef]
36. McCarthy, A.E.; Yoshioka, C.; Mansoor, S.E. Full-Length P2X7 Structures Reveal How Palmitoylation Prevents Channel Desensitization. *Cell* **2019**, *179*, 659–670. [CrossRef]
37. Kasuya, G.; Fujiwara, Y.; Takemoto, M.; Dohmae, N.; Nakada-Nakura, Y.; Ishitani, R.; Hattori, M.; Nureki, O. Structural Insights into Divalent Cation Modulations of ATP-Gated P2X Receptor Channels. *Cell Rep.* **2016**, *14*, 932–944. [CrossRef] [PubMed]
38. Gudipaty, L.; Humphreys, B.D.; Buell, G.; Dubyak, G.R. Regulation of P2X(7) nucleotide receptor function in human monocytes by extracellular ions and receptor density. *Am. J. Physiol. Cell Physiol.* **2001**, *280*, C943–C953. [CrossRef] [PubMed]
39. Falzoni, S.; Munerati, M.; Ferrari, D.; Spisani, S.; Moretti, S.; Di Virgilio, F. The purinergic P2Z receptor of human macrophage cells. Characterization and possible physiological role. *J. Clin. Investig.* **1995**, *95*, 1207–1216. [CrossRef]
40. Gu, B.J.; Zhang, W.Y.; Bendall, L.J.; Chessell, I.P.; Buell, G.N.; Wiley, J.S. Expression of P2X7 purinoceptors on human lymphocytes and monocytes: Evidence for nonfunctional P2X7 receptors. *Am. J. Physiol. Cell Physiol.* **2000**, *279*, C1189–C1197. [CrossRef] [PubMed]
41. Monif, M.; Burnstock, G.; Williams, D.A. Microglia: Proliferation and activation driven by the P2X7 receptor. *Int. J. Biochem. Cell Biol.* **2010**, *42*, 1753–1756. [CrossRef]
42. Qin, J.; Zhang, X.; Tan, B.; Zhang, S.; Yin, C.; Xue, Q.; Zhang, Z.; Ren, H.; Chen, J.; Liu, M.; et al. Blocking P2X7-Mediated Macrophage Polarization Overcomes Treatment Resistance in Lung Cancer. *Cancer Immunol. Res.* **2020**, *8*, 1426–1439. [CrossRef] [PubMed]
43. Raneia E Silva, P.A.; de Lima, D.S.; Mesquita Luiz, J.P.; Camara, N.O.S.; Alves-Filho, J.C.F.; Pontillo, A.; Bortoluci, K.R.; Faquim-Mauro, E.L. Inflammatory effect of Bothropstoxin-I from Bothrops jararacussu venom mediated by NLRP3 inflammasome involves ATP and P2X7 receptor. *Clin. Sci.* **2021**, *135*, 687–701. [CrossRef] [PubMed]
44. Zhou, Y.; Fei, M.; Zhang, G.; Liang, W.C.; Lin, W.; Wu, Y.; Piskol, R.; Ridgway, J.; McNamara, E.; Huang, H.; et al. Blockade of the Phagocytic Receptor MerTK on Tumor-Associated Macrophages Enhances P2X7R-Dependent, STING Activation by Tumor-Derived cGAMP. *Immunity* **2020**, *52*, 357–373. [CrossRef] [PubMed]
45. Gallenga, C.E.; Lonardi, M.; Pacetti, S.; Violanti, S.S.; Tassinari, P.; Di Virgilio, F.; Tognon, M.; Perri, P. Molecular Mechanisms Related to Oxidative Stress in Retinitis Pigmentosa. *Antioxidants* **2021**, *10*, 848. [CrossRef] [PubMed]
46. Xu, S.L.; Lin, Y.; Liu, W.; Zhu, X.Z.; Liu, D.; Tong, M.L.; Liu, L.L.; Lin, L.R. The P2X7 receptor mediates NLRP3-dependent IL-1beta secretion and promotes phagocytosis in the macrophage response to Treponema pallidum. *Int. Immunopharmacol.* **2020**, *82*, 106344. [CrossRef]
47. Janks, L.; Sprague, R.S.; Egan, T.M. ATP-Gated P2X7 Receptors Require Chloride Channels To Promote Inflammation in Human Macrophages. *J. Immunol.* **2019**, *202*, 883–898. [CrossRef] [PubMed]
48. Hempel, C.; Nörenberg, W.; Sobottka, H.; Urban, N.; Nicke, A.; Fischer, W.; Schaefer, M. The phenothiazine-class antipsychotic drugs prochlorperazine and trifluoperazine are potent allosteric modulators of the human P2X7 receptor. *Neuropharmacology* **2013**, *75*, 365–379. [CrossRef]
49. Nurkhametova, D.; Siniavin, A.; Streltsova, M.; Kudryavtsev, D.; Kudryavtsev, I.; Giniatullina, R.; Tsetlin, V.; Malm, T.; Giniatullin, R. Does Cholinergic Stimulation Affect the P2X7 Receptor-Mediated Dye Uptake in Mast Cells and Macrophages? *Front. Cell. Neurosci.* **2020**, *14*, 548376. [CrossRef] [PubMed]
50. Schachter, J.; Motta, A.P.; de Souza, Z.A.; da Silva-Souza, H.A.; Guimaraes, M.Z.; Persechini, P.M. ATP-induced P2X7-associated uptake of large molecules involves distinct mechanisms for cations and anions in macrophages. *J. Cell Sci.* **2008**, *121 Pt 19*, 3261–3270. [CrossRef]
51. Yang, C.; Shi, S.; Su, Y.; Tong, J.S.; Li, L. P2X7R promotes angiogenesis and tumour-associated macrophage recruitment by regulating the, N.F.-κB signalling pathway in colorectal cancer cells. *J. Cell. Mol. Med.* **2020**, *24*, 10830–10841. [CrossRef]
52. Zhang, Y.; Jiang, M.; Cui, B.W.; Jin, C.H.; Wu, Y.L.; Shang, Y.; Yang, H.X.; Wu, M.; Liu, J.; Qiao, C.Y.; et al. P2X7 receptor-targeted regulation by tetrahydroxystilbene glucoside in alcoholic hepatosteatosis: A new strategy towards macrophage-hepatocyte crosstalk. *Br. J. Pharmacol.* **2020**, *177*, 2793–2811. [CrossRef]

53. Dong, X.; Zheng, Z.; Lin, P.; Fu, X.; Li, F.; Jiang, J.; Jiang, J.; Zhu, P. ACPAs promote IL-1-beta production in rheumatoid arthritis by activating the NLRP3 inflammasome. *Cell. Mol. Immunol.* **2020**, *17*, 261–271. [CrossRef]

54. Ousingsawat, J.; Wanitchakool, P.; Kmit, A.; Romao, A.M.; Jantarajit, W.; Schreiber, R.; Kunzelmann, K. Anoctamin 6 mediates effects essential for innate immunity downstream of P2X7 receptors in macrophages. *Nat. Commun.* **2015**, *6*, 6245. [CrossRef]

55. Moreira-Souza, A.C.A.; Almeida-da-Silva, C.L.C.; Rangel, T.P.; Rocha, G.D.C.; Bellio, M.; Zamboni, D.S.; Vommaro, R.C.; Coutinho-Silva, R. The P2X7 Receptor Mediates Toxoplasma gondii Control in Macrophages through Canonical NLRP3 Inflammasome Activation and Reactive Oxygen Species Production. *Front. Immunol.* **2017**, *8*, 1257. [CrossRef]

56. Marques-da-Silva, C.; Chaves, M.M.; Castro, N.G.; Coutinho-Silva, R.; Guimaraes, M.Z. Colchicine inhibits cationic dye uptake induced by ATP in P2X2 and P2X7 receptor-expressing cells: Implications for its therapeutic action. *Br. J. Pharmacol.* **2011**, *163*, 912–926. [CrossRef] [PubMed]

57. Janks, L.; Sharma, C.V.R.; Egan, T.M. A central role for P2X7 receptors in human microglia. *J. Neuroinflammation* **2018**, *15*, 325. [CrossRef]

58. Yaron, J.R.; Gangaraju, S.; Rao, M.Y.; Kong, X.; Zhang, L.; Su, F.; Tian, Y.; Glenn, H.L.; Meldrum, D.R. K$^+$ regulates Ca^{2+} to drive inflammasome signaling: Dynamic visualization of ion flux in live cells. *Cell Death Dis.* **2015**, *6*, e1954. [CrossRef] [PubMed]

59. Pelegrin, P.; Surprenant, A. Pannexin-1 mediates large pore formation and interleukin-1beta release by the ATP-gated P2X7 receptor. *EMBO J.* **2006**, *25*, 5071–5082. [CrossRef]

60. Pfeiffer, Z.A.; Aga, M.; Prabhu, U.; Watters, J.J.; Hall, D.J.; Bertics, P.J. The nucleotide receptor P2X7 mediates actin reorganization and membrane blebbing in, R.A.W 264.7 macrophages via p38 MAP kinase and Rho. *J. Leukoc. Biol.* **2004**, *75*, 1173–1182. [CrossRef] [PubMed]

61. Noguchi, T.; Ishii, K.; Fukutomi, H.; Naguro, I.; Matsuzawa, A.; Takeda, K.; Ichijo, H. Requirement of reactive oxygen species-dependent activation of ASK1-p38 MAPK pathway for extracellular ATP-induced apoptosis in macrophage. *J. Biol. Chem.* **2008**, *283*, 7657–7665. [CrossRef]

62. Le, S.H.; Raymond, M.N. P2X7 receptor-mediated phosphatidic acid production delays ATP-induced pore opening and cytolysis of, R.A.W 264.7 macrophages. *Cell. Signal.* **2007**, *19*, 1909–1918.

63. Thomas, L.M.; Salter, R.D. Activation of macrophages by P2X7-induced microvesicles from myeloid cells is mediated by phospholipids and is partially dependent on, T.L.R4. *J. Immunol.* **2010**, *185*, 3740–3749. [CrossRef]

64. Matty, M.A.; Knudsen, D.R.; Walton, E.M.; Beerman, R.W.; Cronan, M.R.; Pyle, C.J.; Hernandez, R.E.; Tobin, D.M. Potentiation of P2RX7 as a host-directed strategy for control of mycobacterial infection. *Elife* **2019**, *8*, e39123. [CrossRef]

65. Bomfim, C.C.B.; Amaral, E.P.; Cassado, A.D.A.; Salles, A.; do Nascimento, R.S.; Lasunskaia, E.; Hirata, M.H.; Álvarez, J.M.; D'Império-Lima, M.R. P2X7 Receptor in Bone Marrow-Derived Cells Aggravates Tuberculosis Caused by Hypervirulent Mycobacterium bovis. *Front. Immunol.* **2017**, *8*, 435. [CrossRef] [PubMed]

66. Illes, P.; Verkhratsky, A.; Tang, Y. Pathological ATPergic Signaling in Major Depression and Bipolar Disorder. *Front. Mol. Neurosci.* **2019**, *12*, 331. [CrossRef]

67. Virginio, C.; MacKenzie, A.; Rassendren, F.A.; North, R.A.; Surprenant, A. Pore dilation of neuronal P2X receptor channels. *Nat. Neurosci.* **1999**, *2*, 315–321. [CrossRef] [PubMed]

68. Li, M.; Toombes, G.E.; Silberberg, S.D.; Swartz, K.J. Physical basis of apparent pore dilation of ATP-activated P2X receptor channels. *Nat. Neurosci.* **2015**, *18*, 1577–1583. [CrossRef] [PubMed]

69. Pippel, A.; Stolz, M.; Woltersdorf, R.; Kless, A.; Schmalzing, G.; Markwardt, F. Localization of the gate and selectivity filter of the full-length P2X7 receptor. *Proc. Natl. Acad. Sci. USA* **2017**, *114*, E2156–E2165. [CrossRef]

70. Markwardt, F. Human P2X7 receptors-Properties of single ATP-gated ion channels. *Biochem. Pharmacol.* **2021**, *187*, 114307. [CrossRef]

71. Kopp, R.; Krautloher, A.; Ramirez-Fernandez, A.; Nicke, A. P2X7 Interactions and Signaling-Making Head or Tail of It. *Front. Mol. Neurosci.* **2019**, *12*, 183. [CrossRef]

72. Smart, M.L.; Gu, B.; Panchal, R.G.; Wiley, J.; Cromer, B.; Williams, D.A.; Petrou, S. P2X7 receptor cell surface expression and cytolytic pore formation are regulated by a distal C-terminal region. *J. Biol. Chem.* **2003**, *278*, 8853–8860. [CrossRef]

73. Panchin, Y.; Kelmanson, I.; Matz, M.; Lukyanov, K.; Usman, N.; Lukyanov, S. A ubiquitous family of putative gap junction molecules. *Curr. Biol.* **2000**, *10*, R473–R474. [CrossRef]

74. Pelegrin, P.; Surprenant, A. The P2X7 receptor-pannexin connection to dye uptake and IL-1beta release. *Purinergic Signal.* **2009**, *5*, 129–137. [CrossRef] [PubMed]

75. Alberto, A.V.; Faria, R.X.; Couto, C.G.; Ferreira, L.G.; Souza, C.A.; Teixeira, P.C.; Fróes, M.M.; Alves, L.A. Is pannexin the pore associated with the P2X7 receptor? *Naunyn Schmiedebergs Arch. Pharmacol.* **2013**, *386*, 775–787. [CrossRef] [PubMed]

76. Jung, B.C.; Kim, S.H.; Lim, J.; Kim, Y.S. Activation of pannexin-1 mediates triglyceride-induced macrophage cell death. *BMB Rep.* **2020**, *53*, 588–593. [CrossRef]

77. Garre, J.M.; Yang, G.; Bukauskas, F.F.; Bennett, M.V. FGF-1 Triggers Pannexin-1 Hemichannel Opening in Spinal Astrocytes of Rodents and Promotes Inflammatory Responses in Acute Spinal Cord Slices. *J. Neurosci.* **2016**, *36*, 4785–4801. [CrossRef] [PubMed]

78. Perez-Flores, G.; Lévesque, S.A.; Pacheco, J.; Vaca, L.; Lacroix, S.; Pérez-Cornejo, P.; Arreola, J. The P2X7/P2X4 interaction shapes the purinergic response in murine macrophages. *Biochem. Biophys. Res. Commun.* **2015**, *467*, 484–490. [CrossRef]

79. Karasawa, A.; Michalski, K.; Mikhelzon, P.; Kawate, T. The P2X7 receptor forms a dye-permeable pore independent of its intracellular domain but dependent on membrane lipid composition. *Elife* **2017**, *6*, e31186. [CrossRef]
80. Dunning, K.; Martz, A.; Peralta, F.A.; Cevoli, F.; Boue-Grabot, E.; Compan, V.; Gautherat, F.; Wolf, P.; Chataigneau, T.; Grutter, T. P2X7 Receptors and, T.M.EM16 Channels Are Functionally Coupled with Implications for Macropore Formation and Current Facilitation. *Int. J. Mol. Sci.* **2021**, *22*, 6542. [CrossRef]
81. Harkat, M.; Peverini, L.; Cerdan, A.H.; Dunning, K.; Beudez, J.; Martz, A.; Calimet, N.; Specht, A.; Cecchini, M.; Chataigneau, T.; et al. On the permeation of large organic cations through the pore of, ATP-gated P2X receptors. *Proc. Natl. Acad. Sci. USA* **2017**, *114*, E3786–E3795. [CrossRef]
82. Babiychuk, E.B.; Monastyrskaya, K.; Potez, S.; Draeger, A. Blebbing confers resistance against cell lysis. *Cell Death Differ.* **2011**, *18*, 80–89. [CrossRef]
83. Taylor, S.R.; Turner, C.M.; Elliott, J.I.; McDaid, J.; Hewitt, R.; Smith, J.; Pickering, M.C.; Whitehouse, D.L.; Cook, H.T.; Burnstock, G.; et al. P2X7 deficiency attenuates renal injury in experimental glomerulonephritis. *J. Am. Soc. Nephrol.* **2009**, *20*, 1275–1281. [CrossRef] [PubMed]
84. Jiang, L.H.; Rassendren, F.; MacKenzie, A.; Zhang, Y.H.; Surprenant, A.; North, R.A. N-methyl-D-glucamine and propidium dyes utilize different permeation pathways at rat P2X7 receptors. *Am. J. Physiol. Cell Physiol.* **2005**, *289*, C1295–C1302. [CrossRef] [PubMed]
85. Cui, X.; Wan, B.; Yang, Y.; Ren, X.; Guo, L.H.; Zhang, H. Crucial Role of P2X7 Receptor in Regulating Exocytosis of Single-Walled Carbon Nanotubes in Macrophages. *Small* **2016**, *12*, 5998–6011. [CrossRef] [PubMed]
86. Rigato, C.; Swinnen, N.; Buckinx, R.; Couillin, I.; Mangin, J.M.; Rigo, J.M.; Legendre, P.; Le Corronc, H. Microglia proliferation is controlled by P2X7 receptors in a Pannexin-1-independent manner during early embryonic spinal cord invasion. *J. Neurosci.* **2012**, *32*, 11559–11573. [CrossRef]
87. Monif, M.; Reid, C.A.; Powell, K.L.; Drummond, K.J.; O'Brien, T.J.; Williams, D.A. Interleukin-1beta has trophic effects in microglia and its release is mediated by P2X7R pore. *J. Neuroinflammation* **2016**, *13*, 173. [CrossRef] [PubMed]
88. Gheorghe, R.O.; Deftu, A.; Filippi, A.; Grosu, A.; Bica-Popi, M.; Chiritoiu, M.; Chiritoiu, G.; Munteanu, C.; Silvestro, L.; Ristoiu, V. Silencing the Cytoskeleton Protein Iba1 (Ionized Calcium Binding Adapter Protein 1) Interferes with, B.V.2 Microglia Functioning. *Cell. Mol. Neurobiol.* **2020**, *40*, 1011–1027. [CrossRef] [PubMed]
89. Savio, L.E.B.; de Andrade, M.P.; da Silva, C.G.; Coutinho-Silva, R. The P2X7 Receptor in Inflammatory Diseases: Angel or Demon? *Front. Pharmacol.* **2018**, *9*, 52. [CrossRef]
90. Shao, B.Z.; Xu, Z.Q.; Han, B.Z.; Su, D.F.; Liu, C. NLRP3 inflammasome and its inhibitors: A review. *Front Pharmacol.* **2015**, *6*, 262. [CrossRef] [PubMed]
91. Young, C.N.J.; Gorecki, D.C. P2RX7 Purinoceptor as a Therapeutic Target-The Second Coming? *Front. Chem.* **2018**, *6*, 248. [CrossRef]
92. Perregaux, D.G.; Gabel, C.A. Post-translational processing of murine IL-1: Evidence that ATP-induced release of IL-1 alpha and IL-1 beta occurs via a similar mechanism. *J. Immunol.* **1998**, *160*, 2469–2477. [PubMed]
93. Abderrazak, A.; Syrovets, T.; Couchie, D.; El, H.K.; Friguet, B.; Simmet, T.; Rouis, M. NLRP3 inflammasome: From a danger signal sensor to a regulatory node of oxidative stress and inflammatory diseases. *Redox Biol.* **2015**, *4*, 296–307. [CrossRef] [PubMed]
94. Jeong, Y.H.; Walsh, M.C.; Yu, J.; Shen, H.; Wherry, E.J.; Choi, Y. Mice Lacking the Purinergic Receptor P2X5 Exhibit Defective Inflammasome Activation and Early Susceptibility to Listeria monocytogenes. *J. Immunol.* **2020**, *205*, 760–766. [CrossRef] [PubMed]
95. Li, L.H.; Chen, T.L.; Chiu, H.W.; Hsu, C.H.; Wang, C.C.; Tai, T.T.; Ju, T.C.; Chen, F.H.; Chernikov, O.V.; Tsai, W.C.; et al. Critical Role for the NLRP3 Inflammasome in Mediating IL-1beta Production in Shigella sonnei-Infected Macrophages. *Front. Immunol.* **2020**, *11*, 1115. [CrossRef] [PubMed]
96. Munoz-Planillo, R.; Kuffa, P.; Martinez-Colon, G.; Smith, B.L.; Rajendiran, T.M.; Nunez, G. K$^+$ efflux is the common trigger of NLRP3 inflammasome activation by bacterial toxins and particulate matter. *Immunity* **2013**, *38*, 1142–1153. [CrossRef]
97. Brough, D.; Le Feuvre, R.A.; Wheeler, R.D.; Solovyova, N.; Hilfiker, S.; Rothwell, N.J.; Verkhratsky, A. Ca^{2+} stores and Ca^{2+} entry differentially contribute to the release of IL-1 beta and IL-1 alpha from murine macrophages. *J. Immunol.* **2003**, *170*, 3029–3036. [CrossRef]
98. Kahlenberg, J.M.; Dubyak, G.R. Mechanisms of caspase-1 activation by P2X7 receptor-mediated K+ release. *Am. J. Physiol. Cell Physiol.* **2004**, *286*, C1100–C1108. [CrossRef]
99. Hanley, P.J.; Kronlage, M.; Kirschning, C.; Del, R.A.; Di Virgilio, F.; Leipziger, J.; Chessell, I.P.; Sargin, S.; Filippov, M.A.; Lindemann, O.; et al. Transient P2X7 receptor activation triggers macrophage death independent of Toll-like receptors 2 and 4, caspase-1, and pannexin-1 proteins. *J. Biol. Chem.* **2012**, *287*, 10650–10663. [CrossRef]
100. Le Feuvre, R.A.; Brough, D.; Iwakura, Y.; Takeda, K.; Rothwell, N.J. Priming of macrophages with lipopolysaccharide potentiates P2X7-mediated cell death via a caspase-1-dependent mechanism, independently of cytokine production. *J. Biol. Chem.* **2002**, *277*, 3210–3218. [CrossRef]
101. Csoka, B.; Németh, Z.H.; Törö, G.; Idzko, M.; Zech, A.; Koscsó, B.; Spolarics, Z.; Antonioli, L.; Cseri, K.; Erdélyi, K.; et al. Extracellular ATP protects against sepsis through macrophage P2X7 purinergic receptors by enhancing intracellular bacterial killing. *FASEB J.* **2015**, *29*, 3626–3637. [CrossRef] [PubMed]

102. Wang, W.; Hu, D.; Feng, Y.; Wu, C.; Song, Y.; Liu, W.; Li, A.; Wang, Y.; Chen, K.; Tian, M.; et al. Paxillin mediates ATP-induced activation of P2X7 receptor and NLRP3 inflammasome. *BMC Biol.* **2020**, *18*, 182. [CrossRef] [PubMed]
103. Tian, C.; Han, X.; He, L.; Tang, F.; Huang, R.; Lin, Z.; Li, S.; Deng, S.; Xu, J.; Huang, H.; et al. Transient receptor potential ankyrin 1 contributes to the ATP-elicited oxidative stress and inflammation in THP-1-derived macrophage. *Mol. Cell. Biochem.* **2020**, *473*, 179–192. [CrossRef]
104. Wanke, D.; Mauch-Mücke, K.; Holler, E.; Hehlgans, T. Human beta-defensin-2 and -3 enhance pro-inflammatory cytokine expression induced by, T.L.R ligands via ATP-release in a P2X7R dependent manner. *Immunobiology* **2016**, *221*, 1259–1265. [CrossRef] [PubMed]
105. de Torre-Minguela, C.; Barbera-Cremades, M.; Gomez, A.I.; Martin-Sanchez, F.; Pelegrin, P. Macrophage activation and polarization modify P2X7 receptor secretome influencing the inflammatory process. *Sci. Rep.* **2016**, *6*, 22586. [CrossRef]
106. Pelegrin, P.; Surprenant, A. Dynamics of macrophage polarization reveal new mechanism to inhibit IL-1beta release through pyrophosphates. *EMBO J.* **2009**, *28*, 2114–2127. [CrossRef]
107. Lopez-Castejon, G.; Baroja-Mazo, A.; Pelegrin, P. Novel macrophage polarization model: From gene expression to identification of new anti-inflammatory molecules. *Cell. Mol. Life Sci.* **2011**, *68*, 3095–3107. [CrossRef]
108. Neves, A.R.; Castelo-Branco, M.T.; Figliuolo, V.R.; Bernardazzi, C.; Buongusto, F.; Yoshimoto, A.; Nanini, H.F.; Coutinho, C.M.; Carneiro, A.J.; Coutinho-Silva, R.; et al. Overexpression of ATP-activated P2X7 receptors in the intestinal mucosa is implicated in the pathogenesis of Crohn's disease. *Inflamm. Bowel Dis.* **2014**, *20*, 444–457. [CrossRef] [PubMed]
109. Jiang, S.; Zhang, Y.; Zheng, J.H.; Li, X.; Yao, Y.L.; Wu, Y.L.; Song, S.Z.; Sun, P.; Nan, J.X.; Lian, L.H. Potentiation of hepatic stellate cell activation by extracellular ATP is dependent on P2X7R-mediated NLRP3 inflammasome activation. *Pharmacol. Res.* **2017**, *117*, 82–93. [CrossRef] [PubMed]
110. Jiang, M.; Cui, B.W.; Wu, Y.L.; Zhang, Y.; Shang, Y.; Liu, J.; Yang, H.X.; Qiao, C.Y.; Zhan, Z.Y.; Ye, H.; et al. P2X7R orchestrates the progression of murine hepatic fibrosis by making a feedback loop from macrophage to hepatic stellate cells. *Toxicol. Lett.* **2020**, *333*, 22–32. [CrossRef]
111. Wu, X.; Ren, J.; Chen, G.; Wu, L.; Song, X.; Li, G.; Deng, Y.; Wang, G.; Gu, G.; Li, J. Systemic blockade of P2X7 receptor protects against sepsis-induced intestinal barrier disruption. *Sci. Rep.* **2017**, *7*, 4364. [CrossRef]
112. Pereira, J.M.S.; Barreira, A.L.; Gomes, C.R.; Ornellas, F.M.; Ornellas, D.S.; Miranda, L.C.; Cardoso, L.R.; Coutinho-Silva, R.; Schanaider, A.; Morales, M.M.; et al. Brilliant blue, G.; a P2X7 receptor antagonist, attenuates early phase of renal inflammation, interstitial fibrosis and is associated with renal cell proliferation in ureteral obstruction in rats. *BMC Nephrol.* **2020**, *21*, 206. [CrossRef]
113. Koo, T.Y.; Lee, J.G.; Yan, J.J.; Jang, J.Y.; Ju, K.D.; Han, M.; Oh, K.H.; Ahn, C.; Yang, J. The P2X7 receptor antagonist, oxidized adenosine triphosphate, ameliorates renal ischemia-reperfusion injury by expansion of regulatory T cells. *Kidney Int.* **2017**, *92*, 415–431. [CrossRef] [PubMed]
114. Kang, M.J.; Jo, S.G.; Kim, D.J.; Park, J.H. NLRP3 inflammasome mediates interleukin-1beta production in immune cells in response to Acinetobacter baumannii and contributes to pulmonary inflammation in mice. *Immunology* **2017**, *150*, 495–505. [CrossRef] [PubMed]
115. Ye, Q.; Wang, B.; Mao, J. The pathogenesis and treatment of the 'Cytokine Storm' in, C.O.VID-19. *J. Infect.* **2020**, *80*, 607–613. [CrossRef]
116. Wang, C.; Xie, J.; Zhao, L.; Fei, X.; Zhang, H.; Tan, Y.; Nie, X.; Zhou, L.; Liu, Z.; Ren, Y.; et al. Alveolar macrophage dysfunction and cytokine storm in the pathogenesis of two severe, C.O.VID-19 patients. *EBioMedicine* **2020**, *57*, 102833. [CrossRef]
117. Mehta, P.; McAuley, D.F.; Brown, M.; Sanchez, E.; Tattersall, R.S.; Manson, J.J. COVID-19: Consider cytokine storm syndromes and immunosuppression. *Lancet* **2020**, *395*, 1033–1034. [CrossRef]
118. Monteagudo, L.A.; Boothby, A.; Gertner, E. Continuous Intravenous Anakinra Infusion to Calm the Cytokine Storm in Macrophage Activation Syndrome. *ACR Open Rheumatol.* **2020**, *2*, 276–282. [CrossRef] [PubMed]
119. Di Virgilio, F.; Tang, Y.; Sarti, A.C.; Rossato, M. A rationale for targeting the P2X7 receptor in Coronavirus disease 19. *Br. J. Pharmacol.* **2020**, *177*, 4990–4994. [CrossRef] [PubMed]
120. Illes, P. P2X7 Receptors Amplify CNS Damage in Neurodegenerative Diseases. *Int. J. Mol. Sci.* **2020**, *21*, 5996. [CrossRef]
121. Illes, P.; Rubini, P.; Huang, L.; Tang, Y. The P2X7 receptor: A new therapeutic target in Alzheimer's disease. *Expert Opin. Ther. Targets* **2019**, *23*, 165–176. [CrossRef]
122. Francistiova, L.; Bianchi, C.; Di Lauro, C.; Sebastian-Serrano, A.; de Diego-Garcia, L.; Kobolak, J.; Dinnyés, A.; Díaz-Hernández, M. The Role of P2X7 Receptor in Alzheimer's Disease. *Front. Mol. Neurosci.* **2020**, *13*, 94. [CrossRef]
123. Wang, X.H.; Xie, X.; Luo, X.G.; Shang, H.; He, Z.Y. Inhibiting purinergic P2X7 receptors with the antagonist brilliant blue G is neuroprotective in an intranigral lipopolysaccharide animal model of Parkinson's disease. *Mol. Med. Rep.* **2017**, *15*, 768–776. [CrossRef] [PubMed]
124. Oliveira-Giacomelli, A.; Albino, M.; de Souza, H.D.N.; Correa-Velloso, J.; de Jesus Santos, A.P.; Baranova, J.; Ulrich, H. P2Y6 and P2X7 Receptor Antagonism Exerts Neuroprotective/ Neuroregenerative Effects in an Animal Model of Parkinson's Disease. *Front. Cell. Neurosci.* **2019**, *13*, 476. [CrossRef] [PubMed]
125. Apolloni, S.; Amadio, S.; Parisi, C.; Matteucci, A.; Potenza, R.L.; Armida, M.; Popoli, P.; D'Ambrosi, N.; Volonté, C. Spinal cord pathology is ameliorated by P2X7 antagonism in a, S.O.D1-mutant mouse model of amyotrophic lateral sclerosis. *Dis. Models Mech.* **2014**, *7*, 1101–1109.

126. Fabbrizio, P.; Amadio, S.; Apolloni, S.; Volonte, C. P2X7 Receptor Activation Modulates Autophagy in, S.O.D1-G93A Mouse Microglia. *Front. Cell. Neurosci.* **2017**, *11*, 249. [CrossRef]

127. Beaino, W.; Janssen, B.; Kooij, G.; van der Pol, S.M.A.; van Het, H.B.; van Horssen, J.; Windhorst, A.D.; de Vries, H.E. Purinergic receptors P2Y12R and P2X7R: Potential targets for, P.E.T imaging of microglia phenotypes in multiple sclerosis. *J. Neuroinflammation* **2017**, *14*, 259. [CrossRef]

128. Grygorowicz, T.; Struzynska, L. Early P2X7R-dependent activation of microglia during the asymptomatic phase of autoimmune encephalomyelitis. *Inflammopharmacology* **2019**, *27*, 129–137. [CrossRef] [PubMed]

129. Jimenez-Pacheco, A.; Diaz-Hernandez, M.; Arribas-Blazquez, M.; Sanz-Rodriguez, A.; Olivos-Ore, L.A.; Artalejo, A.R.; Alves, M.; Letavic, M.; Miras-Portugal, M.T.; Conroy, R.M.; et al. Transient P2X7 Receptor Antagonism Produces Lasting Reductions in Spontaneous Seizures and Gliosis in Experimental Temporal Lobe Epilepsy. *J. Neurosci.* **2016**, *36*, 5920–5932. [CrossRef]

130. Beamer, E.; Fischer, W.; Engel, T. The ATP-Gated P2X7 Receptor As a Target for the Treatment of Drug-Resistant Epilepsy. *Front. Neurosci.* **2017**, *11*, 21. [CrossRef]

131. Melani, A.; Amadio, S.; Gianfriddo, M.; Vannucchi, M.G.; Volonte, C.; Bernardi, G.; Pedata, F.; Sancesario, G. P2X7 receptor modulation on microglial cells and reduction of brain infarct caused by middle cerebral artery occlusion in rat. *J. Cereb. Blood Flow Metab.* **2006**, *26*, 974–982. [CrossRef] [PubMed]

132. Bartlett, R.; Stokes, L.; Sluyter, R. The P2X7 receptor channel: Recent developments and the use of P2X7 antagonists in models of disease. *Pharmacol. Rev.* **2014**, *66*, 638–675. [CrossRef]

133. Sun, C.; Chu, J.; Singh, S.; Salter, R.D. Identification and characterization of a novel variant of the human P2X7 receptor resulting in gain of function. *Purinergic Signal.* **2010**, *6*, 31–45. [CrossRef] [PubMed]

134. Sluyter, R.; Shemon, A.N.; Wiley, J.S. Glu496 to Ala polymorphism in the P2X7 receptor impairs, ATP-induced IL-1 beta release from human monocytes. *J. Immunol.* **2004**, *172*, 3399–3405. [CrossRef] [PubMed]

135. Gu, B.J.; Zhang, W.; Worthington, R.A.; Sluyter, R.; Dao-Ung, P.; Petrou, S.; Barden, J.A.; Wiley, J.S. A Glu-496 to Ala polymorphism leads to loss of function of the human P2X7 receptor. *J. Biol. Chem.* **2001**, *276*, 11135–11142. [CrossRef]

136. Stokes, L.; Fuller, S.J.; Sluyter, R.; Skarratt, K.K.; Gu, B.J.; Wiley, J.S. Two haplotypes of the P2X7 receptor containing the Ala-348 to Thr polymorphism exhibit a gain-of-function effect and enhanced interleukin-1beta secretion. *FASEB J.* **2010**, *24*, 2916–2927. [CrossRef] [PubMed]

137. Gu, B.J.; Sluyter, R.; Skarratt, K.K.; Shemon, A.N.; Dao-Ung, L.P.; Fuller, S.J.; Barden, J.A.; Clarke, A.L.; Petrou, S.; Wiley, J.S. An Arg307 to Gln polymorphism within the ATP-binding site causes loss of function of the human P2X7 receptor. *J. Biol. Chem.* **2004**, *279*, 31287–31295. [CrossRef]

138. Xiao, J.; Sun, L.; Yan, H.; Jiao, W.; Miao, Q.; Feng, W.; Wu, X.; Gu, Y.; Jiao, A.; Guo, Y.; et al. Metaanalysis of P2X7 gene polymorphisms and tuberculosis susceptibility. *FEMS Immunol. Med. Microbiol.* **2010**, *60*, 165–170. [CrossRef]

139. Britton, W.J.; Fernando, S.L.; Saunders, B.M.; Sluyter, R.; Wiley, J.S. The genetic control of susceptibility to Mycobacterium tuberculosis. *Novartis Found. Symp.* **2007**, *281*, 79–89.

140. Lees, M.P.; Fuller, S.J.; McLeod, R.; Boulter, N.R.; Miller, C.M.; Zakrzewski, A.M.; Mui, E.J.; Witola, W.H.; Coyne, J.J.; Hargrave, A.C.; et al. P2X7 receptor-mediated killing of an intracellular parasite, Toxoplasma gondii, by human and murine macrophages. *J. Immunol.* **2010**, *184*, 7040–7046. [CrossRef]

141. Wiley, J.S.; Sluyter, R.; Gu, B.J.; Stokes, L.; Fuller, S.J. The human P2X7 receptor and its role in innate immunity. *Tissue Antigens* **2011**, *78*, 321–332. [CrossRef] [PubMed]

142. Gu, B.J.; Baird, P.N.; Vessey, K.A.; Skarratt, K.K.; Fletcher, E.L.; Fuller, S.J.; Richardson, A.J.; Guymer, R.H.; Wiley, J.S. A rare functional haplotype of the P2RX4 and P2RX7 genes leads to loss of innate phagocytosis and confers increased risk of age-related macular degeneration. *FASEB J.* **2013**, *27*, 1479–1487. [CrossRef] [PubMed]

143. Jorgensen, N.R. Role of the purinergic P2X receptors in osteoclast pathophysiology. *Curr. Opin. Pharmacol.* **2019**, *47*, 97–101. [CrossRef]

144. Boyle, W.J.; Simonet, W.S.; Lacey, D.L. Osteoclast differentiation and activation. *Nature* **2003**, *423*, 337–342. [CrossRef]

145. Jorgensen, N.R.; Syberg, S.; Ellegaard, M. The role of P2X receptors in bone biology. *Curr. Med. Chem.* **2015**, *22*, 902–914. [PubMed]

146. Varley, I.; Hughes, D.C.; Greeves, J.P.; Fraser, W.D.; Sale, C. SNPs in the vicinity of P2X7R, RANK/RANKL/OPG and Wnt signalling pathways and their association with bone phenotypes in academy footballers. *Bone* **2018**, *108*, 179–185. [CrossRef] [PubMed]

147. McQuillin, A.; Bass, N.J.; Choudhury, K.; Puri, V.; Kosmin, M.; Lawrence, J.; Curtis, D.; Gurling, H.M. Case-control studies show that a non-conservative amino-acid change from a glutamine to arginine in the P2RX7 purinergic receptor protein is associated with both bipolar- and unipolar-affective disorders. *Mol. Psychiatry* **2009**, *14*, 614–620. [CrossRef]

148. Czamara, D.; Müller-Myhsok, B.; Lucae, S. The P2RX7 polymorphism rs2230912 is associated with depression: A meta-analysis. *Prog. Neuropsychopharmacol. Biol. Psychiatry* **2018**, *82*, 272–277. [CrossRef]

149. Green, E.K.; Grozeva, D.; Raybould, R.; Elvidge, G.; Macgregor, S.; Craig, I.; Farmer, A.; McGuffin, P.; Forty, L.; Jones, L.; et al. P2RX7: A bipolar and unipolar disorder candidate susceptibility gene? *Am. J. Med. Genet. B Neuropsychiatr. Genet.* **2009**, *150B*, 1063–1069. [CrossRef]

150. Feng, W.P.; Zhang, B.; Li, W.; Liu, J. Lack of association of P2RX7 gene rs2230912 polymorphism with mood disorders: A meta-analysis. *PLoS ONE* **2014**, *9*, e88575. [CrossRef]

151. Roger, S.; Mei, Z.Z.; Baldwin, J.M.; Dong, L.; Bradley, H.; Baldwin, S.A.; Surprenant, A.; Jiang, L.H. Single nucleotide polymorphisms that were identified in affective mood disorders affect ATP-activated P2X7 receptor functions. *J. Psychiatr. Res.* **2010**, *44*, 347–355. [CrossRef] [PubMed]

152. Aprile-Garcia, F.; Metzger, M.W.; Paez-Pereda, M.; Stadler, H.; Acuna, M.; Liberman, A.C.; Senin, S.A.; Gerez, J.; Hoijman, E.; Refojo, D.; et al. Co-Expression of Wild-Type P2X7R with Gln460Arg Variant Alters Receptor Function. *PLoS ONE* **2016**, *11*, e0151862. [CrossRef] [PubMed]

153. Metzger, M.W.; Walser, S.M.; Dedic, N.; Aprile-Garcia, F.; Jakubcakova, V.; Adamczyk, M.; Webb, K.J.; Uhr, M.; Refojo, D.; Schmidt, M.V.; et al. Heterozygosity for the Mood Disorder-Associated Variant Gln460Arg Alters P2X7 Receptor Function and Sleep Quality. *J. Neurosci.* **2017**, *37*, 11688–11700. [CrossRef] [PubMed]

154. Sluyter, R.; Stokes, L.; Fuller, S.J.; Skarratt, K.K.; Gu, B.J.; Wiley, J.S. Functional significance of P2RX7 polymorphisms associated with affective mood disorders. *J. Psychiatr. Res.* **2010**, *44*, 1116–1117. [CrossRef]

155. Backlund, L.; Nikamo, P.; Hukic, D.S.; Ek, I.R.; Träskman-Bendz, L.; Landén, M.; Edman, G.; Schalling, M.; Frisén, L.; Osby, U. Cognitive manic symptoms associated with the P2RX7 gene in bipolar disorder. *Bipolar Disord.* **2011**, *13*, 500–508. [CrossRef]

156. Gu, B.J.; Field, J.; Dutertre, S.; Ou, A.; Kilpatrick, T.J.; Lechner-Scott, J.; Scott, R.; Lea, R.; Taylor, B.V.; Stankovich, J.; et al. A rare P2X7 variant Arg307Gln with absent pore formation function protects against neuroinflammation in multiple sclerosis. *Hum. Mol. Genet.* **2015**, *24*, 5644–5654. [CrossRef]

International Journal of
Molecular Sciences

Article

Involvement of GPR17 in Neuronal Fibre Outgrowth

Max Braune [1], Nico Scherf [2], Claudia Heine [1], Katja Sygnecka [1], Thanigaimalai Pillaiyar [3], Chiara Parravicini [4], Bernd Heimrich [5], Maria P. Abbracchio [4], Christa E. Müller [3] and Heike Franke [1,*]

1. Rudolf Boehm Institute of Pharmacology and Toxicology, Medical Faculty, University of Leipzig, Härtelstr. 16-18, 04107 Leipzig, Germany; Max.Braune@medizin.uni-leipzig.de (M.B.); claudia.b.heine@t-online.de (C.H.); k.sygnecka@gmail.com (K.S.)
2. Methods and Development Group Neural Data Analysis and Statistical Computing, Max Planck Institute for Human Cognitive and Brain Sciences, Stephanstraße 1A, 04103 Leipzig, Germany; nscherf@cbs.mbp
3. Department of Pharmaceutical & Medicinal Chemistry, Pharmaceutical Institute, University of Bonn, An der Immenburg 4, 53121 Bonn, Germany; thanigai.medchem@gmail.com (T.P.); christa.mueller@uni-bonn.de (C.E.M.)
4. Department of Pharmaceutical Sciences, University of Milan, Via Balzaretti 9, 20133 Milan, Italy; chiara.parravicini@gmail.com (C.P.); mariapia.abbracchio@unimi.it (M.P.A.)
5. Department of Neuroanatomy, Institute of Anatomy and Cell Biology, Center for Basics in NeuroModulation, Faculty of Medicine, University of Freiburg, Albertstr. 23, 79104 Freiburg, Germany; bernd.heimrich@zfn.uni-freiburg.de
* Correspondence: Heike.Franke@medizin.uni-leipzig.de; Tel.: +49-(0)341-9724602; Fax: +49-(0)341-9724609

Citation: Braune, M.; Scherf, N.; Heine, C.; Sygnecka, K.; Pillaiyar, T.; Parravicini, C.; Heimrich, B.; Abbracchio, M.P.; Müller, C.E.; Franke, H. Involvement of GPR17 in Neuronal Fibre Outgrowth. *Int. J. Mol. Sci.* **2021**, 22, 11683. https://doi.org/10.3390/ijms222111683

Academic Editors: Dmitry Aminin and Peter Illes

Received: 4 October 2021
Accepted: 24 October 2021
Published: 28 October 2021

Abstract: Characterization of new pharmacological targets is a promising approach in research of neurorepair mechanisms. The G protein-coupled receptor 17 (GPR17) has recently been proposed as an interesting pharmacological target, e.g., in neuroregenerative processes. Using the well-established *ex vivo* model of organotypic slice co-cultures of the mesocortical dopaminergic system (prefrontal cortex (PFC) and substantia nigra/ventral tegmental area (SN/VTA) complex), the influence of GPR17 ligands on neurite outgrowth from SN/VTA to the PFC was investigated. The growth-promoting effects of Montelukast (MTK; GPR17- and cysteinyl-leukotriene receptor antagonist), the glial cell line-derived neurotrophic factor (GDNF) and of two potent, selective GPR17 agonists (PSB-16484 and PSB-16282) were characterized. Treatment with MTK resulted in a significant increase in mean neurite density, comparable with the effects of GDNF. The combination of MTK and GPR17 agonist PSB-16484 significantly inhibited neuronal growth. qPCR studies revealed an MTK-induced elevated mRNA-expression of genes relevant for neuronal growth. Immunofluorescence labelling showed a marked expression of GPR17 on NG2-positive glia. Western blot and RT-qPCR analysis of untreated cultures suggest a time-dependent, injury-induced stimulation of GPR17. In conclusion, MTK was identified as a stimulator of neurite fibre outgrowth, mediating its effects through GPR17, highlighting GPR17 as an interesting therapeutic target in neuronal regeneration.

Keywords: G protein-coupled receptor 17 (GPR17); neurite outgrowth; montelukast; NG2; *ex vivo* organotypic brain slice co-culture; neurodegeneration and neuroregeneration

1. Introduction

Globally, the number of patients dying from and affected by neurological disorders has increased substantially between 1990 and 2015 [1]. In particular, incidence and prevalence of traumatic brain injury (TBI) increased from 1990 to 2016 [2], as well as the prevalence of Parkinson's disease [3] and dementia [4]. These disorders are the biggest health challenges of the century, posing a serious threat to social and healthcare systems as well as to the future of the global economy [5]. Therefore, new strategies of treatments are pivotal to minimize patients' disabilities promoting better life quality and to reduce costs for society.

The G protein-coupled receptor 17 (GPR17) has recently been proposed as an interesting pharmacological target in neuroinflammatory and neurodegenerative diseases [6–12].

The involvement of GPR17 in other pathophysiological conditions of the brain has been demonstrated, leading to the suggestion that GPR17 acts as a 'sensor of brain damage' [13]. Under normal physiological conditions, GPR17 is almost exclusively expressed in NG2-glia, also known as oligodendrocyte precursor cells (OPC), which give rise to myelin producing oligodendrocytes both during development and throughout adult life [14]. Own previous data indicate that the GPR17 expression was elevated in human brain specimens from neurosurgical and autoptic cases after TBI [15]. Viganò and co-workers showed, that after cerebral damage induced by acute injury or ischemia in mice, GPR17-positive NG2-glia rapidly reacted to the damage, suggesting these cells are a 'reserve pool' of adult progenitors maintained for repair purposes [16]. Subsequent fate mapping studies in the same model of stroke [8] as well as in two different models of demyelination [17] showed that, as a result of injury, GPR17-expressing cells proliferate and markedly accumulate in regions surrounding the lesions, but that only a low percentage of these cells eventually gives rise to mature myelinating oligodendrocytes, due to an unfavourable local inflammatory milieu. These data suggest that GPR17 could be pharmacologically exploited for the benefit of patients with neurological diseases, provided that inflammation is counteracted with appropriate agents.

The pharmacology of GPR17 can be characterized as atypical [18], as GPR17 responds to a diverse set of ligands [9,19,20]. The receptor is phylogenetically related to both purinergic P2Y receptors (P2YRs) and cysteinyl-leukotriene (CysLT) receptors [9]. In 2006, it was discovered that GPR17 is activated by the uracil nucleotides UDP, UDP-glucose and UDP-galactose as well as by the cysteinyl-leukotrienes LTC4 and LTD4 [21]. The known $P2Y_{12}R$ antagonists Cangrelor and Ticagrelor and the cysteinyl-leukotriene receptor (CysLT-R) antagonists Montelukast (MTK) and Pranlukast have also been described as antagonists of GPR17 [7,21]. In 2010, Benned-Jensen and Rosenkilde independently confirmed activation of GPR17 by uracil nucleotides [19], while other groups neither found activation of GPR17 by uracil nucleotides or cysteinyl-leukotrienes, nor inhibition of GPR17 by Cangrelor or Ticagrelor [20,22,23]. These varying effects are likely due to the level of expression of GPR17 in the different transfected cell lines utilized, and in the difficulty of preserving the native pharmacological features of the receptor in artificial recombinant systems.

It has been demonstrated that in animal models of stroke treatment with Cangrelor, and also with MTK and blockade of GPR17 with antisense technology, led to decreased infarct volumes [13,21,24,25]. In rats, treatment with MTK reduced neuroinflammation, elevated hippocampal neurogenesis and improved learning and memory [26]. These effects were mediated by GPR17 as demonstrated by using gene knock down and knock out strategies [26].

Otherwise, in PC12 cells, that natively express GPR17, treatment with UDP-glucose and LTD4 promoted survival and neurite outgrowth [27]. In a neonatal rat model of ischemic periventricular leukomalacia treatment with UDP-glucose improved the thickness of myelin sheaths, motor dysfunction and cognitive functions [28].

Due to its special pharmacology and distinct results, the role of GPR17 in neuroreparative and neuroregenerative mechanisms has not yet been completely elucidated. In particular, it is not clear if and how signalling of GPR17 is involved in repair mechanisms in projection systems after trauma and whether these can be influenced by treatment with ligands of GPR17.

For this reason, the well-established rat *ex vivo* model of organotypic slice co-cultures of the mesocortical dopaminergic system was chosen culturing slices of the substantia nigra/ventral tegmental area (SN/VTA) and the prefrontal cortex (PFC) closely together allowing neuronal fibre outgrowth, to grow from one region to the other [29,30]. Organotypic slice co-cultures largely preserve the tissue architecture of the brain regions that they originate from [31], thereby modelling the *in vivo* situation closely [32]. While the preparation of the co-cultures causes a disruption of the already established connections, this model also strongly correlates with the development of fibre projections of physiological neuronal circuits, thus making the organotypic co-cultures a model of both development and of axonal regrowth after mechanical injury [30]. Concretely, in the present study it was examined (i) whether pharmacological stimulation or inhibition of GPR17 can promote neurite outgrowth; (ii) the dynamics of a small selection of genes involved in growth and differentiation of neurons, myelination and inflammation depending on the treatment with GPR17 antagonists and agonists; (iii) the temporal expression of GPR17 in the co-cultures; and (iv) which cells might be effectors of any observed effects.

2. Results

2.1. Neurite Fibre Outgrowth Modulation

To elucidate if modulation of the GPR17 has any effects on neurite outgrowth in organotypic dopaminergic co-cultures the well-established neurite fibre quantification method in our lab was used [33] (for schematic illustration see Figure 1).

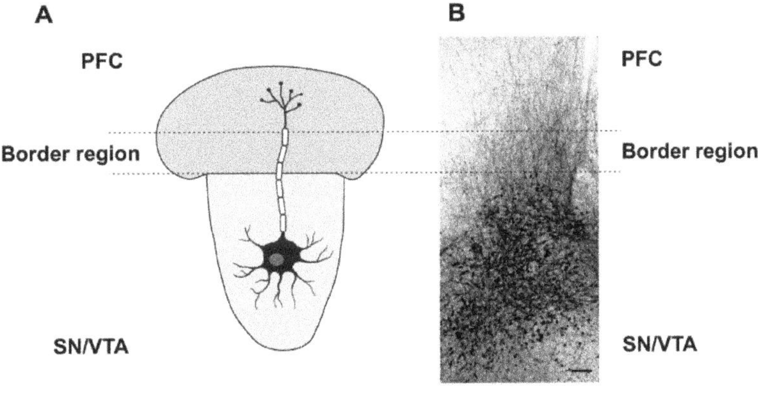

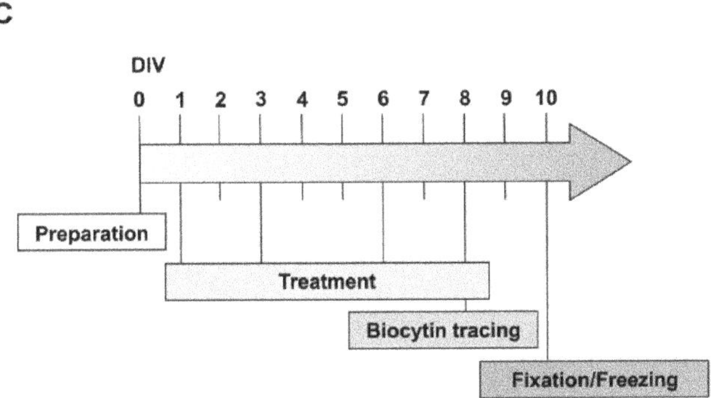

Figure 1. Experimental design. (**A**) Schematic Illustration of the used organotypic *ex vivo* slice co-cultures of the dopaminergic system (SN/VTA and PFC) (Attribution: Free neuron vector from vecteezy: Available online: https://www.vecteezy.com/free-vector/nerve-cell (accessed on 23 July 2021). Nerve Cell Vectors by Vecteezy. (**B**) Overview of the fibre outgrowth visualized by biocytin tracing in an *ex vivo* co-culture system (rat, fixed at DIV 10). The dotted lines characterize the border region, used for quantification of fibre density. (**C**) Timeline with the treatment procedures. Scale bar: **B** = 200 μm.

In initial experiments, the effect of Cangrelor (an anti-platelet agent blocking the $P2Y_{12}Rs$, and reported to act additionally as a non-selective GPR17 antagonist) was investigated, and a significant stimulatory effect on neurite fibre outgrowth in comparison to ACSF (*t*-test with Welch correction t (4.34607) = −3.28534; p = 0.02685; see Figure S1) was observed.

In the next phase of experiments, different concentrations of MTK, the CysLT1R antagonist and proposed GPR17 antagonist, were employed. In the dopaminergic slice co-cultures, in a similar way to Cangrelor, the application of 10 µM of MTK showed a significant stimulation of neurite fibre outgrowth (Figure 2A,C) compared to vehicle (A: 1% ethanol; C: 0.01% DMSO, see Section 4.4.) treated controls (one-way ANOVA F $(2,28)$ = 6.143; p = 0.00614; followed by Tukey's post hoc test p = 0.0062). Neurite fibre outgrowth after application of 1 µM MTK was tendentially increased, but not significantly (Tukey's post hoc test p = 0.57299; Figure 2A).

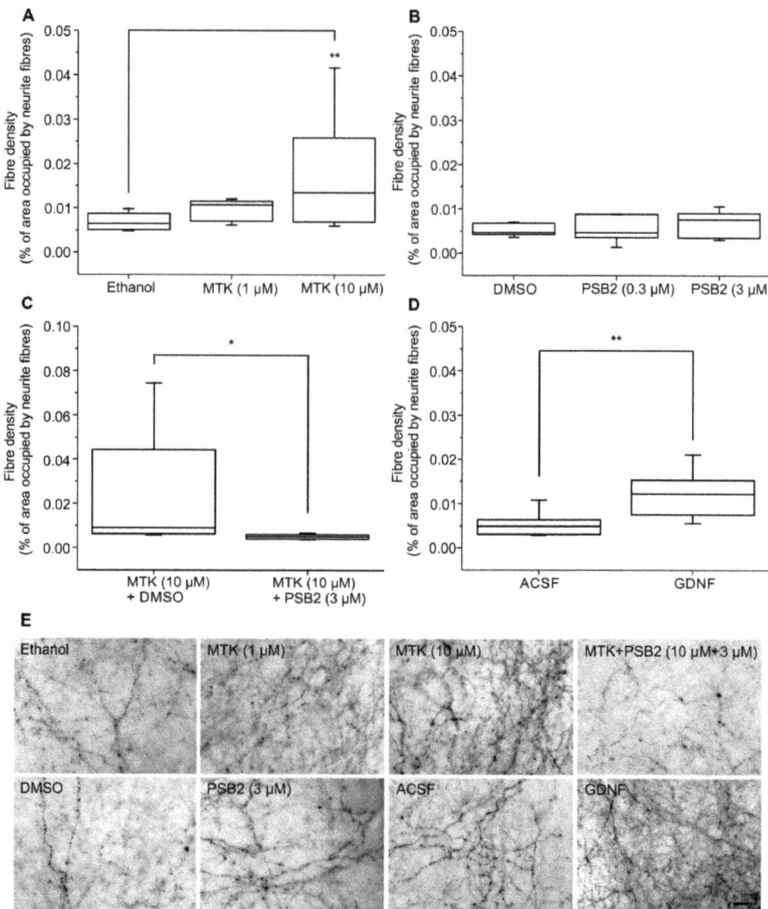

Figure 2. Neurite outgrowth quantification. (**A–D**) Neurite fibre density was quantified after treatment with (**A**) MTK, (**B**) PSB-16484 (PSB2), (**C**) MTK and PSB-16484 (PSB2) and (**D**) GDNF using biocytin-tracing. Data are shown as box plots. The number n of animals being used was for (**A**) $n \geq 9$, (**B**) n = 4, (**C**) $n \geq 5$, (**D**) n = 8. For (**A,B**) ANOVA on Ranks was followed by Tukey's test, for (**C**) Mann–Whitney test was applied and for (**D**) *t*-test was applied; * $p < 0.05$, ** $p < 0.01$. (**E**) Pictures of biocytin-labelled fibres in the border region after application of Ethanol, MTK (1 µM), MTK (10 µM), MTK (10 µM, pre-treatment) + PSB-16484 (PSB2, 3 µM), DMSO, PSB-16484 (PSB2, 3 µM), ACSF and GDNF. Scale bar: 20 µm for all.

Recently, the small molecule 3-(2-carboxy-4,6-dichloro-indol-3-yl)-propionic acid (MDL29,951) was identified to act as a GPR17 agonist [7,20], and optimized analogues of MDL29,951 (which is non-selective since it also interacts with NMDA receptors), were synthesized [34,35] and characterized in functional assays. In a radioligand binding assay, affinity of this class of compounds for GPR17 was confirmed and blockade by MTK was shown [20,34]. Hence, two of the new, selective GPR17 agonists were used to activate the receptor, namely PSB-16282 and PSB-16484, while MTK was employed as a GPR17 antagonist.

The application of the new GPR17 agonists alone, PSB-16484 (PSB2) (one-way ANOVA F $(2,19)$ = 0.29054; p = 0.75113) (Figure 2B) and PSB-16282 (PSB1) (see Figure S2 (one-way ANOVA F $(2,12)$ = 1.04803; p = 0.38063)), did not cause a significant increase or decrease in neurite fibre outgrowth. Thus, for additional studies, to reduce use of animals only PSB-16484 (PSB2) was utilized.

To evaluate if the observed stimulatory effect of MTK could be antagonized by a new, selective synthetic GPR17 agonist, 10 µM MTK was applied in combination with 3 µM PSB-16484 (PSB2) and compared to a group treated only with 10 µM MTK. Neurite fibre density of cultures treated with MTK and PSB-16484 (PSB2) was significantly lower compared to co-cultures treated only with MTK (Mann-Whitney-Test U = 28; p = 0.01732; Figure 2C) indicating that the stimulation of neurite fibre outgrowth after application of MTK is, at least in part, mediated by GPR17.

As a positive control for stimulation of neurite fibre outgrowth, the well-established neurotrophic factor GDNF, which has been shown to provide neuroprotective and neurotrophic effects especially with regard to the mesencephalon, was used [36]. The application of GDNF (50 ng/mL) induced a significant stimulation of neurite fibre outgrowth (t-test t (14) = -3.34886; p = 0.00477) compared to controls, proving its potency as a neurotrophic factor and demonstrating that neurite fibre outgrowth could be stimulated (Figure 2D).

In comparison, treatment with 10 µM MTK showed a 2.6-fold higher mean neurite density compared to controls, while treatment with GDNF led to a 2.3-fold higher mean neurite density compared to controls suggesting that 10 µM MTK is a notable stimulator of neurite fibre outgrowth in the organotypic dopaminergic co-culture system. Examples of neurite fibre outgrowth are given in Figure 2E.

2.2. Gene Expression Analysis of Genes Relevant for Neuronal Growth

To get further insight into the effects caused by application of MTK and PSB-16484 (PSB2), a gene expression analysis was performed using reverse transcription quantitative real-time polymerase chain reaction (RT-qPCR) after day 10 in culture. In addition to GPR17, the expression of five target genes representative of functions in the central nervous system (CNS) or specific cell populations (Figure 3: PFC (A,C,E,G,I,K); SN/VTA (B,D,F,H,J,L)) (Table 1) were investigated.

GAP43 and NFL were chosen because of their role in axonal regeneration [37] and as regeneration associated genes [38]. In brief, GAP43 is known as being connected to neuronal growth, promoting spontaneous formation of new synapses and enhancing sprouting after injury. Additionally, it is linked to neurite outgrowth, nerve-terminal sprouting and long-term potentiation [39]. Its expression in the used *ex vivo* slice co-cultures has been shown immunohistochemically [30]. In the present study, the expression of mRNA of GAP43 (Figure 3A,B) was significantly elevated in the SN/VTA after treatment with MTK compared to controls (ctrl) (one-way ANOVA F $(2,9)$ = 9.73918; p = 0.00561; followed by Tukey's post hoc test p = 0.01067; Figure 3B) supporting the findings in neurite density quantification. Application of PSB-16484 (PSB2) could not block this effect (Tukey's post hoc test p = 0.99838). There was no significant difference between groups in the PFC (one-way ANOVA F $(2,9)$ = 0.92958; p = 0.42954; Figure 3A).

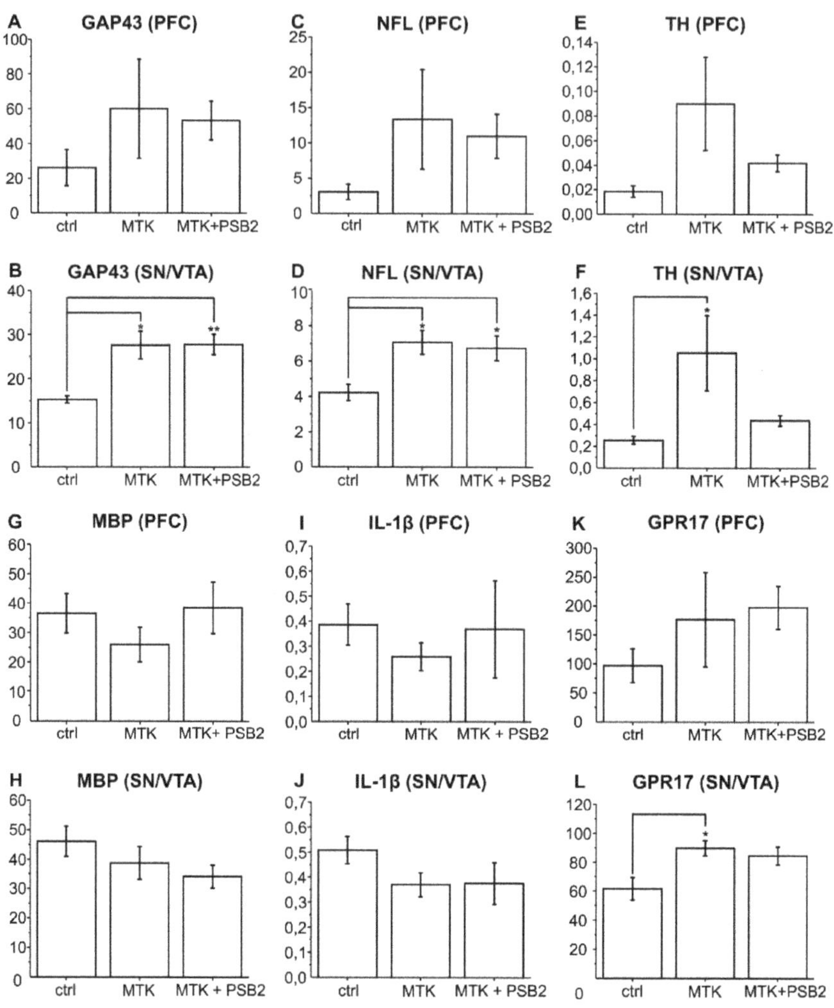

Figure 3. Results of RT-qPCR experiments in the PFC and SN/VTA. Expression of mRNA of GAP43 (**A,B**), NFL (**C,D**), TH (**E,F**), MBP (**G,H**), IL-1β (**I,J**), GPR17 (**K,L**) is shown on *y*-axis as Δ CP. X-Axis represents the different treatments (1% ethanol and 0.01% DMSO as controls (ctrl); 10 μM MTK and 0.01% DMSO; 10 μM MTK and 3 μM PSB-16484 (PSB2)). Statistical analysis was performed in comparison to vehicle control with ANOVA on ranks followed by Tukey's test. Data are shown as bar charts. The number of animals being used was $n = 4$, * $p < 0.05$, ** $p < 0.01$.

Table 1. Target genes and primer sequences.

Accession Number	Target	Forward	Reverse
NM_017195.3	rat GAP43	ACCACTGATAACTCGCCGTC	TGGCTTCATCTACAGCTTCTTTCT
NM_001071777.1	rat GPR17	ACTTGTCCTGTGTGCTGGTC	CCCAAAAGGCCAGTGATTGC
NM_031512.2	rat IL-1β	TAGCAGCTTTCGACAGTGAGG	TCTGGACAGCCCAAGTCAAG
NM_001025293.1	rat MBP	TGTGCCACATGTACAAGGACT	TTCATCTTGGGTCCTCTGCG
NM_001106116.1	rat MrpL32	TTCCGGACCGCTACATAGGTG	CTAGTGCTGGTGCCCACTGAG
NM_031783.2	rat NFL	GCAGCTTACAGGAAACTCTTGG	ACCTGCGAGCTCTGAGAGTA
NM_012740.4	rat TH	TTCTTGAAGGAGCGGACTGG	TGCATTGAAACACGCGGAAG

NFL is particularly abundant in axons where it is essential for the radial growth of axons during development, the maintenance of axon calibre and the transmission of electrical impulses [40]. Our results show that mRNA-expression of NFL (Figure 3C,D) was significantly stimulated by treatment with MTK in the SN/VTA (one-way ANOVA F (2,9) = 6.23136; p = 0.02002; followed by Tukey's post hoc test p = 0.02547; Figure 3D) supporting our findings of higher neurite density in the border region between PFC and SN/VTA after treatment with MTK. The MTK effect was not reduced by PSB-16484 (PSB2) (Tukey's post hoc test p = 0.92805). There was no significant difference between groups in the PFC (one-way ANOVA F (2,9) = 1.45291; p = 0.28391; Figure 3C).

It should be evaluated, if dopaminergic neurons benefit from application of MTK in the co-culture. Therefore, measurement of the mRNA-expression of TH, the marker enzyme of dopaminergic neurons, was performed [41]. TH mRNA-expression (Figure 3E,F) was significantly elevated in the SN/VTA after treatment with MTK (one-way ANOVA F (2,9) = 4.29564; p = 0.04901; followed by Tukey's post hoc test p = 0.0498) and showed non-significantly reduced expression after treatment with PSB-16484 (PSB2) (Tukey's post hoc test p = 0.13171; Figure 3F). Again, in the PFC there was no significant difference between groups (one-way ANOVA F (2,9) = 2.65474; p = 0.1241; Figure 3E).

Additionally, gene expression of myelin basic protein (MBP) was analysed. MBP is an essential part of the myelin sheath insulating axons electrically and thereby allowing saltatory conduction and high conduction velocity [42]. Since GPR17 has been shown to play a role in oligodendrocyte maturation [43,44], it was hypothesized that application of MTK and PSB-16484 (PSB2) could affect expression of the key oligodendrocyte marker MBP. However, expression of mRNA of MBP (Figure 3G,H) showed no significant change after treatment with MTK and PSB-16484 (PSB2) in the PFC (one-way ANOVA F (2,9) = 0.867; p = 0.45255; Figure 3G) and the SN/VTA (one-way ANOVA F (2,9) = 1.48997; p = 0.27609; Figure 3H).

Furthermore, the question was addressed if the observed neuroregenerative effects of MTK could be assigned to a decreased neuroinflammatory milieu in the dopaminergic co-cultures. Because of its central role in neuroinflammation, IL-1β was chosen as an appropriate marker [45]. However, mRNA-expression of IL-1β (Figure 3I,J) showed no significant changes after treatment with MTK and PSB-16484 (PSB2) but was non-significantly decreased in the PFC (one-way ANOVA F (2,9) = 0.30249; p = 0.7462; Figure 3I) and the SN/VTA after treatment with MTK (one-way ANOVA F (2,9) = 0.45195; p = 0.65008; Figure 3J). Treatment with PSB-16484 was able to alleviate this tendency in the PFC, but not in the SN/VTA.

Finally, the expression of mRNA of GPR17 (Figure 3K,L) was monitored, which was found to be significantly elevated after treatment with MTK in the SN/VTA (one-way ANOVA F (2,9) = 5.36593; p = 0.02923; followed by Tukey's post hoc test p = 0.03199), but not in the PFC (one-way ANOVA F (2,9) = 0.94476; p = 0.42418; Figure 3K,L). Treatment with PSB-16484 (PSB2) led to no significant difference in expression of GPR17 compared to treatment with MTK in the PFC (Tukey's post hoc test p = 0.96131) and the SN/VTA (Tukey's post hoc test p = 0.83958).

In conclusion, the data show that the mRNA-expression of all three neuronal markers (GAP43, NFL, TH) as well as of GPR17 are significantly increased after treatment with MTK in SN/VTA, but not in the PFC, supporting a neuroregenerative effect after treatment with MTK and suggesting regional differences in responsiveness to MTK.

2.3. Dynamics of GPR17 Expression in Untreated Organotypic Dopaminergic Co-Cultures
2.3.1. RT-qPCR Analysis of GPR17 Expression

The mRNA expression of GPR17 was characterized at two time points, DIV 3 and DIV 10. In both regions, PFC and SN/VTA, a higher expression was found on DIV 3. The expression of mRNA of GPR17 was significantly reduced after DIV 10 compared to DIV 3 in the PFC (*t*-test t (4) = 3.22794; p = 0.03204) and tendentially reduced in SN/VTA (*t*-test with Welch correction t (2.00119) = 0.97405; p = 0.43272; Figure 4A,B). This corresponds well to the Western blot results, also pointing to an increase in expression of the GPR17 caused by the preparation (injury-induced increase in GPR17 mRNA) (Figure 4C,D).

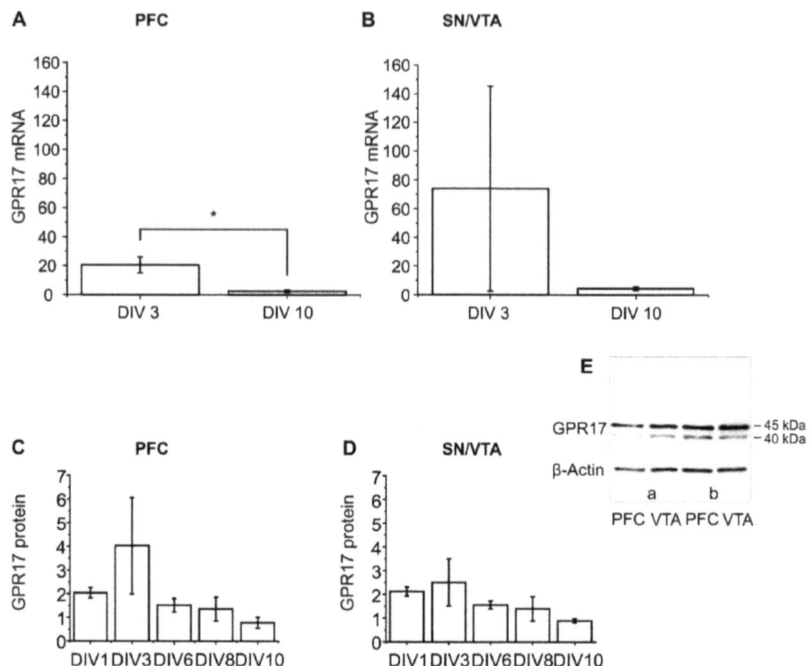

Figure 4. Results of RT-qPCR and Western blot of GPR17 in the *ex vivo* co-cultures. Expression of mRNA of the GPR17 is shown as Δ CP in the PFC (**A**) and SN/VTA (**B**) after DIV 3 and DIV 10. Statistical analysis was performed in comparison to vehicle control with ANOVA on ranks followed by Tukey's test. Data are shown as bar charts. The number of animals was $n = 3$, * $p < 0.05$. Protein expression of GPR17 is shown on DIV 1, 3, 6, 8 and 10 in PFC (**C**) and SN/VTA (**D**), data are shown as bar charts. The number of animals being used was $n \geq 3$. For statistical analysis one-way ANOVA (PFC) and Kruskal–Wallis (SN/VTA) was used. (**E**) Representative examples of Western blot images of samples obtained (DIV 10; (a) ctrl (Ethanol); (b) MTK 10 μM).

2.3.2. Western Blot Analysis

Using the rabbit anti-GPR17 antibody (Sigma-Aldrich, St. Louis, MO, USA), two protein bands were labelled, one at 40 kDa and one at 45 kDa (Figure 4E). These data are in agreement with previously published literature data suggesting that the band at 40 kDa could be a precursor form carrying high mannose oligosaccharide chains [46]. Western blot analysis showed that GPR17 had its peak of expression both in PFC and SN/VTA after DIV 3, followed by a subsequent decrease (Figure 4C,D). There were no significant differences between DIV in the PFC (one-way ANOVA F (1.69403), $p = 0.22724$) and the SN/VTA (Kruskal–Wallis ANOVA chi-square (4) = 6.47059; $p = 0.16665$).

Furthermore, for comparison, native tissue taken from young rats (P2 and P13), from both regions, PFC and SN/VTA, (Figure S3) was investigated. The data showed a slight, but not statistically significant increase in the expression of GPR17 at age P13 compared to P2 in PFC (*t*-test with Welch correction t (3.81358) = −0.62477; $p = 0.56755$) and SN/VTA (*t*-test t (4) = −0.15674; $p = 0.88304$).

In conclusion, the data suggest an injury-induced GPR17 expression in the studied *ex vivo* model. Under *in vivo* conditions no significant changes in GPR17 protein expression between day 2 and day 13 were found (no developmental increase in the investigated

regions). An injury-induced (slice preparation-induced) increase at DIV 3 was postulated. These data are supported by previous experiments using tissue of the complete co-culture (PFC and SN/VTA), indicating an increase at DIV 3 too (data not shown).

2.4. Immunohistochemical Analysis of GPR17 Expression

To investigate which cells express GPR17 in the organotypic dopaminergic co-cultures and could therefore be targeted with ligands of GPR17, the immunolabelling of GPR17 and characteristic cell makers was performed (Table 2).

Table 2. Expression of GPR17 in organotypic slice co-cultures.

Marker	GPR17-Co-Expression
Iba1	+
GFAP	−
TH	−
NeuN	+
βIII-Tubulin	+
NFL	−
NG2	+++
O4	+
CNPase	+
MBP	+

The results show that GPR17 is mainly expressed by NG2-positive glia (Figure 5A–C), but also by neurons (low labelling on NeuN-positive cells; examples are given in Figure 5G,H).

GPR17 is rarely expressed by more developed cells of the oligodendroglial cell lineage (O4-, CNPase-positive cells; an example for O4 is given in Figure 5D–F) and very low in MBP-positive cells, reflecting the peculiar time dependent expression of this receptor during oligodendrocyte maturation [47]. GPR17-labelled cells were observed rather in the proximity of MBP-positive structures (Figure S4A–C). There is also weak expression on microglia (Iba1-positive cells, Figure S4D–F), but no expression on astrocytes (GFAP-positive cells). Other neuronal markers showed only weak (e.g., βIII-Tubulin-positive cells) or no (TH-positive, NFL-positive cells) co-localization with GPR17 (examples are given in Figure 5I,J,K,L). It is important to note that GPR17-positive cells (stars) could be found more in direct proximity of βIII-Tubulin-, TH- and NFL-positive neurons, but not co-expressed on the TH- or NFL-positive structures. It was hypothesized that GPR17 expression was induced in neurons of our co-cultured slices as a result of the experimental procedure, and reflects a neuronal response to the trauma of the slice preparation.

2.5. Toxicological Analysis

To ensure that the applied pharmacological substances had no toxic effects on the organotypic slice co-cultures, the LDH release into incubation medium (IM) was measured. Data showed no significant elevation of LDH activity after treatment with 1 μM and 10 μM MTK compared to untreated control and vehicle-treated control (1% ethanol or 0.01% DMSO; repeated measurement two-way ANOVA with Greenhouse–Geisser correction F (0.29814, 0.60849) = 1.95985; p = 0.2293) e.g., 1% ethanol; Figure 6).

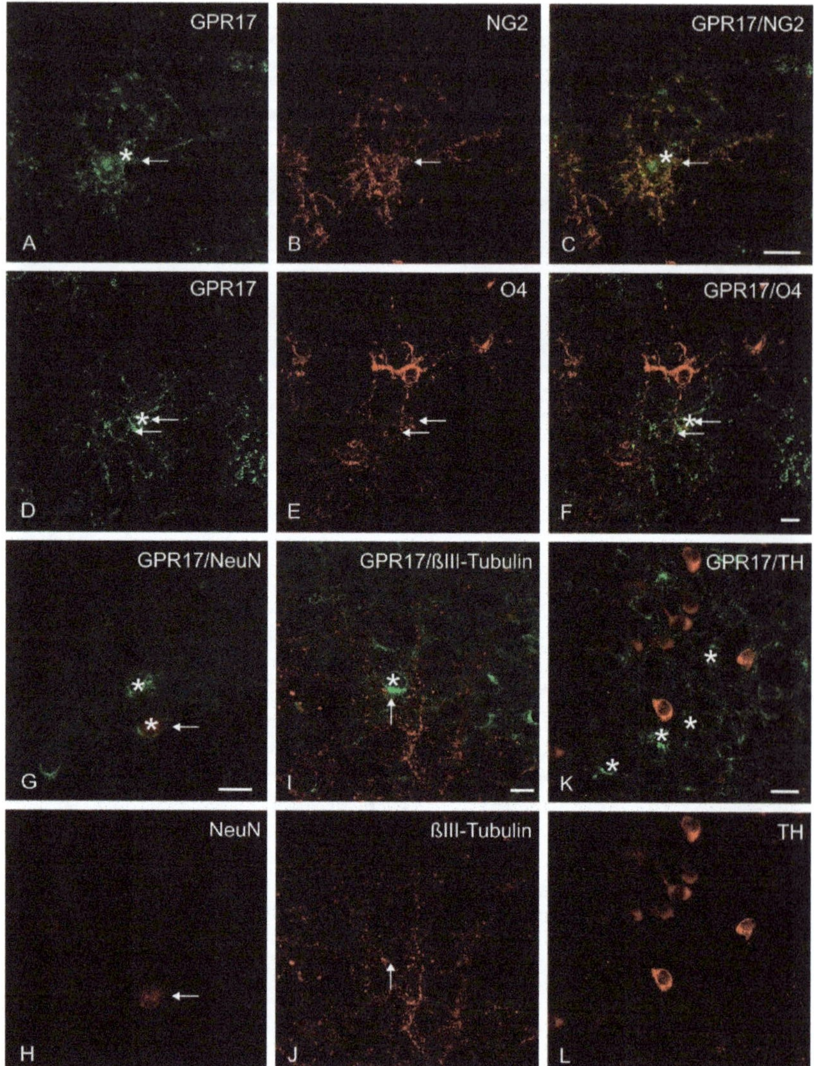

Figure 5. Multiple immunofluorescence study. Representative confocal images of GPR17 expression in organotypic slice co-cultures. At DIV 10 an intense GPR17 immunoreactivity was observed on NG2-positive cells (**A–C**) and on O4-positive cells (**D–F**). A low expression of GPR17 (stars) on a small number of cells was observed on (**G,H**) NeuN-positive cells and (**I,J**) βIII-Tubulin-positive neurons (the thin arrows indicate the co-expression). No co-localization, but a number of GPR17-positive cells (stars) in the proximity were found on (**K,L**) TH-positive cells. Scale bars: (**A–C**) = 20 μm; (**D–J**) = 10 μm; (**K,L**) = 20 μm.

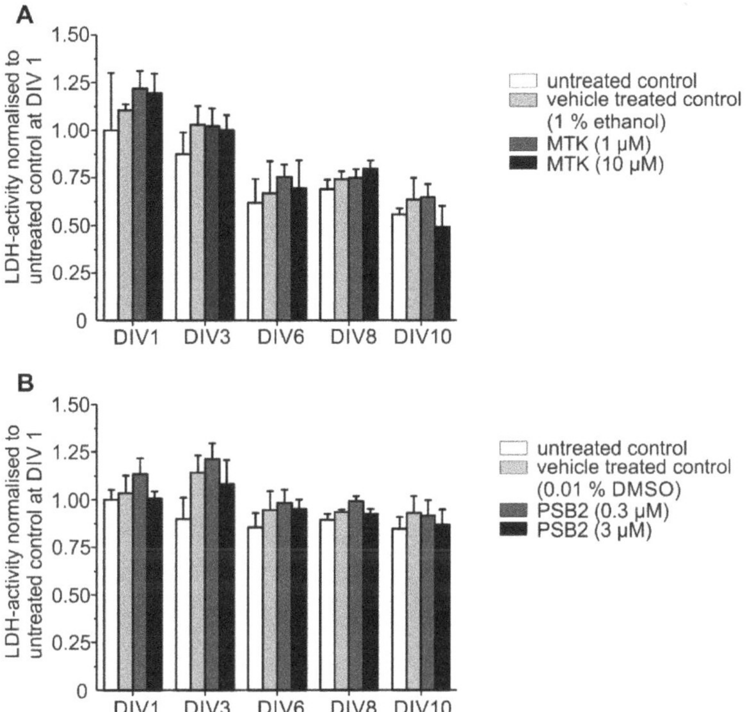

Figure 6. Toxicity testing. (**A**) LDH activity was measured after treatment with MTK (**A**) and PSB-16484 (PSB2) (**B**) compared to vehicle treated control und untreated control. Data have been normalized to the control value at DIV 1 of the respective preparation. Statistical analysis was performed using repeated measurement two-way ANOVA. Each sample represents incubation medium (IM) of one animal. The number of animals was $n = 5$.

After treatment with PSB-16484 (PSB2) the LDH activity was neither elevated compared to untreated control nor to vehicle-treated control (repeated measurement two-way ANOVA with Greenhouse–Geisser correction F (1.6208, 6.48319) = 4.87266; $p = 0.05565$) 0.01% DMSO). Overall LDH activity decreased from DIV 1 to DIV 10 with no significant differences between pharmacological treatment groups and vehicle-treated controls. These data provide proof that neither MTK nor PSB-16484 (PSB2) is toxic to the organotypic dopaminergic slice co-cultures (Figure 6). Both effects, (i) the increase in LDH release (representing the mechanical lesion as a result of the preparation of the co-cultures) and (ii) the decrease in LDH release in the following DIV agrees very well with previous data from our group [48].

2.6. Electron Microscopy

The electron microscopic observation revealed the presence of multi-lamellar myelin-like structures in the slice co-cultures of SN/VTA at DIV 10 of incubation (Figure S5A,B) and also in the early postnatal (P12) ventral midbrain (Figure S5C). High resolution images display n-like lamellar organized membranous structures in the SN/VTA part of the co-culture. Additionally, in the developing fibres bridge from SN/VTA to the PFC, irregularly organized myelin-like lamellae were found ensheathing neuronal profiles (Figure S5B).

For comparison, in the ventral midbrain (P12, rat) cross-sectioned axons of varying diameters could be detected. Only very few axons were enveloped by membranous myelin-like sheaths (Figure S5C).

3. Discussion

The present results confirm the ability of MTK to promote axonal outgrowth and suggest the involvement of GPR17 in these effects. In more detail, the obtained data in the dopaminergic organotypic slice co-cultures indicate that (i) MTK, a non-selective GPR17 antagonist, can promote neurite outgrowth with effects comparable to those induced by the well-known neurotrophic factor GDNF, (ii) treatment with MTK increases mRNA-expression of genes relevant to neuronal growth, (iii) a clear expression of GPR17 on NG2-glia and in some NeuN-positive neurons, (iv) a time-dependent expression of GPR17 in untreated organotypic dopaminergic co-cultures. These results could be of special interest to patients with TBI and other neurological disorders, suggesting that MTK possibly not only attenuates damage, but also promotes neuroregeneration and repair.

3.1. Pharmacological Inhibition of the GPR17 with MTK Can Promote Neurite Outgrowth

The present data show that in the studied dopaminergic organotypic slice co-cultures neuronal outgrowth was stimulated by treatment with the CysLT1-R antagonist and GPR17 antagonist MTK, with effects comparable to those induced by the well-known neurotrophic factor GDNF.

In order to determine if this effect occurred by antagonism of GPR17, two new synthetic GPR17 agonists were applied alone to the co-cultures showing no stimulatory but also no inhibitory effect on neurite outgrowth. Possibly, this reflects the release of endogenous agonists of GPR17 in the slice co-cultures following the preparation procedure, making it difficult to reveal an additional effect caused by the synthetic agonist [49]. Instead, the role of GPR17 was unveiled when the GPR17 agonist PSB-16484 was applied to cultures in the presence of MTK, as shown by abrogation of the MTK induced neurite outgrowth, supporting inhibition of GPR17 as an essential mechanism at the basis of the neuroregenerative effect of MTK under these conditions.

Previous studies in PC12 cells had shown an enhanced neurite outgrowth after treatment with the proposed GPR17 agonist UDP-glucose [27]. While UDP-glucose might have a growth-supporting effect on neurites, its activity as a GPR17 agonist has been questioned [20,22,23]. The different outcomes could also be a result of the chosen models investigating neurite outgrowth, as the organotypic slice co-cultures preserve the architecture and microenvironment of the brain with all its cellular players [30] in contrast to the PC12 cell culture model.

Our initial experiments indicated a growth-promoting effect of the employed $P2Y_{12}R$ and proposed GPR17 antagonist Cangrelor, which may support a beneficial effect of GPR17 antagonists on neurite outgrowth. However, the effect of Cangrelor (playing a central role in the complex processes of activation and aggregation in blood platelets) [50] may be indirect, involving modulation (inhibition) of inflammatory processes by blocking $P2Y_{12}Rs$ e.g., on microglial cells [50–52].

As the specificity and validity of some ligands of GPR17 have been questioned [23,53] the use of new synthetic ligands of GPR17 is a promising way to unveil its effects. A number of recently published studies support the role of MTK as a GPR17 antagonist. MTK was found to inhibit a synthetic GPR17 agonist [^{3}H]2-carboxy-4,6-dichloro-1H-indole-3-propionic acid ([^{3}H]PSB-12150) in a concentration-dependent manner [20]. Furthermore, it was shown that MTK enhanced the growth of neurospheres due to blockade of GPR17 and that knockout of GPR17 led to increased proliferation of neurospheres [26].

3.2. MTK Promotes Neurite Outgrowth by Elevating Neurotrophic Gene Expression

In order to gain some insight into the downstream effects following treatment of the slice co-cultures, RT-qPCR was used to investigate genes involved in neuronal growth of dopaminergic neurons, inflammation and myelination.

The presented RT-qPCR results support a growth-promoting effect of MTK in the co-cultures as mRNA-expression of GAP43, NFL and TH is significantly elevated after treatment with MTK in SN/VTA. GAP43 is a regeneration associated gene [38] and overexpression of GAP43 has been shown to lead to spontaneous formation of new synapses and enhanced sprouting after injury [39]. By stimulating expression of GAP43, treatment with MTK could promote sprouting of neurites. The neurofilament NFL is essential for the radial growth of axons [40]. Treatment with MTK could thus lead to enhanced axonal growth. TH is known to be the key enzyme of dopaminergic neurons and has been used as the characteristic marker for developing dopaminergic fibres from the SN/VTA to the PFC [29,41]. Elevated expression of TH after treatment with MTK underlies MTKs potency to promote growth of dopaminergic neurons in the slice co-cultures.

Therefore, a neurogenic effect of MTK is conceivable as this would also be in line with previously published literature. Transcriptome analysis has shown that GPR17 is specifically highly expressed in adult neural progenitor cells [54]. Recent studies showed that blockade of GPR17 with MTK led to elevated neural stem and progenitor cell proliferation [26,55]. Treatment with MTK resulted in improved cognition of old rats correlating best with enhanced neurogenesis [26]. GPR17 knockdown and knockout in neurospheres induced hyperproliferation and abolished the effects of montelukast [26].

Summarizing, MTK promotes neurite outgrowth in the dopaminergic slice co-culture by enhancing gene expression of neurotrophic genes like GAP43 and NFL. Elevation of mRNA of TH in the SN/VTA after treatment with MTK supports the observed stimulatory effect of neurite fibre outgrowth in the fibre density quantification.

Our data also indicate that GPR17 is not the only mediator of MTKs neurotrophic effects, since MTK induced expression of NFL and GAP43 could not be lowered by treatment with the GPR17 agonist PSB-16484. Only the stimulated mRNA expression of TH was tendentially lowered when co-cultures were treated with PSB-16484, suggesting that elevation of TH could be specifically caused by targeting GPR17 and implying an especially beneficial effect for dopaminergic neurons after blocking GPR17.

3.3. Is Neurite Outgrowth by MTK Stimulated by Modulation of Neuroinflammatory Pathways?

The recent literature data from *in vitro* and *in vivo* experiments suggests that GPR17 is a sensor of damage [13]. It was shown that MTK reduces neuroinflammation, elevates hippocampal neurogenesis and improves learning and memory in 20-month-old rats [26]. By using gene knockdown and knockout approaches, this effect was demonstrated to be mediated through inhibition of GPR17 reducing microglial activation and elevating neurogenesis [26]. After traumatic brain injury in the human brain an increased expression of GPR17 was found followed by a decrease within a few days [15]. In the present study, in untreated dopaminergic *ex vivo* co-cultures, expression of GPR17 was high after DIV 3 in RT-qPCR and Western blot results possibly responding to the tissue damage inflicted by the preparation process with upregulation of GPR17. The Western blot data on native (uninjured) tissue (P2, P13) do not show this effect suggesting that GPR17 expression is a time-dependent dynamic parameter in our model possibly responding to the trauma caused by the preparation of the co-cultures and comparable with the *in vivo* data after TBI.

Inflammatory processes following the preparation procedure (cutting tissue) are conceivable. The authors hypothesized that by treating co-cultures with MTK, a reduction in these neuroinflammatory processes could lead to increased neurite outgrowth. In order to investigate a possible role for neuroinflammatory modulation the mRNA-expression of IL-1β was examined which is known to play a central role in mediating neuroinflammation in pathologies of the CNS [45]. However, in the present study after DIV 10 and treatment with MTK there was only a tendency of reduced expression of mRNA of IL-1β in the dopaminergic slice co-cultures. In the PFC this tendency seemed to be revoked by treatment with PSB-16484 suggesting involvement of GPR17, but differences between the groups were not statistically significant. It cannot be excluded that significant effects on expression of IL-1β measured earlier than on DIV10 might have been missed.

Interestingly, a co-localization of GPR17 with the microglial marker Iba1 was observed. Microglia are known to contribute to neuronal death and neurodegenerative pathologies [56], and microglial inhibition has been shown to be neuroprotective [57]. In a previous study, it has furthermore been demonstrated, that GPR17 mediates ischemia-like neuronal injury via microglial activation [58]. Recently, it was shown that chronically activated microglia and signalling of GPR17 inhibit maturation of OPC and myelination in an optic nerve injury model [59].

Thus, microglial expression of GPR17 and recent literature support an anti-inflammatory effect of treatment with GPR17 antagonists. The known activity of MTK as a CysLTR1 antagonist suggests a role for anti-inflammatory effects. However, in the organotypic slice co-culture model, significant effects on mRNA-expression of IL-1β after treatment with MTK reflecting anti-inflammatory activity were not confirmed but might have been missed. Hence, a relevant anti-inflammatory effect of treatment with MTK cannot be excluded. Additionally, the effect of Cangrelor may partly be explained by an inhibitory role in microglial activation via $P2Y_{12}Rs$ as described above.

3.4. Is Neurite Outgrowth Stimulated by Targeting Oligodendrocytes in the Co-Culture?

Oligodendrocytes produce myelin which allows for saltatory impulse propagation, but they also exert important trophic functions for neurons [60–62]. GPR17 plays an important role in oligodendrocyte differentiation [43,47] and was shown to be a cell-intrinsic timer of myelination [6]. Expression of GPR17 is necessary to start oligodendrocyte differentiation, but must then be downregulated to allow terminal cells' maturation and myelination [18,63,64]. Recently, it was demonstrated that treatment with MTK in a mouse model of stroke increased expression of MBP, numbers of oligodendrocytes and fibre connectivity [65].

On this basis, the expression of GPR17 in the organotypic slice co-culture model using immunohistochemistry was examined. As expected, expression of GPR17 was most abundant on NG2-glia. NG2-glia are OPCs responsible for the generation of mature oligodendrocytes during development and adulthood [14,47].

There is also evidence for NG2-glia maintaining neuronal functions and survival of neurons through regulation of neuroinflammatory pathways [66] and it was shown that NG2-glia are permissive to neurite outgrowth and stabilize sensory axons [67]. Furthermore, participation of the GPR17 to post-acute reactivity of NG2-glia in different injury paradigms was demonstrated [68]. It has been suggested that treatment of NG2-glia with GPR17 ligands could possibly influence their differentiation potential or activate reparative functions [16].

In other models of neurodegeneration, an abnormal increase in GPR17 has invariably been associated with myelin defects and its pharmacological manipulation succeeded in restoring endogenous remyelination. Furthermore, OPCs (isolated from spinal cord of $SOD1^{G93A}$ mice) display defective differentiation compared to control cells, which is rescued by treatment with the GPR17 antagonist MTK [12]. It is concluded that, as a result of either acute injury (e.g., stroke, trauma or demyelination) or genetic defects (as is the case of $SOD1^{G93A}$ mice), GPR17 is initially induced to promote OPC maturation, but its persistence in cells and inability to undergo down regulation unfavourably affects cell's terminal maturation and myelination.

Under these conditions, pharmacological antagonism of GPR17 with MTK alleviates this maturation block, and enables OPCs to resume differentiation, which would in turn possibly promote neurite outgrowth. However, mRNA expression of MBP was not enhanced after treatment with MTK on DIV 10, suggesting that the observed effects of MTK on neurite outgrowth cannot be attributed to a change in myelination in this model. Overall, in the slice co-cultures the degree of myelination was low, as found in the electron microscopic images at DIV 10. An earlier or later onset of myelination after treatment with GPR17 ligands might have been missed since oligodendroglial differentiation and GPR17 expression are coordinated in a temporally complex manner [43]. A longer period of cultivation and shorter intervals of measurement might show differences of myelination caused by treatment with MTK also contributing to increased neurite outgrowth.

In conclusion, the anti-asthmatic drug MTK promotes neurite outgrowth in a dopaminergic slice co-culture system. This effect can be antagonized by treatment with a new potent synthetic agonist of GPR17, pointing to GPR17 as an interesting regulator of neuroregenerative processes. Treatment with MTK results in elevated mRNA-expression of genes relevant for neuronal growth and elevated expression of the dopaminergic marker enzyme tyrosine hydroxylase. GPR17 is most abundantly expressed in NG2-positive glia, suggesting that these cell types are predominantly influenced by treatment with GPR17 ligands.

The present data are in line with previously published findings showing neuroprotective effects associated with the blockade of GPR17 in other disease models characterized by abnormal and prolonged GPR17 upregulation [12,13,24,25,65]. Moreover, the present results support that the neuroregenerative effects induced by MTK after brain injury are, in part, mediated by antagonism of GPR17 but also by additional mechanisms elevating expression of neurotrophic genes.

Repurposing of 'old' drugs to treat both common and rare diseases is increasingly becoming an attractive proposition, because it involves the use of de-risked compounds, with potentially lower overall development costs and shorter development timelines [69]. Intriguingly, MTK is a well-known drug that is currently part of the German guidelines for the therapy of asthma [70].

The shown putative neuroregenerative potential and its easy accessibility makes MTK an interesting drug for patients with TBI and other neurological diseases.

The data and discussions show the need for further research and development of new dual- and multi-target drugs [71], new strategies [72] and new relevant models which will help improving therapeutic strategies.

4. Materials and Methods

4.1. Materials

The following substances and factors were used: artificial cerebrospinal fluid (ACSF, composed of (mM) 126 NaCl; 2.5 KCl; 1.2 NaH_2PO_4; 1.3 $MgCl_2$ and 2.4 $CaCl_2$, pH 7.4; Hospital Pharmacy, University of Leipzig, Germany), biocytin (Sigma-Aldrich Co., St. Louis, MO, USA), dimethyl sulfoxide (DMSO, Applichem GmbH, Darmstadt, Germany), ethanol (VWR Chemicals, Darmstadt, Germany), recombinant human glial derived neurotrophic factor (GDNF; Millipore, Bedford, MA, USA), Montelukast (MTK, Biomol GmbH, Hamburg, Germany), Cangrelor (The Medicines Company, Parsippany-Troy Hills, NJ, USA).

The new potent and selective GPR17 agonists 3-(2-carboxyethyl)-4-fluoro-6-(5-methylhexyloxy)-1*H*-indole-2-carboxylic acid (PSB-16282, (PSB1), EC_{50} 12 nM) and 3-(2-carboxyethyl)-4-fluoro-6-iodo-1*H*-indole-2-carboxylic acid (PSB-16484; (PSB2), EC_{50} 32.1 nM [34]; were synthesized, purified and analysed at the University of Bonn and provided by Professor Dr. C. E. Müller (Pharmaceutical Institute, Pharmaceutical & Medicinal Chemistry, University of Bonn, Germany).

4.2. Animals

Rat breeding was performed in the animal facility of the Rudolf Boehm Institute, Universität Leipzig (see Institutional Review Board Statement). Rats were housed under standard conditions with free access to food and water and a 12 h light–dark cycle (lights on from 7:00 a.m.). Neonatal rat pups (WISTAR RjHan, own breed) of postnatal day 1–3 (P1–3) for the preparation of the organotypic slice co-cultures were used.

4.3. Preparation

The slice co-cultures were prepared from P1–3 neonatal rat pups and cultured following the "static" culture protocol as described previously [29,48,73]; for schematic illustration see Figure 1A,B. In brief, 300 μm coronal sections of mesencephalon and forebrain were cut simultaneously using two vibratomes (Leica VT1200 S; Leica VT1000 S; Nussloch, Germany). After the separation, slices of PFC and SN/VTA were transferred into petri dishes filled with 4 °C preparation medium (PM; Minimum Essential Medium (MEM),

Thermo Fisher Scientific Inc., Waltham, MA, USA), adding glutamine (50 µg/mL; Thermo Fisher Scientific Inc.). The separated slices were then placed side by side as co-cultures (PFC with SN/VTA) on moist translucent membrane inserts (0.4 µm, Millicell-CM, Millipore) in a six-well plate. The wells were filled with incubation medium (IM; MEM (50%), Hank's Balanced Salt Solution (25%), heat-inactivated horse serum (25%) (all from Thermo Fisher Scientific Inc.); supplemented with glutamine to a final concentration of 2 mM and 0.044% sodium bicarbonate (Sigma-Aldrich; pH adjusted to 7.2) and the antibiotic Gentamycin (50 µg/mL, AMRESCO, Solon, OH, USA). The cultures were stored at 37 °C in 5% CO_2, and the medium was changed thrice weekly [73].

4.4. Slice Co-Culture Treatment Procedure

The slice co-cultures were kept for 3 and 10 days *in vitro* (DIV; see Figure 1C). For the pharmacological substance treatments, the slice co-cultures were divided into different experimental groups and treated with the respective substances at different concentrations (see below). The pharmacological treatment was conducted four times on DIV 1, 3, 6 and 8 while changing the IM.

The following compound concentrations were used (selected based on their potencies): 100 pM Cangrelor, 1 µM and 10 µM MTK (solved in 1% ethanol), 0.1 and 1 µM PSB-16282 (dissolved in 0.01% DMSO), 0.3 and 3 µM PSB-16484 (dissolved in 0.01% DMSO). The compounds were applied separately and in combination, to investigate if GPR17 agonists and (proposed) GPR17 antagonist could counteract their effects (to verify receptor inhibition, the antagonist was given first followed by the respective mixture of antagonist and agonist after 15 min).

The glial cell-line derived neurotrophic factor (GDNF; 50 ng/mL; solved in 1% ACSF) was used as a positive control. GDNF is known for its neural growth-promoting properties, especially on dopaminergic neurons [74]. All pharmacological substances were tested in comparison to vehicle-treated control co-cultures.

4.5. Fixation of the Slice Co-Cultures

To perform neurite fibre outgrowth quantification after treatment procedure (Figure 1B) and immunofluorescence labelling, co-cultures were fixed for 2 h in a solution containing 4% paraformaldehyde (Merck, Darmstadt, Germany), 0.1% glutaraldehyde (Serva Electrophoresis GmbH, Heidelberg, Germany), and 0.2% picric acid (Sigma-Aldrich) in 0.1 M phosphate buffer (PB; pH 7.4). Afterwards, co-cultures were rinsed intensively with PB. Finally, the sections were vibratome-cut into 50 µm horizontal sections.

4.6. Neurite Fibre Tracing Procedure

According to the previously described protocol [73], on DIV 8 biocytin crystals were placed on top of SN/VTA, incubating for 2 h allowing the uptake of biocytin and were washed with IM, subsequently.

Then, cultures were re-incubated with IM containing the pharmacological substances as previously described [29,75]. After 48 h (on DIV 10) the co-cultures were immersion-fixed (see above) and cut into 50 µm horizontal sections using the vibratome. The uptake of the anterograde tracer biocytin was labelled using the avidin–biotin complex (1:50, ABC-Elite Kit, Vector Laboratories, Inc., Burlingame, CA, USA) and the nickel/cobalt intensified 3,3′-diaminobenzidine hydrochloride (DAB; Sigma-Aldrich) was used as a chromogen. All dyed sections were transferred on glass slides, dehydrated in a sequence of increasing ethanol concentrations, and covered with Entellan (Merck, Darmstadt, Germany).

4.7. Neurite Fibre Density Quantification

It has been shown that dopaminergic neurons in the mesocortical projection system develop their typical innervation pattern in the organotypic slice co-cultures [29]. For quantifying neurite fibre outgrowth from SN/VTA to PFC previously described protocols have been applied [33,48,73]. Slices were used for analysis only if they fulfilled defined criteria, e.g., the slice cultures should have maintained their cellular organization, the tracer should be placed correctly on the SN/VTA, there had to be a dense network of labelled cell bodies in the SN/VTA, there had to be no labelled cell bodies in the PFC [75].

Image analysis: Raw images from the border region (where the two initially separated brain slices were attached) were taken in 40-fold magnification with an AxioCam ICc 1 camera (Carl Zeiss Jena, Germany) on a light microscope (Axioskop 50; Zeiss Oberkochen, Germany). For quantifying the fibre density, an automated image analysis according to a previously described technique was used [76].

After pre-processing and image binarization, the area occupied by neurite fibres was analysed. In detail, the number of pixels occupied by neurite fibres was divided by the number of pixels of the whole image, giving the percentage of the area occupied by neurite fibres called the neurite fibre density. In the shown data, one sample corresponds to the average value of the neurite fibre density ($n = 4$ slices) measured from one animal (4–9 samples (animals) were used per substance; for details see the legend of Figure 2).

4.8. Multiple Immunofluorescence Labelling

The free-floating slices (50 µm) were pre-incubated with a blocking solution (0.05 M Tris-buffered saline (TBS), pH 7.6), supplemented with foetal calf serum (FCS, 5%) and Triton X-100 (TX-100; 0.3%). After 30 min the slices were incubated in a mixture of primary antibodies diluted in the blocking solution for 48 h at 4 °C. The following primary antibodies were used: goat anti-glial fibrillary acidic protein (GFAP; 1:300; Santa Cruz Biotechnology, Inc., Heidelberg, Germany), mouse anti-GFAP (1:1000; Sigma), rabbit anti-GPR17 (1:100; Cayman Chemical, Ann Arbor, Michigan, USA), rabbit anti-GPR17 (1:1000; Sigma-Aldrich), goat anti-ionized calcium binding adaptor molecule 1 (Iba1; 1:100; abcam, Cambridge, UK), mouse anti-microtubule associated protein 2 (MAP2; 1:200; Chemicon International, CA, USA), goat anti-MAP2 (1:100; Santa Cruz), rat anti-myelin basic protein (MBP; 1:200; Millipore), rabbit anti-NG2 chondroitin sulphate proteoglycan (NG2; 1:200; Millipore), mouse anti-neurofilament (160 kD, NFL Medium, 1:400; abcam), mouse anti-neuronal nuclei (NeuN) (1:100; Chemicon International), mouse anti-tyrosine hydroxylase (TH; 1:1000; Chemicon International) and mouse anti-βIII-Tubulin (1:400; Promega, Fitchburg, WI, USA).

After washing the slices with TBS three times for 5 min, secondary antibodies were applied with blocking solution and incubated for 2 h. For the simultaneous visualization of the different primary antisera a mixture of the following secondary antibodies was used, specific for the appropriate species IgG (rabbit, mouse, goat). Carbocyanine (Cy2- (1:400), Cy3- (1:800), Cy5- (1:100)) conjugated IgGs; all Jackson ImmunoResearch, West Grove, PA, USA) diluted in the blocking solution were applied for 2 h at room temperature. Finally, Hoechst 33342 (Hoe, final concentration 40 mg/mL, Molecular Probes, Leiden, Netherlands) was added for nuclear staining for 5 min in TBS at room temperature. After intensive washing and mounting on glass slides, sections were dehydrated and covered with Entellan (Merck).

No immunofluorescence was observed when slices were incubated in TBS without the primary antibody.

Image analysis: Multiple immunofluorescence was investigated by using a confocal laser scanning microscope (LSM 510 Meta, Zeiss, Oberkochen, Germany) working with excitation wavelengths of 488 nm (argon, yellow-green Cy2-immunofluorescence), 543 nm (helium/neon1, red Cy3-immunofluorescence), and 633 nm (helium/neon2, blue Cy5-immunofluorescence). An ultraviolet laser (362 nm) was used to excite the blue-cyan Hoe 33342 fluorescence.

4.9. Analysis of mRNA-Expression

The tissue of the dopaminergic slice co-cultures was obtained after DIV 10. After putting the membrane inserts in 4 °C cold phosphate buffer solution (PBS; pH 7.3–7.4) PFC and SN/VTA were separated with a scalpel. Out of each well, four slices of PFC and four slices of SN/VTA were put into one tube and considered as one sample (the sample preparation was replicated four times, each using an individual animal). Then, TRIzol reagent (Life Technologies, Gaithersburg, MD, USA) was used to isolate RNA following the manufacturer's protocol. After the first centrifugation, ethanol and GlycoBlue (Life Technologies) were added to the samples for optimal visibility of the pellets. RNA integrity and concentration were analysed with NanoDrop 1000 (NanoDrop Technologies, Wilmington, DE, USA).

The cDNA-synthesis was accomplished using the RevertAid^{TM} H Minus First Strand cDNA Synthesis Kit (Thermo Fisher Scientific Inc.) and a thermal cycler (MJ Research Inc., St. Bruno, QC, Canada). Subsequently, the samples were diluted 20-fold using distilled and RNAse/DNAse free water. Afterwards, 5 µL SYBR Green qPCR Master Mix (2X; Thermo Fisher Scientific Inc.) and 1 µL primer dilution (5 µM each) were added to 4 µL sample dilution. Then, qPCR was performed using a StepOnePlus^{TM} Real-Time PCR System (Thermo Fisher Scientific Inc.). As reference housekeeping gene (HKG) mitochondrial ribosomal protein L32 (Mrpl32) was chosen.

The expression of the following target genes was analysed using primer sequences of growth associated protein 43 (GAP43), GPR17, interleukin-1β (IL-1β), MBP, neurofilament light chain (NFL), TH (Table 1). The following primers were supplied by Eurofins Genomics (Ebersberg, Germany): rat GAP43, rat GPR17, rat MBP, rat MrpL32, rat NFL and rat TH. Rat IL-1β was supplied by Sigma-Aldrich.

The Hot Start Polymerase was activated by a 15 min pre-incubation at 95 °C, followed by 55 amplification cycles at 95 °C for 10 s, 60 °C for 10 s and 72 °C for 10 s.

A melting curve analysis was performed to verify correct qPCR products. The following appropriate controls have been used: no template control (water) and "reverse-transcription-minus control", in order to exclude the presence of genomic DNA in the samples. Quantification of gene expression was performed by the ΔCP method with MrpL32 serving as reference housekeeping gene. Expression levels of the respective receptors are expressed as ΔCP with the HKG MrpL32. Data are shown as bar charts (each sample *n* is equal to four slices of either PFC or SN/VTA of the respective animal).

4.10. Western Blot

At first, PFC and SN/VTA of cultured or native tissue were transferred into one tube, respectively. Then, 5 µL of Triton-X lysis buffer (NaCl (120 mM)), Tris (25 mM), EDTA (1 mM), TX-100 (1%), 1 mM phenylmethylsulfonyl fluoride (PMSF; Sigma-Aldrich) and protease inhibitor mix (0.5%; Sigma-Aldrich) were added to the samples at pH 7.4. The samples were homogenized, 15 min incubated on ice and then centrifuged at $5000 \times g$ at 4 °C. The supernatants were collected, and protein determination was performed using Pierce^{TM} BCA Protein Assay Kit (Thermo Fisher Scientific Inc.). The samples were diluted with sample buffer (1.4-dithiothreitol (DTT) (500 mM), Tris (312.5 mM), glycerol (25%), sodium dodecyl sulphate (SDS, 10%), bromophenol blue (0.005%), pyronin Y (0.005%)), pH 6.8.

This solution was cooked for 5 min at 95 °C. For performing SDS-PAGE Mini-PROTEAN Tetra Cells (Bio-Rad Laboratories GmbH, Munich, Germany) were used. At first, the 10% separating SDS polyacrylamide gel was placed in the mini cells. Before the 5% polyacrylamide stacking gel followed, 200 µL isopropanol were placed on top of the separating gel for 30 min for an ideal flat surface and were then sluiced down. The electrophoresis cells were filled with 1x Laemmli buffer (10x Laemmli buffer: glycin (1.92 M), Tris (250 mM), SDS (1%)). The calculated sample volume and 10 µL molecular weight marker (ColorPlus^{TM} Prestained Protein Ladder, Broad Range; New England Biolabs Inc., USA) were pipetted into the notches. The electrophoresis cells were connected to a power source and the gels were run for 2 h at 50 V and 400 mA.

After the gel electrophoresis, the separating gels were transferred into cathode buffer. Proteins were then transferred onto a polyvinylidene difluoride (PVDF) membrane (Westran; pore size 0.45 µm). The blotting was performed for 90 min at 1.75 mA/cm^2. Subsequently, the PVDF membrane was stained with Ponceau S dye (Carl Roth GmbH + Co. KG; Karlsruhe, Germany) in order to make the protein bands visible.

The PVDF membrane was incubated with a blocking solution for 1 h (5% milk powder, Tris-buffered saline with Tween20 (TBST)). As primary antibody the rabbit anti-GPR17 antibody (Sigma-Aldrich; 1:1000) was used, which was normalized to β-Actin (Sigma-Aldrich; 1:5000). The primary antibody was applied in a blocking solution (5% milk powder, 0.1% NaN_3) and incubated for 24 h at 4 °C. The secondary antibody was incubated in a blocking solution (5% milk powder, TBST) for 1 h at room temperature.

A chemiluminescent solution was applied (Super Signal West Femto Maximum Sensitivity Substrate; Thermo Fisher Scientific Inc.) and after 60 s images of the membrane were taken with a CCD camera (Diana II, Raytest, Isotopenmeßgeräte GmBH, Straubenhardt, Germany) and by using ImageJ (open source, Rasband: https://imagej.nih.gov/ij/; accessed on 8 June 2021).

4.11. Analysis of Cell Injury

As described previously [48], for all pharmacological treatments it was tested if they were toxic to the organotypic co-cultures by measuring the release of lactate dehydrogenase (LDH) in the IM. Briefly, the samples of IM were collected at DIV 1, 3, 5, 8 and 10 (before the cultures were fixed) and stored at −20 °C for short time. When after thawing the samples reached room temperature LDH activity was measured following the protocol described by [76]. The samples were applied to a 96-well plate format using a filter based micro plate reader device (POLAR®star Omega, BMG LABTECH GmbH, Ortenberg, Germany). Briefly, a sample volume of 40 μL was pipetted into the wells followed by 80 μL reaction reagent (40 μL phosphate buffer, 20 μL sodium pyruvate (1.9 mM; Sigma-Aldrich), 20 μL NADH (166.67 μg/mL; Sigma-Aldrich). Then, LDH activity was calculated by measuring the decrease in absorption of NADH at 340 nm (every 10 s for 3 min).

To exclude variations due to temperature changes between individual sets of measurements, LDH activity of all samples has been normalized to the control sample (untreated) at DIV 1 of the respective preparation.

4.12. Electron Microscopy

Fixation: For electron microscopy, rats (three 12-day-old rats) were transcardially perfused using 0.1 M PB containing 4% PFA and 1.5% glutaraldehyde (high purity, Serva). The tissue slice co-cultures were fixed with 0.1 M phosphate buffer (PB) containing 4% paraformaldehyde and 0.05% glutaraldehyde (high purity, Serva).

Fixed brains and co-cultures were further processed for electron microscopic analysis of presence of myelin-like elements. Using a Leica vibratome, horizontal sections (50 μm) of the mesencephalon were cut and the region of interest (PFC, SN/VTA) dissected. From the co-cultures of PFC and SN/VTA complex vibratome sections containing the connecting bridge between both tissue parts were performed. Selected sections were osmicated in 0.5% OsO_4 in 0.1 M PB for 30 min, block-stained with 1% uranyl acetate, dehydrated and flat-embedded in resin (Durcupan, Fluka, Buchs, Switzerland) on glass slides. Ultrathin sections were cut by an Ultracut (Leica) and collected on single-slot Formvar-coated copper grids. Digital images of myelin-like structures/sheets in the regions of interest were taken by a transmission electron microscope Leo 906 E (Zeiss).

4.13. Statistics

Data are presented as mean ± S.E.M. in this study. All data were tested for normality and homogeneity of variance. The applied statistical tests are specified in the figure legends and the results (e.g., ANOVA on ranks). The probability level of 0.05 or less was considered to reflect a statistically significant difference. Significance is given as * $p < 0.05$, ** $p < 0.01$. All quantitative data have been analysed with Origin® statistical analysis program (Origin® 2018b, www.originlab.com/2018b, accessed on 23 July 2021).

Supplementary Materials: All data are available online at https://www.mdpi.com/article/10.3390/ijms222111683/s1.

Author Contributions: Conceptualization: H.F., C.H., M.P.A. and C.E.M.; formal analysis: M.B., H.F., C.H., M.P.A. and C.E.M.; investigation: M.B., K.S., B.H., N.S., C.P., T.P. and C.H.; writing—original draft: M.B. and H.F.; writing—review & editing: all authors; supervision: H.F.; project administration: H.F., M.P.A. and C.E.M.; funding acquisition: M.B., C.H. and M.P.A. All authors have read and agreed to the published version of the manuscript.

Funding: This research was funded by an internal stipend of the Medical Faculty Leipzig to M.B.; by funding from the German Federal Ministry of Education and Research (BMBF, PtJ-Bio, 0315883, 0313909) to C.H. and K.S.; by founding from FISM-Fondazione Italiana Sclerosi Multipla (2017/R/1) to M.P.A. and by founding from the Italian Ministry of University and Research (MUR), PRIN—Progetti di Ricerca di Interesse Nazionale (M.P.A.; Grant no. 2017NSXP8J) to M.P.A.

Institutional Review Board Statement: The study was conducted according to the guidelines of the Declaration of Helsinki, according to the ARRIVE guidelines and German guidelines for the welfare of experimental animals and were approved by the local authorities (Landesdirektion Leipzig (Regional office Saxony); T21/13; T05/16). All possible efforts were made to replace, reduce and refine animal experiments.

Informed Consent Statement: Not applicable.

Data Availability Statement: Data are contained within the article.

Acknowledgments: The authors thank Katrin Becker, Kathrin Krause (Rudolf Boehm Institute of Pharmacology and Toxicology, Leipzig) and Sigrun Nestle (Dept. Neuroanatomy, Institute of Anatomy and Cell Biology, Faculty of Medicine, Albert-Ludwigs-University, Freiburg) for their skilful technical assistance in animal handling, and in the laboratory. The authors are grateful to Jennifer Ding (DAAD RISE Summer Research Internship, 2013).

Conflicts of Interest: The authors declare no conflict of interest. The founders had no role in the design of the study; in the collection, analysis or interpretation of data; in writing of the manuscript; or in the decision on publish the results.

Abbreviations

ACSF	artificial cerebrospinal fluid
CysLT-R	cysteinyl leukotriene receptor
GAP43	growth associated protein 43
GDNF	glial cell line-derived neurotrophic factor
GPR17	G-protein coupled receptor 17
IL-1β	interleukin-1β
LDH	lactate dehydrogenase
LTC/D4	Leukotriene C/D4
MBP	myelin basic protein
MTK	Montelukast
NFL	neurofilament light chain
NG2	neural/glial antigen 2
NeuN	neuronal nuclei antigen
OPC	oligodendrocyte precursor cell
PFC	prefrontal cortex
PSB-16282	(PSB1) 3-(2-Carboxyethyl)-4-fluoro-6-(5-methylhexyloxy)-1H-indole-2-carboxylic acid
PSB-16484	(PSB2) 3-(2-Carboxyethyl)-4-fluoro-6-iodo-1H-indole-2-carboxylic acid
RT-qPCR	reverse transcription-quantitative real-time PCR
SDS	sodium dodecyl sulfate
SN/VTA	substantia nigra/ventral tegmental area
TH	tyrosine hydroxylase

References

1. Feigin, V.L.; Abajobir, A.A.; Abate, K.H. Global, regional, and national burden of neurological disorders during 1990–2015: A systematic analysis for the global burden of disease study 2015. *Lancet Neurol.* **2017**, *16*, 877–897. [CrossRef]
2. James, S.L.; Theadom, A.; Ellenbogen, R.G.; Bannick, M.S.; Montjoy-Venning, W.; Lucchesi, L.R.; Abbasi, N.; Abdulkader, R.; Abraha, H.N.; Adsuar, J.C.; et al. Global, regional, and national burden of traumatic brain injury and spinal cord injury, 1990–2016: A systematic analysis for the global burden of disease study 2016. *Lancet Neurol.* **2019**, *18*, 56–87. [CrossRef]
3. Dorsey, E.R.; Elbaz, A.; Nichols, E.; Abd-Allah, F.; Abdelalim, A.; Adsuar, J.C.; Ansha, M.G.; Brayne, C.; Choi, J.-Y.J.; Collado-Mateo, D.; et al. Global, regional, and national burden of parkinson's disease, 1990–2016: A systematic analysis for the global burden of disease study 2016. *Lancet Neurol.* **2018**, *17*, 939–953. [CrossRef]
4. Nichols, E.; Szoeke, C.E.I.; Vollset, S.E.; Abbasi, N.; Abd-Allah, F.; Abdela, J.; Aichour, M.T.E.; Akinyemi, R.O.; Alahdab, F.; Asgedom, S.W.; et al. Global, regional, and national burden of alzheimer's disease and other dementias, 1990–2016: A systematic analysis for the global burden of disease study 2016. *Lancet Neurol.* **2019**, *18*, 88–106. [CrossRef]
5. Olesen, J.; Gustavsson, A.; Svensson, M.; Wittchen, H.U.; Jönsson, B. The economic cost of brain disorders in europe. *Eur. J. Neurol.* **2012**, *19*, 155–162. [CrossRef]
6. Chen, Y.; Wu, H.; Wang, S.; Koito, H.; Li, J.; Ye, F.; Hoang, J.; Escobar, S.S.; Gow, A.; Arnett, H.A.; et al. The oligodendrocyte-specific g protein-coupled receptor gpr17 is a cell-intrinsic timer of myelination. *Nat. Neurosci.* **2009**, *12*, 1398–1406. [CrossRef] [PubMed]
7. Hennen, S.; Wang, H.; Peters, L.; Merten, N.; Simon, K.; Spinrath, A.; Blättermann, S.; Akkari, R.; Schrage, R.; Schröder, R.; et al. Decoding signaling and function of the orphan g protein-coupled receptor gpr17 with a small-molecule agonist. *Sci. Signal.* **2013**, *6*, ra93. [CrossRef] [PubMed]
8. Bonfanti, E.; Gelosa, P.; Fumagalli, M.; Dimou, L.; Viganò, F.; Tremoli, E.; Cimino, M.; Sironi, L.; Abbracchio, M.P. The role of oligodendrocyte precursor cells expressing the gpr17 receptor in brain remodeling after stroke. *Cell Death Dis.* **2017**, *8*, e2871. [CrossRef] [PubMed]
9. Fumagalli, M.; Lecca, D.; Coppolino, G.T.; Parravicini, C.; Abbracchio, M.P. Pharmacological properties and biological functions of the gpr17 receptor, a potential target for neuro-regenerative medicine. *Adv. Exp. Med. Biol.* **2017**, *1051*, 169–192. [PubMed]
10. Seyedsadr, M.S.; Ineichen, B.V. Gpr17, a player in lysolecithin-induced demyelination, oligodendrocyte survival, and differentiation. *J. Neurosci. Off. J. Soc. Neurosci.* **2017**, *37*, 2273–2275. [CrossRef] [PubMed]
11. Alavi, M.S.; Karimi, G.; Roohbakhsh, A. The role of orphan g protein-coupled receptors in the pathophysiology of multiple sclerosis: A review. *Life Sci.* **2019**, *224*, 33–40. [CrossRef]
12. Bonfanti, E.; Bonifacino, T.; Raffaele, S.; Milanese, M.; Morgante, E.; Bonanno, G.; Abbracchio, M.P.; Fumagalli, M. Abnormal upregulation of gpr17 receptor contributes to oligodendrocyte dysfunction in sod1 g93a mice. *Int. J. Mol. Sci.* **2020**, *21*, 2395. [CrossRef]
13. Lecca, D.; Trincavelli, M.L.; Gelosa, P.; Sironi, L.; Ciana, P.; Fumagalli, M.; Villa, G.; Verderio, C.; Grumelli, C.; Guerrini, U.; et al. The recently identified p2y-like receptor gpr17 is a sensor of brain damage and a new target for brain repair. *PLoS ONE* **2008**, *3*, e3579. [CrossRef] [PubMed]
14. Viganò, F.; Dimou, L. The heterogeneous nature of ng2-glia. *Brain Res.* **2016**, *1638*, 129–137. [CrossRef] [PubMed]
15. Franke, H.; Parravicini, C.; Lecca, D.; Zanier, E.R.; Heine, C.; Bremicker, K.; Fumagalli, M.; Rosa, P.; Longhi, L.; Stocchetti, N.; et al. Changes of the gpr17 receptor, a new target for neurorepair, in neurons and glial cells in patients with traumatic brain injury. *Purinergic Signal.* **2013**, *9*, 451–462. [CrossRef]
16. Viganò, F.; Schneider, S.; Cimino, M.; Bonfanti, E.; Gelosa, P.; Sironi, L.; Abbracchio, M.P.; Dimou, L. Gpr17 expressing ng2-glia: Oligodendrocyte progenitors serving as a reserve pool after injury. *Glia* **2016**, *64*, 287–299. [CrossRef]
17. Coppolino, G.T.; Marangon, D.; Negri, C.; Menichetti, G.; Fumagalli, M.; Gelosa, P.; Dimou, L.; Furlan, R.; Lecca, D.; Abbracchio, M.P. Differential local tissue permissiveness influences the final fate of gpr17-expressing oligodendrocyte precursors in two distinct models of demyelination. *Glia* **2018**, *66*, 1118–1130. [CrossRef] [PubMed]
18. Lecca, D.; Raffaele, S.; Abbracchio, M.P.; Fumagalli, M. Regulation and signaling of the gpr17 receptor in oligodendroglial cells. *Glia* **2020**, *68*, 1957–1967. [CrossRef]
19. Benned-Jensen, T.; Rosenkilde, M.M. Distinct expression and ligand-binding profiles of two constitutively active gpr17 splice variants. *Br. J. Pharmacol.* **2010**, *159*, 1092–1105. [CrossRef] [PubMed]
20. Köse, M.; Ritter, K.; Thiemke, K.; Gillard, M.; Kostenis, E.; Müller, C.E. Development of (3)h2-carboxy-4,6-dichloro-1h-indole-3-propionic acid ((3)hpsb-12150): A useful tool for studying gpr17. *ACS Med. Chem. Lett.* **2014**, *5*, 326–330. [CrossRef]
21. Ciana, P.; Fumagalli, M.; Trincavelli, M.L.; Verderio, C.; Rosa, P.; Lecca, D.; Ferrario, S.; Parravicini, C.; Capra, V.; Gelosa, P.; et al. The orphan receptor gpr17 identified as a new dual uracil nucleotides/cysteinyl-leukotrienes receptor. *EMBO J.* **2006**, *25*, 4615–4627. [CrossRef] [PubMed]
22. Qi, A.-D.; Harden, T.K.; Nicholas, R.A. Is gpr17 a p2y/leukotriene receptor? Examination of uracil nucleotides, nucleotide sugars, and cysteinyl leukotrienes as agonists of gpr17. *J. Pharmacol. Exp. Ther.* **2013**, *347*, 38–46. [CrossRef] [PubMed]
23. Simon, K.; Merten, N.; Schröder, R.; Hennen, S.; Preis, P.; Schmitt, N.-K.; Peters, L.; Schrage, R.; Vermeiren, C.; Gillard, M.; et al. The orphan receptor gpr17 is unresponsive to uracil nucleotides and cysteinyl leukotrienes. *Mol. Pharmacol.* **2017**, *91*, 518–532. [CrossRef] [PubMed]

24. Ceruti, S.; Viganò, F.; Boda, E.; Ferrario, S.; Magni, G.; Boccazzi, M.; Rosa, P.; Buffo, A.; Abbracchio, M.P. Expression of the new p2y-like receptor gpr17 during oligodendrocyte precursor cell maturation regulates sensitivity to atp-induced death. *Glia* **2011**, *59*, 363–378. [CrossRef] [PubMed]

25. Ceruti, S.; Villa, G.; Genovese, T.; Mazzon, E.; Longhi, R.; Rosa, P.; Bramanti, P.; Cuzzocrea, S.; Abbracchio, M.P. The p2y-like receptor gpr17 as a sensor of damage and a new potential target in spinal cord injury. *Brain J. Neurol.* **2009**, *132*, 2206–2218. [CrossRef] [PubMed]

26. Marschallinger, J.; Schäffner, I.; Klein, B.; Gelfert, R.; Rivera, F.J.; Illes, S.; Grassner, L.; Janssen, M.; Rotheneichner, P.; Schmuckermair, C.; et al. Structural and functional rejuvenation of the aged brain by an approved anti-asthmatic drug. *Nat. Commun.* **2015**, *6*, 8466. [CrossRef]

27. Daniele, S.; Lecca, D.; Trincavelli, M.L.; Ciampi, O.; Abbracchio, M.P.; Martini, C. Regulation of pc12 cell survival and differentiation by the new p2y-like receptor gpr17. *Cell. Signal.* **2010**, *22*, 697–706. [CrossRef]

28. Mao, F.-X.; Li, W.-J.; Chen, H.-J.; Qian, L.-H.; Buzby, J.S. Periventricular leukomalacia long-term prognosis may be improved by treatment with udp-glucose, gdnf, and memantine in neonatal rats. *Brain Res.* **2012**, *1486*, 112–120. [CrossRef]

29. Franke, H.; Schelhorn, N.; Illes, P. Dopaminergic neurons develop axonal projections to their target areas in organotypic co-cultures of the ventral mesencephalon and the striatum/prefrontal cortex. *Neurochem. Int.* **2003**, *42*, 431–439. [CrossRef]

30. Heine, C.; Sygnecka, K.; Franke, H. Purines in neurite growth and astroglia activation. *Neuropharmacology* **2016**, *104*, 255–271. [CrossRef]

31. Cho, S.; Wood, A.; Bowlby, M.R. Brain slices as models for neurodegenerative disease and screening platforms to identify novel therapeutics. *Curr. Neuropharmacol.* **2007**, *5*, 19–33. [CrossRef]

32. Humpel, C. Organotypic brain slice cultures: A review. *Neuroscience* **2015**, *305*, 86–98. [CrossRef]

33. Heine, C.; Sygnecka, K.; Scherf, N.; Grohmann, M.; Bräsigk, A.; Franke, H. P2y(1) receptor mediated neuronal fibre outgrowth in organotypic brain slice co-cultures. *Neuropharmacology* **2015**, *93*, 252–266. [CrossRef] [PubMed]

34. Baqi, Y.; Pillaiyar, T.; Abdelrahman, A.; Kaufmann, O.; Alshaibani, S.; Rafehi, M.; Ghasimi, S.; Akkari, R.; Ritter, K.; Simon, K.; et al. 3-(2-carboxyethyl)indole-2-carboxylic acid derivatives: Structural requirements and properties of potent agonists of the orphan g protein-coupled receptor gpr17. *J. Med. Chem.* **2018**, *61*, 8136–8154. [CrossRef]

35. Baqi, Y.; Alshaibani, S.; Ritter, K.; Abdelrahman, A.; Spinrath, A.; Kostenis, E.; Müller, C.E. Improved synthesis of 4-/6-substituted 2-carboxy-1h-indole-3-propionic acid derivatives and structure–activity relationships as gpr17 agonists. *Med. Chem. Commun.* **2014**, *5*, 86–92. [CrossRef]

36. af Bjerkén, S.; Boger, H.A.; Nelson, M.; Hoffer, B.J.; Granholm, A.-C.; Strömberg, I. Effects of glial cell line-derived neurotrophic factor deletion on ventral mesencephalic organotypic tissue cultures. *Brain Res.* **2007**, *1133*, 10–19. [CrossRef]

37. Wang, H.; Wu, M.; Zhan, C.; Ma, E.; Yang, M.; Yang, X.; Li, Y. Neurofilament proteins in axonal regeneration and neurodegenerative diseases. *Neural Regen. Res.* **2012**, *7*, 620–626.

38. Ma, T.C.; Willis, D.E. What makes a rag regeneration associated? *Front. Mol. Neurosci.* **2015**, *8*, 43. [CrossRef] [PubMed]

39. Benowitz, L.I.; Routtenberg, A. Gap-43: An intrinsic determinant of neuronal development and plasticity. *Trends Neurosci.* **1997**, *20*, 84–91. [CrossRef]

40. Yuan, A.; Rao, M.V.; Nixon, R.A. Neurofilaments at a glance. *J. Cell Sci.* **2012**, *125*, 3257–3263. [CrossRef] [PubMed]

41. Bademci, G.; Vance, J.M.; Wang, L. Tyrosine hydroxylase gene: Another piece of the genetic puzzle of parkinson's disease. *CNS Neurol. Disord. Drug Targets* **2012**, *11*, 469–481. [CrossRef] [PubMed]

42. Fumagalli, M.; Lecca, D.; Abbracchio, M.P. Cns remyelination as a novel reparative approach to neurodegenerative diseases: The roles of purinergic signaling and the p2y-like receptor gpr17. *Neuropharmacology* **2016**, *104*, 82–93. [CrossRef]

43. Ou, Z.; Sun, Y.; Lin, L.; You, N.; Liu, X.; Li, H.; Ma, Y.; Cao, L.; Han, Y.; Liu, M.; et al. Olig2-targeted g-protein-coupled receptor gpr17 regulates oligodendrocyte survival in response to lysolecithin-induced demyelination. *J. Neurosci. Off. J. Soc. Neurosci.* **2016**, *36*, 10560–10573. [CrossRef] [PubMed]

44. Mendiola, A.S.; Cardona, A.E. The il-1\textgreek{b} phenomena in neuroinflammatory diseases. *J. Neural Transm.* **2018**, *125*, 781–795. [CrossRef]

45. Fratangeli, A.; Parmigiani, E.; Fumagalli, M.; Lecca, D.; Benfante, R.; Passafaro, M.; Buffo, A.; Abbracchio, M.P.; Rosa, P. The regulated expression, intracellular trafficking, and membrane recycling of the p2y-like receptor gpr17 in oli-neu oligodendroglial cells. *J. Biol. Chem.* **2013**, *288*, 5241–5256. [CrossRef]

46. Fumagalli, M.; Daniele, S.; Lecca, D.; Lee, P.R.; Parravicini, C.; Fields, R.D.; Rosa, P.; Antonucci, F.; Verderio, C.; Trincavelli, M.L.; et al. Phenotypic changes, signaling pathway, and functional correlates of gpr17-expressing neural precursor cells during oligodendrocyte differentiation. *J. Biol. Chem.* **2011**, *286*, 10593–10604. [CrossRef] [PubMed]

47. Sygnecka, K.; Heine, C.; Scherf, N.; Fasold, M.; Binder, H.; Scheller, C.; Franke, H. Nimodipine enhances neurite outgrowth in dopaminergic brain slice co-cultures. *Int. J. Dev. Neurosci. Off. J. Int. Soc. Dev. Neurosci.* **2015**, *40*, 1–11. [CrossRef]

48. Boccazzi, M.; Lecca, D.; Marangon, D.; Guagnini, F.; Abbracchio, M.P.; Ceruti, S. A new role for the p2y-like gpr17 receptor in the modulation of multipotency of oligodendrocyte precursor cells in vitro. *Purinergic Signal.* **2016**, *12*, 661–672. [CrossRef] [PubMed]

49. Gachet, C. P2y(12) receptors in platelets and other hematopoietic and non-hematopoietic cells. *Purinergic Signal.* **2012**, *8*, 609–619. [CrossRef] [PubMed]

50. Maeda, M.; Tsuda, M.; Tozaki-Saitoh, H.; Inoue, K.; Kiyama, H. Nerve injury-activated microglia engulf myelinated axons in a p2y12 signaling-dependent manner in the dorsal horn. *Glia* **2010**, *58*, 1838–1846. [CrossRef]

51. Zhan, T.-W.; Tian, Y.-X.; Wang, Q.; Wu, Z.-X.; Zhang, W.-P.; Lu, Y.-B.; Wu, M. Cangrelor alleviates pulmonary fibrosis by inhibiting gpr17-mediated inflammation in mice. *Int. Immunopharmacol.* **2018**, *62*, 261–269. [CrossRef]
52. Maekawa, A.; Balestrieri, B.; Austen, K.F.; Kanaoka, Y. Gpr17 is a negative regulator of the cysteinyl leukotriene 1 receptor response to leukotriene d4. *Proc. Natl. Acad. Sci. USA* **2009**, *106*, 11685–11690. [CrossRef] [PubMed]
53. Maisel, M.; Herr, A.; Milosevic, J.; Hermann, A.; Habisch, H.-J.; Schwarz, S.; Kirsch, M.; Antoniadis, G.; Brenner, R.; Hallmeyer-Elgner, S.; et al. Transcription profiling of adult and fetal human neuroprogenitors identifies divergent paths to maintain the neuroprogenitor cell state. *Stem Cells* **2007**, *25*, 1231–1240. [CrossRef]
54. Huber, C.; Marschallinger, J.; Tempfer, H.; Furtner, T.; Couillard-Despres, S.; Bauer, H.-C.; Rivera, F.J.; Aigner, L. Inhibition of leukotriene receptors boosts neural progenitor proliferation. *Cell. Physiol. Biochem. Int. J. Exp. Cell. Physiol. Biochem. Pharmacol.* **2011**, *28*, 793–804. [CrossRef]
55. Kabba, J.A.; Xu, Y.; Christian, H.; Ruan, W.; Chenai, K.; Xiang, Y.; Zhang, L.; Saavedra, J.M.; Pang, T. Microglia: Housekeeper of the central nervous system. *Cell. Mol. Neurobiol.* **2017**, *38*, 53–71. [CrossRef] [PubMed]
56. Podbielska, M.; Das, A.; Smith, A.W.; Chauhan, A.; Ray, S.K.; Inoue, J.; Azuma, M.; Nozaki, K.; Hogan, E.L.; Banik, N.L. Neuron-microglia interaction induced bi-directional cytotoxicity associated with calpain activation. *J. Neurochem.* **2016**, *139*, 440–455. [CrossRef] [PubMed]
57. Zhao, B.; Wang, H.; Li, C.-X.; Song, S.-W.; Fang, S.-H.; Wei, E.-Q.; Shi, Q.-J. Gpr17 mediates ischemia-like neuronal injury via microglial activation. *Int. J. Mol. Med.* **2018**, *42*, 2750–2762. [CrossRef] [PubMed]
58. Wang, J.; He, X.; Meng, H.; Li, Y.; Dmitriev, P.; Tian, F.; Page, J.C.; Lu, Q.R.; He, Z. Robust myelination of regenerated axons induced by combined manipulations of gpr17 and microglia. *Neuron* **2020**, *108*, 876–886.e874. [CrossRef]
59. Nave, K.-A. Myelination and support of axonal integrity by glia. *Nature* **2010**, *468*, 244–252. [CrossRef] [PubMed]
60. Fünfschilling, U.; Supplie, L.M.; Mahad, D.; Boretius, S.; Saab, A.S.; Edgar, J.; Brinkmann, B.G.; Kassmann, C.M.; Tzvetanova, I.D.; Möbius, W.; et al. Glycolytic oligodendrocytes maintain myelin and long-term axonal integrity. *Nature* **2012**, *485*, 517–521. [CrossRef] [PubMed]
61. Lee, Y.; Morrison, B.M.; Li, Y.; Lengacher, S.; Farah, M.H.; Hoffman, P.N.; Liu, Y.; Tsingalia, A.; Jin, L.; Zhang, P.-W.; et al. Oligodendroglia metabolically support axons and contribute to neurodegeneration. *Nature* **2012**, *487*, 443–448. [CrossRef]
62. Fumagalli, M.; Bonfanti, E.; Daniele, S.; Zappelli, E.; Lecca, D.; Martini, C.; Trincavelli, M.L.; Abbracchio, M.P. The ubiquitin ligase mdm2 controls oligodendrocyte maturation by intertwining mtor with g protein-coupled receptor kinase 2 in the regulation of gpr17 receptor desensitization. *Glia* **2015**, *63*, 2327–2339. [CrossRef]
63. Simon, K.; Hennen, S.; Merten, N.; Blättermann, S.; Gillard, M.; Kostenis, E.; Gomeza, J. The orphan g protein-coupled receptor gpr17 negatively regulates oligodendrocyte differentiation via g\textgreek{a}i/o and its downstream effector molecules. *J. Biol. Chem.* **2016**, *291*, 705–718. [CrossRef] [PubMed]
64. Gelosa, P.; Bonfanti, E.; Castiglioni, L.; Delgado-Garcia, J.M.; Gruart, A.; Fontana, L.; Gotti, M.; Tremoli, E.; Lecca, D.; Fumagalli, M.; et al. Improvement of fiber connectivity and functional recovery after stroke by montelukast, an available and safe anti-asthmatic drug. *Pharmacol. Res.* **2019**, *142*, 223–236. [CrossRef] [PubMed]
65. Nakano, M.; Tamura, Y.; Yamato, M.; Kume, S.; Eguchi, A.; Takata, K.; Watanabe, Y.; Kataoka, Y. Ng2 glial cells regulate neuroimmunological responses to maintain neuronal function and survival. *Sci. Rep.* **2017**, *7*, 42041. [CrossRef] [PubMed]
66. Busch, S.A.; Horn, K.P.; Cuascut, F.X.; Hawthorne, A.L.; Bai, L.; Miller, R.H.; Silver, J. Adult ng2+ cells are permissive to neurite outgrowth and stabilize sensory axons during macrophage-induced axonal dieback after spinal cord injury. *J. Neurosci. Off. J. Soc. Neurosci.* **2010**, *30*, 255–265. [CrossRef]
67. Boda, E.; Viganò, F.; Rosa, P.; Fumagalli, M.; Labat-Gest, V.; Tempia, F.; Abbracchio, M.P.; Dimou, L.; Buffo, A. The gpr17 receptor in ng2 expressing cells: Focus on in vivo cell maturation and participation in acute trauma and chronic damage. *Glia* **2011**, *59*, 1958–1973. [CrossRef] [PubMed]
68. Pushpakom, S.; Iorio, F.; Eyers, P.A.; Escott, K.J.; Hopper, S.; Wells, A.; Doig, A.; Guilliams, T.; Latimer, J.; McNamee, C.; et al. Drug repurposing: Progress, challenges and recommendations. *Nat. Rev. Drug Discov.* **2018**, *18*, 41–58. [CrossRef] [PubMed]
69. Buhl, R.; Bals, R.; Baur, X.; Berdel, D.; Criée, C.P.; Gappa, M.; Gillissen, A.; Greulich, T.; Haidl, P.; Hamelmann, E.; et al. S2k-leitlinie zur diagnostik und therapie von patienten mit asthma. *Pneumologie* **2017**, *71*, 849–919.
70. Pillaiyar, T.; Funke, M.; Al-Hroub, H.; Weyler, S.; Ivanova, S.; Schlegel, J.; Abdelrahman, A.; Müller, C.E. Design, synthesis and biological evaluation of suramin-derived dual antagonists of the proinflammatory g protein-coupled receptors p2y2 and gpr17. *Eur. J. Med. Chem.* **2020**, *186*, 111789. [CrossRef] [PubMed]
71. Parravicini, C.; Lecca, D.; Marangon, D.; Coppolino, G.T.; Daniele, S.; Bonfanti, E.; Fumagalli, M.; Raveglia, L.; Martini, C.; Gianazza, E.; et al. Development of the first in vivo gpr17 ligand through an iterative drug discovery pipeline: A novel disease-modifying strategy for multiple sclerosis. *PLoS ONE* **2020**, *15*, e0231483. [CrossRef] [PubMed]
72. Heine, C.; Franke, H. Organotypic slice co-culture systems to study axon regeneration in the dopaminergic system ex vivo. *Methods Mol. Biol.* **2014**, *1162*, 97–111. [PubMed]
73. Hudson, J.; Granholm, A.C.; Gerhardt, G.A.; Henry, M.A.; Hoffman, A.; Biddle, P.; Leela, N.S.; Mackerlova, L.; Lile, J.D.; Collins, F. Glial cell line-derived neurotrophic factor augments midbrain dopaminergic circuits in vivo. *Brain Res. Bull.* **1995**, *36*, 425–432. [CrossRef]
74. Heine, C.; Wegner, A.; Grosche, J.; Allgaier, C.; Illes, P.; Franke, H. P2 receptor expression in the dopaminergic system of the rat brain during development. *Neuroscience* **2007**, *149*, 165–181. [CrossRef]

75. Heine, C.; Sygnecka, K.; Scherf, N.; Berndt, A.; Egerland, U.; Hage, T.; Franke, H. Phosphodiesterase 2 inhibitors promote axonal outgrowth in organotypic slice co-cultures. *Neuro-Signals* **2013**, *21*, 197–212. [CrossRef] [PubMed]
76. Koh, J.Y.; Choi, D.W. Quantitative determination of glutamate mediated cortical neuronal injury in cell culture by lactate dehydrogenase efflux assay. *J. Neurosci. Methods* **1987**, *20*, 83–90. [CrossRef]

International Journal of
Molecular Sciences

Review

Astroglial and Microglial Purinergic P2X7 Receptor as a Major Contributor to Neuroinflammation during the Course of Multiple Sclerosis

Marta Sidoryk-Węgrzynowicz * and Lidia Strużyńska *

Laboratory of Pathoneurochemistry, Department of Neurochemistry, Mossakowski Medical Research Institute, 02-106 Warsaw, Poland
* Correspondence: msidoryk@imdik.pan.pl (M.S.-W.); lidkas@imdik.pan.pl (L.S.)

Abstract: Multiple sclerosis (MS) is an autoimmune inflammatory disease of the central nervous system that leads to the progressive disability of patients. A characteristic feature of the disease is the presence of focal demyelinating lesions accompanied by an inflammatory reaction. Interactions between autoreactive immune cells and glia cells are considered as a central mechanism underlying the pathology of MS. A glia-mediated inflammatory reaction followed by overproduction of free radicals and generation of glutamate-induced excitotoxicity promotes oligodendrocyte injury, contributing to demyelination and subsequent neurodegeneration. Activation of purinergic signaling, in particular P2X7 receptor-mediated signaling, in astrocytes and microglia is an important causative factor in these pathological processes. This review discusses the role of astroglial and microglial cells, and in particular glial P2X7 receptors, in inducing MS-related neuroinflammatory events, highlighting the importance of P2X7R-mediated molecular pathways in MS pathology and identifying these receptors as a potential therapeutic target.

Keywords: purinergic receptors; neuroinflammation; autoimmune disease; microglia; astroglia

Citation: Sidoryk-Węgrzynowicz, M.; Strużyńska, L. Astroglial and Microglial Purinergic P2X7 Receptor as a Major Contributor to Neuroinflammation during the Course of Multiple Sclerosis. *Int. J. Mol. Sci.* **2021**, *22*, 8404. https://doi.org/10.3390/ijms22168404

Academic Editors: Dmitry Aminin and Peter Illes

Received: 28 June 2021
Accepted: 1 August 2021
Published: 5 August 2021

Publisher's Note: MDPI stays neutral with regard to jurisdictional claims in published maps and institutional affiliations.

1. Introduction

Multiple sclerosis (MS) is a chronic immune-mediated inflammatory disease of the central nervous system (CNS) with unknown etiology that mainly affects young adults, leading to the progressive physical disability and mental stress of patients. This multifactorial disease is characterized by infiltration of activated peripheral immune cells into the brain and spinal cord with subsequent immune-mediated demyelination and neurodegeneration accompanied by inflammation. The presence of focal demyelinating lesions in the white and grey matter of brain and spinal cord of MS patients, connected with reactive micro- and astrogliosis, is a main diagnostic hallmark of the disease. The pathological basis of these lesions is a selective and primary disruption of myelin sheaths with subsequent axonal degeneration and loss of neurons. Development of lesions in cerebral cortex characterizes the evolution from an early relapsing/remitting phase (RR) into a secondary progressive (SP) phase of the disease [1].

The clinical symptoms of the disease, such as imbalance, muscle weakness, motor dyscoordination, pain, paralysis, cognitive impairment, and depression [2], are correlated with the injured anatomical region of the CNS. The MS etiology is still unknown, although genetic, hormonal, and environmental factors are indicated as significantly influencing the development of the disease. A recent study using single-nucleus RNA sequencing demonstrated region-specific transcriptomic alternations associated with selective damage of neurons in cortical layers associated with the upregulation of stress pathway genes and long non-coding RNAs. This study revealed the vulnerability of oligodendrocytes, reactive astrocytes, and activated microglia which appear most frequently in the areas surrounding MS lesions [3].

Activated glial cells, both microglia and astroglia, substantially contribute to MS pathogenesis and progression by driving and accelerating inflammatory reaction, generating free radicals and releasing cytotoxic excitatory amino acids. All these factors contribute to oligodendrocyte injury, myelin damage, and axonal degeneration, finally leading to the death of neurons. Extracellular ATP (eATP) and plethora of its receptors are of importance in neuron–astrocyte–microglia intercommunication in pathological mechanisms occurring in MS. Recent studies demonstrated that glial cells expressing P2X7 receptors (P2X7Rs) are involved in eATP-dependent signaling involved in the crucial pathological mechanisms at early and progressive stages of MS and experimental autoimmune encephalomyelitis (EAE) [4].

This review focuses on the role of microglial and astroglial cells in MS pathology based on clinical and experimental data, and discusses the involvement of P2X7 purinergic receptors into glia-mediated neuroinflammation.

2. Multiple Sclerosis and Neuroinflammation

2.1. Neuroinflammation during the Course of MS

Focal demyelinated lesions are present in the white matter, as well as in the grey matter of cortex, the basal ganglia, brain stem, and spinal cord [5]. The inflammatory reaction is initiated around post-capillary venoles and veins, spreading further into the surrounding normal-appearing white and grey matter [6]. Perivascular inflammatory infiltrates are composed of T cells, B cells, and immunoglobulin-positive plasma cells, whereas active demyelinating plaques consists mainly of macrophages and activated microglia. CD3+, CD4+, and CD8+ T cells outnumber CD20-positive B cells and plasma cells [7,8], although a subset of inflammatory immune cells differs in composition depending on the brain region (perivascular area vs. parenchymal space) and the stage of the disease (initial vs. advanced) [5]. Activated autoreactive CD4+ cells release pro-inflammatory mediators, such as IFN-γ and/or IL-17, thus driving inflammation. Monocytes recruited from the blood in response to chemokine signaling also constitute a significant component of myelin lesions. Their concentration in lesions has been shown to be positively correlated with the severity of neurological deficits during EAE, whereas depletion of monocytes delays the onset and severity of the disease [9,10]. Activated by cytokines/chemokines, including GM-CSF, INF-γ, and TNF-α, monocytes are transformed into a pro-inflammatory M1 phenotype, expressing MHC II antigens and producing pro-inflammatory factors, proteases, as well as reactive oxygen (ROS) and nitrogen (NRS) species [11]. Thus, migration of immune cells via the blood–brain barrier (BBB), in order to colonize the CNS is a critical step of the disease, contributes to the development of inflammatory cascade. During the course of the disease, in a progressive phase, massive inflammation is observed and mediated by CD4+, CD8+ cells, B cells, and monocytes, and amplified by activated glial cells in response to the presence of disturbed myelin and damaged tissues [12]. The spreading inflammatory process is characterized by upregulation of cytokines, chemokines, and adhesion molecules of both a pro- and anti-inflammatory nature [13]. This situation is aggravated by dysfunction of regulatory T lymphocytes (Tregs) that physiologically suppress autoimmune processes but are impaired in MS patients [14].

MS is considered as an inflammatory disease with a neurodegenerative component, in which an autoimmune inflammatory reaction is accompanied by degeneration of the demyelinated nerve fibers. However, the issue has not been resolved within this sequence of events [15]. Oxidative stress [16] and glutamate-induced excitotoxicity [17] are the main mechanisms, contributing significantly to neurodegeneration in MS pathology, in which glial cells are strongly involved.

2.2. Microglia as Contributors to MS-Related Neuroinflammation

Microglia are resident immune competent cells of the CNS, guarding the homeostasis and protecting nervous tissue against various pathological stimuli [18]. Under "resting" conditions, they in fact function as dynamic sensors that continuously "scan" their environment [19]. In the homeostatic state, they express specific surface markers such as transmembrane protein 119 (TMEM119) and purinergic receptor P2Y12 (reviewed in: Guerrero and Sicotte, 2020, [20]). Depending on the insult, they exhibit different features, either protecting the tissue or exacerbating the injury. When activated under pathological conditions, microglial cells respond rapidly to neuronal distress changing their morphology from ramified to amoeboid. Quiescent state is characterized by small soma, multiple delicate processes, flattened nucleus, and small Golgi apparatus. While transforming to the active state, microglial cells enlarge the soma, retract processes, and overexpress immunomodulatory factors [21]. Two phenotypes of microglia have been classically distinguished: proinflammatory (M1) and anti-inflammatory (M2). This distinction is currently not obvious, as it has been suggested that microglial phenotypes are transient and demonstrate temporal and spatial profiles of transformation following an active response to the changes in the tissue microenvironment [22].

Microglia are implicated in multiple inflammatory and neurodegenerative diseases, including MS. These cells are a key player in the mechanisms underlying MS pathology exhibiting complex roles linked to the stage of the disease. They are critical for antigen presentation to T cells, development and exacerbation of inflammation, and subsequent synaptic loss. However, they can also modulate the inflammatory response and provide the protective functions by phagocytosis of tissue remnants and participation in the processes of tissue repair [23,24]. Moreover, phagocytosis of myelin debris in MS lesions, and expression of anti-inflammatory and protective factors by activated microglia, are essential processes to promote remyelination and enhance neuronal survival [25,26]. Protective factors released from microglia include neurotrophins such as nerve growth factor (NGF), brain-derived neurotrophic factor (BDNF), and basic fibroblast growth factor (bFGF) [21].

As immune competent cells of the CNS, microglia participate in immune response via interactions with other immune cells. There is a cross-talk between microglia and peripheral immune cells that are both present in MS lesions. Activated microglia, presenting MHCs antigens class I and II recognized by T cells, participate in the further recruitment of adaptive immune Th1 and Th17 cells into the CNS [27]. While interacting with cytotoxic T-lymphocyte-associated antigen 4 (CTLA4), microglia induce apoptosis of T cells, whereas the binding of microglia-derived molecules B7-1 and B7-2 to CD28 antigen stimulates the proliferation of T cells and release of cytokines [28].

In an early stage of MS, the initial pool of phagocytic cells in lesions is mainly comprised of microglia, as measured by their specific marker TMEM119 [29]. Increased expression of this marker was observed in MS patients [30]. As the disease progresses, peripheral macrophages are increasingly recruited [31]. Active demyelinating plaques in MS-diseased persons are occupied by phagocytic cells of both activated microglia and macrophages origin that contain products of myelin degradation and tissue debris, and highly expressed NADPH oxidase indicating oxidative stress-related damage [32]. Active demyelination is usually associated with a proinflammatory type of microglia, as indicated by the dominant expression of proinflammatory markers such as CD68 (involved in phagocytosis) and p22phox (involved in the production of reactive oxygen species), as well as CD86 and class II MHC antigens [29]. In the later stages of the disease, microglia/macrophages switches to an intermediate phenotype and co-expresses pro- and anti-inflammatory markers. Inactive lesions mainly consist of CD206-, CD163- and ferritin-positive microglia, indicating anti-inflammatory type [33]. In turn, the animal models of MS show that the presence of myelin debris in microglia is connected with a pro-regenerative phenotype expressing arginase-1, CD206, and insulin-like growth factor-1(IGF-1) necessary for remyelination [20,34].

Experimental evidence obtained using an EAE model of MS also confirms the significant impact of microglial pool of cells over the course of the disease. As reported, activation of microglia occurs in brains of EAE rats very early at a asymptomatic phase of the disease, as evidenced by highly increased immunoreactivity of microglia/macrophage-specific protein Iba-1 and morphological characteristics of microgliosis [35]. Regulation of microglia activity influences the outcome of the disease. The progression and the severity of neurological symptoms significantly declines after inhibition of macrophages/microglia at the developmental phase of the EAE, and the onset of the neurological deficits is delayed [36,37]. Targeting microglia activation by drugs such as minocycline, interferon-β, or fingolimod has been shown to have beneficial effects in reducing inflammation in both clinical and experimental studies of MS (reviewed in [38]). Finally, genetic factors found to be of importance for susceptibility to MS are more frequently related to microglia than other glial cells or neurons [39]. Over-expression of these risk genes highlights the significant contribution of microglia to the mechanisms underlying MS pathogenesis.

2.3. Astroglia as Contributors to MS-Related Neuroinflammation

Astrocytes are the most abundant type of glial cells population of the CNS. These cells maintain the optimal microenvironment for neuronal function [40] and play a crucial role in a variety of processes related to the normal neuronal development, synaptogenesis [41], brain microcirculation, propagation of action potentials, or immunomodulation [42]. Astroglial processes surround neuronal synapses acting functionally as a tripartite synapse wherein they regulate ion and neurotransmitter homeostasis, support neurons metabolically, as well as control and regulate synaptic activity [43].

Astrocytes express a variety of, often specific, anion channels [44], hemichannels [40], and receptors, including ionotropic purinergic receptors [45]. It is well defined that the release of gliotransmitters (amino acids, ATP and peptides) into the extracellular space appears via Ca2+-dependent exocytosis [46]. It should also be emphasized that astrocytes act as an integrated system sensitive to signals derived from all cells that form the syncytium.

A great amount of attention is paid to understanding of the mechanistic processes related to the astroglia dysfunction in response to immune attack, chronic neurodegenerative diseases, or brain injury. It is well accepted that, upon stimulation, morphologically and physiologically abnormal reactive astrocytes may interact with other cells in the CNS in a beneficial or negative way [47].

The characteristics of astrocytes and their perivascular location allow them to play a crucial role during lesion formation and to provide the access of peripheral immune cells into the CNS. In active lesions, astroglia acquire a hypertrophic morphology, as expressed by massive enlargement of the cell soma and reduced processes density that are most likely initiated by the failure of the astrocyte–oligodendrocyte network [9]. It is well established that, in the EAE experimental model of the disease, astrocytes become activated in the developing lesions before significant immune cell infiltration into the parenchyma, suggesting fundamental role of these cells in the lesion development [48]. Astrocyte activation associated with alterations in gene expression and cell hypertrophy was found to be followed by long-lasting scar formation and, in more advanced stages of the disease, with rearrangement of tissue structure.

In the injured CNS, reactive astrocytes form a glial scar and are considered to be detrimental for axonal regeneration. Astrocyte activation appears via canonical signaling cascades, represented by NF-κB pathway that is crucial for neuroinflammation [49] and regulation of innate and adaptive immunity processes. Astrocytic NF-κB signaling is directly activated upon stimulation with the pro-inflammatory cytokines TNF-α and IL-1β, through TLR signaling and various other agents including myelin, mitogens, and free radicals [50]. The downstream NF-κB pathways in astrocytes are involved both in the initiation and exacerbation of inflammatory state in the CNS. A study using transgenic mouse model with astrocyte-specific disruption of NF-κB demonstrated a significant improvement of tissue damage and amelioration of clinical symptoms of EAE and spinal cord injury (SCI).

One of the mechanisms by which astrocytes are activated in MS is a signal transducer and activator of transcription 3 (STAT3)-mediated pathway. It is well established that upregulation of STAT3 activity occurs after pathological stimulation followed by inflammation and injury of the CNS [51]. Pro- and anti-inflammatory pathways activate STAT3 signaling in astrocyte. Great examples are IFN-γ and IL-6 family cytokines that induce STAT3 phosphorylation by binding to the gp130 cell-surface receptor [52]. STAT3-mediated signaling in astrocytes was also characterized as a major player in inhibition of CNS inflammation in astrocyte-specific STAT3 knockouts. Herrmann et al. [53] showed that STAT3 deletion rendered astrocyte resistant to activation and astroglial scar formation, and resulted in active demyelination in the spinal cord lesions after SCI. Outlined changes were associated with the spread of inflammation and an increased volume of SCI-induced lesions. These findings clearly demonstrate that STAT3 signaling is a fundamental mediator of astrogliosis and provide additional evidence that scar-forming astrocytes inhibit the spread of inflammatory signals upon SCI [51]. Furthermore, EAE mice with astrocyte-specific deletion of the STAT3-activated gp130 receptor showed severe disease symptoms, as well as increased infiltration of reactive T-lymphocytes in the areas of demyelination [54].

It is worth noting that disruption of astrocyte–neuron integrity is a common hallmark of various human neurodegenerative diseases [55]. A recent study revealed that classically activated pro-inflammatory microglia secreting Il-1α, TNF, and C1q may induce reactivity of the astrocytes (known as a A1). The major sign of this astrocytic transformation is a lack of the neuroprotective features or gaining novel neurotoxic properties. An in vitro study using retinal ganglion cells co-cultured with A1 demonstrated the inhibition of the synapses development compared to those grown with control astrocytes. Furthermore, A1 astrocytes were found to secrete a soluble toxin that rapidly kills a subset of CNS neurons and mature oligodendrocytes via induction of the apoptosis [47]. Studies using in vivo approaches demonstrated that morphologically changed A1 astrocytes with numerous highly branched processes are mainly localized in the gray matter. Reactive astrogliosis is characterized by a range of functional changes that cause astrocytes unable to respond properly under pathological conditions. Interestingly, microglia-mediated activation of astrocytes and changes of astrocytic phenotype are abundant in almost all human neurodegenerative disorders, including MS.

As outlined above, an increasing amount of evidence points towards the potential of reactive astrogliosis to play either primary or secondary roles in disease progression by disrupting normal glial functions or acquiring negative properties. It is also worth noting that astrocytes in pathological processes associated with MS can acquire both neurotoxic features related to the inflammatory signaling (regulation of leukocyte trafficking) and neuroprotective properties represented by the promotion of tissue repair.

3. P2X7R-Mediated Signaling in Glial Cells

3.1. Purinergic P2X7 Receptor—General Characteristic

Mammalian P2X7Rs are the members of the P2X purinoceptor family consisting of seven subtypes: P2X1, 2, 3, 4, 5, 6, and 7. These ligand-gated ion channels open in response to the extracellular agonist which is adenosine triphosphate (ATP). The P2X7R is a trimer composed of subunits, all of which share the common structure of two transmembrane domains, N- and C-terminal regions and large extracellular loop [56]. Structurally, P2X7R differs from other classes of the family in its long intracellular domain responsible for the pore formation. This specific receptor is activated by high concentrations of ATP in the millimolar range, whereas other members of P2XRs are stimulated by micromolar concentrations of ATP [57]. C-terminus of P2X7Rs, which is longer than in other P2XRs, has been identified as responsible for regulation of the receptor's functions, including signaling pathways, cellular localization, protein–protein interactions, and post-translational modification [58,59].

Prolonged stimulation of P2X7Rs by ATP triggers formation of a non-selective pore, which allows the passage of molecules of up to 900 Da, Na$^+$, and Ca^{2+} influx, and K$^+$ efflux, resulting in changes in the ionic homeostasis of the cell [60]. In addition, the P2X7R may also mediate the large-scale release of intracellular ATP via its intrinsic pore or in connection with pannexin hemichannels thereby augmenting purinergic signaling and inflammation [61]. It is well known that P2X7-pannexin PANX1 pore complex critically determines spreading of depolarization followed by activation of neuroinflammatory machinery [62].

Inflammation-related events associated with P2X7R downstream signaling include the release of inflammatory mediators such as interleukin-1β and TNF-α. It is also becoming increasingly apparent that ATP-dependent signaling via the P2X7R plays a crucial role in astrocyte–neuron communications in a variety of pathological processes that occur in the central and peripheral nervous systems [4]. This prominent function of P2X7 receptors was strongly confirmed in a number of studies related to the neuronal degeneration, as well as behavioral or cognitive disorders.

The strong expression of P2X7R was identified predominantly in cells of haematopoietic linage such as monocytes, macrophages, mast cells, and microglia [63]. Within the CNS, this receptor is also expressed in neuronal processes [64], Müller cells [65], Schwann cells [66], oligodendrocyte precursor cells, and astrocytes [48]. The role of P2X7R-mediated signaling in glial cells, particularly in the context of stimulation and progression of the MS-associated pathology, will be discussed below in detail.

3.2. P2X7R-Mediated Signaling in Microglia

Extracellular ATP is a potent signaling molecule, important in cell–cell communication in the CNS [67,68], which acts through an array of purinergic receptors, including the ATP-gated ion channels, the P2X7R. This specific type of purinergic receptors, widely expressed in the brain [69], has been shown to be substantially engaged in a variety of CNS pathologies, including MS [70]. However, the role of this receptor in the pathomechanisms and theclinical course of MS should be further clarified.

The microglial expression of both quiescent and activated P2X7R in the nervous system has been reported [71,72]. As shown in primary hippocampal cultures, the overexpression of P2X7R is crucial for driving activation and proliferation of microglia [73]. It has also been confirmed that the activation of this type of receptor is strongly involved in the development and propagation of inflammatory reaction [74,75] by releasing a variety of proinflammatory cytokines such as interleukins: IL-1β, IL-18, IL-6, and tumor necrosis factor (TNF-α) [76].

P2X7R activation connected with pore formation [77] results in the outward blebbing of the microglial plasma membrane and the production of extracellular vesicles containing the proinflammatory cytokine, IL-1β, as well as the diffusion of ROS through the plasma membrane [78]. Recent data indicate that lysosomal exocytosis may be involved in the process of IL-1β release, as the lysosomal co-expression of IL-1β and P2X7R has been demonstrated [77]. Upon activation of the receptor, depolarization-induced K$^+$ efflux and downstream signaling via a caspase 1-dependent mechanism leads to the activation of the NLRP3 inflammasome, a protein complex that participates in a proteolytic cleavage of an inactive form of interleukin 1β (pre-IL-1β) and the release of the active cytokine [79]. This primary event drives a self-propagating cycle via an autocrine mechanism and stimulates astrocytes via a paracrine action, thereby initiating the subsequent inflammatory cascade [21,80]. Stimulation of P2X7R is additionally linked to the activation of transcriptional factor NF-kB that up-regulates the IL-1β gene [81].

Microglia that express pore-forming P2X7R exhibit enhanced vesicular exocytosis and IL-1β release which then accelerates the trophic responses in microglia [77] and promotes activation and proliferation of these cells [82] with concomitant induction of neurotoxicity by stimulating the production of TNF-α and other toxic molecules such as IL-6 or reactive oxygen species [74]. P2X7R-dependent expression of ATP-induced pro-inflammatory tumor necrosis factor-α (TNF-α) is regulated in microglial cells by extracellular signal-regulated Kinase (ERK) and c-Jun N-terminal kinase (JNK) [74]. Further, strong evidence indicates that protein kinase δ (PKC δ) acts as an upstream regulator of ERK and JNKGGC [83]. It has also been shown that microglial P2X7R is important in regulating the expression and the release of not only proinflammatory cytokines, but also chemokines. ATP activates P2X7R in microglia with subsequent overexpression of mRNA and release of CXCL2 chemokine that facilitates neurotrophil infiltration. Calcineurin-dependent nuclear factor of activated T cells (NFAT) and protein kinase C (PKC)/MAP kinase (MAPK) are downstream pathways of P2X7R activation that are implicated in this chemokine release [84]. Moreover, via the intercellular communication, CXCL2 may potentiate the expression of other chemokines such as monocyte chemoattractant protein 1 (MCP-1, CCL2), 10 kDa interferon-induced protein (IP-10, CXCL10), and CCL5 in astrocytes [85]. The simplified scheme illustrating the contribution of microglia to the mechanisms underlying MS pathology is presented in Figure 1.

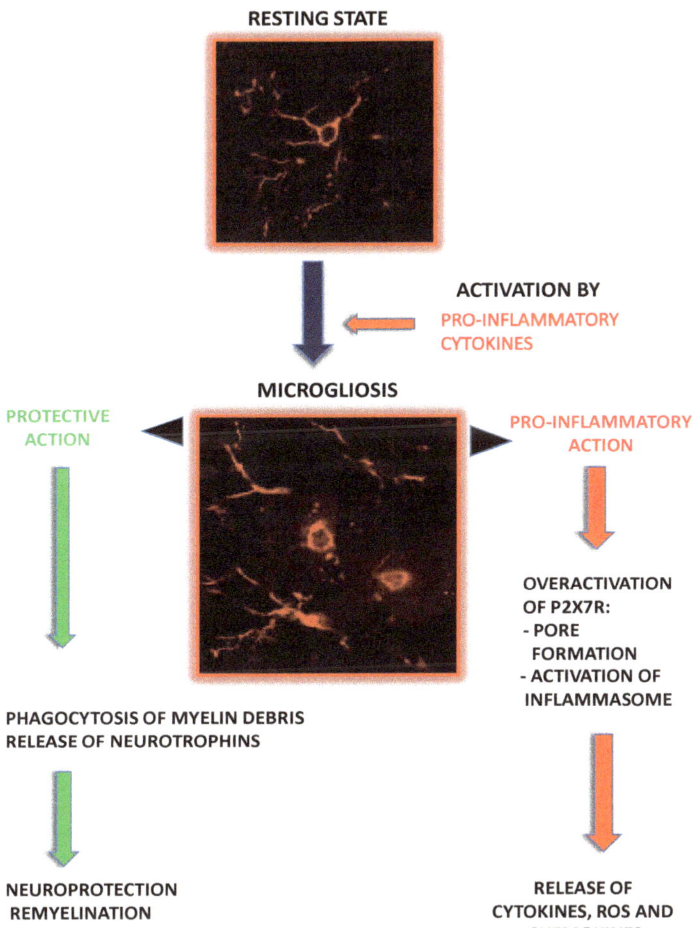

Figure 1. Schematic representation of the protective and negative roles of microglia and their P2XR in MS pathology. Activated microglial cells overexpress P2X7R whose overactivation, combined with the pore formation, mediates activation of inflammasome and the release of proinflammatory cytokines, chemokines, and reactive oxygen species (ROS) that exacerbate neuroinflammation. In parallel, protective functions of reactive microglia may be activated, such as phagocytosis of myelin debris, which subsequently triggers the release of neurotrophins and promotes remyelination (green arrows indicate protective effects, red arrows indicate negative effects). See text for details.

3.3. P2X7R-Mediated Signaling in Astroglia

Extracellular purine (ATP, ADP)-mediated signaling has been recognized as a dominant form of intercellular communication between astrocytes in the CNS, due to the ability of these cells to release purines to modulate neuronal activity and interact with other types of glia [86,87]. Moreover, purine-mediated astroglial signaling is disrupted in pathological states. It is well established that activation of astroglial P2X7Rs are closely associated with processes that initiate neuroinflammation and neuronal dysfunction. These events strongly depend on the pathological changes of astroglia that are related to the overproduction and uncontrolled release of cytokines, glutamate, and reactive oxygen species from astrocytes, all of which modulate astrocyte–neuron integrity and promote demyelination contributing to the neurodegenerative processes [88].

Numerous studies have established the capacity of astrocyte to maintain neuronal network homeostasis via a P2X7R-dependent purinergic signaling system. In the nervous system, ATP is released from neuronal axon terminals during the process of neurotransmission, as well as from astroglia. It is widely accepted that ATP is involved in synaptic transmission in many brain regions [89]. ATP can be stored and released either on its own or together with other neurotransmitters such as glutamate, GABA, noradrenaline or acetylcholine (ACh). Furthermore, activation of P2X7 receptors also triggers the release of gliotransmitters via exocytosis associated with the formation of P2X7-associated transmembrane pore channels (e.g., hemichannels, pannexins, volume sensitive anion channels). For instance, prolonged activation of P2X7Rs leads to a sustained glutamate release by hippocampal astrocytes [90]. High concentrations of ATP, acting through P2X7Rs, were also shown to significantly elevate production of endocannabinoid 2-arachidonoylglycerol in primary cultures of astrocytes [91]. Furthermore, stimulation of P2X7Rs in the same experimental condition has been found to act via different pathways including the release of TNF-α, stimulation of nitric oxide (NO) production, elevation of AKT, and p38MAPK/ERK1/ERK2 phosphorylation, or activation of transmembrane transport of NADH [92–95]. In addition, P2X7-mediated Ca^{2+} signaling was found to increase the production of lipid mediators of inflammatory cysteinyl leukotrienes [96]. Implication of P2X7 receptors in the processes related to the glutamate-glutamine pathway was observed in the RBA-2 astroglial cell line which was represented by a rapid decrease in glutamate uptake by the Na^{+}-dependent transporter system and a decrease in the expression and activity of glutamine syntase. Furthermore, it is well established that P2X7 receptors are crucial in controlling expression of other purinoceptors and channels. For example, P2X7R stimulation increases expression of P2Y2 receptors and decreases expression of aquaporin-4 in primary cultured astrocytes [97].

It has been revealed that P2X7Rs expressed by primary cultured astrocytes may be activated in the absence of any exogenous stimuli, while knock-down of P2X7Rs decreases pore activity of the astrocytic receptor. Astrocytes incubated with an inhibitor of F-actin polymerization (CytD), an effective blocker of the phagocytosis, markedly reduced beads uptake by P2X7R in a manner dependent on actin filaments rearrangement [98]. Another study demonstrated that ATP stimulation of P2X7Rs causes a dissociation of MHC-IIa from its complex with P2X7Rs, resulting in decreased phagocytic activity [99]. These data suggest that basal activity of P2X7R expressed by resting astrocytes mainly regulates their engulfing features.

Constitutive activation of the astrocytic P2XRs is implicated in the clearance of the metabolic product of signaling molecule which is adenosine [100,101]. Proinflammatory stimulation with extracellular ATP via P2XRs at microlomar concentration significantly activates the release of purine nucleoside phosphorylase (PNP) by astrocytes in vitro. Interestingly, PNP release from glial cells partially occurred through the activation of the lysosomal pathway [102].

The simplified scheme illustrating contribution of astroglia to the mechanisms underlying MS pathology is presented in Figure 2.

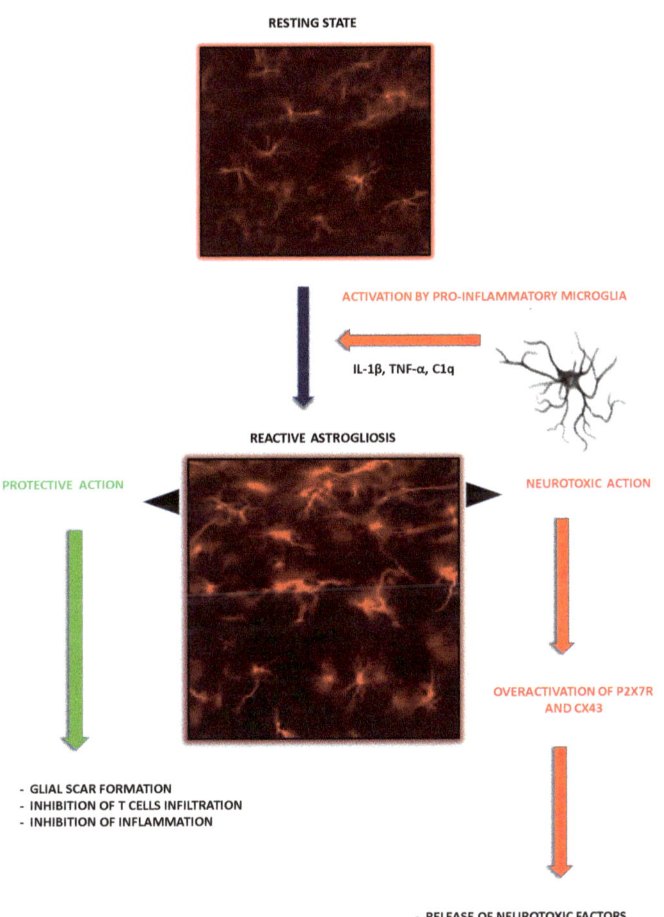

Figure 2. Schematic representation of the protective and negative roles of astroglia and their P2XR in MS pathology. Activated astrocytes overexpress P2X7R whose overactivation mediates efflux of neurotoxic factors such as extracellular ATP, calcium ions (Ca^{2+}) and glutamate, as well as proinflammatory cytokines and chemokines that exacerbate neuroinflammation. In parallel, protective functions of reactive astroglia may be activated that are relevant to the processes of glial scar formation and remyelination. (green arrows indicate protective effects, red arrows indicate negative effects). See text for details.

4. The Involvement of Glial P2X7R-Dependent Signaling into MS Pathology

4.1. Microglial P2X7R-Mediated Signaling in MS/EAE

In the context of MS/EAE pathology, P2XR has been first described to express in oligodendrocytes and myelin sheaths and to induce oligodendroglial cell death in vitro and in vivo. Functional P2X7R contributed to neuronal deficits in EAE animals and was also noticed in MS before lesion formation [103]. In post-mortem-collected MS tissue, increased expression of P2X7R in demyelinating plaques adjacent to blood vessels was shown within activated microglial cells/macrophages [104] where they released IL-1β via the induction of cyclooxygenase-2 and downstream pathogenic mediators [71]. Interestingly, in contrast to these results, in samples of frontal cortex obtained from secondary progressive form of MS, P2X7R expression was not detected either on resting or activated microglia [104].

Evidence from experimental studies using the EAE model of MS indicate that microglia are involved in inducing neuroinflammation via a P2X7R-dependent mechanism. P2X7R deficiency was shown to reduce the development of the disease in mice, inhibit inflammatory reaction, prevent ATP excitotoxicity in oligodendrocytes, and decline the axonal injury [103,105,106]. However, opposite results were also reported, describing exacerbation of the disease in P2X7R knockout mice [107]. Results of our studies showed that pharmacological blockade of P2X7R by its selective antagonist brilliant blue G (BBG) delays the onset of the disease and alleviates clinical symptoms in EAE rats. Moreover, we observed substantially inhibited activation and proliferation of microglia, as shown by decreased Iba-1 immunoreactivity and the morphological characteristics of microglial cells. Concomitantly observed lowered protein expression of proinflammatory cytokines, IL-1β, IL-6, and TNF-α, indicated inhibition of neuroinflammation. The P2X7R-dependent release of the proinflammatory cytokines was constantly decreased during the entire course of the disease after inhibition of the receptor with BBG [35]. Increased expression of P2X7R in microglial cells with concomitant up-regulation of several inflammatory genes associated with the activation of the NLRP3-inflammasome and the polarization of microglia to a pro-inflammatory phenotype was also observed in cuprizone model of demyelination [108]. Importantly, cuprizone-induced demyelination does not fully reflect the pattern of immune cell-related demyelination present in MS/EAE. However, P2X7R knockouts subjected to cuprizone toxicity showed attenuated micro- and astrogliosis, as well as the down-regulation of pro-inflammatory genes. The simplified summary illustrating the functional role of PX2R in different experimental models of MS pathology is presented in Table 1.

Table 1. Functional role of P2X7R during MS; results from experimental animal models.

P2X7 Targeting Approach	Model	Effect	Reference
Brilliant blue G (antagonist)	Acute EAE rats	Reduced onset of the disease; reduced astrogliosis and microglia proliferation	[35] [109]
Brilliant blue G (antagonist)	Chronic EAE mice	Reduced demyelination; ameliorated neurological abnormalities	[103]
Oxidized ATP (oxATP antagonist)	Chronic EAE mice	Inhibition of clinical symptoms and demyelination; reduce antigen T cell	[106]
Cuprizone (demyelination inducer)	P2X7 null mice	Inhibition of astrogliosis and microglia activation	[108]

As mentioned, migration of activated peripheral immune cells, including monocytes, into the CNS is a critical step in the development of neuroinflammation in MS/EAE. Monocytes/macrophages are known to highly express P2X7R. The P2X7R-dependent release of CXCL2 chemokine was presented as an important factor in facilitating neutrophil infiltration in an experimental model of MS [110]. P2X7R protein expression was found to be diminished in monocytes in an acute phase of the disease, both in patients and EAE animals. Moreover, the protein levels of the receptor decreased in healthy monocytes subjected to pro-inflammatory stimuli in vitro. In MS tissue the receptor was lost on both CD14/CD68- or CD14/MHC-II-positive cells near the endothelium of the blood vessels [104]. The authors hypothesized that upregulation of P2X7R might be detrimental to monocytes, therefore secondary autocrine/paracrine down-regulation of P2X7R is triggered to support their survival and invasion into the CNS, thereby contributing to the induction and propagation of neuroinflammation.

4.2. Astroglial P2X7R

In MS patients, P2X7R was found to be upregulated in the parenchymal astrocytes of frontal cortex from SP type of the disease [104]. Several pathological and morphological changes related to the reactive astrogliosis, such as hypertrophy, proliferation, and overlapping of cellular processes, resulting in the disruption of specific astrocytic domains, were revealed in immunochemical studies. In addition, P2X7R-positive astrocytes mediated glial scar formation in the white matter of chronic lesions. The same study revealed co-localization of the astrocytic P2X7R with the monocyte chemoattractant protein 1 (MCP-1), previously found to be responsible for the leukocyte recruitment during the progression of MS [111].

While modeling MS, it has been revealed that mice deficient in P2X7R function are more resistant to EAE than wild-type mice and exhibit reduction in the CNS inflammation-associated processes. Furthermore, within the CNS, astroglia-dependent axonal damage was present, while the opposite effect was observed in the P2X7R null mice. Furthermore, pharmacological inhibition of the receptor significantly abolished astrogliosis in rat EAE and reversed neurological symptoms [105]. These evidences strongly suggest the crucial role of astrocyte in MS pathology, which is associated with gaining new properties of the cells negatively affecting neuronal function.

Studies from our laboratory focusing on the role of P2X7R-mediated signaling during the course of EAE revealed strong P2X7R expression within the frontal motor and somatosensory cortical brain regions, especially in the five and six layers of the cortex neighboring to the cingulum brain area. Interestingly, these changes were associated with the peak of neurological symptoms in immunized rats [109]. Another study using similar experimental conditions revealed the appearance of the astrogliosis in rat forebrains at an early stage of EAE. Overexpression of the specific astroglial markers occurred at fourth day post-immunization. Interestingly, at the same time, astrocyte overexpressed connexin 43 and P2X7R, while inhibition of the P2X7R signaling with BBG abolished activation of the astrocytes. Notably, administration of P2X7R antagonist partially reversed neurological symptoms developed during the disease progression. Given that astrocytes play an important role in the pathogenesis of CNS by releasing several potentially neurotoxic factors (e.g., ATP via purinergic system or glutamate), the dysfunction of activated astroglia suggests pathological involvement of glia cells in MS/EAE starting from an early stage of the disease [112].

As reported, overexpression of the astrocytic P2X7R in MS might be dependent on the disorder progression. Moreover, upregulation of the P2X7R activity seems to be associated not only with inflammatory reaction, but also with a variety of other processes related to the MS pathology, including the removal of glutamate excess, modulation of Ca^+ and ATP efflux, glia scar formation, and lymphocyte homeostasis [113].

It is well accepted that activation of various astroglial receptors causes a transient increase in the intracellular pool of Ca^{2+} within the astroglia [114,115]. A consequence of the Ca^{2+} accumulation is a failure of astroglia to rapidly interact with neighboring cells in the CNS in physiological and pathological conditions [116]. Considerable evidence exists, supporting the role of Ca^{2+} signaling in astrocyte physiology. Synaptically released neurotransmitters mediate Ca^{2+} signaling in astrocytes, and action potentials along axons mediate the efflux of ATP and the intercellular propagation of astroglial Ca^{2+} signals. In turn, astrocytes amplify this initial signal by transmission of the extracellular Ca^{2+} wave to neighboring glia. Notably, propagation of these pathways was not observed in P2X7 knock-out mouse, providing the evidence that gliotransmitter-mediated signal propagation and amplification is strongly dependent on the P2X7 receptors [116]. The evidence also emerged that disruption of this pathway, negatively affecting the neuronal function, is present in a variety of pathological conditions such as neurodegenerative processes.

The recent findings regarding the involvement of astroglia P2X7R in MS pathology revealed the rapid elevation of Ca^{2+} in primary culture of astrocytes upon addition of the isolated CNS-infiltrated immune cells (CNS IICs), predominantly represented by CD4+ T cells, recruited from the periphery to the CNS of EAE rats [117]. Interestingly, CNS IICs-stimulated Ca^{2+} elevation in astrocytes was markedly abolished by the specific block of P2X7 receptors, and was mimicked by the stimulation of this glial receptor with a low concentration of agonist. These results suggest that P2X7 receptor-dependent signaling primarily involves CNS IICs–astrocyte interaction. Furthermore, activation of P2X7 receptors appeared mainly in astroglia, while inhibition of the hemichannel-dependent ATP release in astrocytes declined Ca^{2+} accumulation, which was mediated after the addition of CNS IICs. Although this review focuses on the PX72R, it should be mentioned that other receptors abundantly expressed in glia cells represented by ionotropic P2X4, G protein-coupled P2Y$_1$, and P2Y$_2$ receptors are also involved in the regulation of intracellular Ca^{2+} concentration (more detail in [118]). Taken together, presented data suggest that rapid changes in the content of Ca^{2+} mediated by autoreactive immune cells involve astroglial purinergic signaling and lead to the long-lasting morphological and physiological changes of astroglia in EAE.

5. Conclusions

A growing body of evidence suggests heterogeneity of the processes related to MS pathology in which glial cells are strongly involved. The sensitivity of glial cells to different pathological stimuli and the potentiality to play dual role, positive or negative, point out their importance for neuronal functioning during the disease. The loss of glia protective functions, that is guarding and supporting of neuronal homeostasis, seems to be crucial for pathological processes running in MS-affected CNS.

Recent studies indicate the importance of purinoreceptor-mediated signaling in the glia–neuron cellular network. There is tempting evidence that suggests that activation of the P2X7R is commonly present during MS development and P2X7R-mediated purinergic signaling pathways, which drive and sustain neuroinflammation, significantly contributing to MS/EAE pathology. Taken together, numerous experimental and clinical observations indicate that both pools of glial cells that express P2X7R, microglia and astroglia, are involved in pathological mechanisms operating at early and progressive stages of the disease and should be considered as equally important in the pathogenesis of MS/EAE. Therefore, future studies will have to account for the potential role of purinergic P2X7R as a target for the promising therapeutic interventions in MS pathology.

Author Contributions: Conceptualization, L.S.; Writing—Original Draft Preparation, M.S.-W. and L.S.; Writing—Review & Editing, L.S. and M.S.-W.; Visualization, L.S.; Funding Acquisition, L.S. All authors have read and agreed to the published version of the manuscript.

Funding: This manuscript is financed from the statutable funds provided by the Ministry of Education and Science for Mossakowski Medical Research Institute Polish Academy of Sciences, Warsaw, Poland (9/2021).

Institutional Review Board Statement: Not applicable.

Informed Consent Statement: Not applicable.

Data Availability Statement: Not applicable.

Conflicts of Interest: The authors declare no conflict of interest.

References

1. Kuhlmann, T.; Ludwin, S.; Prat, A.; Antel, J.; Brück, W.; Lassmann, H. An updated histological classification system for multiple sclerosis lesions. *Acta Neuropathol.* **2017**, *133*, 13–24. [CrossRef] [PubMed]
2. Boissy, A.R.; Cohen, J.A. Multiple sclerosis symptom management. *Expert Rev. Neurother.* **2007**, *7*, 1213–1222. [CrossRef]
3. Schirmer, L.; Velmeshev, D.; Holmqvist, S.; Kaufmann, M.; Werneburg, S.; Jung, D.; Vistnes, S.; Stockley, J.H.; Young, A.; Steindel, M.; et al. Neuronal vulnerability and multilineage diversity in multiple sclerosis. *Nature* **2019**, *573*, 75–82. [CrossRef] [PubMed]
4. Illes, P. P2X7 receptors amplify cns damage in neurodegenerative diseases. *Int. J. Mol. Sci.* **2020**, *21*, 5996. [CrossRef] [PubMed]
5. Lassmann, H. Multiple sclerosis pathology. *Cold Spring Harb. Perspect. Med.* **2018**, *8*, a028936. [CrossRef] [PubMed]
6. Frischer, J.M.; Weigand, S.D.; Guo, Y.; Kale, N.; Parisi, J.E.; Pirko, I.; Mandrekar, J.; Bramow, S.; Metz, I.; Brück, W.; et al. Clinical and pathological insights into the dynamic nature of the white matter multiple sclerosis plaque. *Ann. Neurol.* **2015**, *78*, 710–721. [CrossRef]
7. Hayashi, T.; Morimoto, C.; Burks, J.S.; Kerr, C.; Hauser, S.L. Dual-label immunocytochemistry of the active multiple sclerosis lesion: Major histocompatibility complex and activation antigens. *Ann. Neurol.* **1988**, *24*, 523–531. [CrossRef]
8. Franciotta, D.; Salvetti, M.; Lolli, F.; Serafini, B.; Aloisi, F. B cells and multiple sclerosis. *Lancet Neurol.* **2008**, *7*, 852–858. [CrossRef]

9. Brosnan, C.F.; Bornstein, M.B.; Bloom, B.R. The effects of macrophage depletion on the clinical and pathologic expression of experimental allergic encephalomyelitis. *J. Immunol.* **1981**, *126*, 614–620.
10. Ajami, B.; Bennett, J.L.; Krieger, C.; McNagny, K.M.; Rossi, F.M.V. Infiltrating monocytes trigger EAE progression, but do not contribute to the resident microglia pool. *Nat. Neurosci.* **2011**, *14*, 1142–1150. [CrossRef]
11. Nally, F.K.; De Santi, C.; McCoy, C.E. Nanomodulation of Macrophages in Multiple Sclerosis. *Cells* **2019**, *8*, 543. [CrossRef] [PubMed]
12. Henderson, A.P.D.; Barnett, M.H.; Parratt, J.D.E.; Prineas, J.W. Multiple sclerosis: Distribution of inflammatory cells in newly forming lesions. *Ann. Neurol.* **2009**, *66*, 739–753. [CrossRef]
13. Matthews, P.M. Chronic inflammation in multiple sclerosis—Seeing what was always there. *Nat. Rev. Neurol.* **2019**, *15*, 582–593. [CrossRef] [PubMed]
14. Kleinewietfeld, M.; Hafler, D.A. Regulatory T cells in autoimmune neuroinflammation. *Immunol. Rev.* **2014**, *259*, 231–244. [CrossRef]
15. Barnett, M.H.; Prineas, J.W. Relapsing and Remitting Multiple Sclerosis: Pathology of the Newly Forming Lesion. *Ann. Neurol.* **2004**, *55*, 458–468. [CrossRef] [PubMed]
16. Pegoretti, V.; Swanson, K.A.; Bethea, J.R.; Probert, L.; Eisel, U.L.M.; Fischer, R. Inflammation and Oxidative Stress in Multiple Sclerosis: Consequences for Therapy Development. *Oxid. Med. Cell. Longev.* **2020**, *2020*, 7191080. [CrossRef]
17. Groom, A.J.; Smith, T.; Turski, L. Multiple sclerosis and glutamate. *Ann. N. Y. Acad. Sci.* **2003**, *993*, 229–275. [CrossRef] [PubMed]
18. Mallucci, G.; Peruzzotti-Jametti, L.; Bernstock, J.D.; Pluchino, S. The role of immune cells, glia and neurons in white and gray matter pathology in multiple sclerosis. *Prog. Neurobiol.* **2015**, *127*, 1–22. [CrossRef] [PubMed]
19. Nimmerjahn, A.; Kirchhoff, F.; Helmchen, F. Neuroscience: Resting microglial cells are highly dynamic surveillants of brain parenchyma in vivo. *Science* **2005**, *308*, 1314–1318. [CrossRef] [PubMed]
20. Guerrero, B.L.; Sicotte, N.L. Microglia in Multiple Sclerosis: Friend or Foe? *Front. Immunol.* **2020**, *11*, 374. [CrossRef]
21. Monif, M.; Burnstock, G.; Williams, D.A. Microglia: Proliferation and activation driven by the P2X7 receptor. *Int. J. Biochem. Cell Biol.* **2010**, *42*, 1753–1756. [CrossRef]
22. Mathys, H.; Davila-Velderrain, J.; Peng, Z.; Gao, F.; Mohammadi, S.; Young, J.Z.; Menon, M.; He, L.; Abdurrob, F.; Jiang, X.; et al. Single-cell transcriptomic analysis of Alzheimer's disease HHS Public Access. *Nature* **2019**, *570*, 332–337. [CrossRef]
23. Lassmann, H.; Van Horssen, J.; Mahad, D. Progressive multiple sclerosis: Pathology and pathogenesis. *Nat. Rev. Neurol.* **2012**, *8*, 647–656. [CrossRef] [PubMed]
24. Voß, E.V.; Škuljec, J.; Gudi, V.; Skripuletz, T.; Pul, R.; Trebst, C.; Stangel, M. Characterisation of microglia during de- and remyelination: Can they create a repair promoting environment? *Neurobiol. Dis.* **2012**, *45*, 519–528. [CrossRef] [PubMed]
25. Luo, C.; Jian, C.; Liao, Y.; Huang, Q.; Wu, Y.; Liu, X.; Zou, D.; Wu, Y. The role of microglia in multiple sclerosis. *Neuropsychiatr. Dis. Treat.* **2017**, *13*, 1661–1667. [CrossRef] [PubMed]
26. Napoli, I.; Neumann, H. Protective effects of microglia in multiple sclerosis. *Exp. Neurol.* **2010**, *225*, 24–28. [CrossRef] [PubMed]
27. Ransohoff, R.M.; Perry, V.H. Microglial physiology: Unique stimuli, specialized responses. *Annu. Rev. Immunol.* **2009**, *27*, 119–145. [CrossRef]
28. Shi, R. Polyethylene glycol repairs membrane damage and enhances functional recovery: A tissue engineering approach to spinal cord injury. *Neurosci. Bull.* **2013**, *29*, 460–466. [CrossRef]
29. Satoh, J.-I.; Kino, Y.; Asahina, N.; Takitani, M.; Miyoshi, J.; Ishida, T.; Saito, Y. TMEM119 marks a subset of microglia in the human brain. *Neuropathology* **2016**, *36*, 39–49. [CrossRef]
30. Öhrfelt, A.; Axelsson, M.; Malmeström, C.; Novakova, L.; Heslegrave, A.; Blennow, K.; Lycke, J.; Zetterberg, H. Soluble TREM-2 in cerebrospinal fluid from patients with multiple sclerosis treated with natalizumab or mitoxantrone. *Mult. Scler. J.* **2016**, *22*, 1587–1595. [CrossRef]
31. Zrzavy, T.; Hametner, S.; Wimmer, I.; Butovsky, O.; Weiner, H.L.; Lassmann, H. Loss of "homeostatic" microglia and patterns of their activation in active multiple sclerosis. *Brain* **2017**, *140*, 1900–1913. [CrossRef]
32. Fischer, M.T.; Sharma, R.; Lim, J.L.; Haider, L.; Frischer, J.M.; Drexhage, J.; Mahad, D.; Bradl, M.; Van Horssen, J.; Lassmann, H. NADPH oxidase expression in active multiple sclerosis lesions in relation to oxidative tissue damage and mitochondrial injury. *Brain* **2012**, *135*, 886–899. [CrossRef]
33. Haider, L.; Zrzavy, T.; Hametner, S.; Höftberger, R.; Bagnato, F.; Grabner, G.; Trattnig, S.; Pfeifenbring, S.; Brück, W.; Lassmann, H. The topograpy of demyelination and neurodegeneration in the multiple sclerosis brain. *Brain* **2016**, *139*, 807–815. [CrossRef]
34. Lloyd, A.F.; Davies, C.L.; Miron, V.E. Microglia: Origins, homeostasis, and roles in myelin repair. *Curr. Opin. Neurobiol.* **2017**, *47*, 113–120. [CrossRef]
35. Grygorowicz, T.; Strużyńska, L. Early P2X7R-dependent activation of microglia during the asymptomatic phase of autoimmune encephalomyelitis. *Inflammopharmacology* **2019**, *27*, 129–137. [CrossRef]
36. Heppner, F.L.; Greter, M.; Marino, D.; Falsig, J.; Raivich, G.; Hövelmeyer, N.; Waisman, A.; Rülicke, T.; Prinz, M.; Priller, J.; et al. Experimental autoimmune encephalomyelitis repressed by microglial paralysis. *Nat. Med.* **2005**, *11*, 146–152. [CrossRef]
37. Bhasin, M.; Wu, M.; Tsirka, S.E. Modulation of microglial/macrophage activation by macrophage inhibitory factor (TKP) or tuftsin (TKPR) attenuates the disease course of experimental autoimmune encephalomyelitis. *BMC Immunol.* **2007**, *8*, 112. [CrossRef]
38. O'Loughlin, E.; Madore, C.; Lassmann, H.; Butovsky, O. Microglial phenotypes and functions in multiple sclerosis. *Cold Spring Harb. Perspect. Med.* **2018**, *8*, a028993. [CrossRef] [PubMed]

39. Patsopoulos, N.A.; Baranzini, S.E.; Santaniello, A.; Shoostari, P.; Cotsapas, C.; Wong, G.; Beecham, A.H.; James, T.; Replogle, J.; Vlachos, I.S.; et al. Multiple sclerosis genomic map implicates peripheral immune cells and microglia in susceptibility. *Science* **2019**, *365*, eaav7188. [CrossRef]

40. Cotrina, M.L.; Lin, J.H.C.; Alves-Rodrigues, A.; Liu, S.; Li, J.; Azmi-Ghadimi, H.; Kang, J.; Naus, C.C.G.; Nedergaard, M. Connexins regulate calcium signaling by controlling ATP release. *Proc. Natl. Acad. Sci. USA* **1998**, *95*, 15735–15740. [CrossRef] [PubMed]

41. Pfrieger, F.W.; Barres, B.A. Synaptic efficacy enhanced by glial cells in vitro. *Science* **1997**, *277*, 1684–1687. [CrossRef] [PubMed]

42. Hertz, L.; Gibbs, M.E. What learning in day-old chickens can teach a neurochemist: Focus on astrocyte metabolism. *J. Neurochem.* **2009**, *109*, 10–16. [CrossRef]

43. Perea, J.R.; López, E.; Díez-Ballesteros, J.C.; Ávila, J.; Hernández, F.; Bolós, M. Extracellular Monomeric Tau Is Internalized by Astrocytes. *Front. Neurosci.* **2019**, *13*, 442. [CrossRef]

44. Morales, H.P.; Schousboe, A. Volume regulation in astrocytes: A role for taurine as an osmoeffector. *J. Neurosci. Res.* **1988**, *20*, 505–509. [CrossRef]

45. Duan, S.; Anderson, C.M.; Keung, E.C.; Chen, Y.; Chen, Y.; Swanson, R.A. P2X7 receptor-mediated release of excitatory amino acids from astrocytes. *J. Neurosci.* **2003**, *23*, 1320–1328. [CrossRef] [PubMed]

46. Parpura, V.; Basarsky, T.A.; Liu, F.; Jeftinija, K.; Jeftinija, S.; Haydon, P.G. Glutamate-mediated astrocyte-neuron signalling. *Nature* **1994**, *369*, 744–747. [CrossRef] [PubMed]

47. Liddelow, S.A.; Guttenplan, K.A.; Clarke, L.E.; Bennett, F.C.; Bohlen, C.J.; Schirmer, L.; Bennett, M.L.; Münch, A.E.; Chung, W.-S.; Peterson, T.C.; et al. Neurotoxic reactive astrocytes are induced by activated microglia HHS Public Access. *Ted M Dawson* **2017**, *541*, 481–487. [CrossRef]

48. Wang, L.Y.; Cai, W.Q.; Chen, P.H.; Deng, Q.Y.; Zhao, C.M. Downregulation of P2X7 receptor expression in rat oligodendrocyte precursor cells after hypoxia ischemia. *Glia* **2009**, *57*, 307–319. [CrossRef] [PubMed]

49. Shih, R.H.; Wang, C.Y.; Yang, C.M. NF-kappaB signaling pathways in neurological inflammation: A mini review. *Front. Mol. Neurosci.* **2015**, *8*, 77. [CrossRef]

50. Hayden, M.S.; Ghosh, S. NF-κB, the first quarter-century: Remarkable progress and outstanding questions. *Genes Dev.* **2012**, *26*, 203–234. [CrossRef]

51. Nicolas, C.S.; Amici, M.; Bortolotto, Z.A.; Doherty, A.; Csaba, Z.; Fafouri, A.; Dournaud, P.; Gressens, P.; Collingridge, G.L.; Peineau, S. The role of JAK-STAT signaling within the CNS. *JAK-STAT* **2013**, *2*, e22925. [CrossRef]

52. Choi, W.H.; Ji, K.A.; Jeon, S.B.; Yang, M.S.; Kim, H.; Min, K.J.; Shong, M.; Jou, I.; Joe, E.H. Anti-inflammatory roles of retinoic acid in rat brain astrocytes: Suppression of interferon-γ-induced JAK/STAT phosphorylation. *Biochem. Biophys. Res. Commun.* **2005**, *329*, 125–131. [CrossRef] [PubMed]

53. Herrmann, J.E.; Imura, T.; Song, B.; Qi, J.; Ao, Y.; Nguyen, T.K.; Korsak, R.A.; Takeda, K.; Akira, S.; Sofroniew, M.V. STAT3 is a critical regulator of astrogliosis and scar formation after spinal cord injury. *J. Neurosci.* **2008**, *28*, 7231–7243. [CrossRef]

54. Haroon, F.; Drögemüller, K.; Händel, U.; Brunn, A.; Reinhold, D.; Nishanth, G.; Mueller, W.; Trautwein, C.; Ernst, M.; Deckert, M.; et al. Gp130-Dependent Astrocytic Survival Is Critical for the Control of Autoimmune Central Nervous System Inflammation. *J. Immunol.* **2011**, *186*, 6521–6531. [CrossRef]

55. Sidoryk-Węgrzynowicz, M.; Struzyńska, L. Astroglial contribution to tau-dependent neurodegeneration. *Biochem. J.* **2019**, *476*, 3493–3504. [CrossRef] [PubMed]

56. Torres, G.E.; Egan, T.M.; Voigt, M.M. Hetero-oligomeric assembly of P2X receptor subunits. Specificities exist with regard to possible partners. *J. Biol. Chem.* **1999**, *274*, 6653–6659. [CrossRef]

57. Gordon, J.L. Extracellular ATP: Effects, sources and fate. *Biochem. J.* **1986**, *233*, 309–319. [CrossRef]

58. Hattori, M.; Gouaux, E. Molecular mechanism of ATP binding and ion channel activation in P2X receptors. *Nature* **2012**, *485*, 207–212. [CrossRef]

59. Karasawa, A.; Kawate, T. Structural basis for subtype-specific inhibition of the P2X7 receptor. *eLife* **2016**, *5*, e22153. [CrossRef] [PubMed]

60. Coutinho-Silva, R.; Persechini, P.M. P2Z purinoceptor-associated pores induced by extracellular ATP in macrophages and J774 cells. *Am. J. Physiol. Cell Physiol.* **1997**, *273*, C1793–C1800. [CrossRef]

61. Pelegrin, P.; Surprenant, A. Pannexin-1 mediates large pore formation and interleukin-1β release by the ATP-gated P2X7 receptor. *EMBO J.* **2006**, *25*, 5071–5082. [CrossRef]

62. Chen, S.P.; Qin, T.; Seidel, J.L.; Zheng, Y.; Eikermann, M.; Ferrari, M.D.; Van Den Maagdenberg, A.M.J.M.; Moskowitz, M.A.; Ayata, C.; Eikermann-Haerter, K. Inhibition of the P2X7-PANX1 complex suppresses spreading depolarization and neuroinflammation. *Brain* **2017**, *140*, 1643–1656. [CrossRef]

63. Di Virgilio, F.; Sanz, J.M.; Chiozzi, P.; Falzoni, S. The P2Z/P2X7 receptor of microglial cells: A novel immunomodulatory receptor. *Prog. Brain Res.* **1999**, *120*, 355–368. [CrossRef]

64. Deuchars, S.A.; Atkinson, L.; Brooke, R.E.; Musa, H.; Milligan, C.J.; Batten, T.F.C.; Buckley, N.J.; Parson, S.H.; Deuchars, J. Neuronal P2X7 receptors are targeted to presynaptic terminals in the central and peripheral nervous systems. *J. Neurosci.* **2001**, *21*, 7143–7152. [CrossRef]

65. Pannicke, T.; Fischer, W.; Biedermann, B.; Schädlich, H.; Grosche, J.; Faude, F.; Wiedemann, P.; Allgaier, C.; Illes, P.; Burnstock, G.; et al. P2X7 receptors in Muller glial cells from the human retina. *J. Neurosci.* **2000**, *20*, 5965–5972. [CrossRef]

66. Colomar, A.; Amédée, T. ATP stimulation of P2X7 receptors activates three different ionic conductances on cultured mouse Schwann cells. *Eur. J. Neurosci.* **2001**, *14*, 927–936. [CrossRef] [PubMed]
67. Cotrina, M.L.; Lin, J.H.C.; López-García, J.C.; Naus, C.C.G.; Nedergaard, M. ATP-mediated glia signaling. *J. Neurosci.* **2000**, *20*, 2835–2844. [CrossRef]
68. Inoue, K.; Koizumi, S.; Tsuda, M. The role of nucleotides in the neuron-glia communication responsible for the brain functions. *J. Neurochem.* **2007**, *102*, 1447–1458. [CrossRef] [PubMed]
69. Sperlágh, B.; Vizi, E.S.; Wirkner, K.; Illes, P. P2X7 receptors in the nervous system. *Prog. Neurobiol.* **2006**, *78*, 327–346. [CrossRef] [PubMed]
70. Sperlágh, B.; Illes, P. P2X7 receptor: An emerging target in central nervous system diseases. *Trends Pharmacol. Sci.* **2014**, *35*, 537–547. [CrossRef] [PubMed]
71. Yiangou, Y.; Facer, P.; Durrenberger, P.; Chessell, I.P.; Naylor, A.; Bountra, C.; Banati, R.R.; Anand, P. COX-2, CB2 and P2X7-immunoreactivities are increased in activated microglial cells/macrophages of multiple sclerosis and amyotrophic lateral sclerosis spinal cord. *BMC Neurol.* **2006**, *6*, 12. [CrossRef] [PubMed]
72. Kobayashi, K.; Takahashi, E.; Miyagawa, Y.; Yamanaka, H.; Noguchi, K. Induction of the P2X7 receptor in spinal microglia in a neuropathic pain model. *Neurosci. Lett.* **2011**, *504*, 57–61. [CrossRef]
73. Monif, M.; Reid, C.A.; Powell, K.L.; Smart, M.L.; Williams, D.A. The P2X7 receptor drives microglial activation and proliferation: A trophic role for P2X7R pore. *J. Neurosci.* **2009**, *29*, 3781–3791. [CrossRef]
74. Suzuki, T.; Hide, I.; Ido, K.; Kohsaka, S.; Inoue, K.; Nakata, Y. Production and Release of Neuroprotective Tumor Necrosis Factor by P2X 7 Receptor-Activated Microglia. *J. Neurosci.* **2004**, *24*, 1–7. [CrossRef]
75. Inoue, K. Microglial activation by purines and pyrimidines. *Glia* **2002**, *40*, 156–163. [CrossRef] [PubMed]
76. la Sala, A.; Ferrari, D.; Di Virgilio, F.; Idzko, M.; Norgauer, J.; Girolomoni, G. Alerting and tuning the immune response by extracellular nucleotides. *J. Leukoc. Biol.* **2003**, *73*, 339–343. [CrossRef]
77. Monif, M.; Reid, C.A.; Powell, K.L.; Drummond, K.J.; O'Brien, T.J.; Williams, D.A. Interleukin-1β has trophic effects in microglia and its release is mediated by P2X7R pore. *J. Neuroinflamm.* **2016**, *13*, 173. [CrossRef]
78. Illes, P.; Rubini, P.; Ulrich, H.; Zhao, Y.; Tang, Y. Regulation of Microglial Functions by Purinergic Mechanisms in the Healthy and Diseased CNS. *Cells* **2020**, *9*, 1108. [CrossRef]
79. Mariathasan, S.; Weiss, D.S.; Newton, K.; McBride, J.; O'Rourke, K.; Roose-Girma, M.; Lee, W.P.; Weinrauch, Y.; Monack, D.M.; Dixit, V.M. Cryopyrin activates the inflammasome in response to toxins and ATP. *Nature* **2006**, *440*, 228–232. [CrossRef]
80. Mingam, R.; De Smedt, V.; Amédée, T.; Bluthé, R.M.; Kelley, K.W.; Dantzer, R.; Layé, S. In vitro and in vivo evidence for a role of the P2X7 receptor in the release of IL-1β in the murine brain. *Brain. Behav. Immun.* **2008**, *22*, 234–244. [CrossRef] [PubMed]
81. Korcok, J.; Raimundo, L.N.; Ke, H.Z.; Sims, S.M.; Dixon, S.J. Extracellular nucleotides act through P2X7 receptors to activate NF-κB in osteoclasts. *J. Bone Miner. Res.* **2004**, *19*, 642–651. [CrossRef]
82. Schroder, K.; Tschopp, J. The Inflammasomes. *Cell* **2010**, *140*, 821–832. [CrossRef]
83. Bradford, M.D.; Soltoff, S.P. P2X7 receptors activate protein kinase D and p42/p44 mitogen-activated protein kinase (MAPK) downstream of protein kinase C. *Biochem. J.* **2002**, *366*, 745–755. [CrossRef]
84. Shiratori, M.; Tozaki-Saitoh, H.; Yoshitake, M.; Tsuda, M.; Inoue, K. P2X7 receptor activation induces CXCL2 production in microglia through NFAT and PKC/MAPK pathways. *J. Neurochem.* **2010**, *114*, 810–819. [CrossRef]
85. Luo, Y.; Fischer, F.R.; Hancock, W.W.; Dorf, M.E. Macrophage Inflammatory Protein-2 and KC Induce Chemokine Production by Mouse Astrocytes. *J. Immunol.* **2000**, *165*, 4015–4023. [CrossRef]
86. Butt, A.M. ATP: A ubiquitous gliotransmitter integrating neuron—Glial networks. *Semin. Cell Dev. Biol.* **2011**, *22*, 205–213. [CrossRef] [PubMed]
87. Fields, R.D.; Burnstock, G. Purinergic signalling in neuron—Glia interactions. *Nat. Rev. Neurosci.* **2006**, *7*, 423–436. [CrossRef] [PubMed]
88. Burnstock, G.; Knight, G.E. The potential of P2X7 receptors as a therapeutic target, including inflammation and tumour progression. *Purinergic Signal.* **2018**, *14*, 1–18. [CrossRef]
89. Abbracchio, M.P.; Burnstock, G.; Verkhratsky, A.; Zimmermann, H. Purinergic signalling in the nervous system: An overview. *Trends Neurosci.* **2009**, *32*, 19–29. [CrossRef]
90. Fellin, T.; Pozzan, T.; Carmignoto, G. Purinergic receptors mediate two distinct glutamate release pathways in hippocampal astrocytes. *J. Biol. Chem.* **2006**, *281*, 4274–4284. [CrossRef] [PubMed]
91. Walter, L.; Dinh, T.; Stella, N. ATP induces a rapid and pronounced increase in 2-arachidonoylglycerol production by astrocytes, a response limited by monoacylglycerol lipase. *J. Neurosci.* **2004**, *24*, 8068–8074. [CrossRef]
92. Murakami, K.; Nakamura, Y.; Yoneda, Y. Potentiation by ATP of lipopolysaccharide-stimulated nitric oxide production in cultured astrocytes. *Neuroscience* **2003**, *117*, 37–42. [CrossRef]
93. Gendron, F.P.; Neary, J.T.; Theiss, P.M.; Sun, G.Y.; Gonzalez, F.A.; Weisman, G.A. Mechanisms of P2X7 receptor-mediated ERK1/2 phosphorylation in human astrocytoma cells. *Am. J. Physiol. Cell Physiol.* **2003**, *284*, C571–C581. [CrossRef]
94. Jacques-Silva, M.C.; Rodnight, R.; Lenz, G.; Liao, Z.; Kong, Q.; Tran, M.; Kang, Y.; Gonzalez, F.A.; Weisman, G.A.; Neary, J.T. P2X 7 receptors stimulate AKT phosphorylation in astrocytes. *Br. J. Pharmacol.* **2004**, *141*, 1106–1117. [CrossRef] [PubMed]
95. Kucher, B.M.; Neary, J.T. Bi-functional effects of ATP/P2 receptor activation on tumor necrosis factor-alpha release in lipopolysaccharide-stimulated astrocytes. *J. Neurochem.* **2005**, *92*, 525–535. [CrossRef]

96. Potolicchio, I.; Chitta, S.; Xu, X.; Fonseca, D.; Crisi, G.; Horejsi, V.; Strominger, J.L.; Stern, L.J.; Raposo, G.; Santambrogio, L. Conformational Variation of Surface Class II MHC Proteins during Myeloid Dendritic Cell Differentiation Accompanies Structural Changes in Lysosomal MIIC. *J. Immunol.* **2005**, *175*, 4935–4947. [CrossRef]

97. Lee, M.H.; Lee, S.J.; Choi, H.J.; Jung, Y.W.; Frøkiær, J.; Nielsen, S.; Kwon, T.H. Regulation of AQP4 protein expression in rat brain astrocytes: Role of P2X7 receptor activation. *Brain Res.* **2008**, *1195*, 1–11. [CrossRef] [PubMed]

98. Yamamoto, M.; Kamatsuka, Y.; Ohishi, A.; Nishida, K.; Nagasawa, K. P2X7 receptors regulate engulfing activity of non-stimulated resting astrocytes. *Biochem. Biophys. Res. Commun.* **2013**, *439*, 90–95. [CrossRef]

99. Gu, B.J.; Saunders, B.M.; Jursik, C.; Wiley, J.S. The P2X7-nonmuscle myosin membrane complex regulates phagocytosis of nonopsonized particles and bacteria by a pathway attenuated by extracellular ATP. *Blood* **2010**, *115*, 1621–1631. [CrossRef] [PubMed]

100. Nagasawa, K.; Escartin, C.; Swanson, R.A. Astrocyte cultures exhibit P2X7 receptor channel opening in the absence of exogenous ligands. *Glia* **2009**, *57*, 622–633. [CrossRef]

101. Verkhrasky, A.; Krishtal, O.A.; Burnstock, G. Purinoceptors on neuroglia. *Mol. Neurobiol.* **2009**, *39*, 190–208. [CrossRef] [PubMed]

102. Peña-Altamira, L.E.; Polazzi, E.; Giuliani, P.; Beraudi, A.; Massenzio, F.; Mengoni, I.; Poli, A.; Zuccarini, M.; Ciccarelli, R.; Di Iorio, P.; et al. Release of soluble and vesicular purine nucleoside phosphorylase from rat astrocytes and microglia induced by pro-inflammatory stimulation with extracellular ATP via P2X7 receptors. *Neurochem. Int.* **2018**, *115*, 37–49. [CrossRef] [PubMed]

103. Matute, C.; Torre, I.; Pérez-Cerdá, F.; Pérez-Samartín, A.; Alberdi, E.; Etxebarria, E.; Arranz, A.M.; Ravid, R.; Rodríguez-Antigüedad, A.; Sánchez-Gómez, M.V.; et al. P2X7 receptor blockade prevents ATP excitotoxicity in oligodendrocytes and ameliorates experimental autoimmune encephalomyelitis. *J. Neurosci.* **2007**, *27*, 9525–9533. [CrossRef] [PubMed]

104. Amadio, S.; Parisi, C.; Piras, E.; Fabbrizio, P.; Apolloni, S.; Montilli, C.; Luchetti, S.; Ruggieri, S.; Gasperini, C.; Laghi-Pasini, F.; et al. Modulation of P2X7 receptor during inflammation in multiple sclerosis. *Front. Immunol.* **2017**, *8*, 1529. [CrossRef] [PubMed]

105. Sharp, A.J.; Polak, P.E.; Simonini, V.; Lin, S.X.; Richardson, J.C.; Bongarzone, E.R.; Feinstein, D.L. P2x7 deficiency suppresses development of experimental autoimmune encephalomyelitis. *J. Neuroinflamm.* **2008**, *5*, 33. [CrossRef] [PubMed]

106. Lang, P.A.; Merkler, D.; Funkner, P.; Shaabani, N.; Meryk, A.; Krings, C.; Barthuber, C.; Recher, M.; Brück, W.; Häussinger, D.; et al. Oxidized ATP inhibits T-cell-mediated autoimmunity. *Eur. J. Immunol.* **2010**, *40*, 2401–2408. [CrossRef]

107. Chen, L.; Brosnan, C.F. Exacerbation of Experimental Autoimmune Encephalomyelitis in P2X 7 R $-/-$ Mice: Evidence for Loss of Apoptotic Activity in Lymphocytes. *J. Immunol.* **2006**, *176*, 3115–3126. [CrossRef]

108. Bernal-Chico, A.; Manterola, A.; Cipriani, R.; Katona, I.; Matute, C.; Mato, S. P2x7 receptors control demyelination and inflammation in the cuprizone model. *Brain Behav. Immun. Health* **2020**, *4*, 100062. [CrossRef]

109. Grygorowicz, T.; Struzyńska, L.; Sulkowski, G.; Chalimoniuk, M.; Sulejczak, D. Temporal expression of P2X7 purinergic receptor during the course of experimental autoimmune encephalomyelitis. *Neurochem. Int.* **2010**, *57*, 823–829. [CrossRef]

110. Kerstetter, A.E.; Padovani-Claudio, D.A.; Bai, L.; Miller, R.H. Inhibition of CXCR2 signaling promotes recovery in models of multiple sclerosis. *Exp. Neurol.* **2009**, *220*, 44–56. [CrossRef]

111. Simpson, J.E.; Newcombe, J.; Cuzner, M.L.; Woodroofe, M.N. Expression of monocyte chemoattractant protein-1 and other β-chemokines by resident glia and inflammatory cells in multiple sclerosis lesions. *J. Neuroimmunol.* **1998**, *84*, 238–249. [CrossRef]

112. Grygorowicz, T.; Wełniak-Kamińska, M.; Struzyńska, L. Early P2X7R-related astrogliosis in autoimmune encephalomyelitis. *Mol. Cell. Neurosci.* **2016**, *74*, 1–9. [CrossRef] [PubMed]

113. Sofroniew, M.V.; Vinters, H.V. Astrocytes: Biology and pathology. *Acta Neuropathol.* **2010**, *119*, 7–35. [CrossRef]

114. Fumagalli, M.; Brambilla, R.; D'Ambrosi, N.; Volonté, C.; Matteoli, M.; Verderio, C.; Abbracchio, M.P. Nucleotide-mediated calcium signaling in rat cortical astrocytes: Role of P2X and P2Y receptors. *Glia* **2003**, *43*, 218–230. [CrossRef] [PubMed]

115. Hamilton, N.; Vayro, S.; Kirchhoff, F.; Verkhratsky, A.; Robbins, J.; Gorecki, D.C.; Butt, A.M. Mechanisms of ATP- and glutamate-mediated calcium signaling in white matter astrocytes. *Glia* **2008**, *56*, 734–749. [CrossRef] [PubMed]

116. Bazargani, N.; Attwell, D. Astrocyte calcium signaling: The third wave. *Nat. Neurosci.* **2016**, *19*, 182–189. [CrossRef] [PubMed]

117. Bijelić, D.D.; Milićević, K.D.; Lazarević, M.N.; Miljković, D.M.; Pristov, J.J.B.; Savić, D.Z.; Petković, B.B.; Andjus, P.R.; Momčilović, M.B.; Nikolić, L.M. Central nervous system-infiltrated immune cells induce calcium increase in astrocytes via astroglial purinergic signaling. *J. Neurosci. Res.* **2020**, *98*, 2317–2332. [CrossRef] [PubMed]

118. Abbott, N. Inflammatory mediators and modulation of blood—Brain barrier permeability. *Cell. Mol. Neurobiol.* **2000**, *20*, 131–147. [CrossRef]

International Journal of
Molecular Sciences

Review

Neuroprotective Effects of Guanosine in Ischemic Stroke—Small Steps towards Effective Therapy

Karol Chojnowski [1,†] , Mikolaj Opielka [2,3,*,†] , Wojciech Nazar [1] , Przemyslaw Kowianski [4,5] and Ryszard T. Smolenski [2,*]

1   Faculty of Medicine, Medical University of Gdańsk, Marii Skłodowskiej-Curie 3a, 80-210 Gdańsk, Poland; karole97@gumed.edu.pl (K.C.); wojciech.nazar@gumed.edu.pl (W.N.)
2   Department of Biochemistry, Medical University of Gdansk, 1 Debinki St., 80-211 Gdansk, Poland
3   International Research Agenda 3P—Medicine Laboratory, Medical University of Gdańsk, 3A Sklodowskiej-Curie Street, 80-210 Gdansk, Poland
4   Department of Anatomy and Neurobiology, Medical University of Gdansk, 1 Debinki Street, 80-211 Gdańsk, Poland; przemyslaw.kowianski@gumed.edu.pl
5   Institute of Health Sciences, Pomeranian University of Słupsk, Bohaterów Westerplatte 64, 76-200 Słupsk, Poland
*   Correspondence: mopielka@gumed.edu.pl (M.O.); ryszard.smolenski@gumed.edu.pl (R.T.S.)
†   These authors contributed equally to this work.

Abstract: Guanosine (Guo) is a nucleotide metabolite that acts as a potent neuromodulator with neurotrophic and regenerative properties in neurological disorders. Under brain ischemia or trauma, Guo is released to the extracellular milieu and its concentration substantially raises. In vitro studies on brain tissue slices or cell lines subjected to ischemic conditions demonstrated that Guo counteracts destructive events that occur during ischemic conditions, e.g., glutaminergic excitotoxicity, reactive oxygen and nitrogen species production. Moreover, Guo mitigates neuroinflammation and regulates post-translational processing. Guo asserts its neuroprotective effects via interplay with adenosine receptors, potassium channels, and excitatory amino acid transporters. Subsequently, guanosine activates several prosurvival molecular pathways including PI3K/Akt (PI3K) and MEK/ERK. Due to systemic degradation, the half-life of exogenous Guo is relatively low, thus creating difficulty regarding adequate exogenous Guo distribution. Nevertheless, in vivo studies performed on ischemic stroke rodent models provide promising results presenting a sustained decrease in infarct volume, improved neurological outcome, decrease in proinflammatory events, and stimulation of neuroregeneration through the release of neurotrophic factors. In this comprehensive review, we discuss molecular signaling related to Guo protection against brain ischemia. We present recent advances, limitations, and prospects in exogenous guanosine therapy in the context of ischemic stroke.

Keywords: guanosine; stroke; neuroprotection; neuroinflammation; purinergic signaling

Citation: Chojnowski, K.; Opielka, M.; Nazar, W.; Kowianski, P.; Smolenski, R.T. Neuroprotective Effects of Guanosine in Ischemic Stroke—Small Steps towards Effective Therapy. *Int. J. Mol. Sci.* **2021**, 22, 6898. https://doi.org/10.3390/ijms22136898

Academic Editors: Dmitry Aminin and Peter Illes

Received: 31 May 2021
Accepted: 23 June 2021
Published: 27 June 2021

Publisher's Note: MDPI stays neutral with regard to jurisdictional claims in published maps and institutional affiliations.

1. Introduction

Stroke is one of the top causes of death worldwide and the leading cause of permanent disability in developed countries, affecting approximately one in six people in their lifetime worldwide [1–3]. In 2019, on a global scale, ischemic stroke and hemorrhagic stroke accounted for 77.2 million (76%) and 29.1 million (24%) cases of stroke, respectively [4]. Ischemic stroke is characterized by a transient or permanent reduction of cerebral blood flow, causing the depletion of oxygen and glucose levels, uncontrolled depolarization, energy deficit, excitotoxicity, and tissue infarct, thus disturbing physiological cellular function [5]. The central nervous system (CNS) is specifically sensitive to ischemia; thus, a thrombolytic or mechanical restoration of blood flow in a narrow time window remains the treatment of choice for limiting postischemic brain injury [3]. The ischemic core is surrounded by a transition zone—an ischemic penumbra composed of both normal and functionally impaired cells. Ischemic tissue damage develops slower in the penumbra;

therefore, recent studies prompted the creation of therapeutic protocols, which enable clinicians to use thrombolytic therapy and thrombectomy in selected patients during extended time windows. For instance, thrombectomy is advised up to 24 h in eligible patients who have a mismatch between clinical deficits and infarct volume [6–8]. However, clinical evidence demonstrates that in the majority of stroke patients, slow brain injury evolution is observed in hours-to-days time intervals, which may be caused by reperfusion injury and activation of immunoinflammatory mechanisms [9]. Therefore, a neuroprotective therapy protecting neurons in the penumbra against ischemic and reperfusion injury is still highly demanded [10]. Despite a plethora of proposed neuroprotective drugs, including antioxidants, neuropeptides, microRNAs, anti-inflammatory drugs, and antiexcitatory drugs, their status is far from being clinically well established; thus, the question about pleiotropic neuroprotectant(s) in stroke therapy remains unanswered.

In recent years, guanosine (Guo), a part of a guanine-based purinergic system, emerged as a novel neuroprotectant and neuromodulator in CNS disorders. In this comprehensive review, we describe in depth the role and effects of extracellular guanosine in in vitro and in vivo models of ischemic stroke. The main objective of our study was to introduce the concept of Guo as a new potential neuroprotectant and/or neurotrophic agent supplementing the currently available ischemic stroke treatment methods. Starting from the observations of endogenous guanosine release under ischemia, we explore the concept of the therapeutic potential of exogenous guanosine. We discuss the molecular targets in relation to the neuroprotective and neurotrophic effects of extracellular guanosine. We also present the future directions in approaches to guanosine application in ischemic stroke.

We performed a literature search of Medline, Scopus, and Embase databases published between 1 January 1990 and May 2021 to identify studies addressing the role of guanosine in ischemic stroke management in both in vitro and in vivo ischemic stroke models. To the best of our knowledge, there is currently no available experimental data regarding the neuroprotective effects of Guo in hemorrhagic stroke; thus, the scope of the study was limited to ischemic stroke models. Included studies comprised both original and review articles. Searches were independently conducted by two of the authors. We searched Medical Subject Headings (MeSH) terms limited to the English language in multiple combinations, including brain ischemia/ischemic stroke, oxygen glucose deprivation, guanosine, neuroprotective agent, and neuroprotection. Additionally, references from included studies were screened for relevant studies.

2. The Physiological Role and Signaling Targets of Endogenous Guanosine in Central Nervous System

2.1. Cellular Location, Release, and Metabolism of Guanine Derivatives in the Brain

Guanine derivatives (GDs) include guanosine 5′-triphosphate (GTP), guanosine 5′-diphosphate (GDP guanosine 5′-monophosphate (GMP), and guanosine. GTP and GDP have mostly been recognized as intracellular modulators of G-protein activity [11]. However, apart from the regulation of G proteins, GDs are also found to be involved in the extracellular signaling in the CNS [12]. Thus, analogously to the adenosine-based purinergic system, a signaling system based on GDs has been proposed [13].

GTP is co-stored in synaptic vesicles and released with ATP, suggesting the role of GTP in neurotransmission [14,15]. The pool of GDs in the brain is located mainly within the astrocytes [12]. In basal conditions, astrocytes release GDs producing a constant concertation of Guo in the extracellular milieu. Of note, the spontaneous release of GDs from the astrocytes is significantly higher than their adenine-based counterparts [16]. The transport of GDs (including Guo) into an extracellular compartment is mediated by equilibrative nucleoside transporters (ENT) [17]. Moreover, guanine nucleotides can undergo extracellular hydrolysis by membrane-bound ectonucleotidases, providing a secondary source of Guo in the extracellular milieu [18]. After the release into the extracellular compartment, Guo can be transformed into guanine (GUA) [19]. The phosphorolytic breakdown of guanosine to guanine is catalyzed by purine nucleoside phosphorylase (PNP), which is constitutively released by glial cells and neurons into the extracellular space [20,21]. Ultimately, guanine is deaminated by guanine deaminase forming xanthine [22,23].

2.2. Neurotrophic Effects of Guoanosine in CNS—Role in Neurogenesis, Neuritogenesis, and Cell Differentiation

GDs induce proliferative effects, emphasized by an increase in the number of neurons, and proliferation markers in both in vitro and in vivo studies [12,24–27]. These effects can be attributed to GD-induced soluble factor release (neuronal growth factor (NGF), transforming growth factor (TGF), fibroblast growth factor-2 (FGF-2), brain derived neutrophic factor (BDNF), and Guo-induced adenosine release [12,28–30]. In a study performed in vivo on a Parkinson's disease model, Guo treatment increased progenitor/stem cell proliferation in the subventricular zone (SVZ) [24]. A subsequent study, performed on stem cells isolated from 1-day-old healthy rats, corroborated these observations, also reporting Guo-mediated neural stem cell (NSC) proliferation [28]. A more recent study presented that Guo is also able to induce the proliferation of NSCs sampled from adult mice [31].

MAPK and PI3K cascades are well studied molecular pathways involved mainly in neuronal proliferation, differentiation, and survival [32]. The protein kinase C (PKC) pathway takes part in synaptogenesis and, together with protein kinase A (PKA), regulates extracellular matrix (ECM) proteins [33–35]. A different signaling pathway, Ca2+/calmodulin-dependent protein kinase II (CaMKII), participates in Ca2+ signaling and mediates neuronal development and plasticity [36].

The molecular mechanism which would explain Guo-mediated trophic effects has not been fully unraveled. However, some studies indicate that the aforementioned molecular pathways are involved in Guo effect mediation [26]. Moreover, neuronal receptors, such as A2A, NMDA, and non-NMDA receptors, also contribute to the Guo-mediated proliferative and prosurvival effects observed in in vitro studies [25].

Guo stimulates also neuritogenesis—sprouting the neurites from the cell body forming connections between neurons. GTP and Guo coincubated with NGF were able to enhance the stimulatory action of NGF on neurite growth via distinct mechanisms [30,37]. More in-depth studies observed an increase of cAMP and activation of HO-1 and PRK1 in neurons in models of neuronal neurite arborization outgrowth [30,38,39]. HO-1 is an enzyme with antioxidant properties, whereas PRK1 is involved in actin cytoskeleton regulation and neuronal differentiation [40]. In SH-SY5Y neuroblastoma cells, a model of neuronal differentiation, Guo halted the cell cycle in neuroblastoma and promoted differentiation marker expression [41]. In line with this, a different study presented the Guo-elicited proliferation of NSC, followed by differentiation toward neurons [31].

GMP or Guo promotes the reorganization of astrocytic ECM proteins (laminin and fibronectin) in the neuron-astrocyte coculture model. The Guo/GMP-mediated ECM modulation is involved in the neuron-astrocyte interaction process and in trophic effect mediation. Furthermore, this study reported that Guo/GMP did not affect the process of neurogenesis. However, the number of neurons in cocultures increased, putatively due to increased neuronal viability or the neuritogenetic properties of laminin [26].

2.3. Guanine Derivatives and Neuroprotection

The first investigations into the role of GDs in neuropathologies found that the level of GDs is persistently elevated for 7 days after an ischemic injury [42]. Moreover, other studies showed that GDs are physiologically present in cerebrospinal fluid in nanomolar concentrations and rise by three- to fivefold within 30 min of hypoxic/hypoglycemic conditions [16,43]. Furthermore, it was demonstrated that under hypoxic conditions, the extracellular concentrations and activity of PNP decreased, which subsequently prolonged the presence of Guo in the extracellular compartment [21].

These discoveries prompted a new approach in establishing the role of GDs in CNS pathophysiology. It was suggested that GDs constitute an endogenous restorative system that activates after an injury, and its role is to prevent further damage and to re-establish tissue function [19]. Aside from the ischemic stroke, GDs also present a protective effect in in vivo models of other CNS disorders, such as epileptic seizures, Parkinson's disease, Alzheimer's disease, spinal cord injury, sepsis, gliomas, and hepatic encephalopathy [44–53]. It is important to highlight the fact that most of the neuroprotective effects observed in animal models were achieved using exogenously administered guanosine. Therefore, amongst other GDs, guanosine is the most promising potential therapeutic agent [45–52,54].

2.4. Guanosine-Specific Targets: Receptors and Binding Sites

It is still under debate if guanosine has its own specific receptor. Interestingly, as it was demonstrated in the animal models, guanosine probably has some kind of its own G-protein coupled receptor (GPCR). Guo binds strongly to this binding site, and naturally occurring purines (GDP, GMP, ATP, adenosine, xanthine, hypoxanthine, caffeine, theophylline) cannot displace guanosine from this location [55–57]. Moreover, it was demonstrated that this yet unknown binding site is different from well-characterized ARs [57,58]. Possible GPCR candidates include GPR174/LPS3 [59], which is highly homologous with P2Y10, and GPR23/LPA4, which shows high homology with human P2Y5 receptor [57,60]. Moreover, A_1Rs and $A_{2a}Rs$ form receptor complexes with members of the P2Y receptor family. Additionally, P2Y5 receptors share high homology with GPR23. Thus, it is reasonable to hypothesize the occurrence of molecular interactions between GPR23 and $A_1R \backslash A_{2a}Rs$ [57,61]. Nevertheless, despite promising studies, the guanosine-specific receptor has not yet been fully characterized, and thus Guo remains an orphan neuromodulator.

Furthermore, Guo was identified as a weak A_1 receptor (A_1R) and A_{2A} receptor ($A_{2a}R$) agonist [62–64]. It is still unknown to what extent these receptors are responsible for the neuroprotective effect of guanosine. Moreover, some studies show that the antioxidant effect is A_1R dependent but $A_{2A}R$ independent [62,65–67]. This is contrary to the results of Dal-Cim et al., who indicate that some neuroprotective effects may be conducted via $A_{2A}R$ but not A_1R [68]. Some studies suggest that it is the interplay between coexpressed receptors and the fine-tuning mechanism that are responsible for the neuroprotective effect [64,69].

Guo can target not only certain receptors but also potassium channels. Kir 4.1 is an inwardly rectifying K^+ channel commonly found in glial cells in the CNS. This channel plays a role in the sustainment of extracellular K^+ homeostasis, resting membrane potential, and regulation of glutamate uptake [70]. Chronic exposure to Guo in rat cortical astrocytes in vivo promoted the expression of Kir4.1. Of note, inhibition of the translational process prevented the Guo-induced upregulation of Kir 4.1, suggesting that the Kir 4.1 upregulation stimulated by Guo is achieved through de novo protein synthesis [71]. Moreover, Guo acts via large-conductance Ca^{2+}-activated K^+ (BK) channels [27,72]. In this case, guanosine binds to the alfa subunit of K^+ channels and increases K^+ conductance [73]. Moreover, Guo modulates NMDA receptors and probably glutamate transporters, including GLT-1. [74–77].

Overall, guanosine-mediated neuroprotective effects may be mediated by the interplay between known adenosine receptors and/or potential guanosine binding site as well as BK channels.

3. Key Pathophysiological Events of Ischemic Stroke and Targets for Guanosine

The main contributory factors involved in the pathophysiology of ischemic stroke are oxygen and glucose deprivation (OGD), reperfusion injury, glutamate excitotoxicity, and neuroinflammation. The brain is extremely vulnerable to ischemic damage, due to its high metabolic rate, limited energy storage capacity, and sole dependence on glucose as an energy substrate [78]. An area characterized by irreversible neuronal damage due to energy depletion is called an ischemic core. In the ischemic core, within minutes, cells undergo necrosis and excitotoxic cell death. Around the ischemic core is a functionally compromised but structurally intact tissue called the penumbra. In this area, cell death occurs at a slower rate, mostly through oxidative stress-mediated processes like apoptosis and inflammation [79]. As both of these mechanisms are triggered in a relatively orderly fashion, this opens up much more possibilities for therapeutic targeting compared to necrotic cells [80].

Reperfusion injury usually occurs in the course of poststroke thrombolytic therapy or thrombectomy, due to blood flow restoration in the previously occluded blood vessels. Counterintuitively, therapeutically achieved blood flow restoration brings detrimental consequences for the peri-infarct region as it contributes to a secondary burst of ROS generation [81].

Prolonged energy deficiency promotes a rise in glutamate (Glu) levels due to the increased release of Glu into the synaptic cleft and impaired Glu reuptake. Subsequently, excessive stimulation of NMDA receptors by glutamate results in Ca2+ and Na+ influx, which causes cell swelling and excitotoxic cell death [82]. The latter is a result of neuronal overstimulation and subsequent mitochondrial dysfunction, uncontrolled production of reactive oxygen species (ROS), and activation of proapoptotic pathways [83].

Shortly after ischemic stroke onset, the neuroinflammatory processes unfold [84]. The trigger for the acute phase of inflammation is damage-associated molecular patterns (DAMPs) which are released from dying and necrotic cells in the ischemic core. These molecules then activate local immune cells and perivascular endothelial cells by acting on Toll-like receptors and purinergic receptors [85], subsequently leading to inflammasome activation, which initiates a fully fledged inflammatory response [86]. The increased levels of chemokines cause chemotaxis of circulating leukocytes into the injury site. This process of infiltration is supported by activated microglia that upregulate adhesion molecules on cerebral vasculature [87–89]. Subsequently, the microglia (and other immune cells) produce metalloproteinases (MMPs) which increase BBB permeability, allowing other immune cells easier access to damaged brain areas [90]. Locally activated microglia and infiltrated macrophages can then perform proinflammatory or anti-inflammatory functions, depending on their molecular phenotypes [91].

On a molecular level, ischemic insult causes an upregulation of mitogen-activated protein kinases (MAPK) and an expression of the gene encoding nuclear factor kappa beta (NF-kb) protein complex [92]. NF-kb is a heteromeric transcription factor that most commonly is made up of p50 and p65 [93,94]. After translocation to the nucleus, NF-kb binds to specific sites of a DNA and induces transcription of proinflammatory cytokines and inducible NOS (iNOS), which is an enzyme responsible for increased ROS production and inflammation enhancement [95,96].

Most of the abovementioned aspects of neuroinflammation are detrimental to neuronal tissue and further deepen the ischemic injury. Thus, post stroke inflammation presents itself as a potential target of ischemic stroke treatment.

On a cellular level, the plethora of protective mechanisms of guanosine was studied in cortical slices or neural cells exposed to glucose-free medium and hypoxic atmosphere in a so-called oxygen-glucose deprivation (OGD) protocol [27,97]. OGD protocol is a well-established model of mimicking the most significant aspects of ischemic injury. Frequently, the OGD is followed by the reoxygenation period, which simulates the reperfusion stage of ischemic stroke [98].

3.1. PI3K, MEK, and PKC Are Involved in Guanosine-Mediated Neuroprotection

Although the manner in which Guo exerts its effects is still not fully unraveled, a few signaling pathways have been discovered that play a role in Guo effects mediation. In 2008, Oleskovicz et al. demonstrated that Guo acts via the modulation of PKA, PKC, MAPK, and/or PI-3K pathways. These signaling pathways were blocked by specific inhibitors, resulting in the reduction of guanosine-induced neuroprotection [99]. Later studies further confirmed the involvement of these pathways in Guo effect mediation [27,68,100–103] (Figure 1).

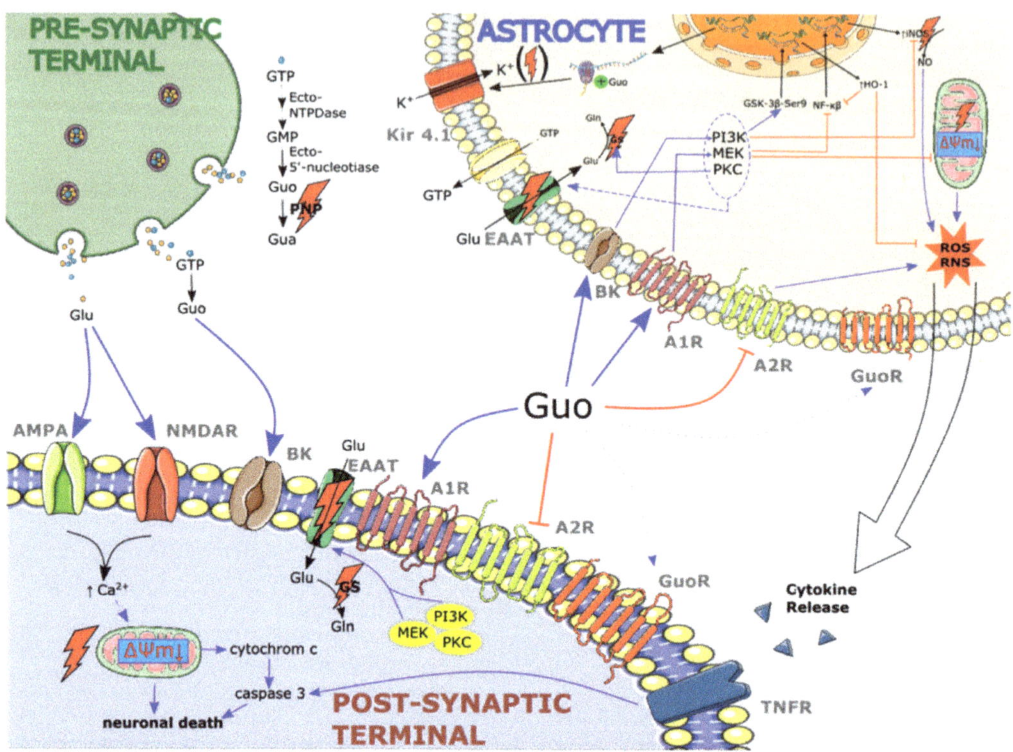

Figure 1. Overview of the most important mechanisms that contribute to guanosine-induced neuroprotection under ischemia: guanosine 5′-triphosphate (GTP) is released from the presynaptic terminal of neurons via synaptic vesicles but can also be transported directly from astrocytes into extracellular space. Extracellular guanosine nucleotides undergo hydrolysis by ectonucleotidases into guanosine 5′-monophosphate (GMP) and subsequently guanosine (Guo). The end product of guanosine nucleotide degradation is guanine. The cleavage of guanosine to guanine (Gua) is catalyzed by purine nucleoside phosphorylase (PNP), the concentration of which is decreased under hypoxic conditions (as marked by a red lightning bolt). Guanosine has a putative specific binding site (GuoR); however, its role has not been fully characterized (the putative Guo–GuoR interaction is marked as dotted arrows). Nevertheless, guanosine can act upon adenosine receptors (A_1R and A_{2A}R) and potassium big conductance channels (BK). Guanosine activates cellular molecular pathways, such as PI3K/Akt (PI3K), MEK/ERK (MEK), and protein kinase C (PKC). Activation of these molecular pathways leads to stimulation of amino acid transporters (EAATs) (depicted as a dashed arrow). Moreover, guanosine acting via the PKC pathway increases glutamine synthase (GS) activity which, combined with EAAT stimulation, protects against glutamate excitotoxicity. The physiological base activity of EAATs and GS is disturbed under ischemic conditions. Guanosine promotes protection against reactive oxygen species (ROS) and reactive nitrogen species (RNS) by activation of MEK and PI3K. These cellular pathways downregulate the expression of iNOS via NF-kb inhibition, upregulate expression of HO-1 via GSK-3β phosphorylation at Ser9 (GSK-3β-Ser9), and prevent loss of mitochondrial membrane potential ($\Delta\Psi$m↓). Notice that some of these effects counteract the detrimental events that occur under ischemic conditions. These include mitochondrial membrane depolarization and iNOS upregulation. The consequence of increased ROS/RNS production is a release of proinflammatory cytokines (marked as an open arrow) some of which then act on tumor necrosis factor receptors (TNFR) promoting apoptosis in neurons. Chronic supplementation of guanosine prevents reduction of inwardly rectifying K^+ channels (Kir 4.1) by promoting de novo Kir4. 1 synthesis. Importantly, expression of Kir 4.1 and/or Kir-mediated currents are reduced up to 14 days after an ischemic injury (marked as red lightning bolt put in braces). Figure designed using image template from Servier Medical Art https://smart.servier.com/image-set-download/, accessed on 12 March 2021.

Guo induces the phosphorylation of protein kinase B (PKB/Akt) via PI3K, which leads to the inactivation of GSK3β through phosphorylation at Ser9 [100]. In a study by Molz et al., the use of PI3K inhibitor (LY294002) prevented Guo-induced GSK3β phosphorylation. Moreover, incubating hippocampal slices with Guo presented a significant rise of GSK3β-Ser9 after 30 min of exposure [101]. In line with this study, blocking PI3K with a different inhibitor (wortmanin) also abolished Guo-induced PKB/Akt phosphorylation. Moreover, in the presence of a BK channel inhibitor (charybdotoxin), the phosphorylation of PKB/Akt promoted by Guo was also blocked [27]. These results indicate that the neuroprotection elicited by Guo involves BK channel activation, a subsequent PI3K-PKB/Akt pathway activation, and phosphorylation of GSK3β, which is a downstream effector of that pathway.

The MAPK/ERK pathway is another crucial pathway that, similar to PI3K, mediates a variety of Guo neuroprotective effects. Its contribution was presented in two studies conducted by Dal-Cim et al. on hippocampal slices and cortical astrocyte cultures. In the presence of MEK inhibitor (PD98059), Guo protection against OGD-induced damage was prevented. Moreover, a similar effect was achieved by blocking A1R. These results show that Guo exerts its neuroprotective effect via a mechanism that involves both the MAPK/ERK pathway and A_1R activation [68,102]. In a different study, using a different MAPK inhibitor (SB203580) in glucose-deprived C6 astroglial cells, a similar effect was observed [103].

Several guanosine effects are also dependent on the PKC pathway. Blocking PKC with BIS II in C6 astroglial cells abolishes some of the neuroprotective effects of Guo [103]. Moreover, Guo coincubated with another PKC inhibitor (chelerythrine) lost the ability to prevent cell death in astrocytes from the murine cerebral cortex [102].

In summary, the most important molecular pathways that participate in Guo effect mediation are PI3K, MEK, and PKC.

3.2. Guanosine-Mediated Neuroprotection Depends on BK Channels Activity

During an ischemic event, the expression of Kir 4.1 and/or Kir-mediated currents are reduced up to 14 days after an injury [104–106]. In a study from 2006 conducted by Befenati et al., chronic exposure to Guo in unexposed to OGD rat cortical astrocytes in vivo promoted the upregulation of Kir4.1 [71]. This study indicates that Guo could potentially counteract the effect of ischemic insult on K+ homeostasis. Nevertheless, further research must be conducted to support this thesis.

In a study performed on hippocampal slices subjected to OGD conditions, the blocking of K+ channels with 4-aminopyridine (4-AP), a voltage-dependent K+ channel blocker, abolished Guo-induced neuroprotection [99]. Moreover, further research showed that Guo effects are mediated by BK channels (BK) [27,72]. BK channels are determined to facilitate membrane potential and activate the PKB/Akt pathway which acts as a cellular defense against oxidative damage [72,107–109]. Charybdotoxin, a BK selective blocker, was able to impede Guo-induced neuroprotection during the reoxygenation period [27]. In line with this study, charybdotoxin also blocked the protective effects of Guo on SH-SY5Y cells against mitochondrial oxidative stress [72]. In the light of the reported evidence, it is conceivable that BK channels are an essential part of the neuroprotective molecular cascade initiated by Guo.

3.3. Guanosine Acts Against Glutamate Excitotoxicity

One of the major mechanisms implicated in guanosine neuroprotection is the stimulation of the glutamate uptake, thus counteracting glutamate excitotoxicity [13,75,110,111]. Primarily, Guo protection against glutamate toxicity was studied based on in vitro glutamate excitotoxicity and seizures models [112–114]. During hypoxia/ischemia, a rapid increase of glutamate occurs mainly due to impairment of the glutamate uptake system, the release of excessive glutamate, the reversal of glutamate transporters activity (reverse uptake), and decreased activity of glutamine synthetase (GS) [115–117]. Using the model of hippocampal slices subjected to OGD with subsequent reoxygenation, it was demonstrated that guanosine administration promoted glutamate uptake by increasing the uptake Vmax and prevented the reversal of uptake induced by excessive glutamate [27,99,111,118]. Notably, the stimulatory effect of guanosine on glutamate uptake was predominantly observed in the reoxygenation but not in the hypoxic period, even when administered up to 3 h in the reoxygenation period [27,97]. This particular effect was caused by ATP level depletion during the ischemic period of OGD, which caused the blockade of ATP-dependent glutamate uptake (relying on Na–K–ATPase activity) [119,120]. The mechanism underlying the ability of guanosine to promote glutamate uptake relies mostly on modulating the glutamate transporter 1 (GLT-1) activity and restoration of GLT-1 expression to basal levels under OGD but also indirectly on the restoration of GS activity to physiological levels, most likely through the activation of the PKC pathway [68,75,102,103,121].

Regarding the intracellular signaling pathways involved in guanosine-mediated glutaminergic modulation, it was demonstrated that guanosine acting through the activation of A1R and the blockade of A2AR receptors or the putative Gi protein-coupled signaling site recruited PI3K/protein kinase B(Akt) and, predominantly, the MEK/ERK and PKC pathways [68,102]. MEK/ERK signaling cascade involvement in astrocytes is particularly important since the activation of ERK1/2 cascade protein is directly related to the activity of glutamate receptors [68,122]. The activation of PI3K/Akt by the Guo signaling cascade was observed only in hippocampal slices subjected to OGD but not in cortical astrocyte cultures, suggesting that the activation of this antiapoptotic, glutamate modulatory pathway by Guo is cell-type specific and may promote PI3K-dependent glutamate uptake in neurons through excitatory amino acid carrier 1 (EAAC1) [102,123].

3.4. Guanosine Prevents Mitochondrial Dysfunction

On the molecular level, the impairment of mitochondria results in increased reactive oxygen species (ROS) production, redox homeostasis disruption, and activation of apoptotic cell death [124–126]. The main events that contribute to mitochondrial dysfunction are excess Ca2+, mitochondrial swelling, loss of mitochondrial membrane potential, impaired oxidative phosphorylation, and accumulation of ROS [127]. Mitochondrial membrane depolarization is a well-known marker of mitochondrial dysfunction [126]. The loss of mitochondrial membrane potential is followed by an increase in ROS production with subsequent oxidative damage, reduced ATP production, and redox homeostasis disruption [128].

Several studies have postulated that Guo can prevent mitochondrial damage by direct ROS scavenging and/or via activation of molecular pathways that induce antioxidant effects [103,107]. More recent studies vastly diminished the role of Guo as a direct antioxidant, showing no or little involvement in nitric oxide (NO) scavenging activity [127,129]. Therefore, the mechanism of action which is responsible for the Guo effect on mitochondria is probably related to its ability to activate molecular pathways which then elicit antioxidant effects. In a study conducted by Dal-Cim et al., performed using hippocampal slices, blocking the MAPK/ERK pathway abolished Guo-induced protection against mitochondrial membrane depolarization. Moreover, blocking A_1R also removed the Guo effect on membrane potential. These results suggest a significant role of A1R and MEK in the mediation of Guo-induced protection against mitochondrial dysfunction during OGD [68].

In the most recent study conducted by Courtes et al. using liver mitochondria, Guo demonstrated a protective effect against Ca2+-induced mitochondrial dysfunction [127]. Interestingly, Guo improved mitochondrial function without any link to the stabilization of mitochondrial membrane potential or direct ROS scavenging. In this study, Guo prevented in vitro Ca2+-induced mitochondrial impairment by reduction of mitochondrial swelling and ROS levels. Moreover, Guo boosted mitochondrial metabolism and helped to establish energy homeostasis [127].

It is important to highlight that the potential differences between liver mitochondria and mitochondria found in CNS could undermine the significance of these data in terms of stroke. Moreover, the heterogeneity of mitochondria within CNS itself also implies caution with the interpretation of data obtained from different brain regions [130–132].

3.5. Guanosine Mediates the Decrease in NO Overproduction

NO is an important CNS messenger and neurotransmitter that actively participates in many pathological processes that occur during an ischemic stroke [133,134]. In an early response to ischemia, NO released from endothelium exerts protective effects by promoting, e.g., vasodilation. However, soon after, NO is massively overproduced by neurons and glia, and its effect becomes deleterious to the surrounding tissues [135]. In neurons, NO is produced mainly by constitutively expressed neuronal nitric oxide synthase (nNOS). Contrarily, in glial cells, the dominant isoform of NOS is iNOS. In contrast to nNOS, iNOS is upregulated in response to OGD, excessive ROS production, and glutamate excitotoxicity [101,136–138].

In line with several papers, Guo exerts its antioxidant effects by inhibition of NOS, downregulating the production of NO and consequently lowering levels of reactive nitrogen species (RNS) and ROS, which in turn ameliorates cell viability [38,68,129].

Studies performed on hippocampal slices and C6 astroglial cells reveal that Guo hinders the expression of iNOS by inhibition of NF-kb, more precisely by preventing it from binding to the promoter sequence of iNOS. Interestingly, the aforementioned Guo effect can be diminished by blocking either A_1R or signaling pathways like MEK or PI3K [68,107]. Moreover, the Guo downregulating effect on NF-kb activation may be mediated by heme oxygenase 1 (HO-1), since blocking this enzyme does not prevent NF-kb from raising in the presence of Guo in OGD conditions [107]. Interestingly, although Guo can elicit iNOS suppression at a concentration of 100 μM, it does not present this ability at concentrations of 30 or 300 μM [101].

It has now been hypothesized that Guo's effects on NO production may be mediated not only by iNOS but also by nNOS. In hippocampal slices subjected to OGD, Guo can in fact decrease iNOS induced by OGD. However, selective blocking of iNOS did not elicit a decline in RNS levels. Contrarily, in the presence of nNOS inhibitors, NO and ONOO- levels significantly decreased [129]. Thus, the modulation of nNOS rather than iNOS activity by Guo can bring antioxidative aid to damage induced by RNS.

More recently, a study on hippocampal slices subjected to OGD followed by reoxygenation presented a modulatory effect of Guo on NO levels. Guo coincubated with the nonselective NOS inhibitor (L-NAME) prevented a decline in ATP production, lactate release, and glutamate uptake in murine brain slices. However, in the presence of NO donors (DETA-NO or SNP), the protective effect of L-NAME or guanosine on bioenergetics and glutamate clearance was abolished [139].

Maintaining sustainable ATP concentration levels within the cell is an important task that can vastly increase the chance of survival in the OGD environment [140]. In CNS, several studies have proven the importance of astrocytes in the maintenance of neuronal energetic equilibrium. In astrocytes subjected to oxygen scarcity, oxygen-independent metabolic pathways such as glycolysis are activated. The end product of glycolysis lactate is then consumed by neurons providing them with energy [141,142]. During the OGD period, extracellular lactate levels decrease due to a putative increase in lactate consumption by neurons enduring hypoxia [143,144]. Guo was able to increase lactate availability and ATP levels in ischemic hippocampal slices. Interestingly, a similar effect was achieved using L-NAME, showing that Guo can alter cellular bioenergetic metabolism putatively via a mechanism involving NO level modulation [139].

In summary, Guo acts through cellular pathways and receptors that modulate NOS enzymes expression and subsequently NO levels. This in turn provides a decrease in ROS and RNS production and prevents the disruption in cell bioenergetics promoted by experimental ischemic stroke models.

3.6. Guanosine Exerts Antioxidative Effects through the Activation of Heme Oxygenase-1

HO-1 is an enzyme involved in the breakdown of pro-oxidant heme into antioxidative bilirubin and biliverdin, consequently shifting the balance in favor of antioxidants [145]. Moreover, HO-1 is also involved in the mediation of anti-inflammatory and antiapoptotic effects [47,146]. Mounting evidence suggests that the activation of HO-1 is one of the cell's rudimentary antioxidant defense [38,147]. Of note, in astrocytes subjected to oxidative or/and inflammatory injury, HO-1 is upregulated, counteracting the insult [107,148]. Considering the fact that an inseparable component of ischemia is an increase in ROS production and neuroinflammation, HO-1 has the potential to be targeted by future neuroprotective drugs [47,79,83,84,107].

The most notable function of HO-1 is intracellular redox environment maintenance, achieved by blocking the activity of iNOS. Moreover, HO-1 can inhibit NF-kB translocation from the cytoplasm to the nucleus, preventing it from inducing the production of inflammation mediators such as IL-1, TNF-alpha, iNOS, and cyclooxygenase-2 (COX2) [107,149]. Azide is a well-known respiratory chain inhibitor and is commonly used to evoke oxidative and nitrosative stress in experimental models [150]. In a study conducted by Quincozes-Santos et al. performed on C6 astroglial cells, Guo presented antioxidative properties by counteracting the detrimental effects of azide-induced ROS production [107]. In line with this, a different study showed that the inhibition of HO-1 by Sn(IV) protoporphyrin-IX dichloride (SnPP) abolishes the protective effects of Guo against mitochondrial stress. Additionally, this study also revealed that HO-1 activation induced by Guo may be preceded by PI3K/Akt/GSK-3beta pathway activation [72].

In summary, guanosine acting via HO-1 modulates the activity of many proinflammatory and pro-oxidant mediators. These include NO, iNOS, TNF-alpha, IL-1, COX2, and NF-kB.

3.7. Guanosine and Post-translational Processes in Ischemia—The Potential Role of SUMOylation

Most recently, guanosine was implicated in modulating the SUMOylation by interacting with A1 and A2A receptors [151]. SUMOylation is defined as a type of post-translational modification mediated by a small ubiquitin-like modifier peptide, which covalently binds to lysine residues of specific proteins, analogously to the ubiquitination process [152,153]. SUMOylation plays important physiological roles including synaptic maturation, regulation, and plasticity [154]. Under hypoxic/ischemic conditions, increased protein SUMOylation takes part in the endogenous neuroprotective system [155–157]. It was clearly demonstrated that extracellular, exogenous guanosine increases global protein SUMOylation in cortical astrocytes and cortical neurons. However, the effect was observed only up to 1 h after guanosine stimulation, thus suggesting that at longer time points, the effect of guanosine is eventually counteracted by deSUMOylating enzymes [151].

Together, this evidence suggests that guanosine is a SUMOylation enhancer, which may partially account for its neuroprotective effects under ischemia.

4. Protective Effects of Guanosine against Ischemic Stroke: Evidence from In Vivo Studies

In recent years, a number of in vivo studies confirmed the neuroprotective effects of guanosine after administration in the acute and chronic phases of ischemic stroke. The authors of these studies undertook different approaches to guanosine administration and studied diversified pathophysiological aspects of the ischemic stroke mechanism. Therefore, it is worth consolidating the existing knowledge to better depict the full spectrum of the neuroprotective effects of guanosine (Table 1).

4.1. Safety and Pharmacokinetics of Exogenous Guanosine in Rodent Models—Implications in Ischemic Stroke

Future neuroprotectants used in stroke treatment should have a wide therapeutic window, be rapidly distributed into the CNS, and target all of the brain components, including neurons, glia, and the BBB. Moreover, any drug's efficacy should be unaffected by sex-specific differences and bring a low risk of side effects and interactions with other drugs (drugs used in the stroke treatment or patients' daily medications).

The drug's bioavailability and the time in which the neuroprotectant reaches the penumbra are largely dependent on the administration route. After intraperitoneal (i.p.)

administration, Guo crosses the BBB via nucleoside-specific transporters located at the endothelium of brain blood vessels, enabling Guo to reach the brain via systemic circulation [158]. Intraperitoneal (i.p) administration of exogenous Guo and GMP elicits a threefold increase of Guo in CSF after 30 min. Additionally, after $5'$-nucleosidase inhibitor (AOPCP) administration, the GMP/Guo relative proportion altered in favor of GMP [159]. In line with this, i.p administration of Guo caused an increase in guanine, a direct Guo metabolite, after 30 min in a spinal cord sample [51]. Moreover, chronic administration of guanosine for 6 weeks resulted in an elevation of Guo metabolite levels in the CSF and plasma. Interestingly, due to Guo-induced adenosine release, the plasma adenosine level was also increased [160]. The results of the aforementioned studies confirm the metabolism of Guo in the central nervous system (CNS) and systemic circulation. Thus, the half-life of Guo is relatively low. An interesting approach to exogenous Guo administration was presented by Ramos et al. in 2016. In this study, Guo was administered intranasally (IN), therefore bypassing systemic circulation and going directly into the brain via olfactory and trigeminal nerves. By partially omitting systemic circulation, a smaller amount of Guo undergoes systemic metabolism. This, in turn, can raise the amount of Guo reaching the brain, consequently increasing the effectiveness of the administered dose. Moreover, when using the IN administration route, beneficial effects, such as the prevention of behavioral impairment and decrease of brain infarct volume, can be achieved with a dose seven times lower compared with using the i.p. administration route [161]. Furthermore, with this route of administration, exogenous Guo can reach the brain within 5 min compared to 15–30 min when administered i.p. [161,162]. A more recent study corroborated these results by outlining the increased time window in IN compared to the systemic route of administration. In the rat ischemia model (thermocoagulation of pial vessels), IN administration of Guo was able to attenuate neurological deficits when administered as late as 3 h after ischemia onset. Contrarily, using the systemic route, Guo-elicited neuroprotection could only be observed when administered immediately after stroke induction [163].

Guo-mediated neuroprotection comprises neurons, glia, and, as recently discovered, the BBB [63,121,160]. However, the main targets of Guo are the astrocytes [102,164]. Moreover, Guo modulates processes, including neuroinflammation and microglia activation, which can be detrimental to virtually all cells found in the brain [84,87,90]. Thus, together with evidence of Guo affecting remote areas from the ischemic lesion, it can be theorized that Guo-mediated neuroprotective effects are expressed globally throughout the CNS [163].

To our knowledge, no studies have reported any major side effects of exogenously administered Guo in in vivo models [74,163,165]. Guo administered i.p did not impair renal function. Indeed, there is evidence of Guo-mediated renoprotection [166]. However, doses higher than 240 mg kg^{-1} caused an elevation in liver function biomarkers. Moreover, Guo significantly decreased barbiturate-induced sleeping time in rodents [74,167]. Interestingly, as opposed to classic NMDAR antagonists (MK-801, ketamine), Guo most probably does not induce psychomimetic effects [54]. Nevertheless, Guo is well known for inducing amnestic effects in rodents [168,169]. This effect is probably the result of the inhibitory action of Guo on the glutaminergic system in the brain limbic structures [170]. Guo presents strong advantages over adenosine, a nucleoside with similar neuroprotective properties, in regard to adverse effects. Compared to adenosine, Guo has much less impact on basal arterial blood pressure [171]. Nevertheless, in the situation where Guo pretreatment is combined with adenosine infusion, the enhancement of the Guo-mediated adenosine's effects can provoke cardiovascular shock [171].

Only one work to date studied the differences in Guo treatment effects between male and female in vivo models, presenting better sensorimotor long-term recovery in female rats [172]. Nevertheless, further research has to be conducted to distinguish the sex-specific differences in response to Guo from neuroprotective effects induced by estrogens [173].

4.2. Neuroprotective and Neurorestorative Effects of Guanosine in Rodent Stroke Models

The neuroprotective effect of guanosine is time and dose dependent. The guanosine administration protocols were evaluated for the first time in models of neonatal hypoxic-ischemic (HI) injury and chronic cerebral hypoperfusion [160,174]. Guanosine administration up to 6 h following HI injury in three consecutive doses was found to modulate the glutamate uptake, thus preventing glutamate excitotoxicity [97,174]. Importantly, only the three-dose administration protocol was sufficient to achieve the protective effects of guanosine, regardless of whether the first dose was given immediately, 3 h, or 6 h after HI, indicating the dose-dependent effect of guanosine [174].

The described protocol was later adopted in ischemic stroke rodent models. The systemic administration of guanosine directly (t = 0 min) or within the first 6 h after permanent focal ischemia or middle cerebral artery occlusion (MCAo) with consecutive doses administered hours to days postischemia was found to increase the postischemic survival of rats and significantly decrease the infarct volume up to 40% [175–178]. The Guo-treated animals achieved progressive improvement in sensorimotor performance and partial restoration of motor dysfunctions assessed by the neurological deficit scale (NGS), indicating both the neuroprotective and neurorestorative properties of guanosine. Interestingly, the Guo-mediated neuroprotective effects differed between sexes. It was demonstrated that the female rats were more sensitive to Guo administration and reached significantly better long-term improvement in terms of sensorimotor deficits, independently of the estrous-cycle phase [172]. A subsequent set of studies demonstrated that the systemic administration of Guo 30 min before MCAo combined with postischemia Guo administration resulted in the most considerable decrease in infarct volume and restoration of neurological function compared to the administration at t = 0 min [176].

The Guo-mediated mechanism of neuroprotection observed in rodent models of the acute phase of ischemic stroke stands in agreement with in vitro studies [60,99,179]. Ex vivo studies performed 24 h after permanent focal ischemia in Guo-treated animals demonstrated the Guo-mediated restoration of decreased GLT-1 expression and increase in glutamine synthetase (GS) activity, thus increasing intracellular glutamate uptake [121]. Guo undeniably protects against glutamate excitotoxicity, but its protective effects observed in vivo also depend on antioxidative and anti-inflammatory properties [47,121,177]. Ischemic stroke triggers a massive production of reactive ROS and RNS, which causes oxidative stress response and an increase in expression of antioxidant enzymes (SOD, CAT) [96,180,181]. However, the activity of antioxidant enzymes drastically decreases, emphasizing the inefficiency of the physiological redox homeostasis system under severe oxidative stress [180,182]. Guo treatment was found to decrease lipid peroxidation and prevent the increase of NO and ROS levels. Moreover, Guo administration fully restored the decreased activity of SOD and CAT, increased the expression of SOD, and partially restored vitamin C levels [177]. A later also study revealed the strong anti-inflammatory properties of Guo. Guo administered in the acute phase of ischemic stroke suppressed the activation and infiltration of microglia to the periphery of the ischemic core. These results were supplemented by a Guo-mediated decrease in proinflammatory cytokines (IL-1, IL-6, TNF-α, IFN-γ) and prevention of a decrease of anti-inflammatory IL-10 cytokine in the CSF and ischemic lesion periphery, consequently restoring the pro-/anti-inflammatory balance [121].

The intranasal administration of Guo was studied in models of permanent focal cerebral ischemia [161,163]. One of the major advantages of IN over the systemic administration route was the rapid penetration to the CNS (5 min post-ischemia), wide distribution in the CSF, decreased influence of systemic metabolism, and lower effective dose. [161]. However, IN Guo administered within 1–3 h postischemia was ineffective in infarct volume reduction. Remarkably IN Guo reduced mitochondrial dysfunction in the penumbra, which positively correlated with neurological outcome [161]. This observation is in line with the correlation of a decrease in oxidative stress markers and sensorimotor recovery observed after systemic administration of Guo [121,178]. Recently, a study by Müller et al. reinforced the safety and long-lasting neuroprotective effects of intranasal Guo. IN Guo administration up to 3 h after ischemic insult improved short-term, poststroke motor deficits and promoted long-term recovery. Using quantitative EEG (qEEG) it was demonstrated that Guo caused a decrease in the global state of synchrony (hyperexcitability) in both hemispheres, thus improving the functional state of the brain affected directly by ischemic insult and its more distant parts. Notably, the authors have shown for the first time that Guo partially prevents the disruption in BBB integrity induced by stroke. The exact nature of the interaction between Guo and BBB is not fully understood but may include the modulation of the survival/apoptotic PI3k/Caspase 3 pathway [163].

Additionally, the effects of Guo were also studied in the rodent model of MCAo followed by 5.5 h reperfusion. Guo administration 5 min before reperfusion or up to 30 min postreperfusion resulted in a dose-dependent decrease in infarct volume and neurological improvement [178]. The 16 mg/kg dose of Guo decreased the infarct volume by approximately 85%, whereas a concentration of 8 mg/kg resulted in a decrease of nearly 60%. Therefore, the dose dependence of Guo in a model of reperfusion is more expressed compared to permanent MCAo models of stroke, where there was no statistical difference between doses of 8 mg/kg and 16mg/kg [176], thus implying the differences in the neuroprotective mechanisms of Guo in permanent ischemia and reperfusion injury. In contrast to the ischemic period, the in vivo neuroprotective effect of Guo in reperfusion injury does not depend on the activation of ER stress pathways, prosurvival modification of calcium homeostasis, or modulation of glutamate uptake [178,183]. Interestingly, there were no observed sustained increases in prosurvival cysteine protease m-calpain levels compared to previous models of permanent MCAo, in which a Guo-mediated increase in m-calpain was implied in protection against necrotic/apoptotic cell death [176,184].

The administration of Guo in a narrow time window directly after or shortly before ischemic insult could be extremely challenging in the clinical environment. Therefore, delayed administration of guanosine after ischemia was studied in a rodent model of photothrombotic stroke (PT) [185]. Administration of guanosine 24 h after ischemic insult induced by PT resulted in remarkable improvement in neurological outcome, starting from 14 days poststroke. However, Guo's treatment failed to reduce the infarct volume measured on day 7. Additionally, no significant neurological improvement in the acute phase was observed. Delayed Guo administration enhanced the endogenous neural progenitor cell proliferation in the peri-infarct region. This observation was also supported by the Guo-mediated major increase in BDNF and VEGF in the ipsilateral hemisphere after PT [185]. BDNF—a major traditional neurotrophic agent—and VEGF—a crucial endothelial growth factor—both play a pivotal role in the poststroke interplay between angiogenesis and neurogenesis by stimulating neuronal plasticity and enhancing neural stem cell (NSCs) migration [130,186,187]. Notably, the events of poststroke neurogenesis and angiogenesis are tightly linked and mutually regulated. In other words, the process of post-stroke vessel formation in the peri-infarct region enhances neuroblast adhesion and migration, while the secretion of VEGF by activated endothelial cells and NSCs acts both as an angiogenesis-stimulating factor and a strong neurotrophin [186]. Together, this evidence suggests that Guo promotes poststroke neurogenesis and angiogenesis, and its effect is neurorestorative rather than neuroprotective when administered in a delayed time interval. This observation also reinforces the concept of the neuroregenerative and neuritogenic effects of extracellular Guo observed in vivo and in neural cell lines [19,31,66,188].

Table 1. The comparison of in vivo studies evaluating the neuroprotective effects of guanosine in ischemic stroke models focused on proposed molecular mechanisms and final outcomes.

Experimental Animal	Experimental Model	Route of Administration	Proposed Mechanism/s of Neuroprotection	Outcome/Guanosine Mediated Effects	Reference
Adult male Wistar Rat	MCAo	I.p.		1. Significantly smaller infarct volume in Guo group compared to control. 2. Major improvement in gait and spontaneous activity in Guo group. 3. No difference in number of cells undergoing apoptosis in penumbra region compared to control.	Chang et al. (2008) [175]
Adult male Sprague Dawley rats	MCAo	I.p. Intracortical Injection	Guo-induced increase in m-calpain level, preventing the necrotic cell death in ischemic area.	1. Significant decrease in infarct volume after 3 days. 2. Significant decrease in infarct volume 6 h after preconditioning with 4mg/kg of Guo. 3. Improvement in motor deficits on day one, two and three after Guo treatment. 4. Increase in m-calpain level in ischemic area. 1. Significant reduction of infarct volume in Guo group.	Rathbone et al. (2011) [176]
Adult male Sprague Dawley rats	MCAo with reperfusion	I.p.	Guo-induced inhibition of proinflammatory events induced by reperfusion. Inhibition of IL-8 release.	1. Time and dose-dependent significant reduction of infarct volume after reperfusion period. 2. Decrease in infarct volume after 24 h following preconditioning 5 min prior to reperfusion. 3. Narrow therapeutic window of Guo administration between 0 and 30 min after reperfusion.	Conell et al. (2013) [178]
Adult male Wistar Rat	Focal thermocoagulation in motor and sensorimotor cortices	I.p.	Guo-induced modulation of oxidative stress response system. Guo-induced glutamate uptake and intracellular conversion to glutamine.	1. Partial recovery of impaired limb function in cylinder test after 24 h, maintained for 15 days. 2. Significant decrease in infarct volume. 3. Prevention of ROS and NO levels increase in ischemic area. 4. Increase in SOD and expression and activity. 5. Increased CAT activity. 6. Restoration of GLT -1 expression. 7. Increased GS activity in ischemic region. 8. Partial reversal of decreased vitamin C level.	Hansel et al. (2014) [177]

Table 1. *Cont.*

Experimental Animal	Experimental Model	Route of Administration	Proposed Mechanism/s of Neuroprotection	Outcome/Guanosine Mediated Effects	Reference
Adult male Wistar Rat	Focal thermocoagulation in motor and sensorimotor cortices	I.p.	Guo-mediated restoration of anti-/proinflammatory balance, prevention of inflammatory cell infiltration.	1. Significant improvement in motor performance in cylinder test, maintained for 15 days. 2. Significant decrease in infarct volume and decrease in number of degenerated cells in penumbra after 24 h. 3. Reduced infiltration of microglia and peripheral immune cells in the periphery of ischemic lesion. 4. Decrease of IL-1, IL-6, TNF-α, and IFN-γ levels. Increase of IL-10 levels.	Hansel et al. (2015) [121]
Adult male Wistar Rat	Focal thermocoagulation in motor and sensorimotor cortices	I.n.	Guo-mediated improvement of mitochondrial status in penumbra.	1. Intranasal Guo administration providing almost immediate (5 min) delivery to the CNS, presenting higher CSF Guo concentrations compared to systemic administration. 2. Significant improvement in symmetry rate in cylinder test. 3. Decrease in infarct volume after 48 h only, when Guo administered immediately after ischemia. 4. Significantly reduced mitochondrial dysfunction in penumbral region after Guo treatment 3 h postischemia. 5. Correlation between mitochondrial status and motor performance of rats.	Ramos et al. (2016) [161]
Adult male C57BL/6J wild-type mice	PT	I.p.	Guo-induced increase in VEGF and BDNF enhancing poststroke angiogenesis and neurogenesis.	1. No decrease in infarct volume after delayed (24 h) administration of Guo. 2. Significant improvement in forelimb function 14 and 28 days postischemia. 3. Proliferation of neural progenitor cells and enhanced differentiation into mature neural cells at all poststroke time intervals. 4. Increased angiogenesis in peri-infarct area. 5. Increased expression of VEGF and BDNF in ischemic brain at 14 days postischemia.	Deng et al. (2017) [185]
Adult female and male Wistar Rat	Focal thermocoagulation in motor and sensorimotor cortices	I.p.		1. Full improvement in forelimb function of female rats in cylinder task, after Guo administration observed earlier in estrogenous group in comparison to only partial (60%) recovery in male subgroup.	Teixeira et al. (2018) [172]

Table 1. *Cont.*

Experimental Animal	Experimental Model	Route of Administration	Proposed Mechanism/s of Neuroprotection	Outcome/Guanosine Mediated Effects	Reference
Adult male Wistar Rat	Focal thermocoagulation in motor and sensorimotor cortices	I.n.	Guo-mediated prevention of disruption in BBB integrity.	1. Partial recovery of impaired limb function in cylinder test directly after ischemia in Guo treated group. 2. Long term improvement in motor deficits. 3. Guo-mediated prevention of apoptotic cell death in ischemic area. 4. Maintenance of BBB integrity after ischemic insult in Guo treated animals. 5. Quantitative EEG study results: partial prevention of global, bilateral state of synchronicity (hyperexcitability) induced by ischemic in Guo treated animals.	Müller et al. (2020) [163]

Abbreviations: Guo—Guanosine; I.p.—Intraperitoneal; I.n.—Intranasal; PT—Photothrombotic Stroke; MCAo—Middle Cerebral Artery Occlusion; ROS—Reactive Oxygen Species; SOD—Superoxide Dismutase; CAT—Catalase; GLT -1—Glutamate Transporter 1; GS—Glutamine Synthetase; VEGF—Vascular Endothelial Growth Factor; BNDF—Brain Derived Neurotrophic Factor; BBB—Blood–Brain Barrier; EEG—Electroencephalography.

5. Current Challenges and Limitations of Guanosine Application in Ischemic Stroke

The pathophysiological processes underlying stroke are driven by the complex interactions between neurons, glial cells, vasculature, immune cells, leukocytes, and matrix components, all participating in the processes of brain injury and neuroregeneration. Currently, the neuroprotective mechanisms of guanosine have been studied mainly concerning pathophysiological processes occurring in neurons and glial cells: for instance, glutamate excitotoxicity or neuroinflammatory response. However, the interactions between guanosine and other components of ischemic stroke are poorly understood. Therefore, we propose to investigate: (i) the effect of Guo on ischemia-induced endothelial activation; (ii) the relationship between Guo and BBB permeability and integrity; (iii) the properties of Guo in the context of peripheral immune cell activation and infiltration to the ischemic area; and (iv) the interactions between guanosine and extracellular purine nucleoside phosphorylase (PNP).

Based on the most recent evidence, the Guo mechanism of action is tightly connected to interactions with cell-surface adenosine receptors. Guo effects are mediated by A1 activation and negative modulation of the A2AR receptor. Moreover, Guo requires both A1R and A2AR coexpression in the form of the A1R-A2AR heteromer [189]. Adenosine receptors are broadly distributed in brain vessels, platelets, and neutrophilic granulocytes and regulate every step of endothelial-related inflammatory processes and vasodilatation [190,191]. Therefore, Guo, through interactions with AR and the proposed guanosine–adenosine interactions, may mitigate the endothelial activation and improve the cerebral microcirculation and thus the oxygen and substrate supply to the ischemic tissue after recanalization. However, this concept requires further evaluation, for instance, in models of in vitro endothelial cell cultures subjected to OGD.

Furthermore, the interaction between Guo and the BBB requires further evaluation. It was demonstrated that Guo can penetrate the BBB after systemic or intranasal administration most probably through equilibrative nucleoside transporters (ENT) [121,163]. Most recently Muller et al. introduced the concept of the in vivo modulation of BBB integrity by Guo [162]. This observation is especially important because ischemic stroke augments the BBB permeability and promotes the entry of immune cells and soluble inflammatory macromolecules into the CNS, thus aggravating the cellular response initiated by ischemia [192]. Moreover, the observed effect is reciprocal to adenosine interactions with BBB, which promote BBB permeability by signaling through A_1 and A_2 ARs [190,191]. Future studies may evaluate in detail the interactions between Guo and BBB and their dependency on Guo-mediated A_1 activation and negative modulation of the A_2AR receptor.

After an acute stroke, multiple immune cells systematically enter the brain parenchyma. Shortly after an ischemic insult, there is a rapid increase in microglia and peripheral immune cells (including dendritic cells, monocytes/macrophages, and neutrophils) that infiltrate within 1–7 days poststroke, resulting in further neuronal damage [193]. Concerning Guo, it was able to suppress the activation of microglia and the infiltration of polymorphonuclear granulocytes and monocytes/macrophages into the ischemic brain region, induced by the breakdown of the BBB [121]. However, the exact nature of this interaction remains elusive, and it is unclear whether it depends on the direct interaction between Guo and the BBB or a receptor-mediated interaction with peripheral immune cells, regarding the fact that neutrophils, macrophages/monocytes, and lymphocytes express a full spectrum of AR, presenting an accessible target for modulation by Guo [190,194]. On the contrary, Guo also activates specific subtypes of Toll-like receptors, TLR 2 and TLR 4, which take part in the activation of the immune system [195,196]. These possibly conflicting mechanisms should be evaluated in the context of interactions between Guo and cellular components of the immune system in ischemic stroke.

Another limitation is that the current understanding of interactions between extracellular PNP and guanosine is incomplete. Astrocytes, microglia, and cerebellar granule neurons constitutively release the intracellular PNP to the extracellular compartment, which rapidly depletes the extracellular pool of guanosine [20]. Apart from this, there is a high activity of

PNP in blood plasma [22]. PNP may be also released from lysed erythrocytes, for instance, during clot lysis in the cerebral artery. If so, the activity of extracellular PNP may reduce the pool of bioavailable exogenously administered Guo, thus hampering Guo from entering the penumbra in the CNS. This issue should be addressed in the future. One of the possible solutions to this issue may include the administration of PNP inhibitors together with Guo.

6. Clinical Perspective

Thrombectomy and thrombolysis are the treatments of choice for ischemic stroke. Both are focused on the elimination of the direct cause of ischemic damage, i.e., the blood clot occluding a specific blood vessel. The most recent advances in this category are extended thrombolysis and thrombectomy protocols, which enable patients to benefit from these treatment options far beyond the standard time window. However, clinical observations show a vast diversity of outcomes found in patients with the same vessel occlusion and treated within a similar time window. These observations indicate that the current treatment methods are often insufficient and that many pathological processes that drive postischemic deterioration are beyond the scope of contemporary stroke patients' care. We propose the concept of Guo-based therapy as a pharmacologically achieved neuroprotection, which would supplement recanalization-oriented treatment. The effects of Guo in such treatment regimens can be bidirectional. First of all, Guo administered in a short-time interval can directly rescue the tissue at risk in the penumbra from the effects of inflammation, oxidative stress, and glutamate toxicity and notably ameliorate the reperfusion injury caused by recanalization. Secondly, Guo can amplify and augment endogenous processes of neuroplasticity and neuroregeneration to support recovery and reduce the rate of poststroke complications.

7. Conclusions

In the present article, we comprehensively summarized the recent advances in the neuroprotective effects of Guo in ischemic stroke. Guo is a potent physiological neuromodulator, which takes part in a "backup" endogenous, restorative system, which protects neural cells from consequences of ischemia/hypoxia. Guanosine is a pleiotropic neuroprotectant in ischemic stroke: in the acute phase of ischemic stroke, Guo-induced neuroprotection is facilitated via antioxidative and anti-inflammatory actions putatively targeted against the oxidative stress and neuroinflammation elicited primarily by hypoxia and secondarily by reperfusion injury. In the chronic phase of ischemic stroke, Guo promotes poststroke neurogenesis and angiogenesis, thus stimulating neuronal plasticity and restoring neuronal function. Guo remains an orphan modulator; thus, its neuroprotective effects are mediated mainly by interactions with AR, glutaminergic receptors/transporters, and ionic channels involved in the process of anoxic depolarization. Furthermore, Guo activates the prosurvival pathways, most notably the MAPK signaling module, which takes part in the regulation of multiple modalities involved in oxygen sensing. Moreover, data collected through studies exploring guanosine's pharmacokinetic properties also look promising. Guo can reach the CNS via systemic, oral, and intranasal routes, causing little to no side effects. This is the major advantage of Guo over adenine-based purines, specifically adenosine, which causes decreased heart rate, blood pressure, and sedation after systemic administration. Nevertheless, due to the short half-life of Guo, the perfect method of administration is still under debate. Overall, based on current evidence, future studies should unravel the precise mechanisms related to the neuroprotective effects of Guo concerning the neurovascular unit and the long-term effect of Guo administration. To address the complex machinery involved in the promotion of ischemic damage in the penumbra, a pleiotropic agent acting on many different pathophysiological processes is greatly desirable. We propose further evaluating the therapeutic potential of Guo in ischemic stroke supportive treatment in human participants.

Author Contributions: Project administration: R.T.S., K.C. and M.O.; Conceptualization: M.O., K.C. and P.K.; Original manuscript preparation: K.C., M.O. and W.N.; Graphic and table preparation: K.C. and M.O.; Supervision: R.T.S. and P.K. All authors have read and agreed to the published version of the manuscript.

Funding: This research received no external funding.

Institutional Review Board Statement: Not applicable.

Informed Consent Statement: Not applicable.

Data Availability Statement: Not applicable.

Acknowledgments: We wish to acknowledge the assistance of Kamil Chwojnicki MD, PhD, Department of Neurology, Clinical Academic Centre, with the clinical aspects of the manuscript.

Conflicts of Interest: The authors declare no conflict of interest. The funders had no role in the design of the study; in the collection, analyses, or interpretation of data; in the writing of the manuscript; or in the decision to publish the results.

References

1. Donkor, E.S. Stroke in the 21st Century: A Snapshot of the Burden, Epidemiology, and Quality of Life. *Stroke Res. Treat.* **2018**, *2018*. [CrossRef]
2. Gorelick, P.B. The global burden of stroke: Persistent and disabling. *Lancet Neurol.* **2019**, *18*, 417–418. [CrossRef]
3. Phipps, M.S.; Cronin, C.A. Management of acute ischemic stroke. *BMJ* **2020**, *368*. [CrossRef] [PubMed]
4. Virani, S.S.; Alonso, A.; Aparicio, H.J.; Benjamin, E.J.; Bittencourt, M.S.; Callaway, C.W.; Carson, A.P.; Chamberlain, A.M.; Cheng, S.; Delling, F.N.; et al. Heart Disease and Stroke Statistics—2021 Update. *Circulation* **2021**, *143*, e254–e743. [CrossRef]
5. Kuriakose, D.; Xiao, Z. Pathophysiology and treatment of stroke: Present status and future perspectives. *Int. J. Mol. Sci.* **2020**, *21*, 7609. [CrossRef]
6. Etherton, M.R.; Gadhia, R.R.; Schwamm, L.H. Thrombolysis beyond 4.5 h in Acute Ischemic Stroke. *Curr. Neurol. Neurosci. Rep.* **2020**, *20*, 1–8. [CrossRef]
7. Baron, J.C. Protecting the ischaemic penumbra as an adjunct to thrombectomy for acute stroke. *Nat. Rev. Neurol.* **2018**, *14*, 325–337. [CrossRef]
8. Nogueira, R.G.; Jadhav, A.P.; Haussen, D.C.; Bonafe, A.; Budzik, R.F.; Bhuva, P.; Yavagal, D.R.; Ribo, M.; Cognard, C.; Hanel, R.A.; et al. Thrombectomy 6 to 24 Hours after Stroke with a Mismatch between Deficit and Infarct. *N. Engl. J. Med.* **2018**, *378*, 11–21. [CrossRef]
9. Back, T.; Hemmen, T.; Schüler, O.G. Lesion evolution in cerebral ischemia. *J. Neurol.* **2004**, *251*, 388–397. [CrossRef]
10. Goenka, L.; Uppugunduri Satyanarayana, C.R.S.S.K.; George, M. Neuroprotective agents in Acute Ischemic Stroke—A Reality Check. *Biomed. Pharm. Ther.* **2019**, *109*, 2539–2547. [CrossRef]
11. Taylor, C.W. The role of G proteins in transmembrane signalling. *Biochem. J.* **1990**, *272*, 1–13. [CrossRef]
12. Ciccarelli, R.; Ballerini, P.; Sabatino, G.; Rathbone, M.P.; D'Onofrio, M.; Caciagli, F.; Di Iorio, P. Involvement of astrocytes in purine-mediated reparative processes in the brain. *Int. J. Dev. Neurosci.* **2001**, *19*, 395–414. [CrossRef]
13. Schmidt, A.P.; Lara, D.R.; Souza, D.O. Proposal of a guanine-based purinergic system in the mammalian central nervous system. *Pharmacol. Ther.* **2007**, *116*, 401–416. [CrossRef]
14. Gualix, J.; Pintor, J.; Miras-Portugal, M.T. Characterization of Nucleotide Transport into Rat Brain Synaptic Vesicles. *J. Neurochem.* **1999**, *73*, 1098–1104. [CrossRef]
15. Santos, T.G.; Souza, D.O.; Tasca, C.I. GTP uptake into rat brain synaptic vesicles. *Brain Res.* **2006**, *1070*, 71–76. [CrossRef]
16. Ciccarelli, R.; Iorio, P.D.I.; Giuliani, P.; Alimonte, I.D.; Ballerini, P.; Caciagli, F.; Rathbone, M.P. Rat Cultured Astrocytes Release Guanine-Based Purines in Basal Conditions and After Hypoxia/Hypoglycemia. *Glia* **1999**, *98*, 93–98. [CrossRef]
17. Dos Santos-Rodrigues, A.; Grañé Boladeras, N.; Bicket, A.; Coe, I.R. Nucleoside transporters in the purinome. *Neurochem. Int.* **2014**, *73*, 229–237. [CrossRef]
18. Zimmermann, H.; Braun, N. Extracellular metabolism of nucleotides in the nervous system. *J. Auton. Pharm. Ther.* **1996**, *16*, 397–400. [CrossRef]
19. Tasca, C.I.; Lanznaster, D.; Oliveira, K.A.; Fernández-dueñas, V.; Ciruela, F.; Jean, R.; Aubert, P. Neuromodulatory Effects of Guanine-Based Purines in Health and Disease. *Front. Cellular Neurosci.* **2018**, *12*, 1–14. [CrossRef]
20. Peña-Altamira, L.E.; Polazzi, E.; Giuliani, P.; Beraudi, A.; Massenzio, F.; Mengoni, I.; Poli, A.; Zuccarini, M.; Ciccarelli, R.; Di Iorio, P.; et al. Release of soluble and vesicular purine nucleoside phosphorylase from rat astrocytes and microglia induced by pro-inflammatory stimulation with extracellular ATP via P2X7 receptors. *Neurochem. Int.* **2018**, *115*, 37–49. [CrossRef]
21. Giuliani, P.; Zuccarini, M.; Buccella, S.; Peña-Altamira, L.E.; Polazzi, E.; Virgili, M.; Monti, B.; Poli, A.; Rathbone, M.P.; Iorio, P.D.; et al. Evidence for purine nucleoside phosphorylase (PNP) release from rat C6 glioma cells. *J. Neurochem.* **2017**, *141*, 208–221. [CrossRef]

22. Giuliani, P.; Zuccarini, M.; Buccella, S.; Rossini, M.; D'Alimonte, I.; Ciccarelli, R.; Marzo, M.; Marzo, A.; Di Iorio, P.; Caciagli, F. Development of a new HPLC method using fluorescence detection without derivatization for determining purine nucleoside phosphorylase activity in human plasma. *J. Chromatogr. B* **2016**, *1009–1010*, 114–121. [CrossRef]
23. Miyamoto, S.; Ogava, H.; Shiraki, H.; Nakagava, H. Guanine Deaminase from Rat Brain. Purification, Characteristics, and Contribution to Ammoniagenesis in the Brain. *J. Biochem.* **1982**, *91*, 167–176. [CrossRef]
24. Su, C.; Elfeki, N.; Ballerini, P.; D'Alimonte, I.; Bau, C.; Ciccarelli, R.; Caciagli, F.; Gabriele, J.; Jiang, S. Guanosine improves motor behavior, reduces apoptosis, and stimulates neurogenesis in rats with parkinsonism. *J. Neurosci. Res.* **2009**, *87*, 617–625. [CrossRef]
25. Decker, H.; Piermartiri, T.C.B.; Nedel, C.B.; Romão, L.F.; Francisco, S.S.; Dal-Cim, T.; Boeck, C.R.; Moura-Neto, V.; Tasca, C.I. Guanosine and GMP increase the number of granular cerebellar neurons in culture: Dependence on adenosine A2A and ionotropic glutamate receptors. *Purinergic Signal.* **2019**, *15*, 439–450. [CrossRef]
26. Decker, H.; Francisco, S.S.; Roma, L.F.; Boeck, C.R.; Moura-neto, V.; Tasca, C.I. Guanine Derivatives Modulate Extracellular Matrix Proteins Organization and Improve Neuron-Astrocyte. *J. Neurosci. Res.* **2007**, *1951*, 1943–1951. [CrossRef]
27. Dal-Cim, T.; Martins, W.C.; Santos, A.R.S.; Tasca, C.I. Guanosine is neuroprotective against oxygen/glucose deprivation in hippocampal slices via large conductance Ca2+-activated K+ channels, phosphatidilinositol-3 kinase/protein kinase B pathway activation and glutamate uptake. *NSC* **2011**, *183*, 212–220. [CrossRef]
28. Su, C.; Wang, P.; Jiang, C.; Ballerini, P.; Caciagli, F.; Rathbone, M.P.; Jiang, S. Guanosine promotes proliferation of neural stem cells through cAMP-CREB pathway. *J. Biol. Regul. Homeost. Agents* **2013**, *27*, 673–680. [PubMed]
29. Middlemiss, P.J.; Gysbers, J.W.; Rathbone, M.P. Extracellular guanosine and guanosine-5′-triphosphate increase: NGF synthesis and release from cultured mouse neopallial astrocytes. *Brain Res.* **1995**, *677*, 152–156. [CrossRef]
30. Gysbers, J.W.; Rathbone, M.P. GTP and guanosine synergistically enhance NGF-induced neurite outgrowth from PC12 cells. *Int. J. Dev. Neurosci. Off. J. Int. Soc. Dev. Neurosci.* **1996**, *14*, 19–34. [CrossRef]
31. Piermartiri, T.C.B.B.C.B.; Santos, B.; Barros-aragão, F.G.Q.Q.G.Q.; Prediger, R.D.D.; dos Santos, B.; Barros-aragão, F.G.Q.Q.G.Q.; Prediger, R.D.D.; Tasca, C.I.; Santos, B.; Barros-aragão, F.G.Q.Q.G.Q.; et al. Guanosine Promotes Proliferation in Neural Stem Cells from Hippocampus and Neurogenesis in Adult Mice. *Mol. Neurobiol.* **2020**, *57*, 3814–3826. [CrossRef]
32. Davis, R.J. The mitogen-activated protein kinase signal transduction pathway. *J. Biol. Chem.* **1993**, *268*, 14553–14556. [CrossRef]
33. Hu, J.Y.; Chen, Y.; Schacher, S. Multifunctional role of protein kinase C in regulating the formation and maturation of specific synapses. *J. Neurosci.* **2007**, *27*, 11712–11724. [CrossRef]
34. Lin, W.; Wang, S.-M.; Huang, T.-F.; Fu, W.-M. Differential regulation of fibronectin fibrillogenesis by protein kinases A and C. *Connect. Tissue Res.* **2002**, *43*, 22–31. [CrossRef]
35. Yang, R.-S.; Tang, C.-H.; Ling, Q.-D.; Liu, S.-H.; Fu, W.-M. Regulation of fibronectin fibrillogenesis by protein kinases in cultured rat osteoblasts. *Mol. Pharm. Ther.* **2002**, *61*, 1163–1173. [CrossRef]
36. Nicole, O.; Pacary, E. Camkiiβ in neuronal development and plasticity: An emerging candidate in brain diseases. *Int. J. Mol. Sci.* **2020**, *21*, 7272. [CrossRef]
37. Gysbers, J.W.; Rathbone, M.P. Guanosine enhances NGF-stimulated neurite outgrowth in PC12 cells. *Neuroreport* **1992**, *3*, 997–1000. [CrossRef]
38. Bau, C.; Middlemiss, P.J.; Hindley, S.; Jiang, S.; Ciccarelli, R.; Caciagli, F.; DiIorio, P.; Werstiuk, E.S.; Rathbone, M.P. Guanosine stimulates neurite outgrowth in PC12 cells via activation of heme oxygenase and cyclic GMP. *Purinergic Signal.* **2005**, *1*, 161–172. [CrossRef]
39. Thauerer, B.; Zur Nedden, S.; Baier-Bitterlich, G. Vital role of protein kinase C-related kinase in the formation and stability of neurites during hypoxia. *J. Neurochem.* **2010**, *113*, 432–446. [CrossRef]
40. Gudi, T.; Chen, J.C.; Casteel, D.E.; Seasholtz, T.M.; Boss, G.R.; Pilz, R.B. cGMP-dependent protein kinase inhibits serum-response element-dependent transcription by inhibiting rho activation and functions. *J. Biol. Chem.* **2002**, *277*, 37382–37393. [CrossRef]
41. Guarnieri, S.; Pilla, R.; Morabito, C.; Sacchetti, S.; Mancinelli, R.; Fanò, G.; Mariggiò, M.A. Extracellular guanosine and GTP promote expression of differentiation markers and induce S-phase cell-cycle arrest in human SH-SY5Y neuroblastoma cells. *Int. J. Dev. Neurosci.* **2009**, *27*, 135–147. [CrossRef]
42. Uemura, Y.; Miller, J.M.; Matson, W.R.; Beal, M.F. Neurochemical analysis of focal ischemia in rats. *Stroke* **1991**, *22*, 1548–1553. [CrossRef] [PubMed]
43. Regner, A.; Crestana, R.E.; Friedman, G.; Chemale, I.; Souza, D. Guanine nucleotides are present in human CSF. *Neuroreport* **2000**, *8*, 3771–3774. [CrossRef] [PubMed]
44. Kovács, Z.; Kékesi, K.A.; Dobolyi, Á.; Lakatos, R.; Juhász, G. Absence epileptic activity changing effects of non-adenosine nucleoside inosine, guanosine and uridine in Wistar Albino Glaxo Rijswijk rats. *Neuroscience* **2015**, *300*, 593–608. [CrossRef] [PubMed]
45. Massari, C.M.; López-Cano, M.; Núñez, F.; Fernández-Dueñas, V.; Tasca, C.I.; Ciruela, F. Antiparkinsonian Efficacy of Guanosine in Rodent Models of Movement Disorder. *Front. Pharm. Ther.* **2017**, *8*, 700. [CrossRef] [PubMed]
46. Lanznaster, D.; Mack, J.M.; Coelho, V.; Ganzella, M.; Almeida, R.F.; Dal-Cim, T.; Hansel, G.; Zimmer, E.R.; Souza, D.O.; Prediger, R.D.; et al. Guanosine Prevents Anhedonic-Like Behavior and Impairment in Hippocampal Glutamate Transport Following Amyloid-β1–40 Administration in Mice. *Mol. Neurobiol.* **2017**, *54*, 5482–5496. [CrossRef] [PubMed]

47. Bellaver, B.; Souza, D.O.D.G.; Bobermin, L.D.; Gonçalves, C.-A.; Souza, D.O.D.G.; Quincozes-Santos, A. Guanosine inhibits LPS-induced pro-inflammatory response and oxidative stress in hippocampal astrocytes through the heme oxygenase-1 pathway. *Purinergic Signal.* **2015**, *11*, 571–580. [CrossRef]
48. Paniz, L.G.; Calcagnotto, M.E.; Pandolfo, P.; Machado, D.G.; Santos, G.F.; Hansel, G.; Almeida, R.F.; Bruch, R.S.; Brum, L.M.; Torres, F.V.; et al. Neuroprotective effects of guanosine administration on behavioral, brain activity, neurochemical and redox parameters in a rat model of chronic hepatic encephalopathy. *Metab. Brain Dis.* **2014**, *29*, 645–654. [CrossRef]
49. Petronilho, F.; Périco, S.R.; Vuolo, F.; Mina, F.; Constantino, L.; Comim, C.M.; Quevedo, J.; Souza, D.O.; Dal-Pizzol, F. Protective effects of guanosine against sepsis-induced damage in rat brain and cognitive impairment. *Brain. Behav. Immun.* **2012**, *26*, 904–910. [CrossRef]
50. Jiang, S.; Khan, M.I.; Lu, Y.; Wang, J.; Buttigieg, J.; Werstiuk, E.S.; Ciccarelli, R.; Caciagli, F.; Rathbone, M.P. Guanosine promotes myelination and functional recovery in chronic spinal injury. *Neuroreport* **2003**, *14*, 2463–2467. [CrossRef]
51. Jiang, S.; Ballerini, P.; Buccella, S.; Giuliani, P.; Jiang, C.; Huang, X.; Rathbone, M.P. Remyelination after chronic spinal cord injury is associated with proliferation of endogenous adult progenitor cells after systemic administration of guanosine. *Purinergic Signal.* **2008**, *4*, 61–71. [CrossRef]
52. Oliveira, K.A.; Dal-Cim, T.A.; Lopes, F.G.; Nedel, C.B.; Tasca, C.I. Guanosine promotes cytotoxicity via adenosine receptors and induces apoptosis in temozolomide-treated A172 glioma cells. *Purinergic Signal.* **2017**, *13*, 305–318. [CrossRef]
53. Molz, S.; Dal-Cim, T.; Tasca, C.I. Guanosine-5'-monophosphate induces cell death in rat hippocampal slices via ionotropic glutamate receptors activation and glutamate uptake inhibition. *Neurochem. Int.* **2009**, *55*, 703–709. [CrossRef]
54. Torres, F.V.; da Silva Filho, M.; Antunes, C.; Kalinine, E.; Antoniolli, E.; Portela, L.V.C.; Souza, D.O.; Tort, A.B.L. Electro-physiological effects of guanosine and MK-801 in a quinolinic acid-induced seizure model. *Exp. Neurol.* **2010**, *221*, 296–306. [CrossRef]
55. Traversa, U.; Bombi, G.; Iorio, P.D.; Ciccarelli, R.; Werstiuk, E.S.; Rathbone, M.P. Specific [3H]-guanosine binding sites in rat brain membranes. *Br. J. Pharm. Ther.* **2002**, *135*, 969–976. [CrossRef]
56. Traversa, U.; Bombi, G.; Camaioni, E.; MacChiarulo, A.; Costantino, G.; Palmieri, C.; Caciagli, F.; Pellicciari, R. Rat brain guanosine binding site: Biological studies and pseudo-Receptor construction. *Bioorganic Med. Chem.* **2003**, *11*, 5417–5425. [CrossRef]
57. Liberto, V.; Di Mudò, G.; Garozzo, R.; Frinchi, M.; Fernandez-dueñas, V.; Iorio, P.; Di Ciccarelli, R.; Caciagli, F.; Condorelli, D.F.; Ciruela, F.; et al. The Guanine-Based Purinergic System: The Tale of An Orphan Neuromodulation. *Front. Pharmacol.* **2016**, *7*, 1–15. [CrossRef]
58. Volpini, R.; Marucci, G.; Buccioni, M.; DalBen, D.; Lambertucci, C.; Lammi, C.; Mishra, R.C.; Thomas, A.; Cristalli, G. Evidence for the existence of a specific gprotein-coupled receptor activated by guanosine. *ChemMedChem* **2011**, *6*, 1074–1080. [CrossRef]
59. Civelli, O.; Reinscheid, R.K.; Zhang, Y.; Wang, Z.; Fredriksson, R.; Schiöth, H.B. G protein-coupled receptor deorphanizations. *Annu. Rev. Pharm. Ther. Toxicol.* **2013**, *53*, 127–146. [CrossRef]
60. Lanznaster, D.; Dal-Cim, T.; Piermartiri, T.C.B.; Tasca, C.I.; Article, R. Guanosine: A Neuromodulator with Therapeutic Potential in Brain Disorders. *Aging Dis.* **2016**, *7*, 657–679. [CrossRef]
61. Borroto-Escuela, D.O.; Brito, I.; Romero-Fernandez, W.; Di Palma, M.; Oflijan, J.; Skieterska, K.; Duchou, J.; Van Craenenbroeck, K.; Suárez-Boomgaard, D.; Rivera, A.; et al. The G protein-coupled receptor heterodimer network (GPCR-HetNet) and its hub components. *Int. J. Mol. Sci.* **2014**, *15*, 8570–8590. [CrossRef]
62. Massari, C.M.; Constantino, L.C.; Marques, N.F.; Binder, L.B.; Valle-León, M.; López-Cano, M.; Fernández-Dueñas, V.; Ciruela, F.; Tasca, C.I. Involvement of adenosine A1 and A2A receptors on guanosine-mediated anti-tremor effects in reserpinized mice. *Purinergic Signal.* **2020**, *16*, 379–387. [CrossRef]
63. Müller, C.E.; Scior, T. Adenosine receptors and their modulators. *Pharm. Acta Helv.* **1993**, *68*, 77–111. [CrossRef]
64. Lanznaster, D.; Massari, C.M.; Marková, V.; Šimková, T.; Duroux, R.; Jacobson, K.A.; Fernández-Dueñas, V.; Tasca, C.I.; Ciruela, F. Adenosine A1-A2A Receptor-Receptor Interaction: Contribution to Guanosine-Mediated Effects. *Cells* **2019**, *8*, 1630. [CrossRef]
65. Almeida, R.F.; Comasseto, D.D.; Ramos, D.B.; Hansel, G.; Zimmer, E.R.; Loureiro, S.O.; Ganzella, M.; Souza, D.O. Guanosine Anxiolytic-Like Effect Involves Adenosinergic and Glutamatergic Neurotransmitter Systems. *Mol. Neurobiol.* **2017**, *54*, 423–436. [CrossRef]
66. Dobrachinski, F.; Gerbatin, R.R.; Sartori, G.; Golombieski, R.M.; Antoniazzi, A.; Nogueira, C.W.; Royes, L.F.; Fighera, M.R.; Porciúncula, L.O.; Cunha, R.A.; et al. Guanosine Attenuates Behavioral Deficits After Traumatic Brain Injury by Modulation of Adenosinergic Receptors. *Mol. Neurobiol.* **2019**, *56*, 3145–3158. [CrossRef]
67. Gerbatin, R.R.; Dobrachinski, F.; Cassol, G.; Soares, F.A.A.; Royes, L.F.F. A 1 rather than A 2A adenosine receptor as a possible target of Guanosine effects on mitochondrial dysfunction following Traumatic Brain Injury in rats. *Neurosci. Lett.* **2019**, *704*, 141–144. [CrossRef] [PubMed]
68. Dal-Cim, T.; Ludka, F.K.K.; Martins, W.C.C.; Reginato, C.; Parada, E.; Egea, J.; López, M.G.; Tasca, C.I.I.; López, M.G.; Tasca, C.I.I. Guanosine controls inflammatory pathways to afford neuroprotection of hippocampal slices under oxygen and glucose deprivation conditions. *J. Neurochem.* **2013**, *126*, 437–450. [CrossRef] [PubMed]
69. Frinchi, M.; Verdi, V.; Plescia, F.; Ciruela, F.; Grillo, M.; Garozzo, R.; Condorelli, D.F.; Iorio, P.; Di Caciagli, F.; Ciccarelli, R.; et al. Guanosine-Mediated Anxiolytic-Like Effect: Interplay with Adenosine A 1 and A 2A Receptors. *Int. J. Mol. Sci.* **2020**, *21*, 9281. [CrossRef] [PubMed]

70. Nwaobi, S.E.; Cuddapah, V.A.; Patterson, K.C.; Randolph, A.C.; Olsen, M.L. The role of glial-specific Kir4. 1 in normal and pathological states of the CNS. *Acta Neuropathol.* **2016**. [CrossRef] [PubMed]

71. Benfenati, V.; Caprini, M.; Nobile, M.; Rapisarda, C.; Ferroni, S. Guanosine promotes the up-regulation of inward rectifier potassium current mediated by Kir4. 1 in cultured rat cortical astrocytes. *J. Neurochem.* **2006**, 430–445. [CrossRef]

72. Dal-cim, T.; Molz, S.; Egea, J.; Parada, E.; Romero, A.; Budni, J.; Martín, M.D.; Saavedra, D.; Tasca, C.I.; López, M.G. Neurochemistry International Guanosine protects human neuroblastoma SH-SY5Y cells against mitochondrial oxidative stress by inducing heme oxigenase-1 via PI3K/Akt/GSK-3 b pathway. *Neurochem. Int.* **2012**, *61*, 397–404. [CrossRef]

73. Chavarria, A.; Perez-H, J.; Garcia, E.; Carrillo-Salgado, C.; Ruiz-Mar, G.; Perez-Tamayo, R. Poster Sessions Tuesday/Wednesday. *J. Neurochem.* **2013**, *125*, 194–280. [CrossRef]

74. Schmidt, A.P.; Böhmer, A.E.; Schallenberger, C.; Antunes, C.; Tavares, R.G.; Wofchuk, S.T.; Elisabetsky, E.; Souza, D.O. Mechanisms involved in the antinociception induced by systemic administration of guanosine in mice. *Br. J. Pharm. Ther.* **2010**, *159*, 1247–1263. [CrossRef]

75. Dal-Cim, T.; Martins, W.C.; Thomaz, D.T.; Coelho, V.; Poluceno, G.G.; Lanznaster, D.; Vandresen-Filho, S.; Tasca, C.I. Neuroprotection Promoted by Guanosine Depends on Glutamine Synthetase and Glutamate Transporters Activity in Hippocampal Slices Subjected to Oxygen/Glucose Deprivation. *Neurotox. Res.* **2016**, *29*, 460–468. [CrossRef]

76. Bettio, L.E.B.; Freitas, A.E.; Neis, V.B.; Santos, D.B.; Ribeiro, C.M.; Rosa, P.B.; Farina, M.; Rodrigues, A.L.S. Guanosine prevents behavioral alterations in the forced swimming test and hippocampal oxidative damage induced by acute restraint stress. *Pharm. Ther. Biochem. Behav.* **2014**, *127*, 7–14. [CrossRef]

77. Bettio, L.E.B.; Cunha, M.P.; Budni, J.; Pazini, F.L.; Oliveira, Á.; Colla, A.R.; Rodrigues, A.L.S. Guanosine produces an antidepressant-like effect through the modulation of NMDA receptors, nitric oxide-cGMP and PI3K/mTOR pathways. *Behav. Brain Res.* **2012**, *234*, 137–148. [CrossRef]

78. Lee, J.M.; Grabb, M.C.; Zipfel, G.J.; Choi, D.W. Brain tissue responses to ischemia. *J. Clin. Investig.* **2000**, *106*, 723–731. [CrossRef]

79. Yoo, A.J.; Hu, R.; Hakimelahi, R.; Lev, M.H.; Nogueira, R.G.; Hirsch, J.A.; González, R.G.; Schaefer, P.W. CT angiography source images acquired with a fast-acquisition protocol overestimate infarct core on diffusion weighted images in acute ischemic stroke. *J. Neuroimaging* **2012**, *22*, 329–335. [CrossRef]

80. Ginsberg, M.D. The new language of cerebral ischemia. *Am. J. Neuroradiol.* **1997**, *18*, 1435–1445.

81. Pan, J.; Konstas, A.-A.; Bateman, B.; Ortolano, G.A.; Pile-Spellman, J. Reperfusion injury following cerebral ischemia: Pathophysiology, MR imaging, and potential therapies. *Neuroradiology* **2007**, *49*, 93–102. [CrossRef] [PubMed]

82. Choi, D.W. Excitotoxic cell death. *J. Neurobiol.* **1992**, *23*, 1261–1276. [CrossRef] [PubMed]

83. Hazell, A.S. Excitotoxic mechanisms in stroke: An update of concepts and treatment strategies. *Neurochem. Int.* **2007**, *50*, 941–953. [CrossRef]

84. Wang, Q.; Tang, X.N.; Yenari, M.A. The inflammatory response in stroke. *J. Neuroimmunol.* **2007**, *184*, 53–68. [CrossRef] [PubMed]

85. Iadecola, C.; Anrather, J. The immunology of stroke: From mechanisms to translation. *Nat. Med.* **2011**, *17*, 796–808. [CrossRef] [PubMed]

86. Martinon, F.; Burns, K.; Tschopp, J. The inflammasome: A molecular platform triggering activation of inflammatory caspases and processing of proIL-beta. *Mol. Cell* **2002**, *10*, 417–426. [CrossRef]

87. Becker, K.J. Inflammation and acute stroke. *Curr. Opin. Neurol.* **1998**, *11*, 45–49. [CrossRef]

88. Stanimirovic, D.; Shapiro, A.; Wong, J.; Hutchison, J.; Durkin, J. The induction of ICAM-1 in human cerebromicrovascular endothelial cells (HCEC) by ischemia-like conditions promotes enhanced neutrophil/HCEC adhesion. *J. Neuroimmunol.* **1997**, *76*, 193–205. [CrossRef]

89. Morioka, T.; Kalehua, A.N.; Streit, W.J. Characterization of microglial reaction after middle cerebral artery occlusion in rat brain. *J. Comp. Neurol.* **1993**, *327*, 123–132. [CrossRef]

90. Danton, G.H.; Dietrich, W.D. Inflammatory Mechanisms after Ischemia and Stroke. *J. Neuropathol. Exp. Neurol.* **2003**, *62*, 127–136. [CrossRef]

91. Hu, X.; Leak, R.K.; Shi, Y.; Suenaga, J.; Gao, Y.; Zheng, P.; Chen, J. Microglial and macrophage polarization—new prospects for brain repair. *Nat. Rev. Neurol.* **2015**, *11*, 56–64. [CrossRef]

92. Baeuerle, P.A.; Henkel, T. Function and activation of NF-kappa B in the immune system. *Annu. Rev. Immunol.* **1994**, *12*, 141–179. [CrossRef]

93. Chariot, A. 20 years of NF-kappaB. *Biochem. Pharm. Ther.* **2006**, *72*, 1051–1053. [CrossRef]

94. Hoffmann, A.; Baltimore, D. Circuitry of nuclear factor kappaB signaling. *Immunol. Rev.* **2006**, *210*, 171–186. [CrossRef]

95. Shirley, R.; Ord, E.N.J.; Work, L.M. Oxidative Stress and the Use of Antioxidants in Stroke. *Antioxidant* **2014**, *3*, 472–501. [CrossRef]

96. Chen, Z.-Q.; Mou, R.-T.; Feng, D.-X.; Wang, Z.; Chen, G. The role of nitric oxide in stroke. *Med. Gas Res.* **2017**, *7*, 194–203. [CrossRef]

97. Thomazi, A.P.P.; Boff, B.; Pires, T.D.D.; Godinho, G.; Battú, C.E.E.; Gottfried, C.; Souza, D.O.O.; Salbego, C.; Wofchuk, S.T.T. Profile of glutamate uptake and cellular viability in hippocampal slices exposed to oxygen and glucose deprivation: Developmental aspects and protection by guanosine. *Brain Res.* **2007**, *8*, 233–240. [CrossRef]

98. Lossi, L.; Merighi, A. Neuronal cell death: Methods and protocols. *Neuronal Cell Death Methods Protoc.* **2014**, *1254*, 1–368. [CrossRef]

99. Oleskovicz, S.P.B.; Martins, W.C.; Leal, R.B.; Tasca, C.I. Mechanism of guanosine-induced neuroprotection in rat hippocampal slices submitted to oxygen–glucose deprivation. *Neurochem. Int.* **2008**, *52*, 411–418. [CrossRef]

100. Pap, M.; Cooper, G.M. Role of glycogen synthase kinase-3 in the phosphatidylinositol 3-Kinase/Akt cell survival pathway. *J. Biol. Chem.* **1998**, *273*, 19929–19932. [CrossRef]

101. Egea, J.; Romero, A.; Barrio, L.; Rodrigues, A.L.S.; Tasca, C.I.; Molz, S.; Dal-Cim, T.; Budni, J.; Martín-de-Saavedra, M.D.; Egea, J.; et al. Neuroprotective effect of guanosine against glutamate-induced cell death in rat hippocampal slices is mediated by the phosphatidylinositol-3 kinase/Akt/ glycogen synthase kinase 3β pathway activation and inducible nitric oxide synthase inhibition. *J. Neurosci. Res.* **2011**, *89*, 1400–1408. [CrossRef]

102. Dal-cim, T.; Poluceno, G.G.; Lanznaster, D.; Oliveira, K.A.; De Nedel, C.B.; Tasca, C.I.; de Oliveira, K.A.; Nedel, C.B.; Tasca, C.I. Guanosine prevents oxidative damage and glutamate uptake impairment induced by oxygen/glucose deprivation in cortical astrocyte cultures: Involvement of A(1) and A(2A) adenosine receptors and PI3K, MEK, and PKC pathways. *Purinergic Signal.* **2019**, *15*, 465–476. [CrossRef] [PubMed]

103. Quincozes-santos, A.; Bobermin, L.D. Gliopreventive effects of guanosine against glucose deprivation in vitro. *Purinergic Signal.* **2013**, 643–654. [CrossRef] [PubMed]

104. Pivonkova, H.; Benesova, J.; Butenko, O.; Chvatal, A.; Anderova, M. Impact of global cerebral ischemia on K+ channel expression and membrane properties of glial cells in the rat hippocampus. *Neurochem. Int.* **2010**, *57*, 783–794. [CrossRef]

105. Steiner, E.; Enzmann, G.U.; Lin, S.; Ghavampour, S.; Hannocks, M.-J.; Zuber, B.; Rüegg, M.A.; Sorokin, L.; Engelhardt, B. Loss of astrocyte polarization upon transient focal brain ischemia as a possible mechanism to counteract early edema formation. *Glia* **2012**, *60*, 1646–1659. [CrossRef]

106. Köller, H.; Schroeter, M.; Jander, S.; Stoll, G.; Siebler, M. Time course of inwardly rectifying K(+) current reduction in glial cells surrounding ischemic brain lesions. *Brain Res.* **2000**, *872*, 194–198. [CrossRef]

107. Quincozes-Santos, A.; Bobermin, L.G.; Souza, D.; Bellaver, B.; Gonçalves, C.-A.; Souza, D. Guanosine protects C6 astroglial cells against azide-induced oxidative damage: A putative role of heme oxygenase 1. *J. Neurochem.* **2014**, *130*. [CrossRef]

108. Gribkoff, V.K.; Starrett, J.E.J.; Dworetzky, S.I.; Hewawasam, P.; Boissard, C.G.; Cook, D.A.; Frantz, S.W.; Heman, K.; Hibbard, J.R.; Huston, K.; et al. Targeting acute ischemic stroke with a calcium-sensitive opener of maxi-K potassium channels. *Nat. Med.* **2001**, *7*, 471–477. [CrossRef]

109. Yamada, K.; Inagaki, N. Neuroprotection by KATP channels. *J. Mol. Cell. Cardiol.* **2005**, *38*, 945–949. [CrossRef]

110. Em, M.; Frizzo, S.; Lara, D.R.; Souza, A.D.; Vargas, C.R.; Salbego, C.G.; Souza, D.O. Guanosine Enhances Glutamate Uptake in Brain Cortical Slices at Normal and Excitotoxic Conditions. *Cell. Molecul. Neurobiol.* **2002**, *22*, 353–363.

111. Em, M.; Schwalm, D.; Frizzo, J.K.K.; Frizzo, M.E.; Schwalm, F.D.; Frizzo, J.K.K.; Soares, F.A.; Souza, D.O. Guanosine Enhances Glutamate Transport Capacity in Brain Cortical Slices. *Cell. Mol. Neurobiol.* **2005**, *25*, 913–921. [CrossRef]

112. Oliveira, D.L.; De Horn, J.F.; Rodrigues, J.M.; Frizzo, M.E.S.; Moriguchi, E.; Souza, D.O.; Wofchuk, S. Quinolinic acid promotes seizures and decreases glutamate uptake in young rats: Reversal by orally administered guanosine. *Brain Res.* **2004**, *1018*, 48–54. [CrossRef]

113. Vinadé, E.R.; Schmidt, A.P.; Frizzo, M.E.S.; Portela, L.V.; Soares, F.A.; Schwalm, F.D.; Elisabetsky, E.; Izquierdo, I.; Souza, D.O. Effects of chronic administered guanosine on behavioral parameters and brain glutamate uptake in rats. *J. Neurosci. Res.* **2005**, *79*, 248–253. [CrossRef]

114. Tavares, R.G.; Schmidt, A.P.; Abud, J.; Tasca, C.I.; Souza, D.O. In vivo quinolinic acid increases synaptosomal glutamate release in rats: Reversal by guanosine. *Neurochem. Res.* **2005**, *30*, 439–444. [CrossRef]

115. Lai, T.W.; Zhang, S.; Wang, Y.T. Excitotoxicity and stroke: Identifying novel targets for neuroprotection. *Prog. Neurobiol.* **2014**, *115*, 157–188. [CrossRef]

116. Jia, M.; Njapo, S.A.N.; Rastogi, V.; Hedna, V.S. Taming glutamate excitotoxicity: Strategic pathway modulation for neuroprotection. *Cns Drugs* **2015**, *29*, 153–162. [CrossRef]

117. Belov Kirdajova, D.; Kriska, J.; Tureckova, J.; Anderova, M. Ischemia-Triggered Glutamate Excitotoxicity from the Perspective of Glial Cells. *Front. Cell. Neurosci.* **2020**, *14*, 1–27. [CrossRef]

118. Nonose, Y.; Pieper, L.Z.; Silva, J.S.; Longoni, A.; Apel, R.V.; Meira-martins, L.A. Guanosine enhances glutamate uptake and oxidation, preventing oxidative stress in mouse hippocampal slices submitted to high glutamate levels. *Brain Res.* **2020**, *1748*, 147080. [CrossRef]

119. Swanson, R.A. Astrocyte glutamate uptake during chemical hypoxia in vitro. *Neurosci. Lett.* **1992**, *147*, 143–146. [CrossRef]

120. Swanson, R.A.; Farrell, K.; Simon, R.P. Acidosis causes failure of astrocyte glutamate uptake during hypoxia. *J. Cereb. Blood Flow Metab.* **1995**, *15*, 417–424. [CrossRef]

121. Hansel, G.; Tonon, A.C.C.; Guella, F.L.L.; Pettenuzzo, L.F.F.; Duarte, T.; Duarte, M.M.M.F.; Oses, J.P.P.; Achaval, M.; Souza, D.O.; Maria, M.; et al. Guanosine Protects Against Cortical Focal Ischemia. Involvement of Inflammatory Response. *Mol. Neurobiol.* **2015**, *52*, 1791–1803. [CrossRef] [PubMed]

122. Gegelashvili, G.; Dehnes, Y.; Danbolt, N.C.; Schousboe, A. The high-affinity glutamate transporters GLT1, GLAST, and EAAT4 are regulated via different signalling mechanisms. *Neurochem. Int.* **2000**, *37*, 163–170. [CrossRef]

123. Sims, K.D.; Straff, D.J.; Robinson, M.B. Platelet-derived growth factor rapidly increases activity and cell surface expression of the EAAC1 subtype of glutamate transporter through activation of phosphatidylinositol 3-kinase. *J. Biol. Chem.* **2000**, *275*, 5228–5237. [CrossRef] [PubMed]

124. Castilho, R.F.; Kowaltowski, A.J.; Vercesi, A.E. 3,5,3′-Triiodothyronine Induces Mitochondrial Permeability Transition Mediated by Reactive Oxygen Species and Membrane Protein Thiol Oxidation. *Arch. Biochem. Biophys.* **1998**, *354*, 151–157. [CrossRef]

125. Jaiswal, M.K.; Zech, W.-D.; Goos, M.; Leutbecher, C.; Ferri, A.; Zippelius, A.; Carrì, M.T.; Nau, R.; Keller, B.U. Impairment of mitochondrial calcium handling in a mtSOD1 cell culture model of motoneuron disease. *Bmc Neurosci.* **2009**, *10*, 64. [CrossRef]

126. Christophe, M.; Nicolas, S. Mitochondria: A target for neuroprotective interventions in cerebral ischemia-reperfusion. *Curr. Pharm. Des.* **2006**, *12*, 739–757. [CrossRef]

127. Courtes, A.A.; Carvalho, N.R.; De Gonçalves, D.F.; Duarte, D.; Carvalho, P.; Dobrachinski, F.; Luis, J.; Onofre, D.; Souza, G.D.; Alexandre, F.; et al. Biomedicine & Pharmacotherapy Guanosine protects against Ca 2 + -induced mitochondrial dysfunction in rats. *Biomed. Pharm. Ther.* **2019**, *111*, 1438–1446. [CrossRef]

128. Wang, W.; Gong, G.; Wang, X.; Wei-LaPierre, L.; Cheng, H.; Dirksen, R.; Sheu, S.-S. Mitochondrial Flash: Integrative Reactive Oxygen Species and pH Signals in Cell and Organelle Biology. *Antioxid. Redox Signal.* **2016**, *25*, 534–549. [CrossRef]

129. Thomaz, D.T.; Dal-cim, T.A.; Martins, W.C.; Cunha, M.P.; Lanznaster, D.; Bem, A.F. De Guanosine prevents nitroxidative stress and recovers mitochondrial membrane potential disruption in hippocampal slices subjected to oxygen/glucose deprivation. *Purinergic Signal.* **2016**, 707–718. [CrossRef]

130. Gao, J.; Wang, L.; Liu, J.; Xie, F.; Su, B.; Wang, X. Abnormalities of Mitochondrial Dynamics in Neurodegenerative Diseases. *Antioxidant* **2017**, *6*, 25. [CrossRef]

131. Brustovetsky, N.; Brustovetsky, T.; Purl, K.J.; Capano, M.; Crompton, M.; Dubinsky, J.M. Increased susceptibility of striatal mitochondria to calcium-induced permeability transition. *J. Neurosci.* **2003**, *23*, 4858–4867. [CrossRef]

132. Janikiewicz, J.; Szymański, J.; Malinska, D.; Patalas-Krawczyk, P.; Michalska, B.; Duszyński, J.; Giorgi, C.; Bonora, M.; Dobrzyn, A.; Wieckowski, M.R. Mitochondria-associated membranes in aging and senescence: Structure, function, and dynamics. *Cell Death Dis.* **2018**, *9*, 332. [CrossRef]

133. Terpolilli, N.A.; Moskowitz, M.A.; Plesnila, N. Nitric oxide: Considerations for the treatment of ischemic stroke. *J. Cereb. Blood Flow Metab. Off. J. Int. Soc. Cereb. Blood Flow Metab.* **2012**, *32*, 1332–1346. [CrossRef]

134. Esplugues, J. V NO as a signalling molecule in the nervous system. *Br. J. Pharm. Ther.* **2002**, *135*, 1079–1095. [CrossRef]

135. Brown, G.C. Nitric oxide and neuronal death. *Nitric Oxide Biol. Chem.* **2010**, *23*, 153–165. [CrossRef]

136. Bolaños, J.; Moro, M.; Lizasoain, I.; Almeida, A. Mitochondria and reactive oxygen and nitrogen species in neurological disorders and stroke: Therapeutic implications. *Adv. Drug Deliv. Rev.* **2009**, *61*, 1299–1315. [CrossRef]

137. Bolanos, J.; Heales, S. Persistent mitochondrial damage by nitric oxide and its derivatives: Neuropathological implications. *Front. Neuroenergetics* **2010**, *2*, 1. [CrossRef]

138. Martín-de-Saavedra, M.D.; del Barrio, L.; Cañas, N.; Egea, J.; Lorrio, S.; Montell, E.; Vergés, J.; García, A.G.; López, M.G. Chondroitin sulfate reduces cell death of rat hippocampal slices subjected to oxygen and glucose deprivation by inhibiting p38, NFκB and iNOS. *Neurochem. Int.* **2011**, *58*, 676–683. [CrossRef]

139. Thomaz, D.T.; Rafognatto, R.; Luisa, A.; Binder, B.; Scheffer, L.; Willms, A.; Fátima, C.; Mena, R.; Silva, B.; Tasca, C.I. Guanosine Neuroprotective Action in Hippocampal Slices Subjected to Oxygen and Glucose Deprivation Restores ATP Levels, Lactate Release and Glutamate Uptake Impairment: Involvement of Nitric Oxide. *Neurochem. Res.* **2020**, *45*, 2217–2229. [CrossRef]

140. Gourdin, M.; Dubois, P. Impact of Ischemia on Cellular Metabolism. *Artery Bypass* **2013**, *54509*, 3–18.

141. Magistretti, P.J.; Pellerin, L.; Rothman, D.L.; Shulman, R.G. Energy on demand. *Science* **1999**, *283*, 496–497. [CrossRef]

142. Bolaños, J.P. Bioenergetics and redox adaptations of astrocytes to neuronal activity. *J. Neurochem.* **2016**, *139* (Suppl. 2), 115–125. [CrossRef]

143. Schurr, A.; Gozal, E. Aerobic production and utilization of lactate satisfy increased energy demands upon neuronal activation in hippocampal slices and provide neuroprotection against oxidative stress. *Front. Pharm. Ther.* **2011**, *2*, 96. [CrossRef]

144. Pellerin, L.; Magistretti, P.J. Glutamate uptake into astrocytes stimulates aerobic glycolysis: A mechanism coupling neuronal activity to glucose utilization. *Proc. Natl. Acad. Sci. USA* **1994**, *91*, 10625–10629. [CrossRef]

145. Stocker, R.; McDonagh, A.F.; Glazer, A.N.; Ames, B.N. Antioxidant activities of bile pigments: Biliverdin and bilirubin. *Methods Enzym. Ther.* **1990**, *186*, 301–309. [CrossRef]

146. Harder, Y.; Amon, M.; Schramm, R.; Rücker, M.; Scheuer, C.; Pittet, B.; Erni, D.; Menger, M.D. Ischemia-induced up-regulation of heme oxygenase-1 protects from apoptotic cell death and tissue necrosis. *J. Surg. Res.* **2008**, *150*, 293–303. [CrossRef]

147. Scapagnini, G.; Butterfield, D.A.; Colombrita, C.; Sultana, R.; Pascale, A.; Calabrese, V. Ethyl ferulate, a lipophilic polyphenol, induces HO-1 and protects rat neurons against oxidative stress. *Antioxid. Redox Signal.* **2004**, *6*, 811–818. [CrossRef]

148. Lu, X.; Gu, R.; Hu, W.; Sun, Z.; Wang, G.; Wang, L.; Xu, Y. Upregulation of heme oxygenase-1 protected against brain damage induced by transient cerebral ischemia-reperfusion injury in rats. *Exp. Ther. Med.* **2018**, *15*, 4629–4636. [CrossRef] [PubMed]

149. Sethi, G.; Sung, B.; Aggarwal, B.B. Nuclear factor-kappaB activation: From bench to bedside. *Exp. Biol. Med.* **2008**, *233*, 21–31. [CrossRef] [PubMed]

150. Tulpule, K.; Dringen, R. Formate generated by cellular oxidation of formaldehyde accelerates the glycolytic flux in cultured astrocytes. *Glia* **2012**, *60*, 582–593. [CrossRef] [PubMed]

151. Zanella, C.A.; Tasca, C.I.; Henley, J.M.; Wilkinson, K.A. Guanosine modulates SUMO2/3-ylation in neurons and astrocytes via adenosine receptors. *Purinergic Signal.* **2020**, 439–450. [CrossRef]

152. Geiss-Friedlander, R.; Melchior, F. Concepts in sumoylation: A decade on. *Nat. Rev. Mol. Cell Biol.* **2007**, *8*, 947–956. [CrossRef]

153. Wilkinson, K.A.; Henley, J.M. Mechanisms, regulation and consequences of protein SUMOylation. *Biochem. J.* **2010**, *428*, 133–145. [CrossRef]
154. Henley, J.M.; Craig, T.J.; Wilkinson, K.A. Neuronal SUMOylation: Mechanisms, physiology, and roles in neuronal dysfunction. *Physiol. Rev.* **2014**, *94*, 1249–1285. [CrossRef]
155. Cimarosti, H.; Ashikaga, E.; Jaafari, N.; Dearden, L.; Rubin, P.; Wilkinson, K.A.; Henley, J.M. Enhanced SUMOylation and SENP-1 protein levels following oxygen and glucose deprivation in neurones. *J. Cereb. Blood Flow Metab.* **2012**, *32*, 17–22. [CrossRef]
156. Lee, Y.J.; Mou, Y.; Klimanis, D.; Bernstock, J.D.; Hallenbeck, J.M. Global SUMOylation is a molecular mechanism underlying hypothermia-induced ischemic tolerance. *Front. Cell. Neurosci.* **2014**, *8*, 1–9. [CrossRef]
157. Zhang, H.; Huang, D.; Zhou, J.; Yue, Y.; Wang, X. SUMOylation participates in induction of ischemic tolerance in mice. *Brain Res. Bull.* **2019**, *147*, 159–164. [CrossRef]
158. Pardridge, W.M. The Blood-Brain Barrier: Bottleneck in Brain Drug Development. *NeuroRx* **2005**, *2*, 3–14. [CrossRef]
159. Soares, F.A.; Schmidt, A.P.; Farina, M.; Frizzo, M.E.S.; Tavares, R.G.; Portela, L.V.C.; Lara, D.R.; Souza, D.O. Anticonvulsant effect of GMP depends on its conversion to guanosine. *Brain Res.* **2004**, *1005*, 182–186. [CrossRef]
160. Ganzella, M.; Dias, E.; Oliveira, A.; De Diniz, D.; Fernanda, C. Effects of chronic guanosine treatment on hippocampal damage and cognitive impairment of rats submitted to chronic cerebral hypoperfusion. *Neurolog. Sci.* **2012**, 985–997. [CrossRef]
161. Ramos, D.B.; Muller, G.C.; Botter, G.; Rocha, M.; Dellavia, G.H.; Almeida, R.F.; Pettenuzzo, L.F.; Loureiro, S.O.; Hansel, G.; Cássio, Â.; et al. Intranasal guanosine administration presents a wide therapeutic time window to reduce brain damage induced by permanent ischemia in rats. *Purinergic Signal.* **2016**, 149–159. [CrossRef]
162. Jiang, S.; Fischione, G.; Guiliani, P.; Romano, S.; Caciagli, F.; Diiorio, P. Metabolism and distribution of guanosine given intraperitoneally: Implications for spinal cord injury. *Nucleosides Nucleotides Nucleic Acids* **2008**, 673–680. [CrossRef] [PubMed]
163. Müller, G.C.; Loureiro, S.O.; Pettenuzzo, L.F.; Almeida, R.F.; Ynumaru, E.Y.; Guazzelli, P.A.; Meyer, F.S.; Pasquetti, M.V.; Ganzella, M.; Calcagnotto, M.E.; et al. Effects of intranasal guanosine administration on brain function in a rat model of ischemic stroke. *Purinergic Signal.* **2021**, *17*, 255–271. [CrossRef] [PubMed]
164. Anderson, C.M.; Swanson, R.A. Astrocyte glutamate transport: Review of properties, regulation, and physiological functions. *Glia* **2000**, *32*, 1–14. [CrossRef]
165. Vinadé, E.R.; Schmidt, A.P.; Frizzo, M.E.S.; Izquierdo, I.; Elisabetsky, E.; Souza, D.O. Chronically administered guanosine is anticonvulsant, amnesic and anxiolytic in mice. *Brain Res.* **2003**, *977*, 97–102. [CrossRef]
166. Kelly, K.J.; Plotkin, Z.; Dagher, P.C. Guanosine supplementation reduces apoptosis and protects renal function in the setting of ischemic injury. *J. Clin. Investig.* **2001**, *108*, 1291–1298. [CrossRef] [PubMed]
167. Schmidt, P.; Paniz, L.; Schallenberger, C.; Bo, A.E.; Wofchuk, S.T.; Elisabetsky, E.; Portela, L.V.C.; Souza, D.O.; Program, G. Guanosine Prevents Thermal Hyperalgesia in a Rat Model of Peripheral Mononeuropathy. *J. Power Sources* **2010**, *11*, 131–141. [CrossRef] [PubMed]
168. Saute, J.A.M.; da Silveira, L.E.; Soares, F.A.; Martini, L.H.; Souza, D.O.; Ganzella, M. Amnesic effect of GMP depends on its conversion to guanosine. *Neurobiol. Learn. Mem.* **2006**, *85*, 206–212. [CrossRef]
169. Vinadé, E.R.; Izquierdo, I.; Lara, D.R.; Schmidt, A.P.; Souza, D.O. Oral administration of guanosine impairs inhibitory avoidance performance in rats and mice. *Neurobiol. Learn. Mem.* **2004**, *81*, 137–143. [CrossRef]
170. Pereira, G.S.; Rossato, J.I.; Sarkis, J.J.F.; Cammarota, M.; Bonan, C.D.; Izquierdo, I. Activation of adenosine receptors in the posterior cingulate cortex impairs memory retrieval in the rat. *Neurobiol. Learn. Mem.* **2005**, *83*, 217–223. [CrossRef]
171. Jackson, E.K.; Mi, Z. The guanosine-adenosine interaction exists in vivo. *J. Pharm. Exp. Ther.* **2014**, *350*, 719–726. [CrossRef]
172. Varaschini, L.; Roberto, T.; Almeida, F.; Rohden, F.; Anderson, L.; Martins, M.; Teixeira, L.V.; Almeida, R.F.; Rohden, F.; Martins, L.A.M.; et al. Neuroprotective Effects of Guanosine Administration on In Vivo Cortical Focal Ischemia in Female and Male Wistar Rats. *Neurochem. Res.* **2018**, *43*, 1476–1489. [CrossRef]
173. Suzuki, S.; Brown, C.M.; Wise, P.M. Neuroprotective effects of estrogens following ischemic stroke. *Front. Neuroendocr. Ther.* **2009**, *30*, 201–211. [CrossRef]
174. Moretto, M.B.; Boff, B.; Lavinsky, D.; Netto, C.A.; Rocha, J.B.T.; Souza, D.O.; Wofchuk, S.T. Importance of schedule of administration in the therapeutic efficacy of guanosine: Early intervention after injury enhances glutamate uptake in model of hypoxia-ischemia. *J. Mol. Neurosci.* **2009**, *38*, 216–219. [CrossRef]
175. Chang, R.; Algird, A.; Dau, C.; Rathbone, M.P.; Jiang, S. Neuroprotective effects of guanosine on stroke models in vitro and in vivo. *Neurosci. Lett.* **2008**, *431*, 101–105. [CrossRef]
176. Rathbone, M.P.; Saleh, T.M.; Connell, B.J.; Chang, R.; Su, C.; Worley, B.; Kim, M.; Jiang, S. Systemic administration of guanosine promotes functional and histological improvement following an ischemic stroke in rats. *Brain Res.* **2011**, *1407*, 79–89. [CrossRef]
177. Souza, G.; Hansel, G.; Ramos, D.B.; Delgado, C.A.; Souza, D.O.; Almeida, F.; Portela, L.V. The Potential Therapeutic Effect of Guanosine after Cortical Focal Ischemia in Rats. *PLoS ONE* **2014**, *9*, 1–10. [CrossRef]
178. Connell, B.J.; Iorio, P.; Di Sayeed, I.; Ballerini, P.; Saleh, M.C.; Giuliani, P.; Saleh, T.M.; Rathbone, M.P.; Su, C.; Jiang, S. Guanosine Protects Against Reperfusion Injury in Rat Brains After Ischemic Stroke. *J. Neurosci. Res.* **2013**, *272*, 262–272. [CrossRef]
179. Bettio, L.E.B.; Gil-mohapel, J.; Rodrigues, A.L.S. Guanosine and its role in neuropathologies. *Purinergic Signal.* **2016**, 411–426. [CrossRef]
180. Allen, C.L.; Bayraktutan, U. Oxidative stress and its role in the pathogenesis of ischaemic stroke. *Int. J. Stroke* **2009**, *4*, 461–470. [CrossRef]

181. Chen, H.; Yoshioka, H.; Kim, G.S.; Jung, J.E.; Okami, N.; Sakata, H.; Maier, C.M.; Narasimhan, P.; Goeders, C.E.; Chan, P.H. Oxidative stress in ischemic brain damage: Mechanisms of cell death and potential molecular targets for neuroprotection. *Antioxid. Redox Signal.* **2011**, *14*, 1505–1517. [CrossRef]

182. Crack, P.J.; Taylor, J.M. Reactive oxygen species and the modulation of stroke. *Free Radic. Biol. Med.* **2005**, *38*, 1433–1444. [CrossRef]

183. Moretto, M.B.; Arteni, N.S.; Lavinsky, D.; Netto, C.A.; Rocha, J.B.T. Hypoxic-ischemic insult decreases glutamate uptake by hippocampal slices from neonatal rats: Prevention by guanosine. *Exp. Neurol.* **2005**, *195*, 400–406. [CrossRef]

184. Wang, Y.; Briz, V.; Chishti, A.; Bi, X.; Baudry, M. Distinct roles for μ-calpain and m-calpain in synaptic NMDAR-mediated neuroprotection and extrasynaptic NMDAR-mediated neurodegeneration. *J. Neurosci.* **2013**, *33*, 18880–18892. [CrossRef]

185. Deng, G.; Qiu, Z.; Li, D.; Fang, Y.U.; Zhang, S. Delayed administration of guanosine improves long-term functional recovery and enhances neurogenesis and angiogenesis in a mouse model of photothrombotic stroke. *Mol. Med. Rep.* **2017**, 3999–4004. [CrossRef]

186. Ruan, L.; Wang, B.; Zhuge, Q.; Jin, K. Coupling of neurogenesis and angiogenesis after ischemic stroke. *Brain Res.* **2015**, *1623*, 166–173. [CrossRef]

187. Chen, S.D.; Wu, C.L.; Hwang, W.C.; Yang, D.I. More insight into BDNF against neurodegeneration: Anti-apoptosis, anti-oxidation, and suppression of autophagy. *Int. J. Mol. Sci.* **2017**, *18*, 545. [CrossRef]

188. Rathbone, M.; Pilutti, L.; Caciagli, F.; Jiang, S. Neurotrophic effects of extracellular guanosine. *Nucleosides. Nucleotides Nucleic Acids* **2008**, *27*, 666–672. [CrossRef]

189. Massari, C.M.; Zuccarini, M.; Di Iorio, P.; Tasca, C.I. Guanosine Mechanisms of Action: Toward Molecular Targets. *Front. Pharm. Ther.* **2021**, *12*, 1–5. [CrossRef]

190. Melani, A.; Pugliese, A.M.; Pedata, F. *Adenosine Receptors in Cerebral Ischemia*, 1st ed.; Elsevier Inc.: Amsterdam, The Netherlands, 2014; Volume 119, ISBN 9780128010228.

191. Bynoe, M.S.; Viret, C.; Yan, A.; Kim, D.G. Adenosine receptor signaling: A key to opening the blood-brain door. *Fluids Barriers Cns* **2015**, *12*, 1–12. [CrossRef]

192. Abdullahi, W.; Tripathi, D.; Ronaldson, P.T. Blood-brain barrier dysfunction in ischemic stroke: Targeting tight junctions and transporters for vascular protection. *Am. J. Physiol. Cell Physiol.* **2018**, *315*, C343–C356. [CrossRef] [PubMed]

193. Jian, Z.; Liu, R.; Zhu, X.; Smerin, D.; Zhong, Y.; Gu, L.; Fang, W.; Xiong, X. The Involvement and Therapy Target of Immune Cells After Ischemic Stroke. *Front. Immunol.* **2019**, *10*, 1–15. [CrossRef] [PubMed]

194. Lymphocytes, S.P.; Pasquini, S.; Vincenzi, F.; Casetta, I.; Laudisi, M. Adenosinergic System Involvement in Ischemic. *Cells* **2020**, *9*, 1072. [CrossRef]

195. Bantia, S.; Choradia, N. Treatment duration with immune-based therapies in Cancer: An enigma 11 Medical and Health Sciences 1107 Immunology. *J. Immunother. Cancer* **2018**, *6*, 4–8. [CrossRef]

196. Shanta, B. Purine Nucleoside Phosphorylase Inhibitors as Novel Immuno-Oncology Agent and Vaccine Adjuvant. *Int. J. Immunol. Immunother.* **2020**, *7*, 1–12. [CrossRef]

International Journal of
Molecular Sciences

MDPI

Review

Purinergic–Glycinergic Interaction in Neurodegenerative and Neuroinflammatory Disorders of the Retina

Laszlo G. Harsing, Jr. [1,*], Gábor Szénási [2], Tibor Zelles [1,3] and László Köles [1,3]

[1] Department of Pharmacology and Pharmacotherapy, Semmelweis University, H-1089 Budapest, Hungary; zelles.tibor@med.semmelweis-univ.hu (T.Z.); koles.laszlo@med.semmelweis-univ.hu (L.K.)

[2] Institute of Translational Medicine, Semmelweis University, H-1089 Budapest, Hungary; szenasi.gabor@med.semmelweis-univ.hu

[3] Department of Oral Biology, Semmelweis University, H-1089 Budapest, Hungary

[*] Correspondence: harsing.laszlo@med.semmelweis-univ.hu; Tel.: +36-1-210-4416

Citation: Harsing, L.G., Jr.; Szénási, G.; Zelles, T.; Köles, L. Purinergic–Glycinergic Interaction in Neurodegenerative and Neuroinflammatory Disorders of the Retina. *Int. J. Mol. Sci.* **2021**, 22, 6209. https://doi.org/10.3390/ijms22126209

Academic Editors: Dmitry Aminin and Peter Illes

Received: 25 May 2021
Accepted: 4 June 2021
Published: 8 June 2021

Abstract: Neurodegenerative–neuroinflammatory disorders of the retina seriously hamper human vision. In searching for key factors that contribute to the development of these pathologies, we considered potential interactions among purinergic neuromodulation, glycinergic neurotransmission, and microglia activity in the retina. Energy deprivation at cellular levels is mainly due to impaired blood circulation leading to increased release of ATP and adenosine as well as glutamate and glycine. Interactions between these modulators and neurotransmitters are manifold. First, P2Y purinoceptor agonists facilitate reuptake of glycine by glycine transporter 1, while its inhibitors reduce reverse-mode operation; these events may lower extracellular glycine levels. The consequential changes in extracellular glycine concentration can lead to parallel changes in the activity of NR1/NR2B type NMDA receptors of which glycine is a mandatory agonist, and thereby may reduce neurodegenerative events in the retina. Second, P2Y purinoceptor agonists and glycine transporter 1 inhibitors may indirectly inhibit microglia activity by decreasing neuronal or glial glycine release in energy-compromised retina. These inhibitions may have a role in microglia activation, which is present during development and progression of neurodegenerative disorders such as glaucomatous and diabetic retinopathies and age-related macular degeneration or loss of retinal neurons caused by thromboembolic events. We have hypothesized that glycine transporter 1 inhibitors and P2Y purinoceptor agonists may have therapeutic importance in neurodegenerative–neuroinflammatory disorders of the retina by decreasing NR1/NR2B NMDA receptor activity and production and release of a series of proinflammatory cytokines from microglial cells.

Keywords: retina; purinergic modulation; glycinergic neurotransmission; microglia; neuroinflammation; neurodegeneration; glycine transporters

1. Introduction

Neurodegenerative disorders occur with a high incidence in the elderly. These pathologies affect the motor system in Parkinson's or Huntington's diseases, memory in Alzheimer's disease, or other additional critically important neural functions after impairments in cerebral blood circulation. The retina, which is commonly considered as an outside part of the central nervous system, is also affected by neurodegenerative disorders. Most of them may lead to loss of neurons in the retinal circuitry and consecutive impaired vision or blindness. Among the neurodegenerative diseases, we mention retinal hypoxia/ischemia, glaucomatous and diabetic neuropathies, and age-related macular degeneration and those that appear less frequently, such as human recessive retinitis pigmentosa or inherited photoreceptor degeneration. These disorders may occur as a sudden pathological event or show a progressively declining clinical course, but vision may be seriously hampered or lost in all disorders. Despite different clinical symptoms, there are a number of common factors in their pathogenesis, like neuroinflammation associated with neurodegeneration. Retinal hypoxia, as a persisting insult, may induce enhanced purinergic modulation and glutamatergic–glycinergic neurotransmission. Activation of the two systems may evoke sustained release of inflammatory mediators from activated microglia and the resultant chronic proinflammatory environment may induce development or exacerbation of retinal neurodegenerative disorders. There are several excellent review articles published on different aspects of the neurodegenerative pathologies of the retina [1–3].

In spite of the extended research effort, the currently used therapeutics only slow down the progression of retinal neurodegenerative disorders, whereas normalization of impaired visual functions can be rarely achieved. It is, therefore, mandatory to develop novel therapeutic interventions. This goal, however, cannot be reached without unfolding the pathophysiology of retinal neurodegenerative diseases. The aim of this review is to highlight a possible interaction between purinergic and glycinergic signaling systems in neuroinflammatory–neurodegenerative disorders of the retina. This interaction is, however, further complicated by the findings that microglia, the resident immune cells of the central nervous system and retina, are involved not only in neuronal cell destructions but also in compensatory neural repair following insulting influences [4,5].

2. Neural Circuitries and Glial Cell Types in the Retina

Figure 1A shows the cell types and the organization of neural circuitry after hematoxylin-eosin staining of the retina [6]. The vertical section of the retina showed the retinal pigment epithelium (PRE) as the outermost layer followed by the photoreceptor layer (PRL) and outer nuclear layer (ONL), the latter containing the cell bodies of the cone and rod photoreceptors. The neurotransmitter in photoreceptors is glutamate, the release of which decreases to light exposition of the retina. The second-degree neurons in the retina are glutamatergic bipolar cells. The cell bodies of the ON and OFF cone bipolar cells and the connecting ganglion cells form the inner nuclear layer (INL) and the ganglion cell layer (GCL) of the retina, respectively. The inner nuclear layer also contains amacrine cells mostly releasing glycine or GABA as neurotransmitters, which, together with the horizontal cells, compose the horizontal visual pathway [7]. Neuronal connection, which transmits signals from rod bipolar cells to cone bipolar cells, is established by the AII cell, a specific type of glycinergic amacrine cell [8]. The outer and inner plexiform layers (OPL, IPL) contain the synaptic connections between the photoreceptors and bipolar cells and the bipolar cells and retinal ganglion cells, respectively (Figure 1B). These layers show characteristic alterations or tapering in various neurodegenerative disorders or injuries of the retina [9,10].

Besides the neural circuits, the retina also contains numerous glia cells termed macroglia (Müller cells and astrocytes) and microglial cells. The principal macroglia Müller cells span all retinal layers, whereas astrocytes are mostly located close to the nerve fiber layer and the ganglion cell layer [3,7,12]. Müller cells are the major source of adenosine triphosphate (ATP) released into the extracellular space, which then modulates both neural and microglial cell activities [7,13]. Müller cells express glycine transporter 1 (GlyT-1), which regulates extracellular glycine concentrations in the retina following its release from macroglia and amacrine cells [14,15]. The glial marker, glial fibrillary acidic protein (GFAP), is expressed in both Müller glia cells and astrocytes [13,16,17].

The other major type of retinal glia cell is microglia. They originate from macrophage or monocyte precursors and are considered as the major immunocompetent cell type of the retina [18]. In healthy retina, microglia cells are characterized by small somata and extensively ramified processes at rest [3]. Microglial cells, which exhibit different morphologies in the retina, are mainly present in the inner and outer plexiform layers, ganglion cell layer, and nerve fiber layer (Figure 1C) [7,11,19,20].

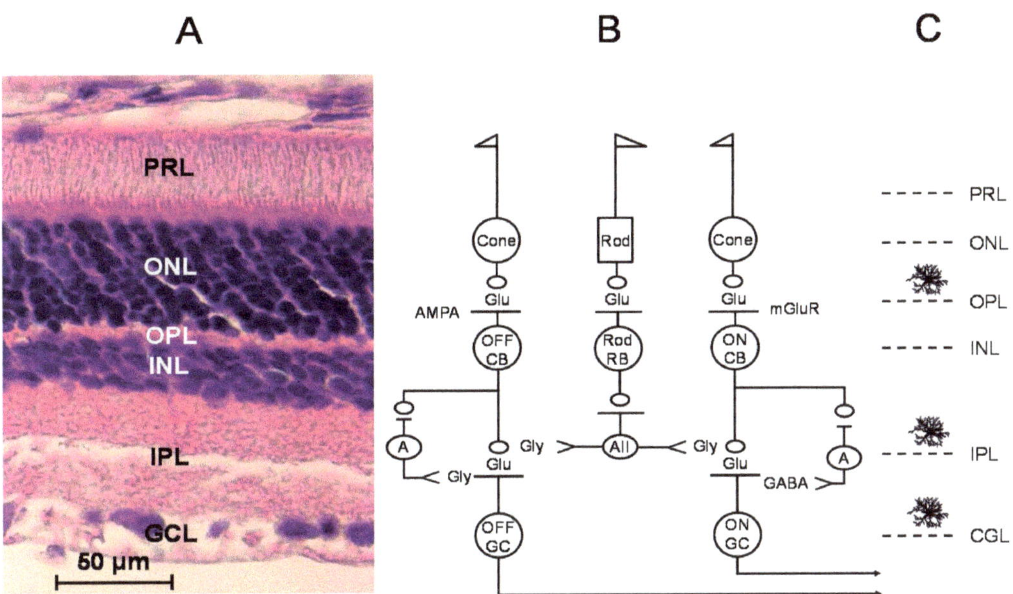

Figure 1. (**A**) Cytoarchitecture of the retina. Vertical section of rat retina stained with hematoxylin-eosine (made with the courtesy of Dr. Mihaly Albert). (**B**) Neural circuitry of the retina. Cone bipolar (CB) axons descend to the inner plexiform layer; the two categories of bipolar cells terminate at different levels: Axons of the OFF bipolar cells send axons to the upper part of the inner plexiform layer and the ON bipolar cells end in the lower part. Rod bipolar (RB) cells release glutamate and synapse to AII amacrine cells. Glycinergic AII cells append to OFF bipolar terminals and OFF ganglion cell dendrites in the upper level of the inner plexiform layer and these interneurons also synapse to ON cone bipolar cells in the lower half of the inner plexiform layer. Retinal ganglion cells receive excitatory glutamatergic innervation from cone bipolar cells and inhibitory influence from amacrine cells. About half of the amacrine cells are glycinergic and the other half is GABAergic. The OFF pathway is under the direct control of glycine released from glycinergic amacrine cells, whereas the ON pathway is under the inhibition of GABAergic amacrine cells. NR1/NR2A-type NMDA receptors are present in the synapses formed between bipolar and ganglion cells, whereas extrasynaptic NR1/NR2B-type NMDA receptors are predominantly expressed in ON ganglion cells. Thus, the primary target for excitotoxicity is the ON ganglion cells in the retina. (**C**) Microglial cells in normal retina are mainly distributed in the outer and inner plexiform layers. In the glaucomatous retina, microglia are present in the ganglion cell layer surrounding the retinal ganglion cell axons and soma. In the diabetic retina, perivascular accumulation of activated microglia cells can be found. These cells are accumulated in the outer nuclear layer and subretinal space in age-related macular degeneration [7,10,11]. PRL, photoreceptor layer; ONL, outer nuclear layer; OPL, outer plexiform layer; INL, inner nuclear layer; IPL, inner plexiform layer; GCL, ganglion cell layer; A, amacrine cell; AII, AII glycinergic interneuron; AMPA, AMPA receptor; mGluR, metabotropic glutamate receptor; CB, cone bipolar cell; RB, rod bipolar cell; GC, ganglion cell.

3. Cellular Biology of Microglia in Neural and Retinal Tissues

Microglia are the resident immune cells of the central nervous system as well as the retina. Microglia exist in different forms. Resting microglia are derived from macrophage-like cells and they can be characterized by small cell bodies and multiple branched processes arising from the cell bodies with down-regulated macrophage-like properties [21]. Microglia processes exhibit continuous movements controlling the immunological environment (surveillance mode) in order to detect signals from stressed neurons or macroglia. Microglial cells also have a function in normal retinal growth, neurogenesis, and retinal blood vessel formation under physiological conditions.

The shape of microglial cells turns to be amoeboid during microglial activation in retinopathological events and cellular hypertrophy and process retraction can be also observed [22]. The amoeboid form of microglia is enlarged in size, and microglia occur either as single cells or in clusters and their phagocytotic activity is enhanced [3]. In response to pathogenic signals such as ischemia or oxidative stress, microglia proliferate and migrate towards the site of tissue injury [23]. Pathological stimuli also activate macroglial cells and cause gliosis of the retina. The gliosis of Müller cells is characterized by cellular hypertrophy, proliferation, and upregulation of GFAP expression [13].

Ramified microglia is a model of resting-like cells, whereas addition of the bacterial cell wall component lipopolysaccharide (LPS) induces activation of microglial cells. This procedure is also called a shift of microglia from an ineffective to an effective (activated) state. Stimulation with LPS is accompanied by altered K^+ channel expression: Resting microglial cells do not express voltage-activated K^+ channels, their expression together with P2Y purinoceptors can be observed following microglia activation [24].

Microglial cells undergo two kinds of activation in response to infection, tissue traumatic or hypoxic injury [10,25]. These two types are designated as M1 and M2 microglial cells, which are also mentioned in the literature as proinflammatory and anti-inflammatory phenotypes [26]. The M1 first state is a neurotoxic phenotype that is characterized with massive inflammatory responses, release of tumor necrosis factor α (TNFα) and other neurotoxic inflammatory mediators/cytokines such as interleukin (IL)-1α, IL-1β, IL-6, and IL-12 [3]. Release of TNFα and IL-6 from microglial cells is toxic and induces neuronal cell death [27]. In certain circumstances, the M1 state may turn to uncontrolled activation of microglial cells, leading to a chronic inflammation in neural tissues. The increased production of the various proinflammatory cytokines may maintain an inflammatory condition and sustained release of these factors also contributes to neuronal cell death [28].

The second microglia phenotype is M2, which can secrete a series of anti-inflammatory mediators (IL-4, IL-10, IL-13, TGF-1β), and neurotrophic factors such as insulin-like growth factor 1 (IGF-1). M2 microglia serve inflammation resolution and promote neuroprotection and neuronal survival [29]. M2 microglia were recognized as neuroprotective in neurodegenerative diseases preventing neuroinflammation both in the brain and in the retina. M2 microglia downregulate inflammatory cells and the released protective and trophic factors induce immunosuppressive responses [30]. During M1 to M2 transition, the acute and prolonged phases of microglia activation may represent a dual role of these cells in neuroinflammatory and neurodegenerative disorders of the retina [26]. Chronic neuroinflammation in neural or retinal tissues may be due to lack of sufficient M2 microglia responses either because of a lower number of microglial cells or a lower secretion of neuroprotective factors [30]. It has been suggested that phenotypic shift of microglia from proinflammatory M1 function to anti-inflammatory M2 function may be beneficial in retinal neurodegenerative disorders [25].

Microglia also become activated in pathological angiogenesis in the retina. Vascular endothelial growth factor (VEGF) released from microglia and monocytes has a crucial role in blood vessel growth. VEGF is also an inflammatory mediator and contributes to inflammatory responses [31]. In a mouse model of macular degeneration, VEGF is released from microglial cells into the subretinal space causing choroidal neovascularization [7]. Increased levels of VEGF and choroidal neovascularization were reported in patients suffering from age-related macular degeneration. Elevated vitreous VEGF levels were also shown in patients with retinal vein occlusion [32]. Microglial cells can produce and the release brain-derived neurotrophic factor (BDNF) and other neurotrophic factors such as nerve growth factor (NGF), inducing trophic cascades but also cell death in developing retina [18,33,34].

4. Purinergic Modulation of Neural Tissues

Ischemia and other pathological conditions like mechanical perturbation and distension, cell swelling, stretching, or increased hydrostatic pressure all lead to ATP release [35]. ATP can be released from neurons into the extracellular space by an external Ca^{2+}-dependent exocytotic mechanism evoked by membrane depolarization. Besides neurons, ATP can also be released from glia cells by Ca^{2+}-dependent and independent means [4]. The release of ATP may also occur though the large transmembrane pannexins hemichannels from neurons and astrocytes [28,36].

ATP, if once released, is metabolized into ADP, AMP, and adenosine in the extracellular space by ecto-5′-nucleotidases (CD73) and ecto-nucleotide triphosphate diphosphohydrolases (NTPDases, CD39) [2,37,38]. Activity of these enzymes can be modified pharmacologically, e.g., ARL67156 is an inhibitor of NTPDases (Table 1). The turnover of the neural or glial ATP released towards the extracellular space may end up by uptake of the formed adenosine into neurons or glial cells. Adenosine reuptake inhibitors increase adenosine concentrations in the extracellular space (Table 1). Adenosine can also be produced intracellularly and effluxed from the intracellular space by adenosine transporter reversal. The mechanism commonly implicated in adenosine release involves bi-directional nucleoside transport. Thus, the Na^+-dependent concentrative nucleoside transporters assure uptake of adenosine from the extracellular space (normal-mode operation), whereas operation of the transporter in the reverse or release-mode operation results in increased adenosine efflux from the cells [39]. The extracellular concentration of adenosine is also determined by the operation of the Na^+-independent equilibrative nucleoside transporter [40]. Adenosine levels increase at the site of tissue damage or in the retina exposed to hypoxia and ischemia-reperfusion [41,42].

Table 1. Drugs used to investigate purinoceptors in neurodegenerative–neuroinflammatory disorders of the central nervous system and retina.

Drugs	Mode of Action
	P2 nucleotide purinoceptors
α,β-Methylene-ATP	P2 agonist
PPADS	P2 antagonist
Suramin	non-specific P2 antagonist
	P2Y nucleotide purinoceptors
2-MeS-ATP	P2 agonist
2-MeS-ADP	$P2Y_{1,12,13}$ agonist
MRS 2365	$P2Y_1$ agonist
MRS 2179	$P2Y_1$ antagonist
MRS 2211	$P2Y_{13}$ antagonist
	P2X nucleotide purinoceptors
β,γ-Methylene-ATP	P2X agonist
BzATP	$P2X_7$ agonist
TNP-ATP	P2X antagonist
NF449	$P2X_1$ antagonist
Brillant Blue G (BBG)	$P2X_7$ antagonist
	P1 nucleoside (adenosine) purinoceptors
R-PIA	A_1 agonist
CCPA	A_1 agonist
CPA	A_1 agonist
DPCPX	A_1 antagonist
CGS21680	A_{2A} agonist
SCH58261	A_{2A} antagonist
ZM-241,385	A_{2A} antagonist
BAY 60-6583	A_{2B} agonist

Table 1. *Cont.*

Drugs	Mode of Action
BzATP	$2',5'$-O-4-benzo-yl)-ATP
CCPA	2-chloro-N6-cyclopentyladenosine
CF101	(N6-(3-iodobenzyl)-$5'$-N-methylcarboxamidoadenosine
CGS 21680	4-[2][6-amino-9-(N-ethyl-β-D-ribofuranuronamidosyl)-9H-purin-2-yl]amino]ethyl] benzeneproprionic acid
CP-532,903	(2S,3S,4R,5R)-3-amino-5-[6-(2,5-dichlorobenzylamino)purin-9-yl]-4 hydroxytetrahydrofuran-2-carboxylic acid methylamide
CPA	N6-cyclopentyladenosine
DPCPX	Dipropylcyclopenthylxanthine
LUF-5835	1-[6-amino-9-[(2R,3R,4S,5R)-3,4-dihydroxy-5-(hydroxymethyl)oxolan-2-yl]purin2-yl]-N-methylpyrazole-4-carboxamide
2-MeS-ATP	2-methylthio-ATP
2-MeSADP	2-methylthio-ADP
MRE 3008F20	N-[2-(2-Furanyl)-8-propyl-8H-pyrazolo[4,3-e][1,2,4]triazolo[1,5-c]pyrimidin 5-yl]-N$'$-(4-methoxyphenyl)urea
MRS 1706	N-(4-acetylphenyl)-2-([4-(2,3,6,7-tetrahydro-2,6-dioxo-1,3-dipropyl-1H-purin-8-yl)phenoxy]acetamide
MRS 2179	N6-methyl-$2'$-deoxyadenosine-$3',5'$-bisphosphate
MRS 2211	pyridoxal-$5'$-phosphate-6-azo(2-chloro-5-nitrophenyl)-2,4-disulfonate
MRS 2365	[[(1R,2R,3S,4R,5S)-4-[6-amino-2-(methylthio)-9H-purin-9-yl]-2,3-dihydroxy bicyclo-[3.1.0]hex-1-yl]methyl] diphosphoric acid monoester
NBMPR	S6-(4-nitrobenzyl)mercaptopurine riboside
NF449	$4,4',4'',4'''$-[Carbonylbis(imino-5,1,3,-benzenetriyl-bis(carbonylimino))]tetrakis-1,3 benzenedisulfonic acid
PPADS	pyridoxalphosphate-6-azaphenyl-2,4-disulfonic acid
R-PIA	R-N6-(2-phenylisopropyl)adenosine
SCH5826	1,7-(2-phenylethyl)-5-amino-2-(2-furyl)-pyrazolo[4,3-e]-1,2,4-triazolo[1,5-c]pyrimidine
TNP-ATP	$2',3'$-(2,4,6,-trinitrophenyl)adenosine-$5'$-triphosphate
ZM241,385	4-(2-[7-amino-2-)2-furyl(triazolo-[1,3,5]triazin-5-ylamino]ethyl)phenol

For further details see Jacobson and Civan [43].

Purinergic influence in the retina is mediated by the release of various endogenous ligands for P1 and P2 purinoceptors from neuronal and non-neuronal cells [44,45]. In the retina, ATP release was found from the pigment epithelium layer, ganglion cells, and also from Müller glia cells and astrocytes [13,46,47]. The presence of pannexin hemichannels and their contribution to ATP release have been demonstrated in the retina [35].

4.1. Heterogeneity of Purinoceptors

Purinergic receptors or purinoceptors are divided into nucleoside P1 and nucleotide P2 types based upon their natural ligands. P2 purinoceptors are further divided into two groups: P2X purinoceptors are ATP-gated ion channels, whereas P2Y purinoceptors are metabotropic and coupled to G proteins. P2 purinoceptors are sensitive to adenine and guanine nucleotides [2]. Released ATP activates P2X and several P2Y purinoceptor subtypes but other adenine or uridine nucleotides ADP, UTP, and UDP are also natural agonist ligands for some P2Y purinoceptor subtypes. P2Y purinoceptors consist of eight different subtypes designated as $P2Y_{1,2,4,6,11,12,13,14}$. The downstream signal transduction linked to P2Y purinoceptors are mostly of two kinds, either Gq ($P2Y_{1,2,4,6,,11}$ receptors) or Gi ($P2Y_{12,13,14}$) proteins are coupled to the receptors. Thus, P2Y purinoceptors with different signal transductions can evoke both excitation and inhibition. In addition, P2Y purinoceptors exhibit different agonist binding profiles: $P2Y_1$, $P2Y_{12}$, and $P2Y_{13}$ purinoceptors are sensitive to adenine nucleotides; $P2Y_2$ and $P2Y_4$ are activated by adenine and uridine nucleotides; and human $P2Y_4$ and $P2Y_6$ purinoceptors are primarily affected by uridine nucleotides [23].

The response of Gq protein-coupled P2Y purinoceptors to ATP is an increase in intracellular inositol trisphosphate (IP_3) level, a rapid transient release of Ca^{2+} from internal stores followed by an increase of intracellular Ca^{2+} concentration, and also a Ca^{2+} influx into the cells [34]. Illes and coworkers [48] proposed that this increase of intracellular Ca^{2+} concentrations results in opening K^+ channels, allowing an outward-directed K^+ current, and membrane hyperpolarization in microglial cells.

The P2X ionotropic purinoceptors are ligand-gated cationic channels and are non-selectively permeable to Na^+, K^+, and Ca^{2+}. P2X purinoceptors have seven subtypes identified as $P2X_1$ to $P2X_7$ and are sensitive to ATP but not to ADP, AMP, and adenosine [4,23]. In response to ATP, P2X purinoceptors mediate fast excitatory neurotransmission as their activation induces an inward-directed cation current and cellular membrane depolarization [34,48]. P2X purinoceptors are also linked to Ca^{2+} signaling via receptor-coupled Ca^{2+}-permeable cationic channels, allowing Ca^{2+} entry into the cells [24].

The other types of purinoceptors are the P1 type [46]. Adenosine is the natural ligand for P1 nucleoside receptors, which activates all four different receptor subtypes, A_1, A_{2A}, A_{2B}, and A_3. Adenosine receptors are only sensitive to the nucleoside adenosine [2] whereas Luongo and coworkers [26] reported that treatment with ATP upregulated adenosine A_1 receptor expression in microglia. Adenosine receptors possess metabotropic signal transduction: A_1 and A_3 receptors are coupled to Gi proteins and A_{2A} and A_{2B} receptors are linked to Gs proteins. Accordingly, activation of adenosine receptors leads to stimulation or inhibition of adenylate cyclase and to an increase or decrease of tissue cAMP levels. Activation of A_1 receptors results in mostly opposite effects to A_{2A} receptors. The P1 purinoceptor agonist adenosine fails to generate a $[Ca^{2+}]i$ response in cultured microglia indicating that ATP action on intracellular Ca^{2+} is mediated only through activation of P2 purinoceptors [24]. Inhibitory response of purinoceptors coupled to Gi proteins (A_1 and A_3 adenosine receptors and some P2Y purinoceptors) is a decrease in cAMP levels leading to Ca^{2+} and K^+ channel dephosphorylation that results in membrane hyperpolarization [26].

4.2. Purinoceptors in the Retina

P2 purinoceptors possess distinct expression profiles in the retina. P2Y metabotropic receptors are expressed in neurons and glial cells and the retinal pigment epithelium [49]. $P2Y_1$ purinoceptors may have a preferential expression in retinal macroglia including Müller glia cells, whereas $P2Y_1$, $P2Y_{12}$, and $P2Y_{13}$ purinoceptors were identified in astroglial cells [50]. Increased purinergic activity induces gliosis in Müller glia cells as was demonstrated by increased GFAP immunoreactivity and the involvement of $P2Y_1$ purinoceptor was suggested to evoke this effect [51]. Microglial cells in the retina express the $P2Y_1$ purinoceptor subtype [51].

P2X purinoceptors are expressed on most classes of neurons in the retina. The presence of P2X receptor immunoreactivity in GABAergic amacrine cells supports previous findings that purinergic signals modulate information processing by GABAergic amacrine cells [49]. Other laboratories identified an abundant expression of $P2X_7$ purinoceptors in the inner and outer plexiform layers, in ganglion cells, Müller glia cells, astrocytes, and microglia in high concentrations [23,52]. Wurm and coworkers [53] reported that, although $P2X_7$ receptors are expressed in human Müller cells, P2X purinoceptors are largely absent in other mammalian Müller glia cells. Depending on the receptor-coupled signal transduction, the effects mediated by P2X and P2Y receptors may be synergistic or antagonistic in retinal circuitry.

Similarly to P2 purinoceptors, P1 purinoceptors are present in retinal pigment epithelial cells and ganglion cells also express multiple adenosine receptors [49]. A_1, A_{2A}, and A_3 adenosine receptor mRNAs were detected in the inner nuclear layer of the retina. A_{2A} and

A_{2B} receptors are present in retinal pigment epithelial cells, Müller glia cells, and they are also expressed in microglia in the retina [2].

A1 receptors are inhibitory in nature and suppress excitatory neurotransmission [46]. Adenosine is a major inhibitor of neural activity in the retina. Increased production of adenosine in the extracellular space activates inhibitory A1 receptors and neuronal activity is decreased [13]. The A_1 receptor agonist CPA decreased the number of apoptotic nuclei and GFAP immunoreactivity in the retina exposed to excessive light to induce neurodegeneration. This effect was reversed by the A_1 receptor antagonist DPCPX [54].

4.3. Purinergic Regulation of Microglia

Microglia express P2Y and P2X purinoceptors in resting and activated states [4]. Further analysis demonstrated that the majority of microglial cells in proliferating (resting) state express P2X purinoceptors and non-proliferating (activated or alerted phenotype) microglial cells express both P2Y and P2X purinoceptors [48]. Färber and Kettenmann [34] reported the presence of all the eight different types of P2Y purinoceptors and six types of P2X purinoceptors ($P2X_{1,2,3,4,6,7,}$) in the microglial cell line. Accordingly, addition of ATP affects several metabotropic and ionotropic transduction systems in microglia [55,56].

ATP and adenosine alter the morphology of microglia by inducing an outgrowth of microglia processes. ATP also regulates the motility of microglial processes and elicits rapid chemotactic responses in the retina [7,57]. Microglial chemotaxis as a response to neuronal injury is regulated by $P2Y_{12}$ receptors. Ogata and coworkers [56] reported that activation of P2Y receptors induces microglia proliferation and accelerates process retraction of the cells. Furthermore, $P2Y_{1,2,4}$ receptors expressed on microglia regulate phagocytosis [58]. Uckermann and coworkers [21] also reported that rabbit retinal microglial cells contain $P2Y_1$ receptors and their activation leads to microglial cell activation. In addition, $P2X_4$ purinoceptors mediate chemotaxis, and $P2X_7$ purinoceptors are involved in transformation of microglia to the proinflammatory phenotype [26]. Of the P1 purinoceptors, A_1 and A_3 adenosine receptors have been negatively implicated into the regulation of microglia morphology and responses, and proliferation and mediator release [11]. A_{2A} receptor upregulation has a role in the regulation of extension and retraction of microglial processes [59].

LPS stimulated TNFα, IL-1β, and IL-6 release and ATP inhibited these effects in cultured microglia obtained from rat spinal cord. In addition, the P2Y purinoceptor agonist 2-MeSATP exerted similar inhibition [56]. It was shown that the $P2Y_2$ and $P2Y_6$ purinoceptor ligands UTP, UDP, and the hydrolysis-resistant ATP analogue ATPγS stimulated the basal and TNFα-induced secretion of the chemotactic factor IL-8 in human retinal pigment epithelium [60]. These observations raise the possibility that P2Y purinoceptors coupled to Gq or Gi proteins may oppositely regulate the production of proinflammatory cytokines from microglia.

Activation of different P2 receptors occurs at a temporal scale. First, ATP activates microglial $P2X_7$ purinoceptors following an ischemic insult inducing an immediate tissue necrosis followed by activation of purinoceptors on astroglia cells and neurons [5]. In addition, ATP stimulates $P2X_7$ purinoceptors at milliomolar concentration, whereas P2Y purinoceptors can be activated in a 10–100 μmol concentration range of ATP [56]. $P2X_7$ purinoceptors exhibit a low sensitivity to ATP resulting in activation only in pathological conditions (tissue damage, neuroinflammation, mechanical stress or injury, trauma, hypoxia, or ischemia) when extracellular ATP concentrations reach high levels.

ATP-activated P2X purinoceptors are also principal regulators of neuroinflammatory responses. P2X purinoceptor activation induces IL-1β release from macrophages and ADP and AMP also act as promoters of LPS-induced IL-1β secretion in microglia cells [36]. $P2X_7$ receptor mediates cytokine release from microglia [52]. A series of studies demonstrated that $P2X_7$ receptor stimulation of immune cells by ATP evokes release of the proinflammatory TNFα and IL-1β from mouse or rat microglial cell lines [61]. This release of TNFα was found to be dependent upon Ca^{2+} influx and MAP kinase cascade activation [62].

In contrast, Shigemito-Mogami and coworkers [63] reported that the production of the proinflammatory cytokine IL-6 was under the control of P2Y rather than P2X purinoceptors in mouse microglial cell lines.

The production and release of proinflammatory cytokines seem to be oppositely regulated by the metabotropic P2Y and the ionotropic P2X receptors in microglial cells [4]. The apparently opposite effects of P2X and P2Y purinoceptors on proinflammatory factors may be explained by the different signal transduction systems utilized by these receptors. Ogata and coworkers [56] also found opposite effects of P2X (stimulation) and P2Y purinoceptor (inhibition) on TNFα release from cultured spinal cord microglia.

Microglial cells express all subtypes of P1 purinoceptors, A_1, $A_{2A//B}$, and A_3 [11]. The various types of adenosine receptors also exhibit different sensitivity to their natural ligand adenosine [39]. Gi protein-coupled A_1 adenosine receptor stimulation decreases and Gs protein-coupled A_{2A} receptors stimulation exerts an opposite effect on microglia activity (Table 2). The A_1 receptor agonist CPA inhibited TNFα and GFAP mRNA expression and the A_1 receptor antagonist DPCPX caused an opposite effect on GFAP immunoreactivity in the retina [54]. CPA also lowered TNFα mRNA expression in light-induced retinal degeneration in rats. Activation of A_1 adenosine receptors inhibited LPS-induced TNFα release in activated retinal microglial cells [2]. Madeira and coworkers [64] reported that blockade of A_{2A} receptor prevents retinal microglia activation and inflammatory responses in the retina. On the contrary, the A_{2A} receptor agonist CGS21680 decreased TNFα release in diabetic retina. CGS21680 also attenuated the expression of inflammatory IL-6 and TNFα. Suppression of the inflammatory cascade C-Raf/ERK by A_{2A} receptor activation may be involved in the latter effect in microglia [2]. Ahmad and coworkers [65] also reported that activation of A_{2A} adenosine receptors attenuated the hypoxia- or LPS-induced TNFα release. At present, conflicting reports can be found in the literature indicating that both stimulation and blockade of A_{2A} receptors regulate the release of microglial proinflammatory bioactive proteins in the same direction [11,39,66]. The A_3 receptor agonist CF101 exerts anti-inflammatory effects and inhibits proinflammatory cytokine release (TNFα, IL-6, IL-12). Downregulation of the NF-kB signaling pathway was suggested to be involved in the mechanism of CF101 [2].

Table 2. Purinoceptors, their signal transductions and endogenous ligands.

Receptor	Signal Transduction	Ligands
P2Y nucleotide purinoceptors		
P2Y purinoceptors subtypes		
$P2Y_{1,2,4,6,11*}$	G_q protein-coupled	ATP, ADP and/or UTP, UDP
P2Y purinoceptors subtypes		
$P2Y_{12,13,14}$	G_i protein-coupled	ADP, UDP
P2X nucleotide purinoceptors		
P2X purinoceptors subtypes		
$P2X_{1,2,3,4,5,6,7}$	cationic ion channel-coupled	ATP
P1 nucleoside purinoceptors		
A_1 adenosine receptor	G_i protein-coupled	adenosine
A_{2A}, A_{2B} adenosine receptor	G_s protein-coupled	adenosine
A_3 adenosine receptor	G_i protein-coupled	adenosine

$P2Y_{1,12,13}$ purinoceptors are sensitive to adenine nucleotides, $P2Y_{2,4}$ purinoceptors are activated by adenine and uridine nucleotides, human $P2Y_{24,6}$ are affected primarily by uridine nucleotides [4,23], $P2Y_{11*}$ purinoceptors also mediate stimulation of adenylyl cyclase activation. Increased cAMP may mobilize intracellular Ca^{2+} via activation of PLC-epsilon [67,68].

5. Glycinergic Transmission in the Retina

Glycine is an inhibitory neurotransmitter that exerts inhibition on various neurons by acting on strychnine-sensitive glycine receptors (GlyRs) in the retina. All the α1-4 subunits of GlyRs were identified in the mammalian retina [69]. Glycine also has a mandatory coagonist role in activation of glutamate N-methyl-D-aspartate (NMDA) receptors both in the central nervous system and in the retina [70–72].

The neural circuit of the retina contains two kinds of glycinergic interneurons. Amacrine cells in the inner nuclear layer are synapsed with the cone pathway and establish feedback inhibition of bipolar cells thereby inhibiting responses of ganglion cells to light [73]. About half of the amacrine cells are glycinergic and they form in- and output innervations with bipolar cells, ganglion cells, and other amacrine cells [74]. Glycinergic amacrine cells synapse primarily with OFF bipolar cells [75] (Figure 1B). The other type of interneuron, which operates with glycine as a neurotransmitter, is the AII amacrine cells [8,76]. These interneurons transfer the light signal from rod bipolar cells to the cone pathways. Of the retinal glial cells, Müller cells and astrocytes are the potential sources of glycine. The cellular location of glycine in the retina indicates that it serves as a neurotransmitter in neurons and a gliotransmitter in glia cells and also participates in microglia activation [77].

The release of glycine from interneurons and macroglia cells occurs with vesicular exocytosis or reverse-mode operation of glycine transporter 1 (GlyT-1). Following neuronal release, glycine may reach the extrasynaptic space by spillover mechanism and affects extrasynaptic GlyR and NMDA receptor activities. In the vicinity of these receptors, glycine concentration is determined by GlyT-1 both in the synaptic and extrasynaptic space of the retina [78]. GlyT-1 is primarily localized to glial cells (Table 3), whereas GlyT-2, which is responsible for refilling presynaptic vesicles with glycine, is absent in the retina [79,80], but see Pena-Rangel et al. [81].

Table 3. Some characteristics of glia cells in the retina.

Glia Cell Type	Locations in the Retina	Receptor Subtype Expression	GlyT Expression
		Macroglia	
Müller cells	Span all retina layers	$P2Y_1$ $P2X_7$ $A_{2A/2B}$	GlyT-1
Astroglia	Ganglion cell layer	$P2Y_{1,12,13}$ $P2X_7$	GlyT-1
		Microglia	
In healthy retina	Inner and outer plexiform layers	$P2X_{4,7}$ $A_{2A/2B}$	?
In retina pathologies	Ganglion cell layer Perivascular accumulation Outer nuclear layer Subretinal space	$P2Y_{1,2,4,12}$ $P2X_7$ $A_{1,2A/2B,3}$	?

?-no data about GlyT in microglia.

Of the macroglia cells, glycine is released from both Müller glia cells and astrocytes in the retina. In energy shortage conditions, declining cellular ATP levels cause failure of ion Na+-K+-ATPase ion pump operation and as a result intracellular Na+ concentration is elevated [1]. This altered ionic milieu favors shifting the operation of both glycine and glutamate transporters into the reverse mode [82]. The consequence of transporter reversal is an increased release of the amino acid neurotransmitters glycine and glutamate as was shown in isolated rat hippocampus and retina as well [15,78]. In response to hypoxia and cellular energy failure, ATP is also released from neuronal or glial sources, although the mechanism of this release differs from that of glutamate or glycine [35].

5.1. Heterogeneity of NMDA Receptors in the Retina

The ionotropic glutamate NMDA receptors, on which glycine acts as a mandatory coagonist, exhibit uneven distribution in the retina: A great number of NMDA receptors can be found in the ganglion cell layer [9,83]. NMDA receptors are expressed in retinal ganglion cells with different subunit compositions: NR1/NR2B NMDA receptors were found on ON ganglion cells in higher density, whereas the presence of NR1/NR2A NMDA receptors was demonstrated both in ON and OFF ganglion cell dendrites [84]. NR1/NR2A receptors are believed to be localized synaptically and extrasynaptic NMDA receptors are primarily composed of NR1NR2B subunits [85]. NR1/NR2A and NR1/NR2B receptors possess different coagonists, D-serine and glycine, respectively [86]. D-Serine may be released from stressed Müller glia cells and astrocytes to potentiate the agonist effect of glutamate on synaptic NMDA receptors [87]. On the other hand, extrasynaptic glycine potentiates the effect of glutamate on extrasynaptic NR1/NR2B NMDA receptors [88]. Glycine, after its release from synapses between amacrine cells and the OFF pathway neurons, diffuses into the extrasynaptic space and influences activity of extrasynaptic NR1/NR2B receptors primarily located on ON ganglion cells [75,89]. Extrasynaptic NR1/NR2B NMDA receptors, which exhibit higher glycine sensitivity, mediate ganglion cell death by activating pro-death intracellular signaling in retinal ischemia/hypoxia [90,91].

5.2. Glycinergic Regulation of Microglia

Glycine bears modulatory features on microglia and is able to influence their pro- and anti-inflammatory functions. Accordingly, Carmans and coworkers [92] reported that glycine activated macrophage functions and stimulated the production of TNFα and other proinflammatory mediators. This effect of glycine was not mediated by GlyRs but rather by activation of the neutral amino acid transporter (NAAT) system. Glycine activates Na^+-dependent NAAT leading to increases in intracellular Na^+ concentrations, inducing membrane depolarization, and enhancement of Ca^{2+} signaling; these events are essential for the release of inflammatory modulators [93]. Increases in extracellular glycine concentrations in response to ischemia of the retina or brain tissues may be high enough to activate the NAAT system in macrophages or microglial cells.

Schilling and Eder [77] also reported GlyR- and GlyT-independent, but NAAT-dependent, glycine-mediated effects in mouse microglia. Metabolic activity of cultured microglia was increased in the presence of glycine, whereas it was suppressed in a glycine-free medium [22]. Glycine is also known to enhance LPS-induced nitric oxide and superoxide productions in microglial cells [93]. Furthermore, glycine enhanced ATP-induced intracellular Ca^{2+} transients in microglial cells and this effect was due to Na^+-coupled NAAT operation [93]. These effects of glycine in microglia point to a depolarization-dependent Ca^{2+} signaling and an increased production of inflammatory mediators. Ca^{2+} signals may, however, be under the control of different signaling systems in microglia. As mentioned above, ATP enhances the glycine-evoked intracellular Ca^{2+} transients in microglial cells possibly via activation of P2X purinoceptors. It was shown by Raouf and coworkers [94] that ATP initiates inflammatory cascades by acting on surface P2X purinoceptors of microglial cells.

Others concluded, however, that glycine has an inhibitory effect on immune cells by reducing the production of proinflammatory cytokines. It has been reported that glycine decreased the production of TNFα and IL-1β in human monocytes and inhibited TNFα production in rat alveolar macrophages [95,96]. These effects of glycine may be mediated by GlyR-dependent mechanisms rather than the NAAT system. These findings indicate that glycine influences macrophages by both GlyR-dependent and GlyR-independent mechanisms.

Hayashi and coworkers [97] reported that transferred microglia- and microglia-conditioned medium potentiated NMDA receptor-mediated synaptic responses in cortical neurons. Analyses with HPLC methods revealed that glycine and serine present in microglia-conditioned medium may be responsible for potentiation of NMDA-induced currents. Released glycine accumulates around glutamatergic synapses and it may build up a concentration sufficient to induce NMDA receptor activity-mediated neurotoxicity. Further in this line, microglia, after activation, have been reported to induce NMDA receptor-mediated neurotoxicity and neuronal death [22].

In energy compromised conditions, when disruption of the balances between normal- and reverse-mode operation of GlyT-1 occurs, GlyT-1 inhibitors decrease extracellular glycine levels [15,78], i.e., this class of compounds may negatively influence microglia activity.

6. Neurodegenerative and Neuroinflammatory Disorders in the Retina

Of the various clinical appearances of retinopathies, hypoxia/ischemia, glaucoma, diabetic neuropathy, and age-related macular degeneration affect a great number of the population. While these disorders exhibit different symptoms, there are several common evens in their pathogeneses. Retinal hypoxia as a persisting insult may induce enhanced purinergic signaling and glutamatergic–glycinergic neurotransmission. Activation of the two systems may lead to a sustained release of inflammatory mediators from activated microglia and the resulting chronic inflammatory environment may initiate or exacerbate retinal neurodegenerative disorders. Ganglion cell destruction can be detected in parallel with abnormally elevated activity of microglia in almost all neurodegenerative disorders of the retina.

6.1. Enhanced Purinergic Signaling in Neurodegenerative/Neuroinflammatory Disorders of the Retina

Purinergic signaling is activated in all neurodegenerative disorders of the retina. ATP is released in excess amount into the extracellular space in hypoxia, inflammation, oxidative stress, or nutrient starvation [4]. Ischemia-like conditions trigger ATP release from rat hippocampal slices and this release likely occurs from the ischemic retina as well [35,98]. Using an in vivo microdialysis technique, co-release of ATP and glutamate was demonstrated from the accumbens nucleus following traumatic injury in the rat and interaction of ATP and glutamate has been proposed [68]. In addition, the non-selective P2 purinoceptor antagonist PPADS and the selective $P2X_7$ purinoceptor antagonist BBG inhibited oxygen and glucose deprivation-induced glutamate release in rat hippocampus [99]. These findings indicate that endogenous ATP, when its release is evoked in energy-compromised conditions, may exert a stimulatory influence on glutamate release. Oxygen-glucose deprivation also evokes glutamate release from isolated rat retina [15] and glutamate released in ischemic retina evokes Ca^{2+}-independent ATP release and transporter-mediated adenosine release [100]. In these experiments, hypoxia was induced by elevated intraocular pressure in rat retina and elevated glutamate release activated purinergic signaling by releasing ATP from glial cells. Thus, a bidirectional facilitation may exist between purinergic signaling and glutamatergic neurotransmission in ischemic neural and retinal tissues.

Experimental observations suggest that purinergic signaling involves P2Y purinoceptors in the mediation of both neurodegenerative and neuroprotective processes. In mice, deletion of P2Y receptors from Müller glia and ganglion cells was associated with increased survival of amacrine cells following retinal ischemia reperfusion, whereas ischemic death of photoreceptor cells appeared to be more pronounced. P2Y purinoceptor activation likely increases cytosolic Ca^{2+}-signaling during ischemia leading to neurodegeneration of amacrine cells [51]. Accordingly, P2Y purinoceptors may support survival of photoreceptors but result in death of amacrine cells. In the zebrafish retina, P2Y1 purinoceptor stimulation by ADP exerts neuroprotection of inner retinal neurons [38]. These findings raise the possibility that neurons express P2Y purinoceptors coupled to Gq or Gi proteins and the excitatory or inhibitory downstream signaling results in neurodegeneration or neuroprotection in the retina [5].

An increase in extracellular ATP concentration also evokes $P2X_7$ purinoceptor stimulation, which opens receptor-coupled ion channels and cation influx into the cells. Activation of $P2X_7$ purinoceptors, which ensures excess Ca^{2+} permeability in neurons, occurs in tissue damage, neuroinflammation, or mechanical stress. Elevated intraocular pressure is also accompanied by increased levels of ATP in the extracellular space [13]. Moreover, $P2X_7$ purinoceptor activation induces release of $TNF\alpha$ and proinflammatory cytokines from microglia and may cause neuronal degeneration and cell death [34]. Stimulation of the $P2X_7$ purinoceptor has a role in ischemic neurodegeneration in the retina and can modulate retinal ganglion cell destruction [52].

Increased adenosine tissue concentration was observed in rat retina following ligation of the central retinal artery [41]. This increase in tissue adenosine levels may be the consequence of increased ATP hydrolysis due to imbalance between energy supply and demand [1]. Adenosine produced from released ATP may contribute to the neuroprotective effect in ischemic retina [41]. Alternatively, energy deprivation provokes enhanced intracellular Na^+ concentrations and equilibrative nucleoside transporter type-1 operating in the reverse-mode increases adenosine efflux [4]. Elevated extracellular adenosine may activate a series of P1 adenosine receptors expressed in retinal neurons or glial cells leading to a short-term antiischemic effect [42,101]. Accordingly, the A_1 receptor agonist R-PIA reduced retinal damage induced by the raise in intraocular pressure and improved electroretinogram after ischemia in the rat eye [102]. On the other hand, the blockade of A_1 receptors increases neurotoxicity caused by hypoxia/ischemia or glutamate [46]. Thus, the Gi protein-coupled A_1 receptor signal transduction can be considered as an endogenous neuroprotective system. This neuroprotective effect may be attributed to inhibition of neuronal Ca^{2+} uptake [39]. In addition, activation of the inhibitory A_1 receptors by adenosine attenuates the release of excitatory amino acids and exerts neuroprotection during glutamatergic neurotoxicity [46].

A_{2A} receptors, which are coupled to Gs proteins, facilitate glutamate release and also facilitate NMDA receptor function at postsynaptic levels [46]. Blockade of A_{2A} receptors with selective antagonists protect ganglion cells in retinal ischemia [11,42,101]. In microglia, inhibition of A_{2A} adenosine receptors reduces the inflammatory response in retinal pigment epithelium and also reduces photoreceptor cell death [64]. In addition, the adenosine A_{2A} receptor antagonist SCH58261 prevented microglia-mediated neuroinflammation and induced neuroprotection in the retina [64]. However, a series of experiments is paradoxical: It has been reported that the A_{2A} receptor agonist CGS21680, similarly to the receptor antagonist, decreased microglial activation and retinal cell death in a mouse traumatic optic neuropathy model [65], but see Santiago et al. [66].

6.2. Enhanced Glutamatergic–Glycinergic Tone in Neurodegenerative/Neuroinflammatory Disorders of the Retina

The concentration of glutamate, the principal neurotransmitter in the retina, is elevated in the vitreous body or aqueosus humor following experimental retinal ischemia, optic nerve neuropathy, and also in glaucoma patients [103]. Increases in extracellular glutamate concentration facilitate glutamate uptake into Müller glia cells, leading to cellular swelling [100]. This finding points to the important role of glia cells in neurodegenerative disorders including excitotoxic injury of the retina [1,103]. In addition, NMDA receptors are overstimulated when extracellular glutamate concentration is increased and this increase plays a pivotal role in neurodegenerative disorders of the retina. The resulting cellular insults can be ameliorated by using various NMDA receptor antagonists [1,104], albeit their clinical usefulness is a subject of debate.

Increased glycine levels have also been demonstrated in neurodegenerative–neuroinflammatory disorders of the central nervous system (amyotrophic lateral sclerosis, multiple sclerosis) and also of the eye [92]. Elevated glycine concentrations were found in the vitreous body in ischemia following ophthalmic artery occlusion [105]. Exocytotic glycine release can also be demonstrated from glycinergic amacrine cells and this release may be directed outward concerning the synaptic cleft by the spillover mechanism. Glycine may be released from Müller glia cells and astrocyte by reverse-mode operation of GlyT-1 [71].

In physiological conditions, the GlyT-1-mediated glycine uptake reduces the possibility of glycine/glutamate-induced neurotoxicity. In oxygen and glucose deprivation-induced neurotoxicity, a decrease in glycine uptake and an increase in glycine release were found to indicate a forced reverse-mode operation of GyT-1 in rat retina [15]. Ischemic conditions evoke simultaneous increase in [^{3}H]glutamate and [^{3}H]glycine release in the retina ensuring excess availability of the agonist and co-agonist at NR1/NR2B-type NMDA receptors [15]. The coagonist role of glycine in NMDA receptor activation indicates that glycine concentrations determined by GyT-1 may have a key role in neurodegenerative disorders of the retina [15,71].

6.3. Activated Microglial Cells in Neurodegenerative Disorders of the Retina

Elevated intraocular pressure, which results in hypoxia in the retina, elicits glaucomatous neuropathy characterized by damage of ganglion cell axons and gradual loss of ganglion cells [100,106]. Reactive microglia have been observed in the retina following increased ocular pressure [11]. Microglial cells are redistributed in the retina as they can be detected in close vicinity of ganglion cell soma and axons [10]. Activation of microglia is showed by the observed increases of TNFα, IL-6, and IL-8 proinflammatory cytokines in aqueous humor obtained from glaucoma patients [11]. These findings indicate that a neuroinflammatory component is present and plays a significant role in glaucomatous retinal neuropathy.

Retinal hypoxia may also be involved in the pathomechanism of diabetic neuropathy of the retina [51]. Tissue hypoxia may result from insufficient microvascular circulation in diabetic neuropathy, which is accompanied by neurodegenerative events in the neural network of the retina. Reactive microglia in diabetic retinopathy are hypertrophic and exhibit an amoeboid shape. The perivascular presence of reactive microglia can be detected in cluster formation in diabetic retina [7]. Increased cytokine levels (TNF, IL-1β, IL-6, and IL-8) have been reported in the vitreous fluid of the eye indicating the involvement of a chronic inflammatory process in diabetic retinopathy [3]. Microcirculatory failure also induces a release of VEGF in diabetic retinopathy [107].

In age-related macular degeneration, pathological changes can be found in the retinal pigment epithelium and loss of photoreceptors is a key finding [64]. Impairment of photoreceptor functions is a consequence of neovascularization as novel blood vessels are formed in the outer retina in the wet-form of this retinal pathology. In age-related macular degeneration, microglia are accumulated in the outer nuclear layer and the subretinal space and around drusen deposits. Choroidal neovascularization and degeneration of the retinal pigment epithelium and photoreceptors are further characteristics of this disorder. Microglia are enlarged and exhibit an amoeboid shape in the retinal pigment epithelium. Activated microglia induce secretion of the proinflammatory cytokine IL-1β and increased cytokine levels were reported in the aqueosus humor [3]. In wet-form of age-related macular degeneration, hypoxia/ischemia induces expression of VEGF, which results in disruption of the blood–retinal barrier and retinal edema underlies the pathology of macular degeneration [31].

7. A Tripartite Interaction: Purinergic–Glycinergic Cross-Talk and Microglia Activation in Neurodegenerative–Neuroinflammatory Disorders of the Retina

We have hypothesized a tripartite interaction in the development of neuroinflammatory–neurodegenerative disorders of the retina with the participation of the purinergic signaling, glutamatergic–glycinergic neurotransmission, and activated microglial cells. This interaction may be synergistic or attenuative depending on whether the participating events strengthen or debilitate each other's effects, creating positive or negative feedback regulation. Whether pathological changes reinforce or weaken the functions of the participants in the tripartite interaction may have importance in microglia activation. Thus, microglia in an M1-activated state may turn into a sustained activation and chronic inflammatory pathologies in the retina, whereas microglia turning into an M2 state induce neuroregeneration, remodeling, and neuronal repair [30].

Activation of the tripartite interaction may be triggered by anoxia/ischemia resulting from disturbances in retinal blood circulation [31,52]. Thromboembolic occlusion of the retinal artery or vein, microcirculatory abnormalities in diabetic neuropathy, increased intraocular pressure, pathological angiogenesis in age-related macular degeneration, or arteriosclerosis of arteries in the optic nerve head all lead to reduced arterial blood flow in the posterior eye [32,105,108]. These disturbances in blood circulation lead to cellular energy shortage, which is accompanied by increased release of ATP and the amino acid neurotransmitters glycine and glutamate from neuronal or glial sources. ATP, when released in response to hypoxia, stimulates P2X receptors expressed on resting microglial cells facilitating their transit to activated state and neuroinflammatory conditions develop. Microglial cell activation by purinergic signaling may result in a sustained pathological state as ATP evokes glutamate release and in turn glutamate stimulates further efflux of ATP [28]. Thus, there might be a self-strengthening regulation between ATP and glutamate release occurring in hypoxic conditions. Further in line with glutamate released from neurons or macroglia cells in high concentrations activates NMDA receptors leading to a condition when excitotoxicity may be triggered.

Hypoxic conditions, however, also evoke glycine release from neurons and astrocytes in the central nervous system and retina. High extracellular glycine concentration stimulates microglia activity and the glycine-induced proinflammatory cytokine release may be an additional factor in neuronal cell damage in retinal circuitry [34,100]. Moreover, high extracellular glycine levels ensure the required coagonist concentration for activation of NMDA receptors. Microglial glycine has been reported to participate in stimulation of NMDA receptors leading to the possibility that activated microglial cells evoke in part overactive NMDA receptor-induced neurodegeneration [22]. Thus, interaction among neurons, macroglia, and microglial cells in glycine release also forms a self-strengthening regulation, which may end up in neurodegeneration.

Besides these self-strengthening mechanisms, attenuative regulations can also be observed in the proposed tripartite interaction. Accordingly, activated microglia are altered in order to express purinoceptors of the P2Y type and their activation decreases the production and release of proinflammatory cytokines and progress of neuroinflammation may slow down [56]. Moreover, P2Y purinoceptors activated by ATP on astrocytes stimulate normal (uptake) mode operation of GlyT-1 leading to a decrease of extracellular glycine concentrations [50]. Recognition of the coagonist role of glycine in glutamate-mediated NMDA receptor activation led to the conclusion that decreases in extracellular glycine concentration may actually result in NMDA receptor hypofunctionality and exert neuroprotection in retinal neurodegenerative disorders. This goal can also be reached by inhibition of the reverse-mode operation of GlyT-1 by its selective inhibitors [15,109]. As shown in Figure 2A, the GlyT-1 inhibitor ACPPB decreased the oxygen-glucose deprivation-induced [^{3}H]glycine release in isolated rat retina and this effect was due to an inhibition of the reverse-mode operation of GlyT-1. Besides inhibition of the reverse mode operation of GlyT-1, stimulation of normal mode operation also leads to a decrease of extracellular glycine concentrations and consequently to inhibition of overactivated NMDA receptors.

Jiménez [50] and coworkers reported that the P2Y$_{1,12,13}$ purinoceptor agonist 2-MeS-ADP stimulated the uptake (normal) mode operation of GlyT-1 in rat brainstem primary neuronal cultures. This effect was partially reversed by the P2Y$_1$ antagonist MRS2179, the P2Y$_{13}$ antagonist MRS2211, and the non-selective P2 purinoceptor antagonist suramin (Table 1). This finding suggests the involvement of the Gq protein-coupled P2Y$_1$ and Gi protein-coupled P2Y$_{13}$ purinoceptors in the regulation of GlyT-1 activity with an apparent P2Y$_1$ purinoceptor dominance. Thus, besides GlyT-1 inhibition, P2 purinoceptor stimulation may also decrease extracellular glycine concentrations by stimulation of the normal-mode operation of GlyT-1. As shown in Figure 2B, the non-selective P2 natural purinoceptor agonist ATP decreased the oxygen-glucose deprivation-induced [^{3}H]glycine release from rat retina. While we consider our finding preliminary and in need of further confirmation using more selective receptor ligands, we speculated that ATP stimulating P2Y purinoceptors forces GlyT-1 operation into the uptake mode. The resulting decrease in extracellular glycine concentrations may then prevent overactivation of NMDA receptors and thereby attenuate the exocytotic cascade. The biological substrate for the cross-talk between purinoceptors and GlyT-1 might be the astrocytes, which express P2 type purinoceptors as well as GlyT-1 immunoreactivity in the retina [68,79].

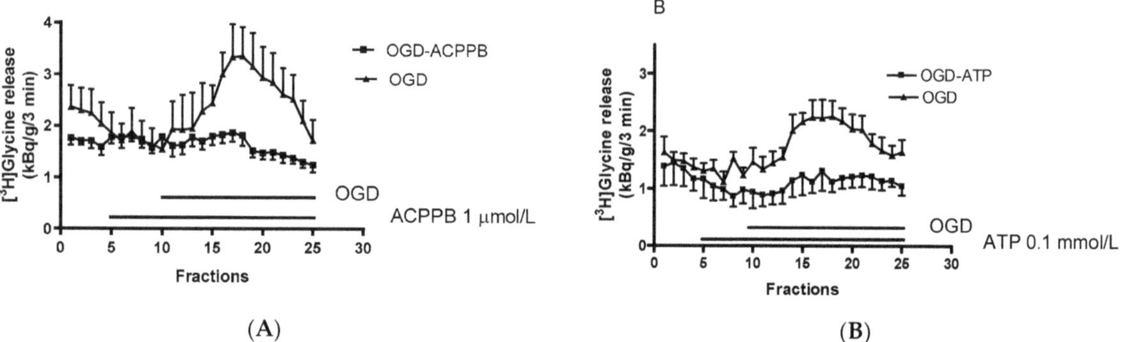

(**A**)   (**B**)

Figure 2. Oxygen and glucose deprivation (OGD)-induced [^{3}H]glycine release from rat retina. This release was reversed by addition of the GlyT-1 inhibitor ACPPB (**A**) and the P2 purinoceptor agonist ligand ATP (**B**). Posterior eyecups containing the retinae were prepared from male Wistar rats, loaded with 10 µCi [^{3}H]glycine for 30 min, and perfused with Krebs-bicarbonate buffer aerated with 95% O_2/5% CO_2 gas mixture for 60 min; then 25 three-min fractions were collected. The perfusion rate was kept at 1 mL/min; [^{3}H]Glycine in the collected fractions and the tissue at the end of superfusion was determined and expressed as kBq/g/3 min. To evoke [^{3}H]glycine release, the eyecups were superfused with glucose-free Krebs-bicarbonate buffer saturated with 95% N_2/5% CO_2 gas mixture added from fraction 10 and maintained through the experiment. The GlyT-1 inhibitor ACPPB, (Merck 13-h, glycine uptake inhibition IC$_{50}$ 1.1×10^{-8} mol/L was determined in rat cerebral cortex synaptosomes) was added in a concentration of 10^{-6} mol/L from fraction 5 and maintained through the experiment [110,111]. ACPPB was synthesized by Professor Dr. Peter Matyus, Semmelweis University, Budapest, Hungary. ATP was added in a concentration of 10^{-4} mol/L from fraction 5 and maintained through the experiment. Data shown as mean±S.E.M., *n* = 3–4. For methodological details see Hanuska et al. [15].

The third participant in this tripartite interaction is microglia. Concerning retinal pathologies, in which glutamatergic–glycinergic neurotransmission and ATP/adenosine-mediated signaling are functional, the presence of activated microglia has also been demonstrated. Induction of proinflammatory cytokine production may be deleterious to injured cells after ischemic insults as was shown in retinal vascular occlusion and reperfusion [112]. Microglial cells in the hypothesized tripartite interaction receive stimulation by glycine released in excess from neurons and astrocytes and by ATP and adenosine via P2X purinoceptors and stimulatory adenosine receptors. On the other hand, a number of experiments point to the inhibitory purinergic influence mediated by ATP and adenosine via P2Y purinoceptors and inhibitory adenosine receptors. The balance between stimulatory and inhibitory influences determines whether activated microglia are shifted to permanent neuroinflammation or remodeling and neuronal repair occur. The hypothesized tripartite interaction between glutamatergic–glycinergic neurotransmission, purinergic modulation, and microglia is shown in Figure 3.

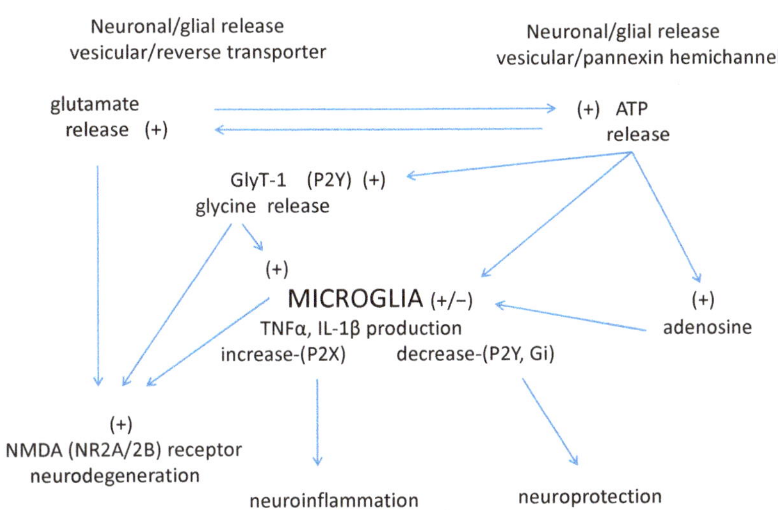

Figure 3. Purinergic–glycinergic cross-talk and microglia activation in neurodegenerative–neuroinflammatory disorders in the retina: A hypothetical model. Impaired microcirculation evokes energy deficiency in the retina leading to increased release of glutamate from neurons or macroglial cells by vesicular exocytosis or reverse-mode operation of excitatory amino acid transporter (EAAT). Cellular energy deficiency also evokes increase in ATP efflux from neurons or glia cells by Ca^{2+}-dependent exocytosis or opening of pannexin hemichannels. There may be a self-strengthening interaction between glutamate and ATP: Glutamate induces ATP release and ATP efflux leads to an increase of glutamate release. In addition, glycine is released from energy-compromised neurons and macroglia by reverse-mode operation of glycine transporter 1 (GlyT-1) in the retina. Excess release of glycine and glutamate overactivates extrasynaptic NR1/NR2B type NMDA receptors, which evokes neurotoxicity. In contrast, synaptic NR1/NR2A type NMDA receptors mediate neuroprotection. Increased glycine release induces further release of glycine from activated microglial cells by altering Ca^{2+} transients and shifts neutral amino acid transporter (NAAT) operation in reverse-mode. Glycine-induced glycine release from microglia, a self-strengthening interaction, facilitates glutamate-induced overstimulation of NMDA receptors, triggering excytotoxicity. This mechanism can be an additional self-strengthening interaction: Glycine induces further release of glycine from activated microglial cells, which participates in extrasynaptic NMDA receptor overstimulation. The evoked NMDA receptor-mediated neurotoxicity primarily damages ON ganglion cells in the retina. Further in the interactions: ATP released by cellular energy deficiency stimulates GlyT-1 normal operation via activation of P2Y purinoceptors, which decreases extracellular glycine levels. GlyT-1 inhibitors inhibit pathological reverse-mode operation of GlyT-1. These two effects on GlyT-1 operation inhibit microglia activity by decreasing extracellular glycine concentrations. ATP released from stressed neurons and glia cells stimulates P2X and P2Y purinoceptors expressed in microglia, which induces stimulation or inhibition of the production and release of TNFα and proinflammatory cytokines (IL-1α, IL-1β, IL-6, IL-12). Adenosine produced by a breakdown of ATP or released from cells stimulates or inhibits microglia activity via adenosine receptors; the participating receptors with opposite effects are the inhibitory A_1 and A_3 and the stimulatory A_{2A} and A_{2B} adenosine receptors. The balance of this tripartite interaction may either be shifted to a neurodegenerative–neuroinflammatory direction or lead to neuroprotection serving inflammation resolution and neuronal survival.

8. Conclusions: Possible Therapeutic Consequences

The tripartite interaction between purinergic modulation, glutamatergic–glycinergic neurotransmission, and microglia activation provides three therapeutic target pathways in neuroinflammatory–neurodegenerative disorders of the retina. First, GlyT-1 inhibitors [16,113] block reverse-mode operation of the transporter in pathological conditions and decrease the concentrations of the coagonist glycine at NR1/NR2B-type NMDA receptors. These effects of GlyT-1 inhibitors may lead to protection of retina ganglion cells. Moreover, decreased extracellular glycine concentrations may also lead to decreased microglia activity with therapeutic consequences.

Second, P2Y purinoceptor agonists and P2X purinoceptor antagonists inhibit micoglial activity and decrease the production and release of TNFα and other proinflammatory biological substances. Thus, modulation of microglial reactivity by P2Y agonists and P2X antagonists has emerged as a possible therapeutic intervention to maintain neurodegenerative courses at a low level. A similar effect might be expected from A_1 adenosine receptor agonists and A_2 adenosine receptor antagonists [43].

Third, microglia activation inhibited by minocycline may suppress the damage of retinal ganglion cells in ischemia and glaucoma [114]. Moreover, decreasing circulated VEGF released from microglia or blockade of VEGF receptor may slow down exacerbation of choroidal neovascularization with a beneficial influence on retina functions in diabetic retinopathy and wet-form of retina macular degeneration [7,31].

Author Contributions: Conceptualization, L.K. and L.G.H.J.; data curation, L.G.H.J., L.K. and T.Z.; writing—original draft preparation, L.G.H.J.; writing—review and editing, L.K., L.G.H.J., G.S., T.Z.; visualization, L.G.H.J., G.S.; funding acquisition, L.K. All authors have read and agreed to the published version of the manuscript.

Funding: This research was funded by the Higher Education Institutional Excellence Program of the Ministry of Human Capacities in Hungary, within the framework of the Neurology thematic program of Semmelweis University (FIKP 2020).

Institutional Review Board Statement: Not applicable.

Informed Consent Statement: Not applicable.

Data Availability Statement: Not applicable.

Acknowledgments: The support of the Higher Education Institutional Excellence Program of the Ministry of Human Capacities in Hungary is greatly acknowledged.

Conflicts of Interest: The authors declare no conflict of interest.

Abbreviations

ATP	adenosine triphosphate
GFAP	glial fibrillary acidic protein
GlyR	glycine receptor
GlyT-1	glycine transporter-1
IL	interleukin
IP_3	inositol trisphosphate
LPS	lipopolysaccharide
NMDA	N-methyl-D-aspartate
NAAT	neutral amino acid transporter
TNFα	tumor necrosis factor α
VEGF	vascular endothelial growth factor

References

1. Osborne, N.N.; Casson, R.J.; Wood, J.P.; Chidlow, G.; Graham, M.; Melena, J. Retinal ischemia: Mechanisms of damage and potential therapeutic strategies. *Prog. Retin. Eye Res.* **2004**, *23*, 91–147. [CrossRef] [PubMed]
2. Guzman-Aranguez, A.; Gasull, X.; Diebold, Y.; Pintor, J. Purinergic receptors in ocular inflammation. *Mediat. Inflamm.* **2014**, *2014*, 320906. [CrossRef]
3. Rashid, K.; Akhtar-Schaefer, I.; Langmann, T. Microglia in Retinal Degeneration. *Front. Immunol.* **2019**, *10*, 1975. [CrossRef] [PubMed]
4. Sperlágh, B. ATP-Mediated Signaling in the Nervous System. In *Handbook of Neurochemistry and Molecular Neurobiology: Neurotransmitter Systems*; Lajtha, A., Vizi, E.S., Eds.; Springer: Boston, MA, USA, 2008; pp. 227–254.
5. Illes, P.; Verkhratsky, A.; Burnstock, G.; Sperlagh, B. Purines in neurodegeneration and neuroregeneration. *Neuropharmacology* **2016**, *104*, 1–3. [CrossRef] [PubMed]

6. Sterling, P.D.; Demb, J.B. Retina. In *The Synaptic Organization of the Brain*; Shepherd, G.M., Ed.; Oxford University Press: Oxford, UK, 2004; pp. 217–270.

7. Madeira, M.H.; Boia, R.; Santos, P.F.; Ambrosio, A.F.; Santiago, A.R. Contribution of microglia-mediated neuroinflammation to retinal degenerative diseases. *Mediat. Inflamm.* **2015**, *2015*, 673090. [CrossRef] [PubMed]

8. Nelson, R.F. Glycinergic neurons process images. *J. Physiol.* **2012**, *590*, 239–240. [CrossRef]

9. Hama, Y.; Katsuki, H.; Tochikawa, Y.; Suminaka, C.; Kume, T.; Akaike, A. Contribution of endogenous glycine site NMDA agonists to excitotoxic retinal damage in vivo. *Neurosci. Res.* **2006**, *56*, 279–285. [CrossRef] [PubMed]

10. Ramirez, A.I.; de Hoz, R.; Salobrar-Garcia, E.; Salazar, J.J.; Rojas, B.; Ajoy, D.; Lopez-Cuenca, I.; Rojas, P.; Trivino, A.; Ramirez, J.M. The Role of Microglia in Retinal Neurodegeneration: Alzheimer's Disease, Parkinson, and Glaucoma. *Front. Aging Neurosci.* **2017**, *9*, 214. [CrossRef]

11. Santiago, A.R.; Baptista, F.I.; Santos, P.F.; Cristovao, G.; Ambrosio, A.F.; Cunha, R.A.; Gomes, C.A. Role of microglia adenosine A(2A) receptors in retinal and brain neurodegenerative diseases. *Mediat. Inflamm.* **2014**, *2014*, 465694. [CrossRef]

12. Bringmann, A.; Pannicke, T.; Grosche, J.; Francke, M.; Wiedemann, P.; Skatchkov, S.N.; Osborne, N.N.; Reichenbach, A. Muller cells in the healthy and diseased retina. *Prog. Retin. Eye Res.* **2006**, *25*, 397–424. [CrossRef]

13. Reichenbach, A.; Bringmann, A. Purinergic signaling in retinal degeneration and regeneration. *Neuropharmacology* **2016**, *104*, 194–211. [CrossRef]

14. Gadea, A.; Lopez, E.; Lopez-Colome, A.M. Characterization of glycine transport in cultured Muller glial cells from the retina. *Glia* **1999**, *26*, 273–279. [CrossRef]

15. Hanuska, A.; Szenasi, G.; Albert, M.; Koles, L.; Varga, A.; Szabo, A.; Matyus, P.; Harsing, L.G., Jr. Some Operational Characteristics of Glycine Release in Rat Retina: The Role of Reverse Mode Operation of Glycine Transporter Type-1 (GlyT-1) in Ischemic Conditions. *Neurochem. Res.* **2016**, *41*, 73–85. [CrossRef]

16. Harsing, L.G., Jr.; Juranyi, Z.; Gacsalyi, I.; Tapolcsanyi, P.; Czompa, A.; Matyus, P. Glycine transporter type-1 and its inhibitors. *Curr. Med. Chem.* **2006**, *13*, 1017–1044. [CrossRef]

17. Van den Eynden, J.; Ali, S.S.; Horwood, N.; Carmans, S.; Brone, B.; Hellings, N.; Steels, P.; Harvey, R.J.; Rigo, J.M. Glycine and glycine receptor signalling in non-neuronal cells. *Front. Mol. Neurosci.* **2009**, *2*, 9. [CrossRef]

18. Cantaut-Belarif, Y.; Antri, M.; Pizzarelli, R.; Colasse, S.; Vaccari, I.; Soares, S.; Renner, M.; Dallel, R.; Triller, A.; Bessis, A. Microglia control the glycinergic but not the GABAergic synapses via prostaglandin E2 in the spinal cord. *J. Cell Biol.* **2017**, *216*, 2979–2989. [CrossRef] [PubMed]

19. Zhang, C.; Lam, T.T.; Tso, M.O. Heterogeneous populations of microglia/macrophages in the retina and their activation after retinal ischemia and reperfusion injury. *Exp. Eye Res.* **2005**, *81*, 700–709. [CrossRef]

20. Okunuki, Y.; Mukai, R.; Nakao, T.; Tabor, S.J.; Butovsky, O.; Dana, R.; Ksander, B.R.; Connor, K.M. Retinal microglia initiate neuroinflammation in ocular autoimmunity. *Proc. Natl. Acad. Sci. USA* **2019**, *116*, 9989–9998. [CrossRef] [PubMed]

21. Uckermann, O.; Uhlmann, S.; Wurm, A.; Reichenbach, A.; Wiedemann, P.; Bringmann, A. ADPbetaS evokes microglia activation in the rabbit retina in vivo. *Purinergic Signal.* **2005**, *1*, 383–387. [CrossRef]

22. Tanaka, J.; Toku, K.; Matsuda, S.; Sudo, S.; Fujita, H.; Sakanaka, M.; Maeda, N. Induction of resting microglia in culture medium devoid of glycine and serine. *Glia* **1998**, *24*, 198–215. [CrossRef]

23. Beamer, E.; Goloncser, F.; Horvath, G.; Beko, K.; Otrokocsi, L.; Kovanyi, B.; Sperlagh, B. Purinergic mechanisms in neuroinflammation: An update from molecules to behavior. *Neuropharmacology* **2016**, *104*, 94–104. [CrossRef]

24. Moller, T.; Kann, O.; Verkhratsky, A.; Kettenmann, H. Activation of mouse microglial cells affects P2 receptor signaling. *Brain Res.* **2000**, *853*, 49–59. [CrossRef]

25. Zhou, T.; Huang, Z.; Sun, X.; Zhu, X.; Zhou, L.; Li, M.; Cheng, B.; Liu, X.; He, C. Microglia Polarization with M1/M2 Phenotype Changes in rd1 Mouse Model of Retinal Degeneration. *Front. Neuroanat.* **2017**, *11*, 77. [CrossRef]

26. Luongo, L.; Guida, F.; Imperatore, R.; Napolitano, F.; Gatta, L.; Cristino, L.; Giordano, C.; Siniscalco, D.; Di Marzo, V.; Bellini, G.; et al. The A1 adenosine receptor as a new player in microglia physiology. *Glia* **2014**, *62*, 122–132. [CrossRef]

27. Viviani, B.; Corsini, E.; Galli, C.L.; Marinovich, M. Glia increase degeneration of hippocampal neurons through release of tumor necrosis factor-alpha. *Toxicol. Appl. Pharmacol.* **1998**, *150*, 271–276. [CrossRef] [PubMed]

28. Eyo, U.B.; Peng, J.; Swiatkowski, P.; Mukherjee, A.; Bispo, A.; Wu, L.J. Neuronal hyperactivity recruits microglial processes via neuronal NMDA receptors and microglial P2Y12 receptors after status epilepticus. *J. Neurosci.* **2014**, *34*, 10528–10540. [CrossRef] [PubMed]

29. Miron, V.E.; Boyd, A.; Zhao, J.W.; Yuen, T.J.; Ruckh, J.M.; Shadrach, J.L.; van Wijngaarden, P.; Wagers, A.J.; Williams, A.; Franklin, R.J.M.; et al. M2 microglia and macrophages drive oligodendrocyte differentiation during CNS remyelination. *Nat. Neurosci.* **2013**, *16*, 1211–1218. [CrossRef]

30. Cherry, J.D.; Olschowka, J.A.; O'Banion, M.K. Neuroinflammation and M2 microglia: The good, the bad, and the inflamed. *J. Neuroinflamm.* **2014**, *11*, 98. [CrossRef]

31. Kaur, C.; Foulds, W.S.; Ling, E.A. Hypoxia-ischemia and retinal ganglion cell damage. *Clin. Ophthalmol.* **2008**, *2*, 879–889. [CrossRef]

32. Whitcup, S.M.; Nussenblatt, R.B.; Lightman, S.L.; Hollander, D.A. Inflammation in retinal disease. *Int. J. Inflamm.* **2013**, *2013*, 724648. [CrossRef]

33. Frade, J.M.; Barde, Y.A. Microglia-derived nerve growth factor causes cell death in the developing retina. *Neuron* **1998**, *20*, 35–41. [CrossRef]
34. Farber, K.; Kettenmann, H. Purinergic signaling and microglia. *Pflüg. Arch.* **2006**, *452*, 615–621. [CrossRef] [PubMed]
35. Reigada, D.; Lu, W.; Zhang, M.; Mitchell, C.H. Elevated pressure triggers a physiological release of ATP from the retina: Possible role for pannexin hemichannels. *Neuroscience* **2008**, *157*, 396–404. [CrossRef] [PubMed]
36. Silverman, W.R.; de Rivero Vaccari, J.P.; Locovei, S.; Qiu, F.; Carlsson, S.K.; Scemes, E.; Keane, R.W.; Dahl, G. The pannexin 1 channel activates the inflammasome in neurons and astrocytes. *J. Biol. Chem.* **2009**, *284*, 18143–18151. [CrossRef]
37. Sperlagh, B.; Szabo, G.; Erdelyi, F.; Baranyi, M.; Vizi, E.S. Homo- and heteroexchange of adenine nucleotides and nucleosides in rat hippocampal slices by the nucleoside transport system. *Br. J. Pharmacol.* **2003**, *139*, 623–633. [CrossRef]
38. Battista, A.G.; Ricatti, M.J.; Pafundo, D.E.; Gautier, M.A.; Faillace, M.P. Extracellular ADP regulates lesion-induced in vivo cell proliferation and death in the zebrafish retina. *J. Neurochem.* **2009**, *111*, 600–613. [CrossRef]
39. Latini, S.; Pedata, F. Adenosine in the central nervous system: Release mechanisms and extracellular concentrations. *J. Neurochem.* **2001**, *79*, 463–484. [CrossRef]
40. Molina-Arcas, M.; Casado, F.J.; Pastor-Anglada, M. Nucleoside transporter proteins. *Curr. Vasc. Pharmacol.* **2009**, *7*, 426–434. [CrossRef] [PubMed]
41. Roth, S.; Rosenbaum, P.S.; Osinski, J.; Park, S.S.; Toledano, A.Y.; Li, B.; Moshfeghi, A.A. Ischemia induces significant changes in purine nucleoside concentration in the retina-choroid in rats. *Exp. Eye Res.* **1997**, *65*, 771–779. [CrossRef]
42. Ghiardi, G.J.; Gidday, J.M.; Roth, S. The purine nucleoside adenosine in retinal ischemia-reperfusion injury. *Vis. Res.* **1999**, *39*, 2519–2535. [CrossRef]
43. Jacobson, K.A.; Civan, M.M. Ocular Purine Receptors as Drug Targets in the Eye. *J. Ocul. Pharmacol. Ther.* **2016**, *32*, 534–547. [CrossRef]
44. Housley, G.D.; Bringmann, A.; Reichenbach, A. Purinergic signaling in special senses. *Trends Neurosci.* **2009**, *32*, 128–141. [CrossRef]
45. Wurm, A.; Erdmann, I.; Bringmann, A.; Reichenbach, A.; Pannicke, T. Expression and function of P2Y receptors on Muller cells of the postnatal rat retina. *Glia* **2009**, *57*, 1680–1690. [CrossRef] [PubMed]
46. Cunha, R.A. Adenosine Neuromodulation and Neuroprotection. In *Handbook of Neurochemistry and Molecular Neurobiology: Neurotransmitter Systems*; Lajtha, A., Vizi, E.S., Eds.; Springer: Boston, MA, USA, 2008; pp. 255–273.
47. Halassa, M.M.; Fellin, T.; Haydon, P.G. The tripartite synapse: Roles for gliotransmission in health and disease. *Trends Mol. Med.* **2007**, *13*, 54–63. [CrossRef] [PubMed]
48. Illes, P.; Norenberg, W.; Gebicke-Haerter, P.J. Molecular mechanisms of microglial activation. B. Voltage- and purinoceptor-operated channels in microglia. *Neurochem. Int.* **1996**, *29*, 13–24. [CrossRef]
49. Sanderson, J.; Dartt, D.A.; Trinkaus-Randall, V.; Pintor, J.; Civan, M.M.; Delamere, N.A.; Fletcher, E.L.; Salt, T.E.; Grosche, A.; Mitchell, C.H. Purines in the eye: Recent evidence for the physiological and pathological role of purines in the RPE, retinal neurons, astrocytes, Muller cells, lens, trabecular meshwork, cornea and lacrimal gland. *Exp. Eye Res.* **2014**, *127*, 270–279. [CrossRef]
50. Jimenez, E.; Zafra, F.; Perez-Sen, R.; Delicado, E.G.; Miras-Portugal, M.T.; Aragon, C.; Lopez-Corcuera, B. P2Y purinergic regulation of the glycine neurotransmitter transporters. *J. Biol. Chem.* **2011**, *286*, 10712–10724. [CrossRef]
51. Pannicke, T.; Frommherz, I.; Biedermann, B.; Wagner, L.; Sauer, K.; Ulbricht, E.; Hartig, W.; Krugel, U.; Ueberham, U.; Arendt, T.; et al. Differential effects of P2Y1 deletion on glial activation and survival of photoreceptors and amacrine cells in the ischemic mouse retina. *Cell Death Dis.* **2014**, *5*, e1353. [CrossRef] [PubMed]
52. Niyadurupola, N.; Sidaway, P.; Ma, N.; Rhodes, J.D.; Broadway, D.C.; Sanderson, J. P2X7 receptor activation mediates retinal ganglion cell death in a human retina model of ischemic neurodegeneration. *Investig. Ophthalmol. Vis. Sci.* **2013**, *54*, 2163–2170. [CrossRef] [PubMed]
53. Wurm, A.; Pannicke, T.; Iandiev, I.; Francke, M.; Hollborn, M.; Wiedemann, P.; Reichenbach, A.; Osborne, N.N.; Bringmann, A. Purinergic signaling involved in Muller cell function in the mammalian retina. *Prog. Retin. Eye Res.* **2011**, *30*, 324–342. [CrossRef] [PubMed]
54. Solino, M.; Lopez, E.M.; Rey-Funes, M.; Loidl, C.F.; Larrayoz, I.M.; Martinez, A.; Girardi, E.; Lopez-Costa, J.J. Adenosine A1 receptor: A neuroprotective target in light induced retinal degeneration. *PLoS ONE* **2018**, *13*, e0198838. [CrossRef]
55. Wang, X.; Franciosi, S.; Bae, J.H.; Kim, S.U.; McLarnon, J.G. Expression of P2y and P2x receptors in cultured human microglia. *Proc. West. Pharmacol. Soc.* **1999**, *42*, 79–81.
56. Ogata, T.; Chuai, M.; Morino, T.; Yamamoto, H.; Nakamura, Y.; Schubert, P. Adenosine triphosphate inhibits cytokine release from lipopolysaccharide-activated microglia via P2y receptors. *Brain Res.* **2003**, *981*, 174–183. [CrossRef]
57. Wang, M.; Wong, W.T. Microglia-Müller cell interactions in the retina. *Adv. Exp. Med. Biol.* **2014**, *801*, 333–338. [CrossRef]
58. Inoue, K. UDP facilitates microglial phagocytosis through P2Y6 receptors. *Cell Adhes. Migr.* **2007**, *1*, 131–132. [CrossRef]
59. Orr, A.G.; Orr, A.L.; Li, X.J.; Gross, R.E.; Traynelis, S.F. Adenosine A(2A) receptor mediates microglial process retraction. *Nat. Neurosci.* **2009**, *12*, 872–878. [CrossRef] [PubMed]
60. Relvas, L.J.; Bouffioux, C.; Marcet, B.; Communi, D.; Makhoul, M.; Horckmans, M.; Blero, D.; Bruyns, C.; Caspers, L.; Boeynaems, J.M.; et al. Extracellular nucleotides and interleukin-8 production by ARPE cells: Potential role of danger signals in blood-retinal barrier activation. *Investig. Ophthalmol. Vis. Sci.* **2009**, *50*, 1241–1246. [CrossRef] [PubMed]

61. Ferrari, D.; Villalba, M.; Chiozzi, P.; Falzoni, S.; Ricciardi-Castagnoli, P.; Di Virgilio, F. Mouse microglial cells express a plasma membrane pore gated by extracellular ATP. *J. Immunol.* **1996**, *156*, 1531–1539. [PubMed]
62. Hide, I.; Tanaka, M.; Inoue, A.; Nakajima, K.; Kohsaka, S.; Inoue, K.; Nakata, Y. Extracellular ATP triggers tumor necrosis factor-alpha release from rat microglia. *J. Neurochem.* **2000**, *75*, 965–972. [CrossRef]
63. Shigemoto-Mogami, Y.; Koizumi, S.; Tsuda, M.; Ohsawa, K.; Kohsaka, S.; Inoue, K. Mechanisms underlying extracellular ATP-evoked interleukin-6 release in mouse microglial cell line, MG-5. *J. Neurochem.* **2001**, *78*, 1339–1349. [CrossRef]
64. Madeira, M.H.; Rashid, K.; Ambrosio, A.F.; Santiago, A.R.; Langmann, T. Blockade of microglial adenosine A2A receptor impacts inflammatory mechanisms, reduces ARPE-19 cell dysfunction and prevents photoreceptor loss in vitro. *Sci. Rep.* **2018**, *8*, 2272. [CrossRef]
65. Ahmad, S.; Fatteh, N.; El-Sherbiny, N.M.; Naime, M.; Ibrahim, A.S.; El-Sherbini, A.M.; El-Shafey, S.A.; Khan, S.; Fulzele, S.; Gonzales, J.; et al. Potential role of A2A adenosine receptor in traumatic optic neuropathy. *J. Neuroimmunol.* **2013**, *264*, 54–64. [CrossRef]
66. Santiago, A.R.; Madeira, M.H.; Boia, R.; Aires, I.D.; Rodrigues-Neves, A.C.; Santos, P.F.; Ambrosio, A.F. Keep an eye on adenosine: Its role in retinal inflammation. *Pharmacol. Ther.* **2020**, *210*, 107513. [CrossRef]
67. Bucheimer, R.E.; Linden, J. Purinergic regulation of epithelial transport. *J. Physiol.* **2004**, *555*, 311–321. [CrossRef] [PubMed]
68. Franke, H.; Krugel, U.; Illes, P. P2 receptors and neuronal injury. *Pflüg. Arch.* **2006**, *452*, 622–644. [CrossRef]
69. Wassle, H.; Heinze, L.; Ivanova, E.; Majumdar, S.; Weiss, J.; Harvey, R.J.; Haverkamp, S. Glycinergic transmission in the Mammalian retina. *Front. Mol. Neurosci.* **2009**, *2*, 6. [CrossRef] [PubMed]
70. Johnson, J.W.; Ascher, P. Glycine potentiates the NMDA response in cultured mouse brain neurons. *Nature* **1987**, *325*, 529–531. [CrossRef] [PubMed]
71. Harsing, L.G., Jr.; Albert, M.; Matyus, P.; Szenasi, G. Inhibition of hypoxia-induced [(3)H]glycine release from chicken retina by the glycine transporter type-1 (GlyT-1) inhibitors NFPS and Org-24461. *Exp. Eye Res.* **2012**, *94*, 6–12. [CrossRef]
72. Eulenburg, V.; Knop, G.; Sedmak, T.; Schuster, S.; Hauf, K.; Schneider, J.; Feigenspan, A.; Joachimsthaler, A.; Brandstatter, J.H. GlyT1 determines the glycinergic phenotype of amacrine cells in the mouse retina. *Brain Struct. Funct.* **2018**, *223*, 3251–3266. [CrossRef]
73. Masland, R.H. The fundamental plan of the retina. *Nat. Neurosci.* **2001**, *4*, 877–886. [CrossRef]
74. Dumitrescu, O.N.; Protti, D.A.; Majumdar, S.; Zeilhofer, H.U.; Wassle, H. Ionotropic glutamate receptors of amacrine cells of the mouse retina. *Vis. Neurosci.* **2006**, *23*, 79–90. [CrossRef] [PubMed]
75. Ivanova, E.; Muller, U.; Wassle, H. Characterization of the glycinergic input to bipolar cells of the mouse retina. *Eur. J. Neurosci.* **2006**, *23*, 350–364. [CrossRef] [PubMed]
76. Lee, E.-J.; Mann, L.B.; Rickman, D.W.; Lim, E.-J.; Chun, M.-H.; Grzywacz, N.M. AII amacrine cells in the distal inner nuclear layer of the mouse retina. *J. Comp. Neurol.* **2006**, *494*, 651–662. [CrossRef] [PubMed]
77. Schilling, T.; Eder, C. A novel physiological mechanism of glycine-induced immunomodulation: Na+-coupled amino acid transporter currents in cultured brain macrophages. *J. Physiol.* **2004**, *559*, 35–40. [CrossRef]
78. Harsing, L.G., Jr.; Matyus, P. Mechanisms of glycine release, which build up synaptic and extrasynaptic glycine levels: The role of synaptic and non-synaptic glycine transporters. *Brain Res. Bull.* **2013**, *93*, 110–119. [CrossRef] [PubMed]
79. Zafra, F.; Aragon, C.; Olivares, L.; Danbolt, N.C.; Gimenez, C.; Storm-Mathisen, J. Glycine transporters are differentially expressed among CNS cells. *J. Neurosci.* **1995**, *15*, 3952–3969. [CrossRef]
80. Oyama, M.; Kuraoka, S.; Watanabe, S.; Iwai, T.; Tanabe, M. Electrophysiological evidence of increased glycine receptor-mediated phasic and tonic inhibition by blockade of glycine transporters in spinal superficial dorsal horn neurons of adult mice. *J. Pharmacol. Sci.* **2017**, *133*, 162–167. [CrossRef]
81. Pena-Rangel, M.T.; Riesgo-Escovar, J.R.; Sanchez-Chavez, G.; Salceda, R. Glycine transporters (glycine transporter 1 and glycine transporter 2) are expressed in retina. *NeuroReport* **2008**, *19*, 1295–1299. [CrossRef]
82. Thomsen, C. Glycine transporter inhibitors as novel antipsychotics. *Drug Discov. Today Ther. Strateg.* **2006**, *3*, 539–545. [CrossRef]
83. Fletcher, E.L.; Kalloniatis, M. Neurochemical architecture of the normal and degenerating rat retina. *J. Comp. Neurol.* **1996**, *376*, 343–360. [CrossRef]
84. Kalbaugh, T.L.; Zhang, J.; Diamond, J.S. Coagonist release modulates NMDA receptor subtype contributions at synaptic inputs to retinal ganglion cells. *J. Neurosci.* **2009**, *29*, 1469–1479. [CrossRef] [PubMed]
85. Liu, Y.; Wong, T.P.; Aarts, M.; Rooyakkers, A.; Liu, L.; Lai, T.W.; Wu, D.C.; Lu, J.; Tymianski, M.; Craig, A.M.; et al. NMDA receptor subunits have differential roles in mediating excitotoxic neuronal death both in vitro and in vivo. *J. Neurosci.* **2007**, *27*, 2846–2857. [CrossRef] [PubMed]
86. Papouin, T.; Ladepeche, L.; Ruel, J.; Sacchi, S.; Labasque, M.; Hanini, M.; Groc, L.; Pollegioni, L.; Mothet, J.P.; Oliet, S.H. Synaptic and extrasynaptic NMDA receptors are gated by different endogenous coagonists. *Cell* **2012**, *150*, 633–646. [CrossRef] [PubMed]
87. Stevens, E.R.; Esguerra, M.; Kim, P.M.; Newman, E.A.; Snyder, S.H.; Zahs, K.R.; Miller, R.F. D-serine and serine racemase are present in the vertebrate retina and contribute to the physiological activation of NMDA receptors. *Proc. Natl. Acad. Sci. USA* **2003**, *100*, 6789–6794. [CrossRef] [PubMed]
88. Reed, B.T.; Sullivan, S.J.; Tsai, G.; Coyle, J.T.; Esguerra, M.; Miller, R.F. The glycine transporter GlyT1 controls N-methyl-D-aspartic acid receptor coagonist occupancy in the mouse retina. *Eur. J. Neurosci.* **2009**, *30*, 2308–2317. [CrossRef]

89. Eggers, E.D.; Lukasiewicz, P.D. Multiple pathways of inhibition shape bipolar cell responses in the retina. *Vis. Neurosci.* **2011**, *28*, 95–108. [CrossRef]

90. Hardingham, G.E.; Fukunaga, Y.; Bading, H. Extrasynaptic NMDARs oppose synaptic NMDARs by triggering CREB shut-off and cell death pathways. *Nat. Neurosci.* **2002**, *5*, 405–414. [CrossRef]

91. Parsons, M.P.; Raymond, L.A. Extrasynaptic NMDA receptor involvement in central nervous system disorders. *Neuron* **2014**, *82*, 279–293. [CrossRef]

92. Carmans, S.; Hendriks, J.J.; Thewissen, K.; Van den Eynden, J.; Stinissen, P.; Rigo, J.M.; Hellings, N. The inhibitory neurotransmitter glycine modulates macrophage activity by activation of neutral amino acid transporters. *J. Neurosci. Res.* **2010**, *88*, 2420–2430. [CrossRef]

93. Van den Eynden, J.; Notelaers, K.; Brone, B.; Janssen, D.; Nelissen, K.; Sahebali, S.; Smolders, I.; Hellings, N.; Steels, P.; Rigo, J.M. Glycine enhances microglial intracellular calcium signaling. A role for sodium-coupled neutral amino acid transporters. *Pflüg. Arch.* **2011**, *461*, 481–491. [CrossRef] [PubMed]

94. Raouf, R.; Chabot-Dore, A.J.; Ase, A.R.; Blais, D.; Seguela, P. Differential regulation of microglial P2X4 and P2X7 ATP receptors following LPS-induced activation. *Neuropharmacology* **2007**, *53*, 496–504. [CrossRef]

95. Spittler, A.; Reissner, C.M.; Oehler, R.; Gornikiewicz, A.; Gruenberger, T.; Manhart, N.; Brodowicz, T.; Mittlboeck, M.; Boltz-Nitulescu, G.; Roth, E. Immunomodulatory effects of glycine on LPS-treated monocytes: Reduced TNF-alpha production and accelerated IL-10 expression. *FASEB J.* **1999**, *13*, 563–571. [CrossRef]

96. Wheeler, M.D.; Thurman, R.G. Production of superoxide and TNF-alpha from alveolar macrophages is blunted by glycine. *Am J. Physiol.* **1999**, *277*, L952–L959. [CrossRef]

97. Hayashi, Y.; Ishibashi, H.; Hashimoto, K.; Nakanishi, H. Potentiation of the NMDA receptor-mediated responses through the activation of the glycine site by microglia secreting soluble factors. *Glia* **2006**, *53*, 660–668. [CrossRef] [PubMed]

98. Juranyi, Z.; Sperlagh, B.; Vizi, E.S. Involvement of P2 purinoceptors and the nitric oxide pathway in [3H]purine outflow evoked by short-term hypoxia and hypoglycemia in rat hippocampal slices. *Brain Res.* **1999**, *823*, 183–190. [CrossRef]

99. Sperlagh, B.; Zsilla, G.; Baranyi, M.; Illes, P.; Vizi, E.S. Purinergic modulation of glutamate release under ischemic-like conditions in the hippocampus. *Neuroscience* **2007**, *149*, 99–111. [CrossRef]

100. Uckermann, O.; Wolf, A.; Kutzera, F.; Kalisch, F.; Beck-Sickinger, A.G.; Wiedemann, P.; Reichenbach, A.; Bringmann, A. Glutamate release by neurons evokes a purinergic inhibitory mechanism of osmotic glial cell swelling in the rat retina: Activation by neuropeptide Y. *J. Neurosci. Res.* **2006**, *83*, 538–550. [CrossRef] [PubMed]

101. Li, B.; Rosenbaum, P.S.; Jennings, N.M.; Maxwell, K.M.; Roth, S. Differing roles of adenosine receptor subtypes in retinal ischemia-reperfusion injury in the rat. *Exp Eye Res* **1999**, *68*, 9–17. [CrossRef]

102. Larsen, A.K.; Osborne, N.N. Involvement of adenosine in retinal ischemia. Studies on the rat. *Investig. Ophthalmol. Vis. Sci.* **1996**, *37*, 2603–2611.

103. Schmidt, K.G.; Bergert, H.; Funk, R.H. Neurodegenerative diseases of the retina and potential for protection and recovery. *Curr. Neuropharmacol.* **2008**, *6*, 164–178. [CrossRef]

104. Chidlow, G.; Wood, J.P.; Casson, R.J. Pharmacological neuroprotection for glaucoma. *Drugs* **2007**, *67*, 725–759. [CrossRef] [PubMed]

105. Wakabayashi, Y.; Yagihashi, T.; Kezuka, J.; Muramatsu, D.; Usui, M.; Iwasaki, T. Glutamate levels in aqueous humor of patients with retinal artery occlusion. *Retina* **2006**, *26*, 432–436. [CrossRef]

106. Osborne, N.N.; Melena, J.; Chidlow, G.; Wood, J.P. A hypothesis to explain ganglion cell death caused by vascular insults at the optic nerve head: Possible implication for the treatment of glaucoma. *Br. J. Ophthalmol* **2001**, *85*, 1252–1259. [CrossRef]

107. Ozawa, Y.; Kamoshita, M.; Narimatsu, T.; Ban, N.; Toda, N.; Okamoto, D.; Yuki, K.; Miyake, S.; Tsubota, K. Neuroinflammation and Neurodegenerative Disorders of the Retina. *Endocrinol. Metab. Synd.* **2013**, *2*. [CrossRef]

108. Hayreh, S.S. Retinal and optic nerve head ischemic disorders and atherosclerosis: Role of serotonin. *Prog. Retin. Eye Res.* **1999**, *18*, 191–221. [CrossRef]

109. Harsing, L.G.; Zsilla, G.; Matyus, P.; Nagy, K.M.; Marko, B.; Gyarmati, Z.; Timar, J. Interactions between glycine transporter type 1 (GlyT-1) and some inhibitor molecules—Glycine transporter type 1 and its inhibitors (review). *Acta Physiol. Hung.* **2012**, *99*, 1–17. [CrossRef] [PubMed]

110. Lindsley, C.W.; Wolkenberg, S.E.; Kinney, G.G. Progress in the preparation and testing of glycine transporter type-1 (GlyT1) inhibitors. *Curr. Top. Med. Chem.* **2006**, *6*, 1883–1896. [CrossRef] [PubMed]

111. Harsing, L.G., Jr.; Albert, M.; Mátyus, P.; Szénási, G. Glycine transporter 1 inhibitors may exert neuroprotective effects in hypoxic retina. *Intrinsic Act.* **2016**, *4* (Suppl. 3), A4.19. [CrossRef]

112. NHangai, M.; Yoshimura, N.; Yoshida, M.; Yabuuchi, K.; Honda, Y. Interleukin-1 gene expression in transient retinal ischemia in the rat. *Investig. Ophthalmol. Vis. Sci.* **1995**, *36*, 571–578.

113. Porter, R.A.; Dawson, L.A. GlyT-1 Inhibitors: From Hits to Clinical Candidates. In *Small Molecule Therapeutics for Schizophrenia*; Celanine, S., Poli, S., Eds.; Topics in Medicinal Chemistry; Springer: Cham, Swizterland, 2015; Volume 13, pp. 51–99.

114. Cukras, C.A.; Petrou, P.; Chew, E.Y.; Meyerle, C.B.; Wong, W.T. Oral minocycline for the treatment of diabetic macular edema (DME): Results of a phase I/II clinical study. *Investig. Ophthalmol. Vis. Sci.* **2012**, *53*, 3865–3874. [CrossRef]

International Journal of
Molecular Sciences

Review

A$_{2B}$ Adenosine Receptors: When Outsiders May Become an Attractive Target to Treat Brain Ischemia or Demyelination

Elisabetta Coppi [1],*, Ilaria Dettori [1], Federica Cherchi [1], Irene Bulli [1], Martina Venturini [1], Daniele Lana [2], Maria Grazia Giovannini [2], Felicita Pedata [1] and Anna Maria Pugliese [1]

[1] Department of Neuroscience, Psychology, Drug Research and Child Health (NEUROFARBA), Section of Pharmacology and Toxicology, University of Florence, 50139 Florence, Italy; ilaria.dettori@unifi.it (I.D.); federica.cherchi@unifi.it (F.C.); irene.bulli@unifi.it (I.B.); martina.venturini@unifi.it (M.V.); felicita.pedata@unifi.it (F.P.); annamaria.pugliese@unifi.it (A.M.P.)

[2] Department of Health Sciences, Section of Clinical Pharmacology and Oncology, University of Florence, 50139 Florence, Italy; daniele.lana@unifi.it (D.L.); mariagrazia.giovannini@unifi.it (M.G.G.)

* Correspondence: elisabetta.coppi@unifi.it

Received: 25 November 2020; Accepted: 16 December 2020; Published: 18 December 2020

Abstract: Adenosine is a signaling molecule, which, by activating its receptors, acts as an important player after cerebral ischemia. Here, we review data in the literature describing A$_{2B}$R-mediated effects in models of cerebral ischemia obtained in vivo by the occlusion of the middle cerebral artery (MCAo) or in vitro by oxygen-glucose deprivation (OGD) in hippocampal slices. Adenosine plays an apparently contradictory role in this receptor subtype depending on whether it is activated on neuro-glial cells or peripheral blood vessels and/or inflammatory cells after ischemia. Indeed, A$_{2B}$Rs participate in the early glutamate-mediated excitotoxicity responsible for neuronal and synaptic loss in the CA1 hippocampus. On the contrary, later after ischemia, the same receptors have a protective role in tissue damage and functional impairments, reducing inflammatory cell infiltration and neuroinflammation by central and/or peripheral mechanisms. Of note, demyelination following brain ischemia, or autoimmune neuroinflammatory reactions, are also profoundly affected by A$_{2B}$Rs since they are expressed by oligodendroglia where their activation inhibits cell maturation and expression of myelin-related proteins. In conclusion, data in the literature indicate the A$_{2B}$Rs as putative therapeutic targets for the still unmet treatment of stroke or demyelinating diseases.

Keywords: cerebral ischemia; oxygen-glucose deprivation; neuroinflammation; A$_{2B}$ receptors; oligodendrocyte differentiation; demyelination; adenosine

---

1. Introduction

Adenosine as a Signaling Molecule

Adenosine is a naturally occurring nucleoside belonging to one of the oldest signaling pathways, the purinergic system, involved in a variety of physiological and pathological processes [1]. In the CNS, adenosine is formed intracellularly from adenosine monophosphate (AMP) degradation, particularly under high energy demand, or extracellularly by the metabolism of released nucleotides operated by membrane-bound ecto-enzymes like CD39 and CD73 [2]. A vesicular mechanism of adenosine release in an excitation–secretion manner has also been postulated [3,4]. Extracellular adenosine is removed by enzymes devoted to its degradation, such as adenosine deaminase (ADA) or adenosine kinase (AK) [5], or taken up by the equilibrative nucleoside transporter (ENT) isoforms ENT1 and

ENT2 [6]. Enhanced extracellular concentrations of adenosine can be considered as a general harm signal, contributing to the recruitment of damage-associated molecular effectors [2].

Adenosine acts through the activation of four different purinergic P1 receptors: A_1, A_{2A}, A_{2B}, and A_3 adenosine receptors (A_1Rs, $A_{2A}Rs$, $A_{2B}Rs$, and A_3Rs, respectively), all belonging to the G-protein coupled, metabotropic receptor family [7]. Adenosine signaling through P1 receptors has long been a target for drug development, with adenosine itself or its derivatives being used clinically since the 1940s. Methylxanthines such as caffeine and theophylline have profound biological effects as antagonists at adenosine receptors [5]. Moreover, drugs such as dipyridamole and methotrexate act by enhancing the activation of adenosine receptors [8,9].

The most widely recognized adenosine signaling is through the activation of A_1Rs, which inhibits adenylyl cyclase (AC) through $G_{i/o}$ protein activation [10]. A_1Rs are widely distributed in most species, and mediate diverse biological effects. They are dominant in the central nervous system (CNS), with high levels being reported in the cerebral cortex, hippocampus, cerebellum, thalamus, brainstem, and spinal cord, where they inhibit neurotransmission by different mechanisms: (1) they decrease glutamate release by inhibiting presynaptic voltage-gated Ca^{2+} channels (VGCC) [11,12] and (2) they stabilize neuronal membrane potential by increasing K^+ and Cl^- conductances at the postsynaptic site [13]. Consistently, A_1R stimulation has been implicated in sedative, anticonvulsant, anxiolytic, and locomotor depressant effects in the CNS, whereas, at cardiovascular levels, they are potent bradycardic agents [14].

The $A_{2A}R$ subtype is known to stimulate AC [10] being coupled to G_s proteins [7]. This receptor subtype was isolated from a human hippocampal cDNA library [15] and, in the brain, it is also highly expressed in the striatum/caudate-putamen nuclei [16], whereas, in the periphery, high expression has been observed in immune tissues [17]. At central level, the functional effect of $A_{2A}R$ activation is at variance from A_1Rs, as they are reported to enhance glutamate release by facilitating Ca^{2+} entry through presynaptic VGCC and inhibiting its uptake [18]. Moreover, $A_{2A}R$ inhibits voltage-dependent K^+ channels, thus promoting cell excitability and neurotransmitter release [19]. Concerning peripheral functions of $A_{2A}Rs$, it is worth noting that adenosine, thanks to its actions on this receptor subtype, is one of the most powerful endogenous anti-inflammatory agents [7]. Indeed, $A_{2A}Rs$ are highly expressed in inflammatory cells including lymphocytes, granulocytes, and monocytes/macrophages, where their activation reduces pro-inflammatory cytokine production, i.e., tumor necrosis factor-alpha (TNFα), interleukin-1β (IL-1 β), and IL-6 [20] and enhances the release of anti-inflammatory mediators, such as IL-10 [21].

The relatively new A_3R subtype, cloned in 1993 [22], is coupled to $G_{i/o}$ proteins and inhibits AC but, under particular conditions or in different cell types, activation of $G_{q/11}$ by A_3R agonists has also been reported [7], with consequent increase in intracellular $[IP_3]$ and $[Ca^{2+}]$. The A_3R shows large interspecies differences, with only 74% sequence homology between rat and human [23]. Its expression is not uniform throughout the body—low levels are found in the brain and spinal cord, whereas a predominance of this receptor subtype is described in peculiar regions at the periphery, i.e., in the testis, lung, kidneys, placenta, heart, brain, spleen, and liver [24]. Interestingly, most of the cell types of the immune system express functional A_3Rs on their surface [25] and its activation is one of the most powerful stimuli for mast cell degranulation [26].

2. A_{2B} Adenosine Receptors ($A_{2B}Rs$)

This adenosine receptor subtype is somewhat the most enigmatic and less studied among the four P1 receptors. Although it was cloned in 1995 [27], a pharmacological and physiological characterization of $A_{2B}Rs$ has long been precluded by the lack of suitable ligands able to discriminate among the other adenosine receptor subtypes [28].

The distribution of $A_{2B}Rs$ in the CNS on neurons and glia is scarce but widespread, whereas in the periphery, abundant expression of $A_{2B}Rs$ is observed in the bronchial epithelium, smooth muscles, mast cells, monocytes, and digestive tracts such as ileum and colon [29]. Functional $A_{2B}Rs$ have been also recognized in vascular beds, fibroblasts, and hematopoietic and neurosecretory cells [30]. The activation of A_{2B}Rs stimulates G_s and, in some cases, $G_{q/11}$ proteins, thus enhancing intracellular [cAMP] or [IP_3], respectively [7]. As mentioned above for the cognate A_{2A}R subtype, in addition to brain cells and endothelial cells, A_{2B}Rs are present on hematic cells, such as lymphocytes and neutrophils, with the highest expression levels on macrophages [31,32]. Here, A_{2B} receptors in most cases are coexpressed with A_{2A}Rs and their activation exerts anti-inflammatory effects, inhibiting vascular adhesion [32] and migration of inflammatory cells [33]. Thus, attenuation of hypoxia-associated increases of tissue neutrophils due to infiltration in different tissues including the brain, may largely depend on blood cell A_{2B}R signaling [34].

Differently from the high affinity A_1Rs, A_{2A}Rs and A_3Rs, which are activated by physiological levels of extracellular adenosine (low nM and high nM, respectively [35]), the A_{2B}R needs much higher adenosine concentrations (in the μM range) reached only in conditions of tissue damage or injury. Such a low affinity of A_{2B}Rs for the endogenous agonist implies that they represent a good therapeutic target, since they are activated only by high adenosine efflux reached under pathological conditions or injury, when a massive release of adenosine occurs [35,36] or that they can be driven to function by selective agonists

3. A_{2B} Adenosine Receptors (A_{2B}Rs) in the Hippocampus

Similarly to the A_{2A}R subtype, A_{2B}R activation within the CNS is reported to increase glutamate release [37,38]. However, a distinct mechanism has been described. Indeed, Cunha and co-workers demonstrated that the A_{2B}R selective agonist BAY60-6583 attenuates the predominant A_1R-mediated inhibitory control of synaptic transmission in the CA1 hippocampus [37]. These data are consistent with the relatively abundant expression of A_{2B}Rs in hippocampal synaptosome preparations reported by the authors [37]. The facilitatory effect of A_{2B}Rs on glutamatergic neurotransmission was assessed in acute hippocampal slices using the electrophysiological protocol of paired pulse facilitation (PPF), which is known to modulate short-term synaptic plasticity. Our group of research recently confirmed that A_{2B}Rs decreases PPF, thus enhancing glutamate release, in an A_1R-dependent manner. Indeed, the effect of BAY60-6583 was prevented not only by the A_{2B}R antagonists MRS1754 and PSB-603, but also by the A_1R blocker DPCPX [38]. Furthermore, we extended results to a newly synthetized BAY60-6583 analogue, the A_{2B}R-selective agonist P453 recently described [39], which proved to have higher affinity than BAY60-6583 [38].

The fact that neither A_{2B}R agonists [37,38] nor antagonists [40] affect basal hippocampal glutamatergic transmission suggests that the role of A_{2B}Rs might be confined to conditions of synaptic plasticity. Nevertheless, the effect of A_{2B}R activation on PPF should not be underestimated as it could be crucial to cognitive performances and mechanisms related to memory and learning [41]. Indeed, it was demonstrated that compounds able to facilitate a PPF ratio, as A_{2B}R agonists do, may improve cognitive processes and memory performances in behavioral tests in rodents [42].

4. A_{2B}Rs and Oligodendrogliogenesis

We recently and originally demonstrated that A_{2B}Rs are crucially involved in oligodendrocyte progenitor cell (OPC) maturation. We found that the selective A_{2B}R agonists BAY60-6583 (10 μM) and P453 (500 nM) inhibited the differentiation of purified primary OPC cultures, as demonstrated by the reduced expression of myelin-related proteins such as myelin basic protein (MBP) or myelin associated glycoprotein (MAG). We also demonstrated that A_{2B}R activation reversibly inhibits tetraethylammonium- (TEA-) sensitive, sustained I_K, and 4-amynopyridine- (4-AP) sensitive, transient I_A, conductances [43]. As I_K are known to be necessary to OPC maturation [44], this could be one of the mechanisms by which A_{2B}Rs inhibit myelin production. These results are similar to what was observed in cultured OPCs exposed to the A_{2A}R agonist CGS21680, as demonstrated by us in a previous work [45–47]. At variance, the activation of a G_i-coupled, P2Y-like receptor GPR17, recently deorphanized [48,49], produces the opposite effects in OPC cultures, i.e., it increases I_K [50]

and stimulates OPC maturation [50,51]. The fact that both A_{2A}Rs and A_{2B}Rs are Gs-coupled whereas GPR17 is a G_i-coupled receptor, let us hypothesize that cAMP is involved in the myelination process. Indeed, when we applied the AC activator forskolin to patch-clamped cultured OPCs, we observed the same effect as A_2R agonists, i.e., a decrease in I_K. Furthermore, a subsequent BAY60-65683 application in the continuous presence of forskolin was devoid of effects [43], thus confirming that cAMP increase is responsible at least for A_{2B}R-mediated inhibition of I_K and cell maturation. Results that forskolin inhibit I_K and cell maturation were reported in the nineties by Soliven and co-workers on cultured ovine OPCs [52].

Interestingly, an interplay occurs in cultured OPCs between A_{2B}Rs and sphingosine kinase 1 (SphK1), one of the enzymes devoted to the synthesis of sphingosine 1 phosphate (S1P). This bioactive lipid mediator is reported to act as a mitogen in OPC cells [43]. We demonstrated that the A_{2B}R agonist BAY60-6583 activates SphK1, thus rising S1P production, whereas its silencing by small interference RNA (siRNA) increases the expression of S1P lyase, the enzyme catalyzing irreversible S1P degradation inside the cells [43]. This observation led to hypothesize that the anti-differentiating effect exerted by A_{2B}R activation in OPCs is mediated by an increase in S1P intracellular levels, as confirmed by findings that the SphK inhibitors VPC96047 or VPC96091 markedly increased MAG and MBP expression and also significantly reduced I_K currents in cultured OPCs [43].

Data about an inhibitory role of A_{2B}R in myelin formation are consistent with recent findings from Manalo et al. [53] who demonstrated that elevated cochlear adenosine levels in $ADA^{-/-}$ mice is associated with sensorineural hearing loss (SNHL) due to cochlear nerve fiber demyelination and mild hair cell loss. Intriguingly, A_{2B}R-specific antagonists administered in $ADA^{-/-}$ mice significantly restored auditory capacity, nerve fiber density, and myelin compaction [53]. The same authors also provided genetic evidence for A_{2B}R upregulation not only in $ADA^{-/-}$ hearing-impaired mice but also in age-related SNHL [53].

5. Conditions of Tissue Damage: Brain Ischemia

Ischemic stroke is the second leading cause of death in industrialized countries and the major cause of long-lasting disabilities worldwide [54]. Current treatments are confined to promote blood fluidity after the insult by the administration of tissue plasminogen activator (tPA) within the first phases (4–4.5 h) after stroke onset [55] and are efficient only in a restricted time window. The considerable socioeconomic impact of this pathology, together with the lack of effectiveness of current treatments, emphasizes the urgent need for new therapeutic targets able to prevent/repair brain damage.

Brain ischemia results from a permanent or transient reduction in cerebral blood flow mostly due to the occlusion of a brain artery. The consequent reduction of blood and/or oxygen supply to the brain leads to neuronal death caused by excessive glutamate release [56]. This early excitotoxic damage is followed by a secondary chronic phase of neuroinflammation that develops hours and days after ischemia. Either or both of these deleterious processes are important therapeutic targets.

Brain ischemia triggers a strong inflammatory response involving damage to the endothelium and to the blood–brain barrier (BBB), early recruitment of granulocytes, and delayed infiltration into the ischemic areas and the boundary zones by T cells and macrophages. This inflammatory *scenario* is mainly generated by necrotic cells, reactive oxygen species (ROS) generation, and numerous other factors caused by blood flow interruption in the brain. Once activated, these initiators of inflammation lead to numerous responses among which is the activation of microglia, the brain's resident immune cells. Microglia then generate more proinflammatory cytokines [57] which in turn leads to adhesion molecule induction in the cerebral vasculature. These are the conditions under which adenosine is typically released in great amounts and may activate all subtypes of P1 adenosine receptors.

Protracted neuroinflammation is now recognized as the predominant mechanism of secondary ischemic damage [58]. Thus, besides the approved treatment with tPA in the first hours after ischemia, an important strategy to counteract the ischemic damage is to control brain injury progression after ischemia. Among neuromodulators involved in this event, adenosine has been recognized as a front line endogenous mediator for anti-inflammatory responses.

During ischemia, adenosine is released in massive amounts [35,59] and has long been known to act predominantly as an endogenous neuroprotectant agent [2,60]. Indeed, adenosine infusion into the ischemic striatum has been shown to significantly ameliorate neurological outcome and reduce infarct volume after transient focal cerebral ischemia [61]. This well recognized neuroprotection by adenosine during ischemia is principally ascribed to A_1R stimulation [62], as they emerged as potent inhibitors of excitatory synaptic transmission both in vitro [63–65] or in vivo [66], as mentioned above. In particular, A_1Rs counteract overstimulation of N-methyl D-aspartate (NMDA) receptors due to excessive glutamate release and consequent intracellular Ca^{2+} overload [11]. Among favorable consequences of A_1R stimulation during ischemic insults are also reduction in cell metabolism and energy consumption [67], and a moderate hypothermia [68–70]. Unfortunately, the use of adenosine A_1R agonists in ischemia is hampered by profound central and peripheral side effects such as bradycardia and sedation [70–72].

Of note, high levels of the enzyme AK, which recycles and removes adenosine by phosphorylation to form AMP, results in a decrease in the ambient adenosine and thus in reduced P1 receptor activation. The prompt ability of AK to alter adenosine availability has been recently described by Boison and Jarvis to provide a "site and event" specificity to the endogenous protective effects of adenosine in situations of cellular stress [73].

6. $A_{2B}Rs$ and Brain Ischemia

As mentioned above, $A_{2B}Rs$ are activated by µM concentrations of adenosine in tissues that experience ischemia, trauma, inflammation, or other types of stressful insults and, interestingly, mRNA and protein expression of $A_{2B}R$ increase to a greater extent after ischemia-reperfusion than does expression of the other three adenosine receptors [74]. Few works have investigated the role of $A_{2B}Rs$ in brain ischemia up to now because of the low potency of adenosine for the receptors and the few selective ligands developed so far.

Since it is known that $A_{2B}Rs$ enhance glutamate release in the CA1 hippocampus [37,38], one of the brain areas more susceptible to ischemic insults, the block of $A_{2B}Rs$ may be neuroprotective as it counteracts glutamate overload by preserving the inhibitory role of A_1Rs on neurotransmission [37,38,40]. This is indeed the case in an in vitro model of brain ischemia reproduced in rat hippocampal slices by oxygen and glucose deprivation (OGD), as demonstrated by our group of research. We recently showed that the selective block of $A_{2B}Rs$ by the prototypical antagonist PSB-603 (50 nM) and by MRS1754 (200 nM) prevents irreversible synaptic failure produced by a severe, 7 min, OGD event in CA1 hippocampal slices. We also showed, in the same work, that anoxic depolarization (AD), an unequivocal sign of glutamate-induced excitotoxicity during OGD [75], is completely abolished in $A_{2B}R$ antagonist-treated slices exposed to 7 min OGD and is significantly delayed in slices undergoing a 30 min OGD insult. These results were accompanied by immunohistochemical analysis revealing that the $A_{2B}R$ block also counteracts the reduction of neuronal density found in CA1 stratum pyramidale at 3 h after OGD insults, decreases apoptosis, and maintains activated mTOR levels similar to those of controls, thus sparing neurons from the degenerative effects caused by the injury. Moreover, astrocytes significantly proliferate in CA1 stratum radiatum at 3 h after the end of OGD, possibly due to increased glutamate release [40]. $A_{2B}R$ antagonism significantly prevents astrocyte modifications. Of note, neither $A_{2B}R$ antagonist tested protects CA1 neurons from the neurodegeneration induced by exogenous glutamate application, indicating that the antagonistic effect is upstream of glutamate release, in line with data indicating a presynaptic effect of $A_{2B}Rs$ on glutamatergic terminals [37,38].

We consolidated the concept of $A_{2B}Rs$ acting through inhibition of presynaptic A_1R subtype by showing that the application of a brief, reversible, synaptic episode of 2 min duration delays the reduction of evoked field potentials during OGD ascribed to A_1R activation [38].

However, beyond neuroprotection exerted by $A_{2B}R$ antagonists acting at the neuro-glial level, it is worth noting that the bulk of evidence in the literature points to a beneficial role exerted by $A_{2B}R$ agonists acting on the same receptor subtype expressed also on peripheral blood cells [32,34]. Indeed, studies in mice ablated of $A_{2B}Rs$ on bone marrow cells indicate an important contribution of vascular $A_{2B}Rs$ in attenuating vascular leakage during hypoxia [34]. It was also found that activation of $A_{2B}Rs$ in a model of femoral artery injury is vasoprotective as it reduces myocardial infarct size in rabbit and mouse hearts when administered before or at the onset of reperfusion [32].

Of note, post-treatment with intravenous BAY60-6583 (1 mg/kg) reduces lesion volume in the absence or presence of tPA (10 mg/kg) and attenuates brain swelling, blood–brain barrier disruption, and tPA-exacerbated hemorrhagic transformation (HT) at 24 h after ischemia induced by transient (2 h) middle cerebral artery occlusion (tMCAo) [74]. Additionally, in the same work, BAY60-6583 mitigates sensorimotor deficits in the presence of tPA and inhibits tPA-enhanced matrix metalloprotease-9 activation, thus decreasing BBB permeability 24 h after ischemia [74]. Protection from BBB permeability after ischemia might protect from blood cell infiltration, that on their turn promote expansion of the inflammatory response in the ischemic tissue [25].

Our group of research contributed to the field by demonstrating that the chronic treatment with BAY60-6583, administered intraperitoneally twice/day for 7 days at the dose of 0.1 mg/kg, from 4 h after focal ischemia induced by tMCAo, since one day after ischemia protects from neurological deficit. Seven days after ischemia it protects from ischemic brain damage, neuronal death, microglia activation, and astrocyte alteration [76]. Interestingly, 7 days after ischemia, the A_{2B} agonist decreases TNF-α and increases IL-10 levels in the blood [76]. Both cytokines are considered valuable blood markers of the brain damage following an ischemic insult [77]. Among putative mechanism mediating protective effects of A_{2B} agonists, it is worth mentioning that $A_{2B}R$ agonists reduce the expression of TNF-α in primary microglia cultures [78] and increase IL-10 production from murine microglial cells [79] with consequent rescuing of the resting state of microglia. Besides a protection due to a direct agonism of $A_{2B}R$ located on rat microglial and/or astrocytic cells, observation that 2 days after tMCAo, BAY60-6583 significantly reduces granulocyte infiltration in the cortex [76] and supports the idea that $A_{2B}R$ activation on peripheral endothelial and blood cells is involved in counteracting inflammation of brain parenchyma. This possibility is also supported by the evidence that adenosine $A_{2B}R$ knock out (KO) mice show increased basal levels of TNF-α and expression of adhesion molecules in lymphoid cells, resulting in increased leukocyte rolling and adhesion [34]. Actually, increasing evidence indicates a role for $A_{2B}R$ in the modulation of inflammation and immune responses in distinct pathologies like cancer, diabetes, as well renal, lung, and vascular diseases [80]. Indeed, stroke and inflammation are strictly interrelated. Brain ischemia induces profound inflammatory changes in peripheral organs (especially lungs and gut) as early as 2 h after tMCAo in mice as detected by whole body single photon emission computed tomography (SPECT)-based imaging protocols [81]. Such peripheral inflammatory changes, in turn, may contribute to poorer recovery after stroke [81]. Data suggest that the $A_{2B}R$ agonists can be proposed as adjuvant therapy to the accepted pharmacological strategy with tPA in brain ischemia.

Taken together, data point toward the possibility that stimulation of $A_{2B}R$ plays a dual time-related role after ischemia. In the early hours after ischemia, a robust and sustained increase in cerebral extracellular levels of adenosine able to activate low affinity $A_{2B}Rs$ in the brain may contribute to expand excitotoxicity. However, in the hours and days following ischemia, when profound neuroinflammation develops, $A_{2B}Rs$ located on glial, vascular endothelial, and blood cells exert a prevalent immunomodulatory role attenuating the neuroinflammation. Thus, on the whole, it appears that $A_{2B}Rs$ located on any cell type of the brain and on vascular and blood cells partake in either salvage or demise of the tissue after a stroke and represent an important target for drugs that have different therapeutic time-windows after stroke.

Since it is known that an $A_{2B}R$ agonist administered within the first 4 h after brain ischemia is able to reduce neurological deficit measures at 24 h after the insult [76], we could hypothesize that a beneficial effect of $A_{2B}R$ activation could be achieved by administration of a selective agonist starting at 1 day after stroke. Table 1 summarizes $A_{2B}R$-mediated effects in in vitro or in vivo experimental models of brain ischemia.

Table 1. Effects of adenosine $A_{2B}R$ receptor ($A_{2B}R$) activation or blockade in different in vitro or in vivo experimental approaches.

$A_{2B}R$-Mediated Effects	In Vitro Ischemia in Rat Hippocampal Slices (OGD)	In Vivo Ischemia Model (MCAo) in the Rat	In Vivo Model of Multiple Sclerosis (EAE) in the Mouse
$A_{2B}R$ activation	↓ A_1R-mediated inhibition of Glu release [38]	Protects from brain damage, neurological deficit [76], and BBB disruption [74]	Reverses MSC-mediated BBB repair [82]
$A_{2B}R$ block	Protects from AD and synaptic failure [40]	unknown	Protects from myelin loss and neurological damage [83]

Glutamate: Glu. Anoxic depolarization: AD. Oxygen and glucose deprivation: OGD. Middle cerebral artery occlusion: MCAo. Experimental autoimmune encephalomyelitis: EAE. Mesenchymal stem cells: MSCs.

In recent years, a similar dual role of the other subtype of A_2Rs, the $A_{2A}R$, was described. Indeed, antagonists at this receptor proved neuroprotective when applied during in vitro OGD [84,85] or in the acute post-ischemic phase after MCAo in rats [86–89], whereas the same receptor agonists proved protective at later phases after ischemia by decreasing neuroinflammation [35,90]. This "paradoxical" role of the $A_{2A}Rs$ has been discussed in several review papers [35,91].

Hence, it appears that the reciprocal influence of central (neuro-glia) versus peripheral (blood vessels–inflammatory cells) mechanisms involved in brain ischemia are hard to dissect.

What recently and interestingly emerged about the $A_{2B}R$ subtype is that it might be a "sensor" of blood oxygen levels and rescue hypoxic tissues by an additive mechanism involving the release of oxygen from erythrocytes. Indeed, recent findings obtained in humans challenged with altitude indicate that $A_{2B}Rs$ expressed on erythrocytes elevate their O_2 releasing capacity by increasing 2,3-biphosphoglycerate (2,3-BPG) levels in an AMP-activated protein kinase-dependent manner [92]. Of note, the same authors also reported a significant increase in plasma adenosine concentrations and soluble CD73 activity in 21 healthy humans within 2 h of arrival at 5260 m altitude [92]. These findings unveiled a novel mechanism of human adaptation to hypoxia, in this case due to high altitude, and pointed to $A_{2B}Rs$ as possible targets to facilitate O_2 release capacity to peripheral tissues and a potential therapeutic approach for counteracting hypoxia-induced tissue damage.

7. $A_{2B}Rs$ and Demyelinating Diseases

Demyelination occurs in a variety of pathological conditions affecting central or peripheral nervous systems. As an example, myelin disorganization in caudate/putamen striatal nuclei have been reported by us [88] and others [93,94]. Furthermore, chronic demyelinating diseases, such as multiple sclerosis (MS), are highly invalidating pathologies with elevated incidence among the "under 40" population worldwide [95], but an efficacious therapy is still lacking.

It is interesting to note that elevated adenosine levels have been detected in cerebrospinal fluid of MS patients [96,97]. Furthermore, an upregulation of $A_{2A}R$ [98] and $A_{2B}R$ [83] has been reported in peripheral blood leukocytes of MS patients and in the CNS [82] and peripheral lymphoid tissues [83,99] in a mouse model of MS, the experimental autoimmune encephalomyelitis (EAE).

Hence, a crucial role of adenosine, and in particular of $A_{2A}R$ and/or $A_{2B}R$ subtypes, in demyelinating pathologies have been postulated. Under these conditions, excessive signaling by excitatory neurotransmitters like glutamate may be deleterious to neurons and oligodendroglia by exacerbating excitotoxicity and contributing to brain injury. For this reason, the inhibitory effect on glutamate release described above for antagonists at both A_2R subtypes could prove protective. This was indeed the case, as demonstrated by Chen and colleagues [100] and by Wei and co-workers [83] who reported that $A_{2A}R$ and $A_{2B}R$ antagonists, respectively, alleviated the clinical symptoms of EAE and prevented demyelination and CNS damage. Recent data by Liu and co-workers [82] confirmed that $A_{2B}R$ activation seems to participate in EAE-induced damage as BAY60-6583 reverted the protective effects, i.e., reduced inflammatory cell infiltration and demyelination, exerted by mesenchymal stem cell therapy in EAE mice. Of note, the above results demonstrating a deleterious role of $A_{2B}Rs$ in demyelinating diseases are in agreement with our in vitro data demonstrating that $A_{2B}R$ blockade [44], as well as $A_{2A}R$ antagonism [45], facilitates OPC differentiation in vitro.

However, things are probably more complicated as suggested by the fact that, again, A_2R-mediated actions are mainly anti-inflammatory when observed in a longer time-span. Indeed, genetically modified $A_{2A}R^{-/-}$ EAE mice are more prone to EAE-induced damage [45], and the $A_{2A}R$ agonist CGS61680 ameliorates EAE by reducing Th1 lymphocyte activation and cytokine-induced BBB dysfunction [101]. Indeed, adenosine receptors are expressed also by infiltrating lymphocytes, macrophages, and microglial cells and, accordingly, many works suggest that the possible beneficial effects linked to this pathway are mainly related to the immune-modulating consequences rather than to a remyelinating or neuroprotecting effect. The dual role of $A_{2A}Rs$ in demyelinating diseases has been elegantly commented by Rajasundaram who tried to reconcile the apparently paradoxical data reported up to now about $A_{2A}Rs$ in demyelinating diseases [102].

Unfortunately, data about the effects of $A_{2B}R$ ligands in demyelinating pathologies are not that abundant yet, as shown in Table 1.

8. Conclusions

In conclusion, results underlie that after hypoxia/ischemia, brain injury results from a complex sequence of pathophysiological events that evolve over time—a primary acute mechanism of excitotoxicity and periinfarct depolarizations followed by a secondary brain injury activation triggered by protracted neuroinflammation. Information so far acquired indicates that adenosine $A_{2B}Rs$ located on any cell type of the brain and on vascular and blood cells partake in either salvage or demise of the tissue after a stroke, including protracted demyelination.

Thus, they all represent important targets for drugs having different therapeutic time-windows after stroke. The final protective outcome for an agonist versus antagonist compound depends on time of administration and district of activation of the receptor targeted by the drug.

Funding: The study was supported by the University of Florence (Fondi Ateneo Ricerca) to AMP, the Ministry of Education, Universities and Research, Italy: MIUR-PRIN 2017CY3J3W_003 to FP, Fondazione Umberto Veronesi to EC (FUV2020-3299), and by Fondazione Italiana Sclerosi Multipla (FISM): 2019/R-Single/036 to AMP and EC.

Conflicts of Interest: The authors declare no conflict of interest.

Abbreviations

4-AP	4-amynopyridine
A_1Rs	A_1 receptors
$A_{2A}Rs$	A_{2A} receptors
LD	linear dichroism
$A_{2B}Rs$	A_{2B} receptors
A_3Rs	A_3 receptors
AC	adenylyl cyclase
ADA	adenosine deaminase
AK	adenosine kinase
AMP	adenosine monophosphate
BBB	blood brain barrier
CA1	cornus ammonis 1
cAMP	cyclic adenosine monophosphate
CNS	central nervous system

ENT	equilibrative nucleoside transporter
I_K	potassium current
IL-1β	interleukin-1β
IL-6	interleukin-6
IL-10	interleukin-10
IP$_3$	inositol 1,4,5-triphosphate
MAG	myelin associated glycoprotein
MCAo	middle cerebral artery occlusion
MBP	myelin basic protein
NMDA	N-methyl D-aspartate
PPF	paired pulse facilitation
OGD	oxygen-glucose deprivation
ROS	reactive oxygen species
S1P	sphingosine 1 phosphate
siRNA	small interference RNA
SNHL	sensorineural hearing loss
SphK1	sphingosine kinase 1
TNFα	tumor necrosis factor-alpha
TEA	tetraethylammonium
tPA	tissue plasminogen activator
VGCC	voltage-gated Ca^{2+} channels

References

1. Verkhratsky, A.; Burnstock, G. Biology of purinergic signalling: Its ancient evolutionary roots, its omnipresence and its multiple functional significance. *BioEssays* **2014**, *36*, 697–705. [CrossRef]
2. Latini, S.; Pedata, F. Adenosine in the central nervous system: Release mechanisms and extracellular concentrations. *J. Neurochem.* **2008**, *79*, 463–484. [CrossRef] [PubMed]
3. Melani, A.; Corti, F.; Stephan, H.; Müller, C.E.; Donati, C.; Bruni, P.; Vannucchi, M.-G.; Pedata, F. Ecto-ATPase inhibition: ATP and adenosine release under physiological and ischemic in vivo conditions in the rat striatum. *Exp. Neurol.* **2012**, *233*, 193–204. [CrossRef] [PubMed]
4. Corti, F.; Cellai, L.; Melani, A.; Donati, C.; Bruni, P.; Pedata, F. Adenosine is present in rat brain synaptic vesicles. *NeuroReport* **2013**, *24*, 982–987. [CrossRef] [PubMed]
5. Chen, J.-F.; Eltzschig, H.K.; Fredholm, B.B. Adenosine receptors as drug targets—What are the challenges? *Nat. Rev. Drug Discov.* **2013**, *12*, 265–286. [CrossRef] [PubMed]
6. Inoue, K. Molecular Basis of Nucleobase Transport Systems in Mammals. *Biol. Pharm. Bull.* **2016**, *40*, 1130–1138.
7. Antonioli, L.; Blandizzi, C.; Pacher, P.; Haskó, G. The Purinergic System as a Pharmacological Target for the Treatment of Immune-Mediated Inflammatory Diseases. *Pharmacol. Rev.* **2019**, *71*, 345–382. [CrossRef] [PubMed]
8. Cronstein, B.N.; Aune, T.M. Methotrexate and its mechanisms of action in inflammatory arthritis. *Nat. Rev. Rheumatol.* **2020**, *16*, 145–154. [CrossRef]
9. Lana, D.; Ugolini, F.; Melani, A.; Nosi, D.; Pedata, F.; Giovannini, M. The neuron-astrocyte-microglia triad in CA3 after chronic cerebral hypoperfusion in the rat: Protective effect of dipyridamole. *Exp. Gerontol.* **2017**, *96*, 46–62. [CrossRef]
10. Van Calker, D.; Müller, M.; Hamprecht, B. Adenosine regulates via two different types of receptors, the accumulation of cyclic AMP in cultured brain cells. *J. Neurochem.* **1979**, *33*, 999–1005. [CrossRef]
11. Andiné, P. Involvement of adenosine in ischemic and postischemic calcium regulation. *Mol. Chem. Neuropathol.* **1993**, *18*, 35–49. [CrossRef] [PubMed]
12. Corradetti, R.; Conte, G.L.; Moroni, F.; Passani, M.B.; Pepeu, G. Adenosine decreases aspartate and glutamate release from rat hippocampal slices. *Eur. J. Pharmacol.* **1984**, *104*, 19–26. [CrossRef]
13. Choi, D.W. Possible mechanisms limiting N-methyl-D-aspartate receptor overactivation and the therapeutic efficacy of N-methyl-D-aspartate antagonists. *Stroke* **1990**, *21*, 20–22.

14. Deb, P.K.; Deka, S.; Borah, P.; Abed, S.N.; Klotz, K.-N. Medicinal Chemistry and Therapeutic Potential of Agonists, Antagonists and Allosteric Modulators of A1 Adenosine Receptor: Current Status and Perspectives. *Curr. Pharm. Des.* **2019**, *25*, 2697–2715. [CrossRef]
15. Furlong, T.J.; Pierce, K.D.; Selbie, L.A.; Shine, J. Molecular characterization of a human brain adenosine A2 receptor. *Mol. Brain Res.* **1992**, *15*, 62–66. [CrossRef]
16. Peterfreund, R.A.; MacCollin, M.; Gusella, J.; Fink, J.S. Characterization and Expression of the Human A2a Adenosine Receptor Gene. *J. Neurochem.* **2002**, *66*, 362–368. [CrossRef] [PubMed]
17. Yu, L.; Frith, M.C.; Suzuki, Y.; Peterfreund, R.A.; Gearan, T.; Sugano, S.; Schwarzschild, M.A.; Weng, Z.; Fink, J.; Chen, J.-F. Characterization of genomic organization of the adenosine A2A receptor gene by molecular and bioinformatics analyses. *Brain Res.* **2004**, *1000*, 156–173. [CrossRef]
18. Gonçalves, M.; Ribeiro, J.A. Adenosine A2 receptor activation facilitates 45Ca^{2+} uptake by rat brain synaptosomes. *Eur. J. Pharmacol.* **1996**, *310*, 257–261.
19. Lopes, L.V.; Cunha, R.; Kull, B.; Fredholm, B.; Ribeiro, J. Adenosine A2A receptor facilitation of hippocampal synaptic transmission is dependent on tonic A1 receptor inhibition. *Neuroscience* **2002**, *112*, 319–329. [CrossRef]
20. Varani, K.; Padovan, M.; Vincenzi, F.; Targa, M.; Trotta, F.; Govoni, M.; Borea, P.A. A2A and A3 adenosine receptor expression in rheumatoid arthritis: Upregulation, inverse correlation with disease activity score and suppression of inflammatory cytokine and metalloproteinase release. *Arthritis Res. Ther.* **2011**, *13*, R197. [CrossRef]
21. Bortoluzzi, A.; Vincenzi, F.; Govoni, M.; Padovan, M.; Ravani, A.; Borea, P.A.; Varani, K. A2A adenosine receptor upregulation correlates with disease activity in patients with systemic lupus erythematosus. *Arthritis Res.* **2016**, *18*, 192. [CrossRef]
22. Sajjadi, F.G.; Firestein, G.S. cDNA cloning and sequence analysis of the human A3 adenosine receptor. *Biochim. Biophys. Acta Bioenerg.* **1993**, *1179*, 105–107. [CrossRef]
23. Koscsó, B.; Csóka, B.; Pacher, P.; Haskó, G. Investigational A3adenosine receptor targeting agents. *Expert Opin. Investig. Drugs* **2011**, *20*, 757–768. [CrossRef] [PubMed]
24. Gessi, S.; Merighi, S.; Varani, K.; Leung, E.; Mac Lennan, S.; Borea, P.A. The A3 adenosine receptor: An enigmatic player in cell biology. *Pharmacol. Ther.* **2008**, *117*, 123–140. [CrossRef] [PubMed]
25. Haskó, G.; Linden, J.; Cronstein, B.; Pacher, P. Adenosine receptors: Therapeutic aspects for inflammatory and immune diseases. *Nat. Rev. Drug Discov.* **2008**, *7*, 759–770. [CrossRef] [PubMed]
26. Ramkumar, V.; Stiles, G.L.; Beaven, M.A.; Ali, H. TheE A(3) adenosine receptor is the unique adenosine receptor which facilitates release of allergic mediators in mast-cells. *J. Biol. Chem.* **1993**, *268*, 16887–16890. [PubMed]
27. Jacobson, M.A.; Johnson, R.G.; Luneau, C.J.; Salvatore, C.A. Cloning and Chromosomal Localization of the Human A2b Adenosine Receptor Gene (ADORA2B) and Its Pseudogene. *Genome* **1995**, *27*, 374–376. [CrossRef]
28. Popoli, P.; Pepponi, R. Potential therapeutic relevance of adenosine A2B and A2A receptors in the central nervous system. *CNS Neurol. Disord. Drug Targets* **2012**, *11*, 664–674. [CrossRef]
29. Feoktistov, I.; Biaggioni, I.; Polosa, R.; Holgate, S. Adenosine A2B receptors: A novel therapeutic target in asthma? *Trends Pharmacol. Sci.* **1998**, *19*, 148–153.
30. Balakumar, C.; Samarneh, S.; Jaber, A.M.Y.; Kassab, G.; Agrawal, N. Therapeutic Potentials of A2B Adenosine Receptor Ligands: Current Status and Perspectives. *Curr. Pharm. Des.* **2019**, *25*, 2741–2771.
31. Gessi, S.; Varani, K.; Merighi, S.; Cattabriga, E.; Pancaldi, C.; Szabadkai, Y.; Rizzuto, R.; Klotz, K.-N.; Leung, E.; Mac Lennan, S.; et al. Expression, Pharmacological Profile, and Functional Coupling of A2B Receptors in a Recombinant System and in Peripheral Blood Cells Using a Novel Selective Antagonist Radioligand, [3H]MRE 2029-F20. *Mol. Pharmacol.* **2005**, *67*, 2137–2147. [CrossRef]
32. Yang, D.; Zhang, Y.; Nguyen, H.G.; Koupenova, M.; Chauhan, A.K.; Makitalo, M.; Jones, M.R.; Hilaire, C.S.; Seldin, D.C.; Toselli, P.; et al. The A2B adenosine receptor protects against inflammation and excessive vascular adhesion. *J. Clin. Investig.* **2006**, *116*, 1913–1923. [CrossRef] [PubMed]
33. Wakai, A.; Wang, J.H.; Winter, D.C.; Street, J.T.; O'sullivan, R.G.; Redmond, H.P. Adenosine inhibits neutrophil vascular endothelial growth factor release and transendothelial migration via A2B receptor activation. *Shock* **2001**, *15*, 297–301. [PubMed]

34. Eckle, T.; Faigle, M.; Grenz, A.; Laucher, S.; Thompson, L.F.; Eltzschig, H.K. A2B adenosine receptor dampens hypoxia-induced vascular leak. *Blood* **2008**, *111*, 2024–2035. [CrossRef]

35. Pedata, F.; Dettori, I.; Coppi, E.; Melani, A.; Fusco, I.; Corradetti, R.; Pugliese, A.M. Purinergic signalling in brain ischemia. *Neuropharmacology* **2016**, *104*, 105–130. [CrossRef] [PubMed]

36. Latini, S.; Bordoni, F.; Corradetti, R.; Pepeu, G.; Pedata, F. Temporal correlation between adenosine outflow and synaptic potential inhibition in rat hippocampal slices during ischemia-like conditions. *Brain Res.* **1998**, *794*, 325–328. [CrossRef]

37. Goncalves, F.Q.; Pires, J.; Pliassova, A.; Beleza, R.; Lemos, C.; Marques, J.M.; Rodrigues, R.J.; Canas, P.M.; Kofalvi, A.; Cunha, R.A.; et al. Adenosine A(2b) receptors control A(1) receptor-mediated inhibition of synaptic transmission in the mouse hippocampus. *Eur. J. Neurosci.* **2015**, *41*, 876–886.

38. Fusco, I.; Cherchi, F.; Catarzi, D.; Colotta, V.; Varano, F.; Pedata, F.; Pugliese, A.M.; Coppi, E. Functional characterization of a novel adenosine A2B receptor agonist on short-term plasticity and synaptic inhibition during oxygen and glucose deprivation in the rat CA1 hippocampus. *Brain Res. Bull.* **2019**, *151*, 174–180. [CrossRef]

39. Betti, M.; Catarzi, D.; Varano, F.; Falsini, M.; Varani, K.; Vincenzi, F.; Ben, D.D.; Lambertucci, C.; Colotta, V. The aminopyridine-3,5-dicarbonitrile core for the design of new non-nucleoside-like agonists of the human adenosine A2B receptor. *Eur. J. Med. Chem.* **2018**, *150*, 127–139. [CrossRef]

40. Fusco, I.; Ugolini, F.; Lana, D.; Coppi, E.; Dettori, I.; Gaviano, L.; Nosi, D.; Cherchi, F.; Pedata, F.; Giovannini, M.G.; et al. The Selective Antagonism of Adenosine A2B Receptors Reduces the Synaptic Failure and Neuronal Death Induced by Oxygen and Glucose Deprivation in Rat CA1 Hippocampus in Vitro. *Front. Pharmacol.* **2018**, *9*, 399.

41. Li, X.L.; Sun, W.; An, L. Nano-CuO impairs spatial cognition associated with inhibiting hippocampal long-term potentiation via affecting glutamatergic neurotransmission in rats. *Toxicol. Ind. Health* **2018**, *34*, 409–421. [CrossRef] [PubMed]

42. Pereda, D.; Al-Osta, I.; Okorocha, A.E.; Easton, A.; Hartell, N.A. Changes in presynaptic calcium signalling accompany age-related deficits in hippocampal LTP and cognitive impairment. *Aging Cell* **2019**, *18*, e13008. [CrossRef] [PubMed]

43. Coppi, E.; Cherchi, F.; Fusco, I.; Dettori, I.; Gaviano, L.; Magni, G.; Catarzi, D.; Colotta, V.; Varano, F.; Rossi, F.; et al. Adenosine A2B receptors inhibit K+ currents and cell differentiation in cultured oligodendrocyte precursor cells and modulate sphingosine-1-phosphate signaling pathway. *Biochem. Pharmacol.* **2020**, *177*, 113956. [CrossRef] [PubMed]

44. Gallo, V.; Zhou, J.M.; McBain, C.J.; Wright, P.; Knutson, P.L.; Armstrong, R.C. Oligodendrocyte progenitor cell proliferation and lineage progression are regulated by glutamate receptor-mediated K+ channel block. *J. Neurosci.* **1996**, *16*, 2659–2670. [CrossRef]

45. Coppi, E.; Cellai, L.; Maraula, G.; Pugliese, A.M.; Pedata, F. Adenosine A2A receptors inhibit delayed rectifier potassium currents and cell differentiation in primary purified oligodendrocyte cultures. *Neuropharmacology* **2013**, *73*, 301–310. [CrossRef]

46. Coppi, E.; Cellai, L.; Maraula, G.; Dettori, I.; Melani, A.; Pugliese, A.M.; Pedata, F. Role of adenosine in oligodendrocyte precursor maturation. *Front. Cel. Neurosci.* **2015**, *9*, 155. [CrossRef]

47. Cherchi, F.; Pugliese, A.M.; Coppi, E. Oligodendrocyte precursor cell maturation: Role of adenosine receptors. *Neural Regen. Res.* **2020**. Accepted on 9 December.

48. Pugliese, A.M.; Trincavelli, M.L.; Lecca, D.; Coppi, E.; Fumagalli, M.; Ferrario, S.; Failli, P.; Daniele, S.; Martini, C.; Pedata, F.; et al. Functional characterization of two isoforms of the P2Y-like receptor GPR17: S-35 GTP gamma S binding and electrophysiological studies in 1321N1 cells. *Am. J. Physiol. Cell Physiol.* **2009**, *297*, C1028–C1040. [CrossRef]

49. Ciana, P.; Fumagalli, M.; Trincavelli, M.L.; Verderio, C.; Rosa, P.; Lecca, D.; Ferrario, S.; Parravicini, C.; Capra, V.; Gelosa, P.; et al. The orphan receptor GPR17 identified as a new dual uracil nucleotides/cysteinyl-leukotrienes receptor. *EMBO J.* **2006**, *25*, 4615–4627. [CrossRef]

50. Coppi, E.; Maraula, G.; Fumagalli, M.; Failli, P.; Cellai, L.; Bonfanti, E.; Mazzoni, L.; Coppini, R.; Abbracchio, M.P.; Pedata, F.; et al. UDP-glucose enhances outward K plus currents necessary for cell differentiation and stimulates cell migration by activating the GPR17 receptor in oligodendrocyte precursors. *Glia* **2013**, *61*, 1155–1171. [CrossRef]

51. Fumagalli, M.; Daniele, S.; Lecca, D.; Lee, P.R.; Parravicini, C.; Fields, R.D.; Rosa, P.; Antonucci, F.; Verderio, C.; Trincavelli, M.L.; et al. Phenotypic Changes, Signaling Pathway, and Functional Correlates of GPR17-expressing Neural Precursor Cells during Oligodendrocyte Differentiation. *J. Biol. Chem.* **2011**, *286*, 10593–10604. [CrossRef] [PubMed]

52. Soliven, B.; Szuchet, S.; Arnason, B.G.W.; Nelson, D.J. Forskolin and phorbol esters decrease the same K+ conductance in cultured oligodendrocytes. *J. Membr. Biol.* **1988**, *105*, 177–186. [CrossRef] [PubMed]

53. Manalo, J.M.; Liu, H.; Ding, D.; Hicks, J.; Sun, H.; Salvi, R.; Kellems, R.E.; Pereira, F.A.; Xia, Y. Adenosine A2B receptor: A pathogenic factor and a therapeutic target for sensorineural hearing loss. *FASEB J.* **2020**. [CrossRef] [PubMed]

54. Boehme, A.K.; Esenwa, C.; Elkind, M.S.V. Stroke Risk Factors, Genetics, and Prevention. *Circ. Res.* **2017**, *120*, 472–495. [CrossRef]

55. Fonarow, G.C.; Smith, E.E.; Saver, J.L.; Reeves, M.J.; Bhatt, D.L.; Grau-Sepulveda, M.V.; Olson, D.M.; Hernandez, A.F.; Peterson, E.D.; Schwamm, L.H. Timeliness of Tissue-Type Plasminogen Activator Therapy in Acute Ischemic Stroke Patient Characteristics, Hospital Factors, and Outcomes Associated With Door-to-Needle Times Within 60 Minutes. *Circulation* **2011**, *123*, 750–758. [CrossRef]

56. Rossi, D.J.; Oshima, T.; Attwell, D. Glutamate release in severe brain ischaemia is mainly by reversed uptake. *Nat. Cell Biol.* **2000**, *403*, 316–321. [CrossRef]

57. Stoll, G.; Jander, S.; Schroeter, M. Inflammation and glial responses in ischemic brain lesions. *Prog. Neurobiol.* **1998**, *56*, 149–171. [CrossRef]

58. Wang, X.; Xuan, W.; Zhu, Z.-Y.; Li, Y.; Zhu, H.; Zhu, L.; Fu, D.-Y.; Yang, L.-Q.; Li, P.; Yu, W. The evolving role of neuro-immune interaction in brain repair after cerebral ischemic stroke. *CNS Neurosci. Ther.* **2018**, *24*, 1100–1114. [PubMed]

59. Dale, N.; Pearson, T.; Frenguelli, B.G. Direct measurement of adenosine release during hypoxia in the CA1 region of the rat hippocampal slice. *J. Physiol.* **2000**, *526*, 143–155. [CrossRef]

60. Pedata, F.; Melani, A.; Pugliese, A.M.; Coppi, E.; Cipriani, S.; Traini, C. The role of ATP and adenosine in the brain under normoxic and ischemic conditions. *Purinergic Signal.* **2007**, *3*, 299–310. [CrossRef]

61. Kitagawa, H.; Mori, A.; Shimada, J.; Mitsumoto, Y.; Kikuchi, T. Intracerebral adenosine infusion improves neurological outcome after transient focal ischemia in rats. *Neurol. Res.* **2002**, *24*, 317–323. [CrossRef] [PubMed]

62. Kashfi, S.; Ghaedi, K.; Baharvand, H.; Nasr-Esfahani, M.H.; Javan, M. A1 Adenosine Receptor Activation Modulates Central Nervous System Development and Repair. *Mol. Neurobiol.* **2016**, *54*, 8128–8139. [CrossRef] [PubMed]

63. Fowler, J.C. Adenosine antagonists delay hypoxia-induced depression of neuronal activity in hippocampal brain slice. *Brain Res.* **1989**, *490*, 378–384. [CrossRef]

64. Fowler, J.C. Adenosine antagonists alter the synaptic response to invitro ischemia in the rat hippocampus. *Brain Res.* **1990**, *509*, 331–334. [CrossRef]

65. Latini, S.; Bordoni, F.; Pedata, F.; Corradetti, R. Extracellular adenosine concentrations during in vitro ischaemia in rat hippocampal slices. *Br. J. Pharmacol.* **1999**, *127*, 729–739. [CrossRef]

66. Fowler, J.C.; Gervitz, L.M.; Hamilton, M.E.; Walker, J.A. Systemic Hypoxia and the Depression of Synaptic Transmission in Rat Hippocampus after Carotid Artery Occlusion. *J. Physiol.* **2003**, *550*, 961–972. [CrossRef]

67. Greene, R.W.; Haas, H. The electrophysiology of adenosine in the mammalian central nervous system. *Prog. Neurobiol.* **1991**, *36*, 329–341. [CrossRef]

68. Muzzi, M.; Blasi, F.; Masi, A.; Coppi, E.; Traini, C.; Felici, R.; Pittelli, M.; Cavone, L.; Pugliese, A.M.; Moroni, F.; et al. Neurological Basis of AMP-Dependent Thermoregulation and its Relevance to Central and Peripheral Hyperthermia. *Br. J. Pharmacol.* **2012**, *33*, 183–190. [CrossRef]

69. Alexander, V.G. Fever in systemic inflammation: Roles of purines. *Front. Biosci.* **2004**, *9*, 1011–1022.

70. Von Lubitz, D.K.; Lin, R.-S.; Melman, N.; Ji, X.-D.; Carter, M.F.; Jacobson, K.A. Chronic administration of selective adenosine A1 receptor agonist or antagonist in cerebral ischemia. *Eur. J. Pharmacol.* **1994**, *256*, 161–167. [CrossRef]

71. Sollevi, A. Cardiovascular effects of adenosine in man; possible clinical implications. *Prog. Neurobiol.* **1986**, *27*, 319–349. [CrossRef]

72. Sweeney, J.; Hohmann, C.; Oster-Granite, M.; Coyle, J. Neurogenesis of the basal forebrain in euploid and trisomy 16 mice: An animal model for developmental disorders in down syndrome. *Neuroscience* **1989**, *31*, 413–425. [CrossRef]

73. Boison, D.; Jarvis, M.F. Adenosine kinase: A key regulator of purinergic physiology. *Biochem. Pharmacol.* **2020**, 114321. [CrossRef] [PubMed]

74. Li, Q.; Han, X.N.; Lan, X.; Hong, X.H.; Gao, Y.F.; Luo, T.Q.; Yang, Q.W.; Koehler, R.C.; Zhai, Y.; Zhou, J.Y.; et al. Inhibition of tPA-induced hemorrhagic transformation involves adenosine A2b receptor activation after cerebral ischemia. *Neurobiol. Dis.* **2017**, *108*, 173–182. [CrossRef]

75. Colotta, V.; Lenzi, O.; Catarzi, D.; Varano, F.; Squarcialupi, L.; Costagli, C.; Galli, A.; Ghelardini, C.; Pugliese, A.M.; Maraula, G.; et al. 3-Hydroxy-1H-quinazoline-2,4-dione derivatives as new antagonists at ionotropic glutamate receptors: Molecular modeling and pharmacological studies. *Eur. J. Med. Chem.* **2012**, *54*, 470–482. [CrossRef] [PubMed]

76. Dettori, I.; Gaviano, L.; Ugolini, F.; Lana, D.; Bulli, I.; Magni, G.; Rossi, F.; Giovannini, M.G.; Pedata, F. Protective effect of adenosine A2B receptor agonist, BAY60-6583, against transient focal brain ischemia in rat. *Front. Pharmacol.* **2020**, *11*, 1639.

77. Jickling, G.C.; Sharp, F.R. Blood Biomarkers of Ischemic Stroke. *Neurotherapeutics* **2011**, *8*, 349–360.

78. Merighi, S.; Borea, P.A.; Stefanelli, A.; Bencivenni, S.; Castillo, C.A.; Varani, K.; Gessi, S. A(2a) and a(2b) adenosine receptors affect HIF-1 alpha signaling in activated primary microglial cells. *Glia* **2015**, *63*, 1933–1952. [CrossRef]

79. Koscsó, B.; Csóka, B.; Selmeczy, Z.; Himer, L.; Pacher, P.; Virág, L.; Haskó, G. Adenosine Augments IL-10 Production by Microglial Cells through an A2B Adenosine Receptor-Mediated Process. *J. Immunol.* **2011**, *188*, 445–453. [CrossRef]

80. Borea, P.A.; Gessi, S.; Merighi, S.; Vincenzi, F.; Varani, K. Pharmacology of Adenosine Receptors: The State of the Art. *Physiol. Rev.* **2018**, *98*, 1591–1625. [CrossRef]

81. Szigeti, K.; Horváth, I.; Veres, D.S.; Martinecz, B.; Lénárt, N.; Kovács, N.; Bakcsa, E.; Márta, A.; Semjéni, M.; Máthé, D.; et al. A Novel SPECT-Based Approach Reveals Early Mechanisms of Central and Peripheral Inflammation after Cerebral Ischemia. *Br. J. Pharmacol.* **2015**, *35*, 1921–1929. [CrossRef] [PubMed]

82. Liu, Y.; Ma, Y.; Du, B.; Wang, Y.; Yang, G.-Y.; Bi, X. Mesenchymal Stem Cells Attenuated Blood-Brain Barrier Disruption via Downregulation of Aquaporin-4 Expression in EAE Mice. *Mol. Neurobiol.* **2020**, *57*, 3891–3901. [CrossRef] [PubMed]

83. Wei, W.; Du, C.; Lv, J.; Zhao, G.; Li, Z.; Wu, Z.; Haskó, G.; Xie, X. Blocking A2B Adenosine Receptor Alleviates Pathogenesis of Experimental Autoimmune Encephalomyelitis via Inhibition of IL-6 Production and Th17 Differentiation. *J. Immunol.* **2012**, *190*, 138–146. [CrossRef] [PubMed]

84. Maraula, G.; Traini, C.; Mello, T.; Coppi, E.; Galli, A.; Pedata, F.; Pugliese, A.M. Effects of oxygen and glucose deprivation on synaptic transmission in rat dentate gyrus: Role of A2A adenosine receptors. *Neuropharmacology* **2013**, *67*, 511–520. [CrossRef]

85. Maraula, G.; Lana, D.; Coppi, E.; Gentile, F.; Mello, T.; Melani, A.; Galli, A.; Giovannini, M.G.; Pedata, F.; Pugliese, A.M. The Selective Antagonism of P2X7 and P2Y1 Receptors Prevents Synaptic Failure and Affects Cell Proliferation Induced by Oxygen and Glucose Deprivation in Rat Dentate Gyrus. *PLoS ONE* **2014**, *9*, e115273. [CrossRef]

86. Melani, A.; Pantoni, L.; Bordoni, F.; Gianfriddo, M.; Bianchi, L.; Vannucchi, M.G.; Bertorelli, R.; Monopoli, A.; Pedata, F. The selective A2A receptor antagonist SCH 58261 reduces striatal transmitter outflow, turning behavior and ischemic brain damage induced by permanent focal ischemia in the rat. *Brain Res.* **2003**, *959*, 243–250. [CrossRef]

87. Melani, A.; Gianfriddo, M.; Vannucchi, M.G.; Cipriani, S.; Baraldi, P.G.; Giovannini, M.G.; Pedata, F. The selective A2A receptor antagonist SCH 58261 protects from neurological deficit, brain damage and activation of p38 MAPK in rat focal cerebral ischemia. *Brain Res.* **2006**, *1073*, 470–480. [CrossRef]

88. Melani, A.; Cipriani, S.; Vannucchi, M.G.; Nosi, D.; Donati, C.; Bruni, P.; Giovannini, M.G.; Pedata, F. Selective adenosine A2a receptor antagonism reduces JNK activation in oligodendrocytes after cerebral ischaemia. *Brain* **2009**, *132*, 1480–1495. [CrossRef]

89. Melani, A.; Dettori, I.; Corti, F.; Cellai, L.; Pedata, F. Time-course of protection by the selective A2A receptor antagonist SCH58261 after transient focal cerebral ischemia. *Neurol. Sci.* **2015**, *36*, 1441–1448. [CrossRef]

90. Melani, A.; Corti, F.; Cellai, L.; Vannucchi, M.-G.; Pedata, F. Low doses of the selective adenosine A2A receptor agonist CGS21680 are protective in a rat model of transient cerebral ischemia. *Brain Res.* **2014**, *1551*, 59–72. [CrossRef]

91. Pedata, F.; Pugliese, A.M.; Coppi, E.; Dettori, I.; Maraula, G.; Cellai, L.; Melani, A. Adenosine A(2A) Receptors Modulate Acute Injury and Neuroinflammation in Brain Ischemia. *Mediat. Inflamm.* **2014**, *2014*, 805198. [CrossRef] [PubMed]

92. Liu, H.; Zhang, Y.; Wu, H.; D'Alessandro, A.; Yegutkin, G.G.; Song, A.; Sun, K.; Li, J.; Cheng, N.-Y.; Huang, A.; et al. Beneficial Role of Erythrocyte Adenosine A2B Receptor–Mediated AMP-Activated Protein Kinase Activation in High-Altitude Hypoxia. *Circulation* **2016**, *134*, 405–421. [CrossRef] [PubMed]

93. Zhao, H.; Gao, X.; Liu, Z.; Lin, J.; Wang, S.; Wang, D.; Zhang, Y. Effects of the transcription factor Olig1 on the differentiation and remyelination of oligodendrocyte precursor cells after focal cerebral ischemia in rats. *Mol. Med. Rep.* **2019**, *20*, 4603–4611. [CrossRef] [PubMed]

94. Khodanovich, M.Y.; A Kisel, A.; E Akulov, A.; Atochin, D.N.; Kudabaeva, M.S.; Glazacheva, V.Y.; Svetlik, M.V.; A Medvednikova, Y.; Mustafina, L.R.; Yarnykh, V. Quantitative assessment of demyelination in ischemic stroke in vivo using macromolecular proton fraction mapping. *Br. J. Pharmacol.* **2018**, *38*, 919–931.

95. Compston, A.; Coles, A. Multiple sclerosis. *Lancet* **2002**, *359*, 1221–1231. [CrossRef]

96. Lazzarino, G.; Amorini, A.M.; Eikelenboom, M.; Killestein, J.; Belli, A.; Di Pietro, V.; Tavazzi, B.; Barkhof, F.; Polman, C.; Uitdehaag, B.; et al. Cerebrospinal fluid ATP metabolites in multiple sclerosis. *Mult. Scler. J.* **2010**, *16*, 549–554. [CrossRef]

97. Amorini, A.M.; Petzold, A.; Tavazzi, B.; Eikelenboom, J.; Keir, G.; Belli, A.; Giovannoni, G.; Di Pietro, V.; Polman, C.; D'Urso, S.; et al. Increase of uric acid and purine compounds in biological fluids of multiple sclerosis patients. *Clin. Biochem.* **2009**, *42*, 1001–1006. [CrossRef]

98. Vincenzi, F.; Corciulo, C.; Targa, M.; Merighi, S.; Gessi, S.; Casetta, I.; Gentile, M.; Granieri, E.; Borea, P.A.; Varani, K. Multiple sclerosis lymphocytes upregulate A(2A) adenosine receptors that are antiinflammatory when stimulated. *Eur. J. Immunol.* **2013**, *43*, 2206–2216. [CrossRef]

99. Ingwersen, J.; Wingerath, B.; Graf, J.; Lepka, K.; Hofrichter, M.; Schröter, F.; Wedekind, F.; Bauer, A.; Schrader, J.; Hartung, H.-P.; et al. Dual roles of the adenosine A2a receptor in autoimmune neuroinflammation. *J. Neuroinflamm.* **2016**, *13*, 48.

100. Chen, Y.; Zhang, Z.-X.; Zheng, L.-P.; Wang, L.; Liu, Y.-F.; Yin, W.-Y.; Chen, Y.-Y.; Wang, X.-S.; Hou, S.-T.; Chen, J.; et al. The adenosine A2A receptor antagonist SCH58261 reduces macrophage/microglia activation and protects against experimental autoimmune encephalomyelitis in mice. *Neurochem. Int.* **2019**, *129*, 104490. [CrossRef]

101. Liu, Y.; Alahiri, M.; Ulloa, B.; Xie, B.; Sadiq, S.A. Adenosine A2A receptor agonist ameliorates EAE and correlates with Th1 cytokine-induced blood brain barrier dysfunction via suppression of MLCK signaling pathway. *Immun. Inflamm. Dis.* **2017**, *6*, 72–80. [CrossRef] [PubMed]

102. Rajasundaram, S. Adenosine A2A Receptor Signaling in the Immunopathogenesis of Experimental Autoimmune Encephalomyelitis. *Front. Immunol.* **2018**, *9*, 402. [CrossRef] [PubMed]

Publisher's Note: MDPI stays neutral with regard to jurisdictional claims in published maps and institutional affiliations.

International Journal of
Molecular Sciences

Review

Targeting Adenosine Receptors: A Potential Pharmacological Avenue for Acute and Chronic Pain

Fabrizio Vincenzi [1,*], **Silvia Pasquini [1]**, **Pier Andrea Borea [2]** and **Katia Varani [1]**

[1] Department of Translational Medicine and for Romagna, Pharmacology Section, University of Ferrara, 44121 Ferrara, Italy; psqslv@unife.it (S.P.); vrk@unife.it (K.V.)
[2] University of Ferrara, 44121 Ferrara, Italy; bpa@unife.it
* Correspondence: fabrizio.vincenzi@unife.it; Tel.: +39-0532-455214

Received: 12 October 2020; Accepted: 17 November 2020; Published: 18 November 2020

Abstract: Adenosine is a purine nucleoside, responsible for the regulation of multiple physiological and pathological cellular and tissue functions by activation of four G protein-coupled receptors (GPCR), namely A_1, A_{2A}, A_{2B}, and A_3 adenosine receptors (ARs). In recent years, extensive progress has been made to elucidate the role of adenosine in pain regulation. Most of the antinociceptive effects of adenosine are dependent upon A_1AR activation located at peripheral, spinal, and supraspinal sites. The role of A_{2A}AR and A_{2B}AR is more controversial since their activation has both pro- and anti-nociceptive effects. A_3AR agonists are emerging as promising candidates for neuropathic pain. Although their therapeutic potential has been demonstrated in diverse preclinical studies, no AR ligands have so far reached the market. To date, novel pharmacological approaches such as adenosine regulating agents and allosteric modulators have been proposed to improve efficacy and limit side effects enhancing the effect of endogenous adenosine. This review aims to provide an overview of the therapeutic potential of ligands interacting with ARs and the adenosinergic system for the treatment of acute and chronic pain.

Keywords: adenosine; pain; adenosine receptors; antinociception

1. Introduction

Today, although substantial progress has been made, many pathological pain conditions remain poorly understood and resist currently available treatments. There is, therefore, a need for novel molecular targets to develop new therapeutic agents with improved efficacy and tolerability. Many experimental reports have identified adenosine receptors (ARs) as potential targets for the management of acute and chronic pain.

Adenosine is a ubiquitous endogenous autacoid that mediates its physiopathological effects by interacting with four G protein-coupled receptors (GPCR), namely A_1, A_{2A}, A_{2B}, and A_3 ARs [1]. A_1 and A_3AR are coupled with G_i and G_o members of the G protein family, through which they have an inhibitory effect on adenylyl cyclase (AC) activity, while A_{2A}ARs and A_{2B}ARs stimulate it by coupling to G_s proteins. The consequent modulation of cyclic adenosine monophosphate (cAMP) levels activates or inhibits a large variety of signaling pathways depending on the specific type of cell involved. Although there are instances in which adenosine exerts detrimental effects in various pathological conditions, it is generally considered a protective and homeostatic mediator against tissue damages and stress conditions [2,3]. In physiological and unstressed conditions, the extracellular concentrations of adenosine are maintained low as a result of the rapid metabolism and uptake [4]. However, its levels rise considerably during conditions involving increased metabolic demand, hypoxia, inflammation, and tissue injury. In particular, increased levels of extracellular adenosine were observed in pathological conditions such as epilepsy [5,6], ischemia [7,8], cancer [9,10], inflammation [11], and ultimately pain [12,13]. Although adenosine can be produced intracellularly, the main source of adenosine in pathological states is adenosine triphosphate (ATP), released by cells under stressful conditions and dephosphorylated from the combined action of two hydrolyzing enzymes termed ectonucleoside triphosphate diphosphohydrolase (CD39) and ecto-5′-nucleotidase (CD73) [1]. Regarding nociception, these elevated levels of endogenous adenosine can alter pain transmission by actions at spinal, supraspinal, and peripheral sites. The extracellular action of adenosine can then be terminated by its transformation to inosine through adenosine deaminase (ADA) and/or by intracellular uptake via nucleoside transporters [14]. Intracellularly, adenosine is phosphorylated to AMP by adenosine kinase or deaminated to inosine by ADA. Given these regulation mechanisms of adenosine concentration, potential pain management can be obtained not only with specific ligand interacting with ARs but also by manipulating endogenous tissue levels of adenosine by modulating its metabolism or transport [13] (Figure 1).

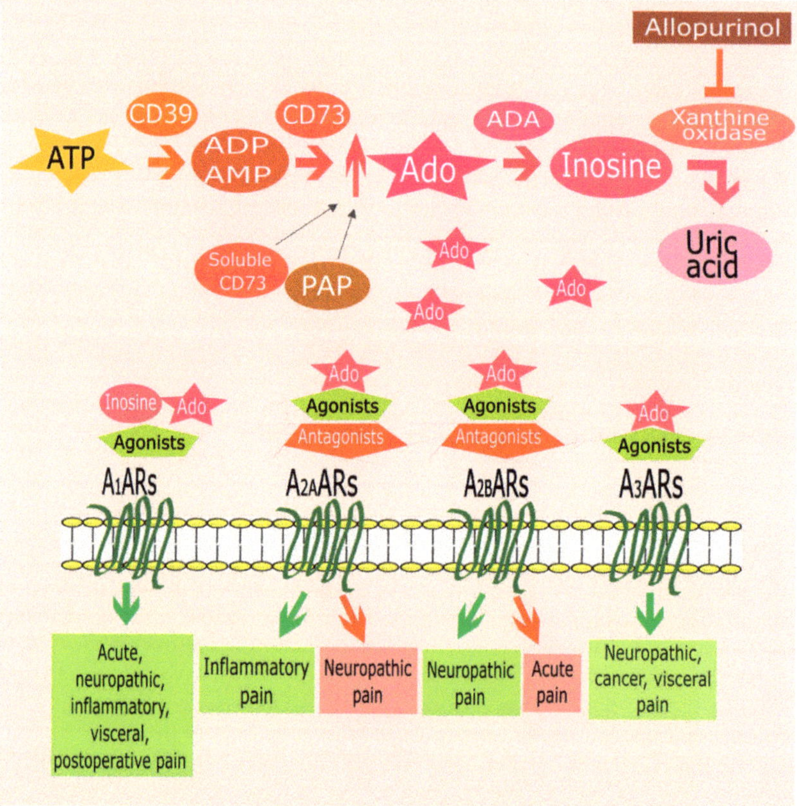

Figure 1. Adenosine (Ado) metabolism and involvement of adenosine receptors (ARs) in pain. The main source of adenosine is adenosine triphosphate (ATP) released from various cell types in response to different stimuli. ATP is dephosphorylated to adenosine diphosphate (ADP)/adenosine monophosphate (AMP) and then to adenosine by two ectonucleotidases (CD39, CD73). In nociception, the elevated levels of adenosine may alter the pain signaling. Thus, the modulation of adenosine metabolisms, increasing its levels, could represent an alternative strategy for pain management. Soluble CD73 provokes long-lasting thermal antihyperalgesic and mechanical antiallodynic effects through A_1AR activation. Prostatic acid phosphatase (PAP), acting as an ectonucleotidase, induces A_1AR-dependent antinociceptive effects in inflammatory and neuropathic pain models. Extracellular adenosine is rapidly

metabolized to inosine by adenosine deaminase (ADA). Inosine is able to bind A_1ARs, with an affinity similar to that of adenosine, inducing antinociceptive effects. Another strategy to promote the accumulation of inosine is represented by the inhibitors of the enzyme xanthine oxidase such as allopurinol. In the extracellular space, adenosine can interact with its receptors. A_1ARs stimulation with adenosine, adenosine metabolites like inosine, or synthetic agonists presents analgesic effects in acute, neuropathic, visceral, postoperative, and inflammatory pain. Activation of $A_{2A}ARs$ by endogenous adenosine or exogenous agonists results in antinociception in case of inflammatory pain. While, $A_{2A}ARs$ blockade shows analgesic effects in neuropathic pain. Regarding $A_{2B}ARs$, their stimulation has antinociceptive effects in neuropathic pain and their blockade is useful for acute pain treatment. Finally, A_3ARs activation gives analgesic effects in different types of pain such as neuropathic, cancer, and visceral pain.

Although adenosine and its receptors represent a clear target for pharmacological treatment of various diseases and pathological states including pain, very few drugs acting on the adenosinergic system have so far reached the market. The reason behind this discrepancy may be partly due to the ubiquitous distribution of ARs in almost every cell and tissue, making it difficult to avoid unwanted side effects. In recent years, many efforts have been made to improve our understanding of the role of adenosine in nociception and identify novel strategies to exploit the therapeutic potential of the adenosinergic system such as selective ligands, partial agonists, allosteric modulators, or adenosine concentration modulating agents.

The focus of the present review is to describe the recent advances in our understanding of the role of ARs in nociception. For each receptor subtype, we will briefly summarize and discuss the preclinical experimental studies that investigated their role and mechanism of action in the modulation of acute and chronic pain.

2. ARs and Pain

2.1. A1ARs

The antinociceptive effect of adenosine has been primarily attributed to the activation of A_1ARs [15] and various A_1AR agonists or positive allosteric modulators have been shown to be effective in several preclinical models of pain (Table 1). The signaling pathway underlying A_1ARs antinociception includes inhibition of cyclic AMP and consequently protein kinase A (PKA) activation, inhibition of Ca^{2+} channels, activation of K^+ currents, and interactions with phospholipase C (PLC), inositol triphosphate (IP3), diacylglycerol (DAG), extracellular signal-regulated kinases (ERK), and β-arrestin pathways [3]. The prominent role of this receptor subtype in analgesic responses is due to its peculiar expression in different sites relevant to pain transmission. A_1ARs are indeed located on peripheral sensory nerve endings in the spinal cord dorsal horn, and at supraspinal pain-processing structures [13,16]. Microglia represent another important localization for the antinociceptive action of A_1ARs, especially for pain states involving glial activation [17]. The peripheral activation of A_1ARs diminished inflammatory hypernociception caused by carrageenan intraplantar administration. Using specific inhibitors, the antinociceptive effect of the A_1AR agonist CPA was shown to be dependent on the nitric oxide (NO)/cyclic guanosine monophosphate (cGMP)/protein kinase G (PKG)/K_{ATP} signaling pathway [18]. The contribution of peripheral A_1ARs to antinociception was further corroborated when the selective A_1AR antagonist DPCPX reversed the antinociceptive effects of locally and systemically administered acetaminophen or tramadol in the formalin test [19]. A proof of the supraspinal antinociceptive action of A_1AR has been reported in a study where the A_1AR agonist 2′-Me-CCPA injected into the intra-periaqueductal grey (PAG) reduced pain behavior in the plantar and formalin tests. When microinjected into the PAG, 2′-Me-CCPA decreased the ongoing activity of the pronociceptive ON cells and increased the ongoing activity of the antinociceptive OFF cell in the rostral ventromedial medulla [20]. In neuropathic pain rats, the A_1AR agonist CPA reduced thermal and mechanical sensitivity, while in naïve rats it decreased hypersensitivity to heat but not to mechanical stimuli. In this study, electrophysiological experiments suggested that spinal application of CPA depressed long-term potentiation of A- and C-fiber evoked field potentials while it depressed the baseline of C-fiber but not A-fiber response. To explain this different response, authors have hypothesized that A_1ARs may be more expressed at C-fiber nerve endings than at A-fiber endings. [21]. In resiniferatoxin-induced neuropathy, the downregulation of A_1ARs was suggested to contribute to nociception, while the intrathecal injection of adenosine attenuated mechanical allodynia, an effect abrogated by A_1AR antagonism [22].

A_1ARs also seem to be involved in visceral antinociception. Centrally injected agonist CPA increased the threshold volume of colonic distension-induced abdominal withdrawal reflex in conscious rats. Besides, the use of the A_1 antagonist DPCPX suggested that adenosinergic signaling via A_1ARs is also involved in the central orexin-induced antinociceptive action against colonic distension [23]. In a subsequent study, the authors suggested that serotonin $5\text{-}HT_{1A}$, $5\text{-}HT_{2A}$, dopamine D1 or cannabinoid CB_1 receptors, and the opioid system might specifically mediate the CPA-induced visceral antinociception [24].

The potential role of A_1ARs in postoperative pain was also investigated. Intrathecal administration of the A_1AR agonist R-PIA decreased nonevoked spontaneous pain behavior and increased withdrawal thresholds after plantar incision. The opening of K_{ATP} channels contributed to this antinociceptive effect [25]. In a mouse model of acute postoperative pain, ankle joint mobilization decreased hyperalgesia through the involvement of peripheral and central A_1ARs [26]. In another report, intrathecal adenosine injection inhibited hyperalgesia in two neuropathic pain models but not in a postoperative pain model represented by the plantar incision. However, in this model A_1AR mRNA and protein expression were decreased suggesting that the lack of antinociceptive effect of adenosine on postoperative pain was due to the decrease in A_1ARs [27].

An intriguing connection has been uncovered between A_1ARs and acupuncture, an invasive practice worldwide used to relieve pain. Many studies report that the antinociceptive effects of acupuncture are dependent upon A_1AR activation. It was shown that extracellular adenosine concentration is increased during acupuncture in mice and A_1AR expression is required for the adenosine-mediated analgesic effect of acupuncture [28]. The involvement of A_1ARs in the reduction in neuropathic pain exerted by electroacupuncture was demonstrated by the intrathecal injection of the A_1AR antagonist DPCPX in a chronic constriction injury (CCI) model. In this report, the effect of A_1ARs was related to the inhibition of astrocyte activation. [29]. Similar results were obtained in a Complete Freund's adjuvant (CFA)-induced inflammatory pain mouse model, corroborating the involvement of A_1ARs in electroacupuncture-mediated antinociception [30]. In another study, the analgesic effect of electroacupuncture was suggested to be mediated by overexpressed A_1ARs in the spinal cord [31].

Different studies suggested that A_1AR activation is required for the antinociceptive action of various natural compounds. Indeed, the A_1AR antagonist DPXPC blocked the effect of norisoboldine, a benzylisoquinoline alkaloid isolated from *Radix Linderae* that diminishes pain response, in the formalin and writhing test [32]. In addition, A_1AR is necessary to the analgesic effect of paeoniflorin, the major active component extracted from *Paeonia lactiflora*. In a study carried out in mice, paeoniflorin increased the mechanical threshold and prolonged the thermal latency after partial sciatic nerve ligation (SCNL), an effect abolished by the A_1AR antagonist CPT or the genetic deletion of A_1ARs [33]. In the hot plate test, the antinociceptive effect of (−)-linalool, a natural occurring enantiomer in essential oils, was blocked by both an A_1 and an $A_{2A}AR$ antagonist [34]. D-Fructose-1,6-bisphosphate is an intermediate in the glycolytic pathway, inhibiting hyperalgesia induced by intraplantar injection of carrageenin and its mechanism of action seems dependent on adenosine accumulation that in turns exerts antinociceptive effects by activating peripheral A_1ARs [35].

Adenosine is rapidly metabolized to inosine by ADA. Interestingly, different studies have identified inosine as a putative endogenous ligand of A_1ARs and demonstrated the A_1-mediated antinociceptive effect of the more stable metabolite of adenosine. In particular, inosine binds to A_1ARs with an affinity resembling that of adenosine and induces antinociceptive, antiallodynic, and antihyperalgesic effects. In rats, both the A_1AR antagonist DPCPX and the $A_{2A}AR$ antagonist ZM241385 reversed the antiallodynic, and antihyperalgesic effects of inosine in models of mechanical and heat hyperalgesia induced by bradykinin and phorbol 12-myristate 13-acetate [36]. In the formalin test, inosine did not induce antinociception in A_1ARs knockout (KO) mice and the A_1AR antagonist DPCPX inhibited its effects [37]. In a subsequent study, DPCPX, but not the $A_{2A}AR$ antagonist SCH58261, abrogated the antinociceptive effect of inosine in the intraplantar glutamate test [38]. A different strategy to promote the accumulation of purines like adenosine or inosine is by using the xanthine oxidase inhibitor, allopurinol. Indeed, it has been reported that intraperitoneal administration of allopurinol increased cerebrospinal fluid concentrations of adenosine and its metabolites inducing antinociceptive effects in different pain models. The selective A_1AR antagonist DPCPX, but not the selective $A_{2A}AR$ antagonist SCH58261, prevented allopurinol-induced anti-nociception [39,40]. Since extracellular adenosine is primarily derived from the hydrolysis of AMP, the antinociceptive effect of a soluble version of the recombinant CD73, the enzyme that converts AMP to adenosine, has been tested in different pain models. The results of this study revealed long-lasting thermal antihyperalgesic and mechanical antiallodynic effects that were dependent on A_1AR activation [41]. Prostatic acid phosphatase (PAP) acts as an ectonucleotidase hydrolyzing extracellular AMP to adenosine in nociceptive dorsal root ganglia neurons [42,43]. Intrathecal injection of a secretory version of human PAP induced A_1AR-dependent antinociceptive effects in inflammatory and neuropathic pain models [44,45]. Furthermore, the injection of PAP into the popliteal fossa—a common acupuncture point—reduces pain responses in mouse models that lasted up to six days after a single injection, an effect dependent upon A_1AR activation [46].

Several papers in the literature proposed a link between opioid-mediated antinociception and A_1ARs. In a rat with spinal cord injury (SCI), it was demonstrated a supra-additive interaction between the adenosine A_1AR agonist R-PIA and morphine in the reduction in mechanical allodynia-like behavior [47]. In spinal cord neuronal nociceptive responses, the antinociceptive effects of the A_1AR agonist CPA were associated with activation of κ-opioid receptors since the reversal of the CPA effect was observed with norbinaltorphimine (a selective κ-opioid receptor antagonist) but not with low doses of μ-opioid antagonist naloxone [48]. While the opioid antagonist naltrexone did not affect the antinociception induced by CPA in the formalin test, the activation of A_1 or $A_{2A}AR$ counteracted the μ-opioid receptor increase induced by formalin in the spinal cord, confirming the interaction between adenosinergic and opioid systems [49]. In a rat model of nerve ligation injury, the intrathecal administration of morphine synergistically enhanced the antiallodynic effect of the A_1AR agonist R-PIA, suggesting an interaction between μ-opioid receptors and A_1ARs at the spinal level [50]. In addition, other works reported that the antiallodynic/antihyperalgesic effect of morphine is reversed in the presence of the selective A_1AR antagonist DPCPX [51] or in A_1ARs KO mice [52]. Beyond opioids, the involvement of A_1ARs has been observed in the antinociceptive effect of non-steroidal anti-inflammatory drugs such as acetaminophen. In the formalin test, when acetaminophen was administered systemically or locally, its antinociceptive effects were reversed by the intraplantar injection of the A_1AR antagonist DPCPX, suggesting a link between activation of peripheral A_1ARs and acetaminophen effects [53]. The contribution of spinal A_1ARs to the action of acetaminophen secondarily to the involvement of descending serotonin pathways and the release of adenosine within the spinal cord was also suggested [54]. The involvement of A_1ARs was also demonstrated in the antinociceptive effects of amitriptyline [55,56], oxcarbazepine [57], levetiracetam [58], and neuropeptide S [59].

Collectively, these preclinical studies provide strong support for the therapeutic potential of A_1AR agonists. However, limited clinical efficacy and relevant cardiovascular and central adverse effects have, to date, hampered the development of A_1AR agonists as analgesic drugs. An alternative approach to increase selectivity and reduce the possibility of adverse effects exploiting the physiological action of endogenous adenosine is the development of A_1AR-positive allosteric modulators [60,61]. These agents enhance the function of receptors activated by endogenous agonists, they are expected to have a much lower side effect potential than an exogenous orthosteric ligand, a low propensity for receptor desensitization, and a high selectivity for a given receptor subtype [62]. T62 was the first A_1AR-positive allosteric modulator to be tested in animal models of pain. Intrathecal or systemic administration of T62 reduced mechanical hypersensitivity induced by spinal nerve ligation (SNL) [63,64], reversed thermal hypersensitivity in carrageenin-inflamed rats [65], and was effective for postoperative hypersensitivity following plantar incision [66]. More recently, TRR469 was characterized as one of the most potent A_1AR-positive allosteric modulators so far synthesized being able to increase adenosine affinity by 33 fold [67–69]. TRR469 effectively inhibited nociceptive behaviors in the formalin and writhing tests, with effects comparable to morphine. Furthermore, it revealed an antiallodynic action in the streptozotocin (STZ)-induced diabetic neuropathic pain model without inducing locomotor or cataleptic side effects as the orthosteric-acting CCPA did [69].

Table 1. A_1AR ligands with antinociceptive effects in preclinical models of pain.

Ligand	Pharmacological Behavior	Pain Model	Species	Route of Administration
2'-Me-CCPA	agonist	formalin test	rat	intra-PAG, i.p. [20]
		plantar test	rat	intra-PAG, i.p. [20]
		tail flick test	rat	intra-PAG, i.p. [20]
CCPA	agonist	formalin test	mouse	i.p. [69]
		writhing test	mouse	i.p. [69]
		STZ-induced mechanical allodynia	mouse	i.p. [69]
		CFA induced-mechanical allodynia and thermal hyperalgesia	mouse	Zusanli acupoint-injection [28]
		SCNL induced-mechanical allodynia and thermal hyperalgesia	mouse	Zusanli acupoint-injection [28]
CPA	agonist	formalin test	mouse	i.p. [49]; i.t. [54]
		CFA-induced-mechanical allodynia and thermal hyperalgesia	mouse	i.m. [30]; i.p. [42]; Weizhong acupoint-injection [46]
		carrageenan-induced mechanical allodynia	rat	i.pl. [18]
		PGE$_2$-induced mechanical allodynia	rat	i.pl. [18]
		SCNL-induced mechanical allodynia and thermal hyperalgesia	rat	i.p. [21]
		colonic distension-induced visceral pain	rat	s.c., i.c. [23,24]
R-PIA	agonist	plantar incision-induced mechanical allodynia	rat	i.t. [25]
		photochemical SCI-induced mechanical and thermal allodynia	rat	i.t. [47]
		SCNL-induced mechanical allodynia	rat	i.t. [50]
		photochemical sciatic nerve injury-induced mechanical and thermal allodynia	rat, mouse	i.t. [52]
		carrageenan-induced mechanical and thermal allodynia	rat, mouse	i.t. [52]
T62	positive allosteric modulator	SNL-induced mechanical allodynia	rat	i.p. [63]; i.t. [63,64]; p.o. [64]
		carrageenan-induced thermal hyperalgesia	rat	i.t. [65]
		plantar incision-induced mechanical allodynia	rat	i.t. [66]
TRR469	positive allosteric modulator	formalin test	mouse	i.p. [69]
		writhing test	mouse	i.p. [69]
		STZ-induced mechanical allodynia	mouse	i.p. [69]

PAG (periaqueductal grey); i.p. (intraperitoneal); STZ (streptozotocin); CFA (Complete Freund's adjuvant); SCNL (sciatic nerve ligation); i.t. (intrathecal); i.m. (intramuscular); i.pl. (intraplantar); s.c. (subcutaneous); i.c. (intracisternal); SCI (spinal cord injury); SNL (spinal nerve ligation); p.o. (per os).

2.2. $A_{2A}ARs$ and Pain

The presence of $A_{2A}ARs$ both on neurons and on glial cells is at the basis of $A_{2A}ARs$ implications in pain [70]. The relation between $A_{2A}ARs$ and pain has been controversial with evidence sustaining either pronociceptive and antinociceptive activity depending on the receptors' localization and the kind of pain (Table 2) [13]. Studies supporting the pronociceptive role of $A_{2A}AR$ report that the selective blockade of this receptor subtype by systemic administration of SCH58261, a selective $A_{2A}AR$ antagonist, is able to counteract nociception; even the administration at the spinal level produced an equal effect [13,54]. These results are supported by experimental models of acute and nerve injury pain in $A_{2A}ARs$ KO which showed a decreased algesic reaction to pain tests and even a reduction in markers of neural activity [71]. Moreover, the administration of caffeine, which is a well-known non-selective antagonist of ADA, avert the sleep deprivation due to hypersensitivity following surgical operation. $A_{2A}AR$ selective blockade with ZM241385 has shown to decrease surgical pain levels and the thermal hyperalgesia caused by sleep deprivation in rats. These results support the hypothesis that $A_{2A}ARs$ are implicated in the regulation of the interplay between sleep and pain [72]. The pronociceptive effect of $A_{2A}AR$ stimulation was further corroborated in a study reporting that carrageenan-induced hyperalgesia was significantly reduced in $A_{2A}AR$ KO mice compared to wild type controls. Interestingly, the $A_{2A}AR$ inverse agonist ZM241385 injected into the hindpaw reduced the nociceptive behavior following carrageenan in female wild type mice, but not in males suggesting a sex difference in response to $A_{2A}AR$ activation in the periphery [73]. In addition, a series of inverse agonists showing two different affinity values for the $A_{2A}ARs$ with the high affinity value in the picomolar/femtomolar range was recently synthesized [74,75] and tested for their antinociceptive properties. In particular, one of these potent inverse agonists, namely TP455, proved to be more potent than morphine in writhing and tail immersion tests in mice [74].

Furthermore, the blockade of $A_{2A}ARs$ could provide protection in cases of neuropathic pain, which is one of the most common kinds of chronic pain, and it is found in different disorders and could lead to nerve dysfunctionalities [76]. Neuropathic pain pathophysiology is extremely intricate because it comprises central and peripheral mechanisms such as changes in ion channel expression, neurotransmitter release, and pain pathways [77]. Even oxidative stress could play an important role in the neuropathic pain origin process [78]. A body of evidence reveals that, after SCI, there are events that trigger reactive oxygen species (ROS) formation pathways such as microglia activation and glutamate release [79,80]. The injury at the sensory nerves level also involves damage to nuclear and mitochondrial DNA, and loss of antioxidant enzymes [81–83]. In fact, numerous studies report that the anti-oxidant or ROS scavengers administration has analgesic effects in many in vivo models of neuropathic pain. Furthermore, neuropathic pain is often a consequence of antitumoral treatments containing platinum because these drugs can provoke peripheral neuropathy and chemotherapy-induced oxidative stress is one of the important pathogenic factors damaging peripheral sensory neurons [84]. Recently, it has been proved that novel $A_{2A}AR$ antagonists featuring antioxidant moieties can reduce pain associated with oxaliplatin treatment in a mouse model of neuropathy reducing ROS level [85,86]. After peripheral nerve injury, $A_{2A}ARs$ stimulation induces both activation and proliferation of microglia and astrocytes responsible for inflammation occurring in neuropathic pain, while genetic deletion of the $A_{2A}ARs$ decreases all the behavioral and histological signs of pain [77,87]. Several studies also showed that systemic and spinal administration of the selective $A_{2A}AR$ antagonist SCH58261 has antinociceptive effects in different preclinical models [54,74].

Notwithstanding the coherence of the studies testifying for a pronociceptive role of $A_{2A}ARs$, in the literature there is evidence even for an antinociceptive role. In particular, since $A_{2A}ARs$ are expressed in immune cells where they exert a potent anti-inflammatory action, their stimulation may be helpful in cases of inflammatory pain [3,13]. $A_{2A}ARs$ KO animals under prolonged inflammatory conditions show an up-regulation of markers of spinal cord neural activation. In these KO mice, the loss of the antinociceptive $A_{2A}ARs$ on immune cells exceeds the decrease in pronociceptive $A_{2A}ARs$ on nerve terminals leading to enhanced pain signaling [88]. It is well known that the stimulation of $A_{2A}ARs$ has anti-inflammatory effects but less is known about $A_{2A}AR$ agonists treatment and chronic inflammatory pain. Different studies report that $A_{2A}ARs$ expression is up-regulated in lymphocytes of rheumatoid arthritis patients, these data should represent a basis for further investigations in this field [89,90].

The selective agonist of $A_{2A}AR$ CGS21680 shows the ability to slow down disease progression in an in vivo model of arthritis [91]. Even in a rat animal model, it has been demonstrated that CGS21680 treatment was very effective in decreasing clinical features in comparison to standard antirheumatic drugs such as methotrexate and etanercept [92]. The treatment with the $A_{2A}AR$ agonist CGS21680 was also able to inhibit the nuclear factor kappa-light-chain-enhancer of activated B cells (NF-κB) activation and to reduce the release of inflammatory cytokines such as tumor necrosis factor-α (TNF-α), IL-1β, and IL-6. Besides, the $A_{2A}AR$ stimulation leads to a decrease in metalloproteinases 1 and 3 [93]. Finally, in another mice model of monoarthritis, a new $A_{2A}AR$ agonist, named LASSBio-1359, showed an important analgesic effect in response to inflammatory pain. This treatment was also able to reduce inflammation by decreasing TNF-α, inducible NO synthase (iNOS) expression, and joint damage [94]. The results of the above-mentioned studies highlight a role for $A_{2A}AR$ agonists as a potential therapeutic tool in the management of inflammatory pain [89,93].

Different reports demonstrated an antinociceptive role of $A_{2A}AR$ activation in models of neuropathic pain. An acute administration of $A_{2A}AR$ agonists, such as ATL313 and CGS21680, leads to an analgesic effect that lasts for many weeks and reverses the mechanical allodynia and thermic hyperalgesia while decreasing the markers of microglia and astrocytes activation [95]. Interestingly, the effect of $A_{2A}ARs$ activation was just specific for nerve injury or sensitized state suggesting a potential role of $A_{2A}AR$ agonists for neuropathic pain. Moreover, the blockade of $A_{2A}ARs$ by using a receptor antagonist in the presence of an anti-IL-10 antibody reverted the effect of ATL313, suggesting that the observed effects were due to the activation of $A_{2A}ARs$ and the simultaneous enhanced IL-10 production [95]. In a subsequent study, ATL313 induced long-lasting protection against allodynia caused by CCI, SNL, and sciatic inflammatory neuropathy (SIN), through a mechanism involving PKA and protein kinase C (PKC) [96]. In a recent study, a single intrathecal injection of the $A_{2A}AR$ agonists CGS21680 reversed mechanical allodynia in a rat model of SCI termed spinal neuropathic avulsion pain for at least 6weeks [97]. In the follow-up work, the peri-sciatic injection of the agonist ATL313 also demonstrated the efficacy of $A_{2A}AR$ activation at the site of nerve injury. These anti-allodynic effects were accompanied by a reduction in interleukin (IL)-1β and NO release, and reduced expression of iNOS and sciatic markers of monocytes/macrophages [98]. These studies revealed that the agonism toward $A_{2A}ARs$ was able to reduce different kinds of neuropathic pain such as inflammatory neuropathic pain and traumatic ones. In all these cases the $A_{2A}ARs$ stimulation averted and reverted the nociceptive stimuli amplification [98]. Additionally, the long time span of the analgesic effect after a single treatment suggests that $A_{2A}AR$ agonists could be useful for central neuropathic pain therapy. It is worth noting that in these studies, the antiallodynic effects of $A_{2A}AR$ agonists were associated with diminished reactive gliosis. Glial cells have a pivotal role in starting and carrying on neuropathic pain, and for this reason, many studies are directed toward the discovery of new strategies in order to defeat the pain expansion directed by glia. In recent years, $A_{2A}AR$ agonists have emerged as possible candidates for glial inhibition thanks to their capability to suppress inflammation in immune cells; consequently, $A_{2A}AR$ agonists represent a promising tool for the treatment of chronic pain of neuroinflammatory origin [99].

The activation of the $A_{2A}AR$ subtype also seems to be involved in the analgesic effect of neuropeptide S observed in the formalin test. Intracerebroventricular administration of this eicosapeptide reduced formalin-induced nociception during both phases 1 and phase 2 of the test, an effect counteracted by the non-selective AR antagonist caffeine or the selective $A_{2A}AR$ antagonist ZM241385 [59]. Besides, an interaction between $A_{2A}ARs$ and the opioid system was reported when the antinociceptive effect exerted by the intracerebroventricular injection of Adonis, an agonist-like monoclonal antibody with high specificity for the $A_{2A}ARs$, was counteracted by naloxone, a non-selective opioid antagonist [100].

Table 2. $A_{2A}AR$ ligands with antinociceptive effects in preclinical models of pain.

Ligand	Pharmacological Behavior	Pain Model	Species	Route of Administration
ATL313	agonist	CCI-induced mechanical allodynia and thermal hyperalgesia	rat	i.t. [95,96]; peri-sciatic nerve injection [98]
		SNL-induced mechanical allodynia	rat	i.t. [96]
		SIN-induced mechanical allodynia	rat	i.t. [96]
		SCI-induced mechanical and thermal allodynia	rat	i.t. [97]
CGS21680	agonist	formalin test (early phase)	mouse	i.p. [49]
		CFA-induced-mechanical allodynia and thermal hyperalgesia	rat	i.p. [92]
		CCI-induced mechanical allodynia and thermal hyperalgesia	rat	i.t. [95,96]
		SCI-induced mechanical and thermal allodynia	rat	i.t. [97]
LASSBio-1359	agonist	formalin test	mouse	i.p. [94]
		carrageenan induced-mechanical allodynia and thermal hyperalgesia	mouse	i.p. [94]
Adonis	agonist-like monoclonal antibody	hot plate test	mouse	i.c.v. [100]
		tail flick test	mouse	i.c.v. [100]
TP455	inverse agonist	writhing test	mouse	i.p. [74]
		tail immersion test	mouse	i.p. [74]
ZM241385	antagonist	writhing test	mouse	i.p. [74]
		tail immersion test	mouse	i.p. [74]
		carrageenan induced-mechanical allodynia	mouse	s.c. [73]
		sleep deprivation-induced thermal hyperalgesia	rat	i.c.v. [72]
		plantar incision-induced mechanical allodynia and thermal hyperalgesia	rat	i.c.v. [72]

CCI (chronic constriction injury); i.t. (intrathecal); SNL (spinal nerve ligation); SIN (sciatic inflammatory neuropathy); SCI (spinal cord injury); i.p. (intraperitoneal); CFA (Complete Freund's adjuvant); i.c.v. (intracerebroventricular); s.c. (subcutaneous).

2.3. $A_{2B}ARs$ and Pain

$A_{2B}ARs$ are expressed both at the central level and in the periphery: among pain-relevant sites, they are localized on immune-inflammatory cells, where they have pro-inflammatory functions, in the spinal cord, and on astrocytes [1,101,102]. Since adenosine presents a lower affinity for $A_{2B}ARs$ in comparison to other AR subtypes, $A_{2B}ARs$ are more involved when adenosine concentration rises, for example in pathological conditions such as hypoxia/ischemia and inflammation [2,103].

Nonetheless, the different functions of $A_{2B}ARs$ in various tissues and their involvement in the pathogenesis of pain are poorly known. As a consequence, more studies are needed in order to clarify their pro or anti-nociceptive actions in different types of pain conditions [11,101].

Unfortunately, studies on the relationship between pain and $A_{2B}ARs$ are limited due to the lack of selective ligands (Table 3). One of the first studies using selective $A_{2B}AR$ antagonists reported an antinociceptive activity of $A_{2B}ARs$ blockade in an acute pain model represented by the hot plate test. One of these ligands, PSB-1115, did not penetrate the blood brain barrier due to its polar sulfonate group, suggesting that peripheral $A_{2B}ARs$ were implicated in the analgesic activity [104]. Interestingly, the efficacy of morphine was enhanced by subeffective doses of these $A_{2B}AR$ antagonists. In a follow-up study, the systemic administration of PSB-1115 decreased the algesic response and edema in both phases of the formalin test [105]. In the same test, the selective blockade of $A_{2B}ARs$ by using alloxazine resulted in a dose-dependent reduction in nociceptive behavior [106]. Moreover, it has been reported that the treatment with $A_{2B}AR$ antagonists, MRS1754 and PSB-1115, was able to decrease pain in visceral hypersensitivity rat models [107,108]. PSB-1115 also reverted the antinociceptive effect of diphenyl diselenide, organoselenium compounds, in the hot plate test in mice [109].

A$_{2B}$ARs seem to be involved even in chronic pain, with evidence highlighting that these receptor subtypes stimulate the interactions between immune cells and neurons. It was reported that high extracellular adenosine levels activate A$_{2B}$ARs on myeloid cells, and that this leads to the activation of pain sensory neurons giving rise to hypersensitivity and chronic pain [110]. Intriguingly, the author demonstrated that A$_{2B}$AR stimulation caused nociceptor hyperexcitability and promoted chronic pain via soluble IL-6 receptor trans-signaling. From these results, it is possible to deduce that the blockade of A$_{2B}$ARs may repress the nociceptive activity.

All these findings seem to testify for a pronociceptive role of A$_{2B}$ARs. However, it has been reported that even the activation of A$_{2B}$ARs, using a selective agonist (BAY606583), presented an analgesic effect in an accredited model of neuropathic pain, in a similar way to A$_{2A}$AR agonists treatment [96]. As it is well known, both A$_{2A}$ARs and A$_{2B}$ARs lead to increased cAMP accumulation and activation of downstream pathways; they also probably have a similar spinal mechanism of action. Normally, A$_{2B}$AR stimulation activates PKA and the pathway of PLC/IP3/DAG leading to changes in gene transcription, while β-arrestins are responsible for the receptor internalization mechanism [13,111].

Table 3. A$_{2B}$AR ligands with antinociceptive effects in preclinical models of pain.

Ligand	Pharmacological Behavior	Pain Model	Species	Route of Administration
BAY606583	agonist	CCI-induced mechanical allodynia	mouse	i.t. [96]
PSB-10	antagonist	formalin test	mouse	i.p. [105]
PSB-36	antagonist	formalin test	mouse	i.p. [105]
PSB-1115	antagonist	formalin test	mouse	i.p. [105]

CCI (chronic constriction injury); i.t. (intrathecal); i.p. (intraperitoneal).

2.4. A$_3$ARs and Pain

A$_3$ARs are present at the peripheral level in many tissues including inflammatory cells; they are less expressed in the central nervous system, nonetheless their activation causes functional effects, in particular, in glial cells [112]. The possibility to exploit A$_3$AR stimulation, using selective agonists, has been studied in different pathologies counting cancer and inflammation [112,113].

A$_3$ARs involvement has also been investigated in relation to pain; the first pieces of evidence reported a pronociceptive role [114]. Further studies, using more selective ligands, overturned previous results showing that A$_3$AR agonists present antinociceptive activity so, they can be useful as analgesics especially for neuropathic pain (Table 4) [113]. In fact, the systemic administration of selective A$_3$AR agonists, such as IB-MECA, Cl-IB-MECA and MRS1898, reduced the mechanical allodynia in a model of neuropathic pain—especially IB-MECA was as efficacious as morphine. The specificity of this effect was demonstrated by blocking A$_3$ARs with the selective antagonist MRS1523, which abrogated the analgesic effect of A$_3$AR agonists [115]. Interestingly, the A$_3$AR agonists have no effects in acute pain models, for instance, hot plate and tail flick tests [116]. Another A$_3$AR selective agonist, named MRS5698, was demonstrated to be able to reduce mechanical allodynia in different models of neuropathic pain. MRS5698 had an analgesic effect in acute pain tests but its activity persisted with repeated administrations [117]. The mechanism of action of this agonist involves GABA signaling: the A$_3$ARs activation normalizes the changes in GABA concentrations caused by nerve damages, thus restoring the GABA inhibitory effect on pain transmission [118]. Moreover, it has been noticed that A$_3$ARs stimulation inhibits N-type calcium channel opening in isolated rat dorsal root ganglion neurons, causing a reduction in the neurotransmitter release and the neuronal excitation [119]. In another model of nerve injury that produces tactile allodynia, the daily administration of IB-MECA averted the appearance of hypersensitivities, the activation of glial cells and the altered transmission of nociceptive stimuli, resulting in an attenuation of neuropathic pain [120]. In a recent study, MRS7476, a prodrug with increased aqueous solubility compared with parent MRS5698, was found to be efficacious in reversing neuropathic pain induced by CCI [121].

Anticancer chemotherapeutic treatments often induce neuropathy as an adverse effect; the stimulation of A_3ARs can help to decrease the pain in these cases. The A_3AR agonist IB-MECA is able to reduce the allodynia and the hyperalgesia induced by different anticancer drugs such as paclitaxel, oxaliplatin and bortezomib without diminishing their antitumoral effectiveness; even other A_3AR agonists, Cl-IB-MECA and MRS1898 present the same effects [115,116]. The pathway involved seems to imply NF-κB, ERK and p38 inhibition and the production of inflammatory cytokines. In particular, the treatment with A_3AR agonists reduces the release of the pro-inflammatory cytokines TNF-α and IL-1β while increases the anti-inflammatory IL-10 [122]. Other mechanisms have been proposed to explain the antinociceptive activity of A_3ARs; among these are the diminished activation of astrocytes, inhibition of cAMP, PKA, the interaction with the PLC/IP3/DAG and phosphoinositide 3-kinase (PI3K)/mitogen-activated protein kinase (MAPK)/ERK/cAMP response element-binding protein (CREB) pathways and the signaling via Gi [123]. Even in the A_3AR subtype, the internalization of the receptor is mediated by β-arrestins [111]. Recently, in a model of cancer chemotherapy-induced neuropathic pain, the A_3AR agonist MRS5698 attenuated pro-inflammatory IL-1β production and promoted anti-inflammatory and neuroprotective IL-10 expression by regulating the nucleotide-binding oligomerization domain-like receptor protein 3 inflammasome [124].

Besides the antinociceptive effect of A_3AR agonists in cancer pain and neuropathic pain related to chemotherapy, they have also found to be potent antitumoral agents in many animal models of different forms of cancer (melanoma, prostate, colon, and hepatocellular carcinoma), where they are able to reduce tumor growth [113,125]. Their therapeutic potential has also been assessed in a model of bone cancer pain in which mammary gland tumoral cells were injected into the tibia [126]. In this model, the repeated administration of Cl-IB-MECA decreased tumor growth, mitigated the nociception and the bone degradation associated with cancer. In addition, the A_3AR agonist was also effective in delaying the onset and the advancement of bone cancer with a major efficacy when the treatment with Cl-IB-MECA was done before the injection of cancer cells [126].

The involvement of A_3ARs in diabetic neuropathy was also investigated. It has been demonstrated that IB-MECA treatment ameliorates mechanical hyperalgesia and thermal hypoalgesia in STZ-treated mice. Moreover, reduced expression or functionality of A_3ARs promoted diabetic neuropathy development [127].

It is well established that long-lasting treatments with opioids lead to hyperalgesia and tolerance to drugs, reducing the analgesic effect of opioids in chronic pain [128,129]. In a rodent model, it has been reported that the opioid adverse effects seem to be linked to reduced A_3ARs signaling. In fact, the stimulation with A_3AR agonists ameliorates pain sensitivity suggesting that selective A_3AR agonists may be used in addition to opioids for chronic pain management [130]. Importantly, it has been reported that the antinociceptive effects of A_3AR agonists persist at least up to 2 weeks of treatment, suggesting that stimulation of A_3ARs does not induce tolerance [87].

A recent study reports the effect of a new A_3AR agonist, AL170, in a rat model of colitis. AL170 mitigates the colonic damage and inflammation, reducing the release of TNF-α, IL-1β, and myeloperoxidase. AL170 was demonstrated to have an efficacy comparable to that of dexamethasone, one of the most used drugs in the colitis treatment and other inflammatory bowel diseases [131]. The activation of A_3ARs resulted able to decrease the infiltration of inflammatory cells and the production of inflammatory mediators thus softening visceral pain [131]. A further study revealed that the treatment with A_3AR agonists is useful in another model of abdominal pain induced by colitis. In this model, Cl-IB-MECA and MRS5980 decreased visceral hypersensitivity in the postinflammatory phase as well as in the and persistence one and showed effectiveness comparable to that of linaclotide, a drug used for the treatment of irritable bowel syndrome [132,133].

The possibility to exploit A_3AR agonists in rheumatic pathologies has been studied, starting from the observation that A_3ARs are up-regulated in synovial tissue and peripheral blood mononuclear cells in rheumatoid arthritis patients. Treatment with A_3AR agonists leads to an improvement of symptoms and clinical signs [90,93,113]. Another potential therapeutic approach for arthritis could be represented by allosteric enhancer, in order to exploit the anti-inflammatory action of A_3ARs and the high adenosine concentrations typical of inflammatory pathologies. LUF6000, an A_3AR allosteric modulator, was showed to reduce inflammation in models of adjuvant- and monoiodoacetate-induced arthritis [134]. Even if the analysis of the nociceptive activity was not comprised in these studies, it is reasonable to hypothesize that a decreased inflammation would be accompanied by a reduction in pain.

Table 4. A_3AR ligands with antinociceptive effects in preclinical models of pain.

Ligand	Pharmacological Behavior	Pain Model	Species	Route of Administration
AR170	A_3AR agonist	colitis-induced visceral hypersensitivity	rat	i.p. [131]
Cl-IB-MECA	A_3AR agonist	chemotherapy-induced mechanical allodynia	mouse	i.p. [115]
		CCI-induced mechanical allodynia	mouse	i.p. [115]
		bone cancer-induced mechanical allodynia	rat	i.p. [126]
		colitis-induced visceral hypersensitivity	rat	i.p. [133]
IB-MECA	A_3AR agonist	chemotherapy-induced mechanical allodynia	mouse/rat	i.p. [115,122]
		CCI-induced mechanical allodynia	mouse	i.p. [115]; i.t. [118]
		STZ-induced mechanical allodynia and thermal hyperalgesia	mouse	i.p. [127]
		opioid-induced thermal hyperalgesia	rat	p.o. [130]
		tibial nerve injury-induced mechanical allodynia	rat	i.p. [120]
MRS1898	A_3AR agonist	chemotherapy-induced mechanical allodynia	mouse	i.p. [115]
		CCI-induced mechanical allodynia	mouse	i.p. [115]
MRS5698	A_3AR agonist	CCI-induced mechanical allodynia	rat	s.c., i.p., i.v. [117]; i.t. [118]
		spared nerve injury-induced mechanical allodynia	rat	s.c., i.p., i.v. [117]
		SNL-induced mechanical allodynia	rat	s.c., i.p., i.v. [117]
		bone cancer-induced mechanical allodynia	rat	s.c., i.p., i.v. [117]
		chemotherapy-induced mechanical allodynia	rat	s.c., i.p., i.v. [117]; i.t. [124]
		opioid-induced thermal hyperalgesia	rat	p.o. [130]
MRS5980	A_3AR agonist	colitis-induced visceral hypersensitivity	rat	i.p. [133]
MRS7422	A_3AR agonist	CCI-induced mechanical allodynia	mouse	p.o. [121]

i.p. (intraperitoneal); CCI (chronic constriction injury); i.t. (intrathecal); STZ (streptozotocin); p.o. (per os); s.c. (subcutaneous); i.v. (intravenous); SNL (spinal nerve ligation).

3. Conclusions

The modulation of ARs, especially their activation, induces potent antinociceptive effects in diverse preclinical models of acute and chronic pain. Nevertheless, the efficacy of AR ligands for the pharmacological treatment of pain in humans is still ambiguous and it has also yet to be determined whether ARs modulation could be exploited to inhibit spontaneous pain. Many hopes were initially placed on A_1AR agonists, but cardiovascular side effects prevented their progress in the clinic, at least with regard to their systemic administration. To get around these obstacles, alternative strategies have been proposed to continue exploiting the huge potential of adenosine and its receptors in pain management. Partial agonist and allosteric modulators of ARs have been tested in preclinical settings revealing potent antinociceptive effects with fewer side effects than conventional full agonists. Furthermore, localized activation of ARs has been proposed as a valid alternative to systemic delivery to maintain efficacy and reduce side effects, especially considering the ubiquitous localization of ARs in the human body. Exogenous ectonucleotidases as well as metabolizing enzyme inhibitors could increase the extracellular concentration of the short-living mediator adenosine, enhancing its nociceptive effects. As reviewed here, these alternative pharmacological approaches have shown promising results in preclinical models of pain and could offer a means to overcome the issues encountered so far by AR ligands in the clinic. Overall, the data summarized in this review highlight the therapeutic potential of ARs as pharmacological targets for the treatment of acute and chronic pain and the need to develop new and more effective strategies to exploit this potential.

Author Contributions: Conceptualization, F.V. and S.P.; writing—original draft preparation, F.V. and S.P.; writing—review and editing, P.A.B. and K.V.; visualization, F.V. and S.P.; supervision, P.A.B. and K.V. All authors have read and agreed to the published version of the manuscript.

Funding: This research received no external funding.

Conflicts of Interest: The authors declare no conflict of interest.

Abbreviations

GPCR	G protein coupled receptors
AR	adenosine receptor
AC	adenylate cyclase
cAMP	cyclic adenosine monophosphate
ATP	adenosine triphosphate
ADA	adenosine deaminase
PKA	protein kinase A
PLC	protein lipase C
IP3	inositol triphosphate
DAG	diacylglycerol
ERK	extracellular signal-regulated kinase
NO	nitric oxide
cGMP	cyclic guanosine monophosphate
PKG	protein kinase G
PAG	periaqueductal gray
CCI	chronic constriction injury
CFA	complete Freund's adjuvant
SCNL	sciatic nerve ligation
KO	knockout
PAP	prostatic acid phosphatase
SCI	spinal cord injury
SNL	spinal nerve ligation
STZ	streptozotocin
ROS	reactive oxygen species
NF-κB	nuclear factor kappa-light-chain-enhancer of activated B cells
TNF-α	tumor necrosis factor-α
iNOS	inducible nitric oxide synthase
SIN	sciatic inflammatory neuropathy
PKC	protein kinase C
IL	interleukin
PI3K	phosphoinositide 3-kinase
MAPK	mitogen-activated protein kinase
CREB	cAMP response element-binding protein

References

1. Borea, P.A.; Gessi, S.; Merighi, S.; Vincenzi, F.; Varani, K. Pharmacology of Adenosine Receptors: The State of the Art. *Physiol. Rev.* **2018**, *98*, 1591–1625. [CrossRef] [PubMed]
2. Borea, P.A.; Gessi, S.; Merighi, S.; Vincenzi, F.; Varani, K. Pathological overproduction: The bad side of adenosine. *Br. J. Pharmacol.* **2017**, *174*, 1945–1960. [CrossRef] [PubMed]

3. Borea, P.A.; Gessi, S.; Merighi, S.; Varani, K. Adenosine as a Multi-Signalling Guardian Angel in Human Diseases: When, Where and How Does it Exert its Protective Effects? *Trends Pharmacol. Sci.* **2016**, *37*, 419–434. [CrossRef] [PubMed]
4. Haskó, G.; Linden, J.; Cronstein, B.; Pacher, P. Adenosine receptors: Therapeutic aspects for inflammatory and immune diseases. *Nat. Rev. Drug Discov.* **2008**, *7*, 759–770. [CrossRef]
5. Gompel, J.J.V.; Bower, M.R.; Worrell, G.A.; Stead, M.; Chang, S.-Y.; Goerss, S.J.; Kim, I.; Bennet, K.E.; Meyer, F.B.; Marsh, W.R.; et al. Increased cortical extracellular adenosine correlates with seizure termination. *Epilepsia* **2014**, *55*, 233–244. [CrossRef]
6. During, M.J.; Spencer, D.D. Adenosine: A potential mediator of seizure arrest and postictal refractoriness. *Ann. Neurol.* **1992**, *32*, 618–624. [CrossRef]
7. Ganesana, M.; Venton, B.J. Early changes in transient adenosine during cerebral ischemia and reperfusion injury. *PLoS ONE* **2018**, *13*, e0196932. [CrossRef]
8. Pedata, F.; Dettori, I.; Coppi, E.; Melani, A.; Fusco, I.; Corradetti, R.; Pugliese, A.M. Purinergic signalling in brain ischemia. *Neuropharmacology* **2016**, *104*, 105–130. [CrossRef]
9. Merighi, S.; Battistello, E.; Giacomelli, L.; Varani, K.; Vincenzi, F.; Borea, P.A.; Gessi, S. Targeting A3 and A2A adenosine receptors in the fight against cancer. *Expert Opin. Ther. Targets* **2019**, *23*, 669–678. [CrossRef]
10. Gessi, S.; Merighi, S.; Sacchetto, V.; Simioni, C.; Borea, P.A. Adenosine receptors and cancer. *Biochim. Biophys. Acta* **2011**, *1808*, 1400–1412. [CrossRef]
11. Antonioli, L.; Csóka, B.; Fornai, M.; Colucci, R.; Kókai, E.; Blandizzi, C.; Haskó, G. Adenosine and inflammation: What's new on the horizon? *Drug Discov. Today* **2014**, *19*, 1051–1068. [CrossRef] [PubMed]
12. Sawynok, J.; Liu, X.J. Adenosine in the spinal cord and periphery: Release and regulation of pain. *Prog. Neurobiol.* **2003**, *69*, 313–340. [CrossRef]
13. Sawynok, J. Adenosine receptor targets for pain. *Neuroscience* **2016**, *338*, 1–18. [CrossRef] [PubMed]
14. Fredholm, B.B. Adenosine—A physiological or pathophysiological agent? *J. Mol. Med.* **2014**, *92*, 201–206. [CrossRef]
15. Varani, K.; Vincenzi, F.; Merighi, S.; Gessi, S.; Borea, P.A. Biochemical and Pharmacological Role of A1 Adenosine Receptors and Their Modulation as Novel Therapeutic Strategy. *Adv. Exp. Med. Biol.* **2017**, *1051*, 193–232. [CrossRef]
16. Schulte, G.; Robertson, B.; Fredholm, B.B.; DeLander, G.E.; Shortland, P.; Molander, C. Distribution of antinociceptive adenosine a1 receptors in the spinal cord dorsal horn, and relationship to primary afferents and neuronal subpopulations. *Neuroscience* **2003**, *121*, 907–916. [CrossRef]
17. Luongo, L.; Guida, F.; Imperatore, R.; Napolitano, F.; Gatta, L.; Cristino, L.; Giordano, C.; Siniscalco, D.; Di Marzo, V.; Bellini, G.; et al. The A1 adenosine receptor as a new player in microglia physiology. *Glia* **2014**, *62*, 122–132. [CrossRef]
18. Lima, F.O.; Souza, G.R.; Verri, W.A.; Parada, C.A.; Ferreira, S.H.; Cunha, F.Q.; Cunha, T.M. Direct blockade of inflammatory hypernociception by peripheral A1 adenosine receptors: Involvement of the NO/cGMP/PKG/KATP signaling pathway. *Pain* **2010**, *151*, 506–515. [CrossRef]
19. Sawynok, J.; Reid, A.R.; Liu, J. Spinal and peripheral adenosine A1 receptors contribute to antinociception by tramadol in the formalin test in mice. *Eur. J. Pharmacol.* **2013**, *714*, 373–378. [CrossRef]
20. Maione, S.; de Novellis, V.; Cappellacci, L.; Palazzo, E.; Vita, D.; Luongo, L.; Stella, L.; Franchetti, P.; Marabese, I.; Rossi, F.; et al. The antinociceptive effect of 2-chloro-2′-C-methyl-N6-cyclopentyladenosine (2′-Me-CCPA), a highly selective adenosine A1 receptor agonist, in the rat. *Pain* **2007**, *131*, 281–292. [CrossRef]
21. Gong, Q.-J.; Li, Y.-Y.; Xin, W.-J.; Wei, X.-H.; Cui, Y.; Wang, J.; Liu, Y.; Liu, C.-C.; Li, Y.-Y.; Liu, X.-G. Differential effects of adenosine A1 receptor on pain-related behavior in normal and nerve-injured rats. *Brain Res.* **2010**, *1361*, 23–30. [CrossRef] [PubMed]
22. Kan, H.-W.; Chang, C.-H.; Lin, C.-L.; Lee, Y.-C.; Hsieh, S.-T.; Hsieh, Y.-L. Downregulation of adenosine and adenosine A1 receptor contributes to neuropathic pain in resiniferatoxin neuropathy. *Pain* **2018**, *159*, 1580–1591. [CrossRef] [PubMed]
23. Okumura, T.; Nozu, T.; Kumei, S.; Takakusaki, K.; Miyagishi, S.; Ohhira, M. Adenosine A1 receptors mediate the intracisternal injection of orexin-induced antinociceptive action against colonic distension in conscious rats. *J. Neurol. Sci.* **2016**, *362*, 106–110. [CrossRef]

24. Okumura, T.; Nozu, T.; Ishioh, M.; Igarashi, S.; Kumei, S.; Ohhira, M. Adenosine A1 receptor agonist induces visceral antinociception via 5-HT1A, 5-HT2A, dopamine D1 or cannabinoid CB1 receptors, and the opioid system in the central nervous system. *Physiol. Behav.* **2020**, *220*, 112881. [CrossRef]

25. Zahn, P.K.; Straub, H.; Wenk, M.; Pogatzki-Zahn, E.M. Adenosine A1 but not A2a receptor agonist reduces hyperalgesia caused by a surgical incision in rats: A pertussis toxin-sensitive G protein-dependent process. *Anesthesiology* **2007**, *107*, 797–806. [CrossRef] [PubMed]

26. Martins, D.F.; Mazzardo-Martins, L.; Cidral-Filho, F.J.; Stramosk, J.; Santos, A.R.S. Ankle joint mobilization affects postoperative pain through peripheral and central adenosine A1 receptors. *Phys. Ther.* **2013**, *93*, 401–412. [CrossRef]

27. Yamaoka, G.; Horiuchi, H.; Morino, T.; Miura, H.; Ogata, T. Different analgesic effects of adenosine between postoperative and neuropathic pain. *J. Orthop. Sci.* **2013**, *18*, 130–136. [CrossRef]

28. Goldman, N.; Chen, M.; Fujita, T.; Xu, Q.; Peng, W.; Liu, W.; Jensen, T.K.; Pei, Y.; Wang, F.; Han, X.; et al. Adenosine A1 receptors mediate local anti-nociceptive effects of acupuncture. *Nat. Neurosci.* **2010**, *13*, 883–888. [CrossRef]

29. Zhang, M.; Dai, Q.; Liang, D.; Li, D.; Chen, S.; Chen, S.; Han, K.; Huang, L.; Wang, J. Involvement of adenosine A1 receptor in electroacupuncture-mediated inhibition of astrocyte activation during neuropathic pain. *Arq. Neuro-Psiquiatr.* **2018**, *76*, 736–742. [CrossRef]

30. Liao, H.-Y.; Hsieh, C.-L.; Huang, C.-P.; Lin, Y.-W. Electroacupuncture Attenuates CFA-induced Inflammatory Pain by suppressing Nav1.8 through S100B, TRPV1, Opioid, and Adenosine Pathways in Mice. *Sci. Rep.* **2017**, *7*, 42531. [CrossRef]

31. Dai, Q.-X.; Huang, L.-P.; Mo, Y.-C.; Yu, L.-N.; Du, W.-W.; Zhang, A.-Q.; Geng, W.-J.; Wang, J.-L.; Yan, M. Role of spinal adenosine A1 receptors in the analgesic effect of electroacupuncture in a rat model of neuropathic pain. *J. Int. Med. Res.* **2020**, *48*, 300060519883748. [CrossRef] [PubMed]

32. Gao, X.; Lu, Q.; Chou, G.; Wang, Z.; Pan, R.; Xia, Y.; Hu, H.; Dai, Y. Norisoboldine attenuates inflammatory pain via the adenosine A1 receptor. *Eur. J. Pain* **2014**, *18*, 939–948. [CrossRef] [PubMed]

33. Yin, D.; Liu, Y.-Y.; Wang, T.-X.; Hu, Z.-Z.; Qu, W.-M.; Chen, J.-F.; Cheng, N.-N.; Huang, Z.-L. Paeoniflorin exerts analgesic and hypnotic effects via adenosine A1 receptors in a mouse neuropathic pain model. *Psychopharmacology* **2016**, *233*, 281–293. [CrossRef] [PubMed]

34. Peana, A.T.; Rubattu, P.; Piga, G.G.; Fumagalli, S.; Boatto, G.; Pippia, P.; De Montis, M.G. Involvement of adenosine A1 and A2A receptors in (-)-linalool-induced antinociception. *Life Sci.* **2006**, *78*, 2471–2474. [CrossRef] [PubMed]

35. Valério, D.A.; Ferreira, F.I.; Cunha, T.M.; Alves-Filho, J.C.; Lima, F.O.; De Oliveira, J.R.; Ferreira, S.H.; Cunha, F.Q.; Queiroz, R.H.; Verri, W.A. Fructose-1,6-bisphosphate reduces inflammatory pain-like behaviour in mice: Role of adenosine acting on A1 receptors. *Br. J. Pharmacol.* **2009**, *158*, 558–568. [CrossRef]

36. Nascimento, F.P.; Figueredo, S.M.; Marcon, R.; Martins, D.F.; Macedo, S.J.; Lima, D.A.N.; Almeida, R.C.; Ostroski, R.M.; Rodrigues, A.L.S.; Santos, A.R.S. Inosine reduces pain-related behavior in mice: Involvement of adenosine A1 and A2A receptor subtypes and protein kinase C pathways. *J. Pharmacol. Exp. Ther.* **2010**, *334*, 590–598. [CrossRef]

37. Nascimento, F.P.; Macedo-Júnior, S.J.; Pamplona, F.A.; Luiz-Cerutti, M.; Córdova, M.M.; Constantino, L.; Tasca, C.I.; Dutra, R.C.; Calixto, J.B.; Reid, A.; et al. Adenosine A1 receptor-dependent antinociception induced by inosine in mice: Pharmacological, genetic and biochemical aspects. *Mol. Neurobiol.* **2015**, *51*, 1368–1378. [CrossRef]

38. de Oliveira, E.D.; Schallenberger, C.; Böhmer, A.E.; Hansel, G.; Fagundes, A.C.; Milman, M.; Silva, M.D.P.; Oses, J.P.; Porciúncula, L.O.; Portela, L.V.; et al. Mechanisms involved in the antinociception induced by spinal administration of inosine or guanine in mice. *Eur. J. Pharmacol.* **2016**, *772*, 71–82. [CrossRef]

39. Schmidt, A.P.; Böhmer, A.E.; Antunes, C.; Schallenberger, C.; Porciúncula, L.O.; Elisabetsky, E.; Lara, D.R.; Souza, D.O. Anti-nociceptive properties of the xanthine oxidase inhibitor allopurinol in mice: Role of A1 adenosine receptors. *Br. J. Pharmacol.* **2009**, *156*, 163–172. [CrossRef]

40. Abad, A.S.S.; Falanji, F.; Ghanbarabadi, M.; Rad, A.; Nazemi, S.; Pejhan, A.; Amin, B. Assessment of anti-nociceptive effect of allopurinol in a neuropathic pain model. *Brain Res.* **2019**, *1720*, 146238. [CrossRef]

41. Sowa, N.A.; Voss, M.K.; Zylka, M.J. Recombinant ecto-5′-nucleotidase (CD73) has long lasting antinociceptive effects that are dependent on adenosine A1 receptor activation. *Mol. Pain* **2010**, *6*, 20. [CrossRef] [PubMed]

42. Zylka, M.J.; Sowa, N.A.; Taylor-Blake, B.; Twomey, M.A.; Herrala, A.; Voikar, V.; Vihko, P. Prostatic acid phosphatase is an ectonucleotidase and suppresses pain by generating adenosine. *Neuron* **2008**, *60*, 111–122. [CrossRef] [PubMed]

43. Street, S.E.; Walsh, P.L.; Sowa, N.A.; Taylor-Blake, B.; Guillot, T.S.; Vihko, P.; Wightman, R.M.; Zylka, M.J. PAP and NT5E inhibit nociceptive neurotransmission by rapidly hydrolyzing nucleotides to adenosine. *Mol. Pain* **2011**, *7*, 80. [CrossRef] [PubMed]

44. Sowa, N.A.; Street, S.E.; Vihko, P.; Zylka, M.J. Prostatic Acid Phosphatase Reduces Thermal Sensitivity and Chronic Pain Sensitization by Depleting Phosphatidylinositol 4,5-Bisphosphate. *J. Neurosci.* **2010**, *30*, 10282–10293. [CrossRef] [PubMed]

45. Zylka, M.J. Pain-relieving prospects for adenosine receptors and ectonucleotidases. *Trends Mol. Med.* **2011**, *17*, 188–196. [CrossRef] [PubMed]

46. Hurt, J.K.; Zylka, M.J. PAPupuncture has localized and long-lasting antinociceptive effects in mouse models of acute and chronic pain. *Mol. Pain* **2012**, *8*, 28. [CrossRef]

47. von Heijne, M.; Hao, J.X.; Sollevi, A.; Xu, X.J.; Wiesenfeld-Hallin, Z. Marked enhancement of anti-allodynic effect by combined intrathecal administration of the adenosine A1-receptor agonist R-phenylisopropyladenosine and morphine in a rat model of central pain. *Acta Anaesthesiol. Scand.* **2000**, *44*, 665–671. [CrossRef]

48. Ramos-Zepeda, G.A.; Herrero-Zorita, C.; Herrero, J.F. Interaction of the adenosine A1 receptor agonist N6-cyclopentyladenosine and κ-opioid receptors in rat spinal cord nociceptive reflexes. *Behav. Pharm.* **2014**, *25*, 741–749. [CrossRef]

49. Borghi, V.; Przewlocka, B.; Labuz, D.; Maj, M.; Ilona, O.; Pavone, F. Formalin-induced pain and μ-opioid receptor density in brain and spinal cord are modulated by A1 and A2a adenosine agonists in mice. *Brain Res.* **2002**, *956*, 339–348. [CrossRef]

50. Hwang, J.-H.; Hwang, G.-S.; Cho, S.-K.; Han, S.-M. Morphine can enhance the antiallodynic effect of intrathecal R-PIA in rats with nerve ligation injury. *Anesth. Analg.* **2005**, *100*, 461–468. [CrossRef]

51. Zhang, Y.; Conklin, D.R.; Li, X.; Eisenach, J.C. Intrathecal morphine reduces allodynia after peripheral nerve injury in rats via activation of a spinal A1 adenosine receptor. *Anesthesiology* **2005**, *102*, 416–420. [CrossRef]

52. Wu, W.-P.; Hao, J.-X.; Halldner, L.; Lövdahl, C.; DeLander, G.E.; Wiesenfeld-Hallin, Z.; Fredholm, B.B.; Xu, X.-J. Increased nociceptive response in mice lacking the adenosine A1 receptor. *Pain* **2005**, *113*, 395–404. [CrossRef] [PubMed]

53. Liu, J.; Reid, A.R.; Sawynok, J. Antinociception by systemically-administered acetaminophen (paracetamol) involves spinal serotonin 5-HT7 and adenosine A1 receptors, as well as peripheral adenosine A1 receptors. *Neurosci. Lett.* **2013**, *536*, 64–68. [CrossRef]

54. Sawynok, J.; Reid, A.R. Caffeine inhibits antinociception by acetaminophen in the formalin test by inhibiting spinal adenosine A1 receptors. *Eur. J. Pharmacol.* **2012**, *674*, 248–254. [CrossRef] [PubMed]

55. Sawynok, J.; Reid, A.R.; Fredholm, B.B. Caffeine reverses antinociception by amitriptyline in wild type mice but not in those lacking adenosine A1 receptors. *Neurosci. Lett.* **2008**, *440*, 181–184. [CrossRef] [PubMed]

56. Liu, J.; Reid, A.R.; Sawynok, J. Spinal serotonin 5-HT7 and adenosine A1 receptors, as well as peripheral adenosine A1 receptors, are involved in antinociception by systemically administered amitriptyline. *Eur. J. Pharmacol.* **2013**, *698*, 213–219. [CrossRef] [PubMed]

57. Sawynok, J.; Reid, A.R.; Fredholm, B.B. Caffeine reverses antinociception by oxcarbazepine by inhibition of adenosine A1 receptors: Insights using knockout mice. *Neurosci. Lett.* **2010**, *473*, 178–181. [CrossRef] [PubMed]

58. Stepanović-Petrović, R.M.; Micov, A.M.; Tomić, M.A.; Ugrešić, N.D. The local peripheral antihyperalgesic effect of levetiracetam and its mechanism of action in an inflammatory pain model. *Anesth. Analg.* **2012**, *115*, 1457–1466. [CrossRef] [PubMed]

59. Victor Holanda, A.D.; Asth, L.; Santos, A.R.; Guerrini, R.; Soares-Rachetti, V.d.P.; Caló, G.; André, E.; Gavioli, E.C. Central adenosine A1 and A2A receptors mediate the antinociceptive effects of neuropeptide S in the mouse formalin test. *Life Sci.* **2015**, *120*, 8–12. [CrossRef]

60. Romagnoli, R.; Baraldi, P.G.; Tabrizi, M.A.; Gessi, S.; Borea, P.A.; Merighi, S. Allosteric enhancers of A1 adenosine receptors: State of the art and new horizons for drug development. *Curr. Med. Chem.* **2010**, *17*, 3488–3502. [CrossRef]

61. Romagnoli, R.; Baraldi, P.G.; Moorman, A.R.; Borea, P.A.; Varani, K. Current status of A1 adenosine receptor allosteric enhancers. *Future Med. Chem.* **2015**, *7*, 1247–1259. [CrossRef] [PubMed]

62. Urwyler, S. Allosteric modulation of family C G-protein-coupled receptors: From molecular insights to therapeutic perspectives. *Pharmacol. Rev.* **2011**, *63*, 59–126. [CrossRef] [PubMed]

63. Pan, H.L.; Xu, Z.; Leung, E.; Eisenach, J.C. Allosteric adenosine modulation to reduce allodynia. *Anesthesiology* **2001**, *95*, 416–420. [CrossRef] [PubMed]

64. Li, X.; Conklin, D.; Ma, W.; Zhu, X.; Eisenach, J.C. Spinal noradrenergic activation mediates allodynia reduction from an allosteric adenosine modulator in a rat model of neuropathic pain. *Pain* **2002**, *97*, 117–125. [CrossRef]

65. Li, X.; Conklin, D.; Pan, H.-L.; Eisenach, J.C. Allosteric adenosine receptor modulation reduces hypersensitivity following peripheral inflammation by a central mechanism. *J. Pharmacol. Exp. Ther.* **2003**, *305*, 950–955. [CrossRef]

66. Obata, H.; Li, X.; Eisenach, J.C. Spinal adenosine receptor activation reduces hypersensitivity after surgery by a different mechanism than after nerve injury. *Anesthesiology* **2004**, *100*, 1258–1262. [CrossRef]

67. Romagnoli, R.; Baraldi, P.G.; Carrion, M.D.; Cara, C.L.; Cruz-Lopez, O.; Salvador, M.K.; Preti, D.; Tabrizi, M.A.; Moorman, A.R.; Vincenzi, F.; et al. Synthesis and biological evaluation of 2-amino-3-(4-chlorobenzoyl)-4-[(4-arylpiperazin-1-yl)methyl]-5-substituted-thiophenes. effect of the 5-modification on allosteric enhancer activity at the A1 adenosine receptor. *J. Med. Chem.* **2012**, *55*, 7719–7735. [CrossRef]

68. Vincenzi, F.; Ravani, A.; Pasquini, S.; Merighi, S.; Gessi, S.; Romagnoli, R.; Baraldi, P.G.; Borea, P.A.; Varani, K. Positive allosteric modulation of A1 adenosine receptors as a novel and promising therapeutic strategy for anxiety. *Neuropharmacology* **2016**, *111*, 283–292. [CrossRef]

69. Vincenzi, F.; Targa, M.; Romagnoli, R.; Merighi, S.; Gessi, S.; Baraldi, P.G.; Borea, P.A.; Varani, K. TRR469, a potent A1 adenosine receptor allosteric modulator, exhibits anti-nociceptive properties in acute and neuropathic pain models in mice. *Neuropharmacology* **2014**, *81*, 6–14. [CrossRef]

70. Gomes, C.; Ferreira, R.; George, J.; Sanches, R.; Rodrigues, D.I.; Gonçalves, N.; Cunha, R.A. Activation of microglial cells triggers a release of brain-derived neurotrophic factor (BDNF) inducing their proliferation in an adenosine A2A receptor-dependent manner: A2A receptor blockade prevents BDNF release and proliferation of microglia. *J. Neuroinflammation* **2013**, *10*, 780. [CrossRef]

71. Hussey, M.J.; Clarke, G.D.; Ledent, C.; Kitchen, I.; Hourani, S.M.O. Genetic deletion of the adenosine A2A receptor in mice reduces the changes in spinal cord NMDA receptor binding and glucose uptake caused by a nociceptive stimulus. *Neurosci. Lett.* **2010**, *479*, 297–301. [CrossRef] [PubMed]

72. Hambrecht-Wiedbusch, V.S.; Gabel, M.; Liu, L.J.; Imperial, J.P.; Colmenero, A.V.; Vanini, G. Preemptive Caffeine Administration Blocks the Increase in Postoperative Pain Caused by Previous Sleep Loss in the Rat: A Potential Role for Preoptic Adenosine A2A Receptors in Sleep–Pain Interactions. *Sleep* **2017**, 40. [CrossRef] [PubMed]

73. Li, L.; Hao, J.X.; Fredholm, B.B.; Schulte, G.; Wiesenfeld-Hallin, Z.; Xu, X.J. Peripheral adenosine A2A receptors are involved in carrageenan-induced mechanical hyperalgesia in mice. *Neuroscience* **2010**, *170*, 923–928. [CrossRef] [PubMed]

74. Varano, F.; Catarzi, D.; Vincenzi, F.; Betti, M.; Falsini, M.; Ravani, A.; Borea, P.A.; Colotta, V.; Varani, K. Design, Synthesis, and Pharmacological Characterization of 2-(2-Furanyl)thiazolo[5,4-d]pyrimidine-5,7-diamine Derivatives: New Highly Potent A2A Adenosine Receptor Inverse Agonists with Antinociceptive Activity. *J. Med. Chem.* **2016**, *59*, 10564–10576. [CrossRef] [PubMed]

75. Varano, F.; Catarzi, D.; Vincenzi, F.; Falsini, M.; Pasquini, S.; Borea, P.A.; Colotta, V.; Varani, K. Structure-activity relationship studies and pharmacological characterization of N5-heteroarylalkyl-substituted-2-(2-furanyl)thiazolo[5,4-d]pyrimidine-5,7-diamine-based derivatives as inverse agonists at human A2A adenosine receptor. *Eur. J. Med. Chem.* **2018**, *155*, 552–561. [CrossRef] [PubMed]

76. Luongo, L.; Salvemini, D. Targeting metabotropic adenosine receptors for neuropathic pain: Focus on A2A. *Brain Behav. Immun.* **2018**, *69*, 60–61. [CrossRef]

77. Bura, A.S.; Nadal, X.; Ledent, C.; Maldonado, R.; Valverde, O. A2A adenosine receptor regulates glia proliferation and pain after peripheral nerve injury. *Pain* **2008**, *140*, 95–103. [CrossRef]

78. Areti, A.; Yerra, V.G.; Naidu, V.; Kumar, A. Oxidative stress and nerve damage: Role in chemotherapy induced peripheral neuropathy. *Redox Biol.* **2014**, *2*, 289–295. [CrossRef]
79. Carrasco, C.; Naziroğlu, M.; Rodríguez, A.B.; Pariente, J.A. Neuropathic Pain: Delving into the Oxidative Origin and the Possible Implication of Transient Receptor Potential Channels. *Front. Physiol.* **2018**, *9*, 95. [CrossRef]
80. Kim, H.K.; Park, S.K.; Zhou, J.-L.; Taglialatela, G.; Chung, K.; Coggeshall, R.E.; Chung, J.M. Reactive oxygen species (ROS) play an important role in a rat model of neuropathic pain. *Pain* **2004**, *111*, 116–124. [CrossRef]
81. Han, Y.; Smith, M.T. Pathobiology of cancer chemotherapy-induced peripheral neuropathy (CIPN). *Front. Pharm.* **2013**, *4*, 156. [CrossRef] [PubMed]
82. Kerckhove, N.; Collin, A.; Condé, S.; Chaleteix, C.; Pezet, D.; Balayssac, D. Long-Term Effects, Pathophysiological Mechanisms, and Risk Factors of Chemotherapy-Induced Peripheral Neuropathies: A Comprehensive Literature Review. *Front. Pharm.* **2017**, *8*, 86. [CrossRef] [PubMed]
83. Naik, A.K.; Tandan, S.K.; Dudhgaonkar, S.P.; Jadhav, S.H.; Kataria, M.; Prakash, V.R.; Kumar, D. Role of oxidative stress in pathophysiology of peripheral neuropathy and modulation by N-acetyl-L-cysteine in rats. *Eur. J. Pain* **2006**, *10*, 573–579. [CrossRef] [PubMed]
84. Shim, H.S.; Bae, C.; Wang, J.; Lee, K.-H.; Hankerd, K.M.; Kim, H.K.; Chung, J.M.; La, J.-H. Peripheral and central oxidative stress in chemotherapy-induced neuropathic pain. *Mol. Pain* **2019**, *15*, 1–11. [CrossRef]
85. Falsini, M.; Catarzi, D.; Varano, F.; Ceni, C.; Dal Ben, D.; Marucci, G.; Buccioni, M.; Volpini, R.; Di Cesare Mannelli, L.; Lucarini, E.; et al. Antioxidant-Conjugated 1,2,4-Triazolo[4,3-a]pyrazin-3-one Derivatives: Highly Potent and Selective Human A2A Adenosine Receptor Antagonists Possessing Protective Efficacy in Neuropathic Pain. *J. Med. Chem.* **2019**, *62*, 8511–8531. [CrossRef]
86. Betti, M.; Catarzi, D.; Varano, F.; Falsini, M.; Varani, K.; Vincenzi, F.; Pasquini, S.; di Cesare Mannelli, L.; Ghelardini, C.; Lucarini, E.; et al. Modifications on the Amino-3,5-dicyanopyridine Core to Obtain Multifaceted Adenosine Receptor Ligands with Antineuropathic Activity. *J. Med. Chem.* **2019**, *62*, 6894–6912. [CrossRef]
87. Jacobson, K.A.; Giancotti, L.A.; Lauro, F.; Mufti, F.; Salvemini, D. Treatment of chronic neuropathic pain: Purine receptor modulation. *Pain* **2020**, *161*, 1425–1441. [CrossRef] [PubMed]
88. Hussey, M.J.; Clarke, G.D.; Ledent, C.; Kitchen, I.; Hourani, S.M.O. Deletion of the adenosine A2A receptor in mice enhances spinal cord neurochemical responses to an inflammatory nociceptive stimulus. *Neurosci. Lett.* **2012**, *506*, 198–202. [CrossRef] [PubMed]
89. Varani, K.; Padovan, M.; Govoni, M.; Vincenzi, F.; Trotta, F.; Borea, P.A. The role of adenosine receptors in rheumatoid arthritis. *Autoimmun. Rev.* **2010**, *10*, 61–64. [CrossRef]
90. Varani, K.; Padovan, M.; Vincenzi, F.; Targa, M.; Trotta, F.; Govoni, M.; Borea, P.A. A2A and A3 adenosine receptor expression in rheumatoid arthritis: Upregulation, inverse correlation with disease activity score and suppression of inflammatory cytokine and metalloproteinase release. *Arthritis Res. Ther.* **2011**, *13*, R197. [CrossRef]
91. Mazzon, E.; Esposito, E.; Impellizzeri, D.; Di Paola, R.; Melani, A.; Bramanti, P.; Pedata, F.; Cuzzocrea, S. CGS 21680, an Agonist of the Adenosine (A2A) receptor, reduces progression of murine type II collagen-induced arthritis. *J. Rheumatol.* **2011**, *38*, 2119–2129. [CrossRef] [PubMed]
92. Vincenzi, F.; Padovan, M.; Targa, M.; Corciulo, C.; Giacuzzo, S.; Merighi, S.; Gessi, S.; Govoni, M.; Borea, P.A.; Varani, K. A(2A) adenosine receptors are differentially modulated by pharmacological treatments in rheumatoid arthritis patients and their stimulation ameliorates adjuvant-induced arthritis in rats. *PLoS ONE* **2013**, *8*, e54195. [CrossRef] [PubMed]
93. Ravani, A.; Vincenzi, F.; Bortoluzzi, A.; Padovan, M.; Pasquini, S.; Gessi, S.; Merighi, S.; Borea, P.A.; Govoni, M.; Varani, K. Role and Function of A2A and A3 Adenosine Receptors in Patients with Ankylosing Spondylitis, Psoriatic Arthritis and Rheumatoid Arthritis. *Int. J. Mol. Sci.* **2017**, *18*, 697. [CrossRef] [PubMed]
94. Montes, G.C.; Hammes, N.; da Rocha, M.D.; Montagnoli, T.L.; Fraga, C.A.M.; Barreiro, E.J.; Sudo, R.T.; Zapata-Sudo, G. Treatment with Adenosine Receptor Agonist Ameliorates Pain Induced by Acute and Chronic Inflammation. *J. Pharmacol. Exp. Ther.* **2016**, *358*, 315–323. [CrossRef]
95. Loram, L.C.; Harrison, J.A.; Sloane, E.M.; Hutchinson, M.R.; Sholar, P.; Taylor, F.R.; Berkelhammer, D.; Coats, B.D.; Poole, S.; Milligan, E.D.; et al. Enduring Reversal of Neuropathic Pain by a Single Intrathecal Injection of Adenosine 2A Receptor Agonists: A Novel Therapy for Neuropathic Pain. *J. Neurosci.* **2009**, *29*, 14015–14025. [CrossRef]

96. Loram, L.C.; Taylor, F.R.; Strand, K.A.; Harrison, J.A.; RzasaLynn, R.; Sholar, P.; Rieger, J.; Maier, S.F.; Watkins, L.R. Intrathecal injection of adenosine 2A receptor agonists reversed neuropathic allodynia through protein kinase (PK)A/PKC signaling. *Brain Behav. Immun.* **2013**, *33*, 112–122. [CrossRef]

97. Kwilasz, A.J.; Ellis, A.; Wieseler, J.; Loram, L.; Favret, J.; McFadden, A.; Springer, K.; Falci, S.; Rieger, J.; Maier, S.F.; et al. Sustained reversal of central neuropathic pain induced by a single intrathecal injection of adenosine A2A receptor agonists. *Brain Behav. Immun.* **2018**, *69*, 470–479. [CrossRef]

98. Kwilasz, A.J.; Green Fulgham, S.M.; Ellis, A.; Patel, H.P.; Duran-Malle, J.C.; Favret, J.; Harvey, L.O.; Rieger, J.; Maier, S.F.; Watkins, L.R. A single peri-sciatic nerve administration of the adenosine 2A receptor agonist ATL313 produces long-lasting anti-allodynia and anti-inflammatory effects in male rats. *Brain Behav. Immun.* **2019**, *76*, 116–125. [CrossRef]

99. Jacobson, K.A.; Tosh, D.K.; Jain, S.; Gao, Z.-G. Historical and Current Adenosine Receptor Agonists in Preclinical and Clinical Development. *Front. Cell Neurosci.* **2019**, *13*, 124. [CrossRef]

100. By, Y.; Condo, J.; Durand-Gorde, J.-M.; Lejeune, P.-J.; Mallet, B.; Guieu, R.; Ruf, J. Intracerebroventricular injection of an agonist-like monoclonal antibody to adenosine A(2A) receptor has antinociceptive effects in mice. *J. Neuroimmunol.* **2011**, *230*, 178–182. [CrossRef]

101. Feoktistov, I.; Biaggioni, I. Chapter 5—Role of Adenosine A2B Receptors in Inflammation. In *Advances in Pharmacology*; Jacobson, K.A., Linden, J., Eds.; Pharmacology of Purine and Pyrimidine Receptors; Academic Press: Cambridge, MA, USA, 2011; Volume 61, pp. 115–144.

102. Popoli, P.; Pepponi, R. Potential Therapeutic Relevance of Adenosine A2B and A2A Receptors in the Central Nervous System. *CNS Neurol. Disord.-Drug Targets (Former. Curr. Drug Targets CNS Neurol. Disord.)* **2012**, *11*, 664–674. [CrossRef]

103. Fredholm, B.B.; IJzerman, A.P.; Jacobson, K.A.; Linden, J.; Müller, C.E. International union of basic and clinical pharmacology. LXXXI. Nomenclature and classification of adenosine receptors—An update. *Pharmacol. Rev.* **2011**, *63*, 1–34. [CrossRef] [PubMed]

104. Abo-Salem, O.M.; Hayallah, A.M.; Bilkei-Gorzo, A.; Filipek, B.; Zimmer, A.; Müller, C.E. Antinociceptive effects of novel A2B adenosine receptor antagonists. *J. Pharmacol. Exp. Ther.* **2004**, *308*, 358–366. [CrossRef] [PubMed]

105. Bilkei-Gorzo, A.; Abo-Salem, O.M.; Hayallah, A.M.; Michel, K.; Müller, C.E.; Zimmer, A. Adenosine receptor subtype-selective antagonists in inflammation and hyperalgesia. *Naunyn. Schmiedebergs. Arch. Pharmacol.* **2008**, *377*, 65–76. [CrossRef]

106. Yoon, M.H.; Bae, H.B.; Choi, J.I.; Kim, S.J.; Chung, S.T.; Kim, C.M. Roles of Adenosine Receptor Subtypes in the Antinociceptive Effect of Intrathecal Adenosine in a Rat Formalin Test. *PHA* **2006**, *78*, 21–26. [CrossRef]

107. Asano, T.; Tanaka, K.-I.; Tada, A.; Shimamura, H.; Tanaka, R.; Maruoka, H.; Takenaga, M.; Mizushima, T. Aminophylline suppresses stress-induced visceral hypersensitivity and defecation in irritable bowel syndrome. *Sci. Rep.* **2017**, *7*, 40214. [CrossRef] [PubMed]

108. Asano, T.; Takenaga, M. Adenosine A2B Receptors: An Optional Target for the Management of Irritable Bowel Syndrome with Diarrhea? *J. Clin. Med.* **2017**, *6*, 104. [CrossRef]

109. Savegnago, L.; Jesse, C.R.; Nogueira, C.W. Caffeine and a selective adenosine A(2B) receptor antagonist but not imidazoline receptor antagonists modulate antinociception induced by diphenyl diselenide in mice. *Neurosci. Lett.* **2008**, *436*, 120–123. [CrossRef]

110. Hu, X.; Adebiyi, M.G.; Luo, J.; Sun, K.; Le, T.-T.T.; Zhang, Y.; Wu, H.; Zhao, S.; Karmouty-Quintana, H.; Liu, H.; et al. Sustained Elevated Adenosine via ADORA2B Promotes Chronic Pain through Neuro-immune Interaction. *Cell Rep.* **2016**, *16*, 106–119. [CrossRef]

111. Chen, J.-F.; Lee, C.; Chern, Y. Chapter One—Adenosine Receptor Neurobiology: Overview. In *International Review of Neurobiology*; Mori, A., Ed.; Adenosine Receptors in Neurology and Psychiatry; Academic Press: Cambridge, MA, USA, 2014; Volume 119, pp. 1–49.

112. Borea, P.A.; Varani, K.; Vincenzi, F.; Baraldi, P.G.; Tabrizi, M.A.; Merighi, S.; Gessi, S. The A3 Adenosine Receptor: History and Perspectives. *Pharm. Rev.* **2015**, *67*, 74–102. [CrossRef]

113. Fishman, P.; Bar-Yehuda, S.; Liang, B.T.; Jacobson, K.A. Pharmacological and therapeutic effects of A3 adenosine receptor agonists. *Drug Discov. Today* **2012**, *17*, 359–366. [CrossRef]

114. Wu, W.-P.; Hao, J.-X.; Halldner-Henriksson, L.; Xu, X.-J.; Jacobson, M.A.; Wiesenfeld-Hallin, Z.; Fredholm, B.B. Decreased inflammatory pain due to reduced carrageenan-induced inflammation in mice lacking adenosine A3 receptors. *Neuroscience* **2002**, *114*, 523–527. [CrossRef]

115. Chen, Z.; Janes, K.; Chen, C.; Doyle, T.; Bryant, L.; Tosh, D.K.; Jacobson, K.A.; Salvemini, D. Controlling murine and rat chronic pain through A3 adenosine receptor activation. *FASEB J.* **2012**, *26*, 1855–1865. [CrossRef] [PubMed]

116. Janes, K.; Symons-Liguori, A.; Jacobson, K.A.; Salvemini, D. Identification of A3 adenosine receptor agonists as novel non-narcotic analgesics. *Br. J. Pharm.* **2016**, *173*, 1253–1267. [CrossRef]

117. Little, J.W.; Ford, A.; Symons-Liguori, A.M.; Chen, Z.; Janes, K.; Doyle, T.; Xie, J.; Luongo, L.; Tosh, D.K.; Maione, S.; et al. Endogenous adenosine A3 receptor activation selectively alleviates persistent pain states. *Brain* **2015**, *138*, 28–35. [CrossRef] [PubMed]

118. Ford, A.; Castonguay, A.; Cottet, M.; Little, J.W.; Chen, Z.; Symons-Liguori, A.M.; Doyle, T.; Egan, T.M.; Vanderah, T.W.; Koninck, Y.D.; et al. Engagement of the GABA to KCC2 Signaling Pathway Contributes to the Analgesic Effects of A3AR Agonists in Neuropathic Pain. *J. Neurosci.* **2015**, *35*, 6057–6067. [CrossRef] [PubMed]

119. Coppi, E.; Cherchi, F.; Fusco, I.; Failli, P.; Vona, A.; Dettori, I.; Gaviano, L.; Lucarini, E.; Jacobson, K.A.; Tosh, D.K.; et al. Adenosine A3 receptor activation inhibits pronociceptive N-type Ca2+ currents and cell excitability in dorsal root ganglion neurons. *Pain* **2019**, *160*, 1103–1118. [CrossRef]

120. Terayama, R.; Tabata, M.; Maruhama, K.; Iida, S. A3 adenosine receptor agonist attenuates neuropathic pain by suppressing activation of microglia and convergence of nociceptive inputs in the spinal dorsal horn. *Exp. Brain Res.* **2018**, *236*, 3203–3213. [CrossRef]

121. Suresh, R.R.; Jain, S.; Chen, Z.; Tosh, D.K.; Ma, Y.; Podszun, M.C.; Rotman, Y.; Salvemini, D.; Jacobson, K.A. Design and in vivo activity of A3 adenosine receptor agonist prodrugs. *Purinergic Signal.* **2020**, *16*, 367–377. [CrossRef]

122. Janes, K.; Esposito, E.; Doyle, T.; Cuzzocrea, S.; Tosh, D.K.; Jacobson, K.A.; Salvemini, D. A3 adenosine receptor agonist prevents the development of paclitaxel-induced neuropathic pain by modulating spinal glial-restricted redox-dependent signaling pathways. *Pain* **2014**, *155*, 2560–2567. [CrossRef]

123. Kim, Y.; Kwon, S.Y.; Jung, H.S.; Park, Y.J.; Kim, Y.S.; In, J.H.; Choi, J.W.; Kim, J.A.; Joo, J.D. Amitriptyline inhibits the MAPK/ERK and CREB pathways and proinflammatory cytokines through A3AR activation in rat neuropathic pain models. *Korean J. Anesth.* **2019**, *72*, 60–67. [CrossRef]

124. Wahlman, C.; Doyle, T.M.; Little, J.W.; Luongo, L.; Janes, K.; Chen, Z.; Esposito, E.; Tosh, D.K.; Cuzzocrea, S.; Jacobson, K.A.; et al. Chemotherapy-induced pain is promoted by enhanced spinal adenosine kinase levels through astrocyte-dependent mechanisms. *Pain* **2018**, *159*, 1025–1034. [CrossRef] [PubMed]

125. Antonioli, L.; Blandizzi, C.; Pacher, P.; Haskó, G. Immunity, inflammation and cancer: A leading role for adenosine. *Nat. Rev. Cancer* **2013**, *13*, 842–857. [CrossRef] [PubMed]

126. Varani, K.; Vincenzi, F.; Targa, M.; Paradiso, B.; Parrilli, A.; Fini, M.; Lanza, G.; Borea, P.A. The stimulation of A3 adenosine receptors reduces bone-residing breast cancer in a rat preclinical model. *Eur. J. Cancer* **2013**, *49*, 482–491. [CrossRef] [PubMed]

127. Yan, H.; Zhang, E.; Feng, C.; Zhao, X. Role of A3 adenosine receptor in diabetic neuropathy. *J. Neurosci. Res.* **2016**, *94*, 936–946. [CrossRef]

128. Hayhurst, C.J.; Durieux, M.E. Differential Opioid Tolerance and Opioid-induced HyperalgesiaA Clinical Reality. *Anesthesiology* **2016**, *124*, 483–488. [CrossRef]

129. Rivat, C.; Ballantyne, J. The dark side of opioids in pain management: Basic science explains clinical observation. *Pain Rep.* **2016**, *1*, e570. [CrossRef]

130. Doyle, T.M.; Largent-Milnes, T.M.; Chen, Z.; Staikopoulos, V.; Esposito, E.; Dalgarno, R.; Fan, C.; Tosh, D.K.; Cuzzocrea, S.; Jacobson, K.A.; et al. Chronic Morphine-Induced Changes in Signaling at the A3 Adenosine Receptor Contribute to Morphine-Induced Hyperalgesia, Tolerance, and Withdrawal. *J. Pharmacol. Exp. Ther.* **2020**, *374*, 331–341. [CrossRef]

131. Antonioli, L.; Lucarini, E.; Lambertucci, C.; Fornai, M.; Pellegrini, C.; Benvenuti, L.; Di Cesare Mannelli, L.; Spinaci, A.; Marucci, G.; Blandizzi, C.; et al. The Anti-Inflammatory and Pain-Relieving Effects of AR170, an Adenosine A3 Receptor Agonist, in a Rat Model of Colitis. *Cells* **2020**, *9*, 1509. [CrossRef]

132. Lucarini, E.; Coppi, E.; Micheli, L.; Parisio, C.; Vona, A.; Cherchi, F.; Pugliese, A.M.; Pedata, F.; Failli, P.; Palomino, S.; et al. Acute visceral pain relief mediated by A3AR agonists in rats: Involvement of N-type voltage-gated calcium channels. *Pain* **2020**. [CrossRef]

133. Grundy, L.; Harrington, A.M.; Castro, J.; Garcia-Caraballo, S.; Deiteren, A.; Maddern, J.; Rychkov, G.Y.; Ge, P.; Peters, S.; Feil, R.; et al. Chronic linaclotide treatment reduces colitis-induced neuroplasticity and reverses persistent bladder dysfunction. *JCI Insight* **2018**, *3*, e121841. [CrossRef] [PubMed]
134. Cohen, S.; Barer, F.; Bar-Yehuda, S.; IJzerman, A.P.; Jacobson, K.A.; Fishman, P. A$_3$ adenosine receptor allosteric modulator induces an anti-inflammatory effect: In vivo studies and molecular mechanism of action. *Mediat. Inflamm.* **2014**, *2014*, 708746. [CrossRef] [PubMed]

Publisher's Note: MDPI stays neutral with regard to jurisdictional claims in published maps and institutional affiliations.

International Journal of
Molecular Sciences

Article

P2Y$_2$ and P2X4 Receptors Mediate Ca^{2+} Mobilization in DH82 Canine Macrophage Cells

Reece Andrew Sophocleous [1,2], Nicole Ashleigh Miles [1,2], Lezanne Ooi [1,2] and Ronald Sluyter [1,2,*]

1   Illawarra Health and Medical Research Institute, Wollongong, NSW 2522, Australia;
    rs256@uowmail.edu.au (R.A.S.); nam993@uowmail.edu.au (N.A.M.); lezanne@uow.edu.au (L.O.)
2   Molecular Horizons and School of Chemistry and Molecular Bioscience, University of Wollongong,
    Wollongong, NSW 2522, Australia
*   Correspondence: rsluyter@uow.edu.au; Tel.: +612-4221-5508

Received: 7 October 2020; Accepted: 12 November 2020; Published: 13 November 2020

Abstract: Purinergic receptors of the P2 subclass are commonly found in human and rodent macrophages where they can be activated by adenosine 5′-triphosphate (ATP) or uridine 5′-triphosphate (UTP) to mediate Ca^{2+} mobilization, resulting in downstream signalling to promote inflammation and pain. However, little is understood regarding these receptors in canine macrophages. To establish a macrophage model of canine P2 receptor signalling, the expression of these receptors in the DH82 canine macrophage cell line was determined by reverse transcription polymerase chain reaction (RT-PCR) and immunocytochemistry. P2 receptor function in DH82 cells was pharmacologically characterised using nucleotide-induced measurements of Fura-2 AM-bound intracellular Ca^{2+}. RT-PCR revealed predominant expression of P2X4 receptors, while immunocytochemistry confirmed predominant expression of P2Y$_2$ receptors, with low levels of P2X4 receptor expression. ATP and UTP induced robust Ca^{2+} responses in the absence or presence of extracellular Ca^{2+}. ATP-induced responses were only partially inhibited by the P2X4 receptor antagonists, 2′,3′-*O*-(2,4,6-trinitrophenyl)-ATP (TNP-ATP), paroxetine and 5-BDBD, but were strongly potentiated by ivermectin. UTP-induced responses were near completely inhibited by the P2Y$_2$ receptor antagonists, suramin and AR-C118925. P2Y$_2$ receptor-mediated Ca^{2+} mobilization was inhibited by U-73122 and 2-aminoethoxydiphenyl borate (2-APB), indicating P2Y$_2$ receptor coupling to the phospholipase C and inositol triphosphate signal transduction pathway. Together this data demonstrates, for the first time, the expression of functional P2 receptors in DH82 canine macrophage cells and identifies a potential cell model for studying macrophage-mediated purinergic signalling in inflammation and pain in dogs.

Keywords: P2Y$_2$ receptor; P2X4 receptor; canine; dog; purinergic signalling; DH82; macrophage; pain; neuroinflammation

1. Introduction

The activation of purinergic receptors by nucleotides such as adenosine 5′-triphosphate (ATP) and uridine 5′-triphosphate (UTP) is crucial for a number of inflammatory processes, including those in the central nervous system (CNS) such as chronic pain [1–4] and remyelination of nerves following injury to the CNS [5,6]. The P2 receptor family consists of seven mammalian ionotropic P2X receptors (P2X1-7) and eight mammalian metabotropic P2Y receptors (P2Y$_{1,2,4,6,11-14}$) that can modulate intracellular Ca^{2+} concentrations through direct ion channel permeation or mobilization of intracellular Ca^{2+} stores, respectively [7]. P2 receptors, such as the P2X4 and P2Y$_2$ receptors, are commonly expressed on human and rodent macrophages and macrophage cell lines [8–14], and have demonstrated roles in signalling pathways that control chronic pain and inflammation in humans or rodents [2,8,15–18]. Despite this, studies on purinergic signalling in canine macrophages are lacking.

The DH82 cell line is a canine macrophage cell line isolated from a 10 year old Golden Retriever with malignant histiocytosis [19]. This cell line has recently been demonstrated as a useful model of canine macrophage physiology, bearing similarities to an M0 macrophage subtype with demonstrated potential for polarisation to either M1 or M2a subtypes through cytokine stimulation [20]. Studies have demonstrated that DH82 cells express a number of macrophage markers, such as CD11c and CD18 [21], and can secrete tumour necrosis factor (TNF)-α and interleukin (IL)-6 similar to that observed in lipopolysaccharide (LPS)-stimulated canine monocytes [22]. Despite its use as an in vitro model of viral and protozoan infection [23–25], knowledge regarding purinergic signalling in DH82 cells is limited to a single report describing ATP- and adenosine-induced cytokine release [26]. Although this study did not investigate any purinergic receptor per se, DH82 cells represent a possible model to study endogenous P2 receptors in canine macrophages for a number of reasons. Firstly, the original study revealed that ATP could alter cytokine expression in LPS-stimulated DH82 cells [26]. Secondly, infection of DH82 cells with canine distemper virus modulates inflammatory signalling pathways [27] that are common to P2X receptor-mediated signalling [28]. Thirdly, despite few studies analysing the expression of P2 receptors in dogs, it has been demonstrated that canine monocytes express P2X7 receptors [29,30]. Lastly, human and rodent macrophage or myeloid cell lines, such as THP-1 and RAW264.7 cells, are well-established models for studying endogenous P2 receptors commonly expressed on human and rodent macrophages [31–38].

The current study aimed to establish a canine macrophage model of P2 receptor signalling. Through investigation of canine P2 receptor expression and functional characterisation of these receptors, this study has identified the $P2Y_2$ receptor and, to a lesser extent, the P2X4 receptor, as the primary functional P2 receptors in DH82 cells, which are responsible for nucleotide-mediated Ca^{2+} mobilization.

2. Results

2.1. DH82 Cells Express Abundant P2RX4 mRNA Compared to Other P2 Receptors

To establish a P2 receptor mRNA expression profile for DH82 cells, cDNA was amplified by RT-PCR using primer pairs (Table S1) designed to genes encoding canine P2X1-7 receptors and canine $P2Y_{1,2,4,6,11-14}$ receptors and amplicons were semi-quantitatively analysed by agarose gel electrophoresis and densitometry. P2X4 receptor mRNA was most abundant in DH82 cells, with relative amounts comparable to glyceraldehyde 3-phosphate dehydrogenase (*GAPDH*; Figure 1) and β-actin (*ACTB*; data not shown). Other P2X receptor mRNAs, including P2X1 and P2X7 receptors were detected, but to a much lesser degree than the P2X4 receptor (Figure 1). Additionally, mRNA from a number of P2Y receptor subtypes were also detected, including $P2Y_1$, $P2Y_2$, $P2Y_6$ and $P2Y_{11}$ receptors, however, as with P2X1 and P2X7 receptors, these were expressed at greatly reduced levels compared to the P2X4 receptor (Figure 1).

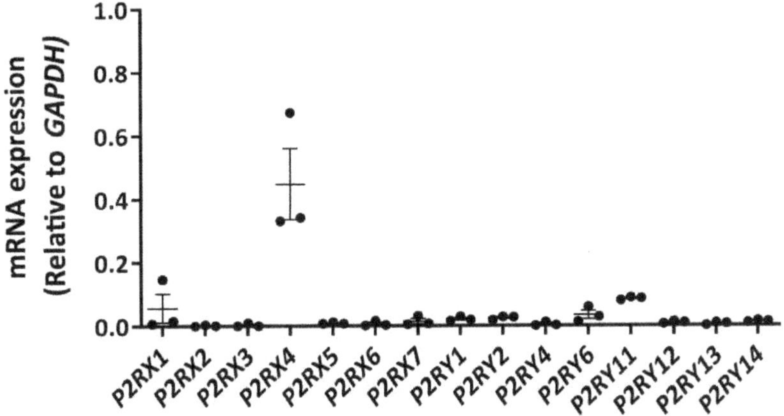

Figure 1. Expression of P2X and P2Y receptor mRNA in DH82 cells. RNA was isolated from DH82 cells and cDNA was synthesized and amplified using primer pairs designed to each respective *P2RX* or *P2RY* genes, with glyceraldehyde 3-phosphate dehydrogenase (*GAPDH*) as a positive control. Amplification in absence of cDNA was conducted for each primer pair to ensure primer specificity. Amplicons were visualized by agarose gel electrophoresis using GelRed and the GelDoc XR+ imaging system and semi-quantitatively analysed by densitometry. Data shown are the mean ± SEM relative to *GAPDH* expression from three independent experiments.

2.2. Nucleotides Mediate Ca^{2+} Responses in DH82 Cells

To establish an agonist profile for functional P2 receptors in DH82 cells, nucleotides which have previously demonstrated activity towards P2 receptors from dogs, humans or rodents [39,40] were utilised to measure changes in intracellular Ca^{2+}. These nucleotides were ATP, 3′-O-(4-benzoyl)benzoyl-ATP (BzATP), adenosine-5′-diphosphate (ADP), 2-methylthio-ADP (2MeSADP), uridine-5′-triphosphate (UTP) and uridine-5′-diphosphate (UDP). ADP was preincubated with hexokinase to remove trace amounts of ATP [41]. Incubation with ATP or UTP induced robust Ca^{2+} responses (ΔF340/380) in DH82 cells, which peaked approximately 15 s after application of nucleotides (Figure 2A,B). Ca^{2+} responses then decayed more slowly, returning to baseline approximately 70–80 s after the initial peak was observed (Figure 2A,B). Incubation with ADP or BzATP resulted in much smaller Ca^{2+} responses compared to ATP and UTP (Figure 2C,D). UDP and 2MeSADP were unable to induce Ca^{2+} responses up to 30 μM and 100 μM, respectively (Figure 2E,F). Decay time, calculated as the time constant (τ), was similar for ATP and UTP (τ = 56.9 ± 4.8 s and 56.7 ± 5.9 s, respectively), however, BzATP (τ = 102.3 ± 17.9 s) had a significantly longer decay time compared to ATP ($p < 0.05$), UTP ($p < 0.05$) or ADP (τ = 23.8 ± 0.4 s; $p < 0.001$), while UDP and 2MeSADP did not respond and as such, decay time could not be calculated (Figure 2G).

As Ca^{2+} responses were observed with a number of nucleotides, including those known to activate mammalian ionotropic P2X (ATP, BzATP) and metabotropic P2Y receptors (ATP, UTP, ADP), both the peak nucleotide-induced Ca^{2+} responses (Figure 2H) and net Ca^{2+} movement (Figure 2I; calculated as area under the curve [AUC]) were used for constructing concentration-response curves to account for potential differences in Ca^{2+} response phenotypes in a model of co-expression of P2X and P2Y receptor subtypes. In DH82 cells, nucleotides induced concentration-dependent Ca^{2+} responses with the rank order of potency of UTP > ATP >> ADP ≈ BzATP, with UDP and 2MeSADP being unresponsive (Figure 2H,I; Table 1). There were no significant differences between the EC_{50} values calculated for net Ca^{2+} movement and peak Ca^{2+} response for any nucleotide (Table 1).

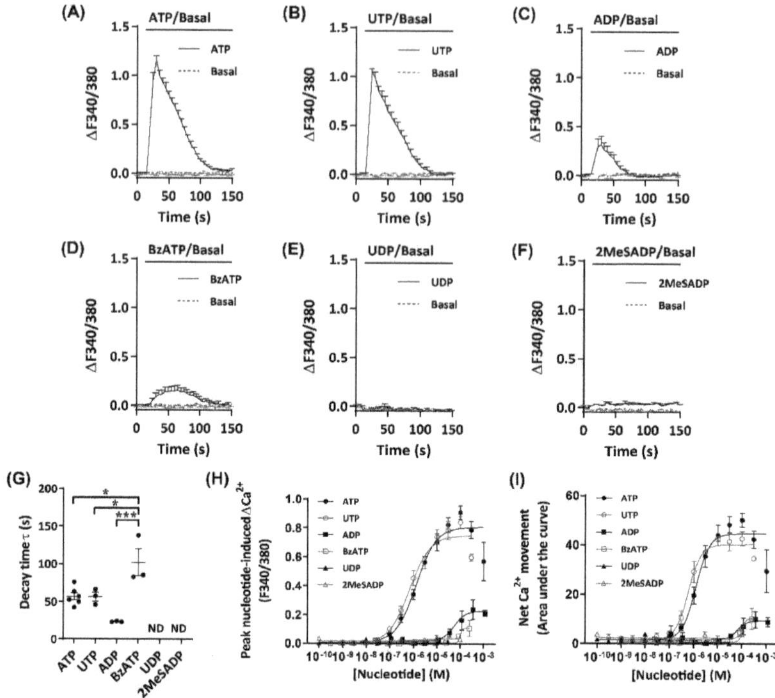

Figure 2. Nucleotide-induced Ca^{2+} response profiles for DH82 cells. (**A–I**) DH82 cells in extracellular Ca^{2+} solution (ECS) were loaded with Fura-2, incubated in the absence (basal) or presence of each nucleotide (as indicated) and Fura-2 fluorescence was recorded. Ca^{2+} traces (ΔF340/380) for (**A**) 100 μM adenosine 5′-triphosphate (ATP) ($n = 6$), (**B**) 30 μM uridine 5′-triphosphate (UTP) ($n = 3$), (**C**) 100 μM adenosine 5′-diphosphate (ADP) (preincubated with 4.5 U/mL hexokinase for 1 h at 37 °C) ($n = 3$), (**D**) 300 μM 3′-O-(4-benzoyl)benzoyl-ATP (BzATP) ($n = 3$), (**E**) 30 μM uridine 5′-diphosphate (UDP) ($n = 3$) and (**F**) 100 μM 2-methylthio-ADP (2MeSADP) ($n = 3$). (**G**) One phase decay time (τ) calculated from the peak of each Ca^{2+} trace, ND = no data. (**H**) Peak nucleotide-induced Ca^{2+} responses and (**I**) net Ca^{2+} movement were fit to the Hill equation to produce concentration-response curves (n values correspond to respective individual traces above). (**A–I**) Data shown are the mean ± SEM from three to six independent experiments as indicated. (**G**) * $p < 0.05$ and *** $p < 0.001$ between nucleotides as indicated analysed using a one-way ANOVA with Bonferroni post hoc test.

Table 1. Nucleotide-induced changes in intracellular Ca^{2+} in DH82 cells as measured by half maximal effective concentration.

Nucleotide	Peak Ca^{2+} Response		Net Ca^{2+} Response	
	pEC$_{50}$	Hill Coefficient	pEC$_{50}$	Hill Coefficient
ATP	5.88 ± 0.05 (100%)	0.99	5.92 ± 0.09 (100%)	1.43
UTP	6.16 ± 0.09 (65.9%)	1.02	6.26 ± 0.12 (69.1%)	1.71
ADP [1]	4.03 ± 0.30 (26.4%) [2]	1.47	4.07 ± 0.21 (19.9%) [2]	1.00
BzATP	<4.00 (12.1%) [3]	2.26	<4.00 (20.5%) [3]	2.66
UDP	ND (<10%)	-	ND (<10%)	-
2MeSADP	ND (<10%)	-	ND (<10%)	-

Abbreviations: AUC, area under the curve; ND, not determined (pEC$_{50}$ not calculated due to lack of response). Values in parentheses indicate the percent of each maximum agonist response compared to 100 μM ATP. [1] ADP in the presence of hexokinase to remove contaminating ATP. [2] $p < 0.05$ compared to the respective pEC$_{50}$ of ATP and UTP (one-way ANOVA). [3] $p < 0.01$ compared to the respective pEC$_{50}$ of ATP and UTP (one-way ANOVA).

2.3. P2X4 Receptors Mediate Minor Changes in Intracellular Ca^{2+} in DH82 Cells

2.3.1. TNP-ATP Partially Reduces ATP-Induced Net Ca^{2+} Movement

To determine if the observed Ca^{2+} responses were mediated by P2X receptors, DH82 cells were preincubated with the non-selective P2X receptor antagonist, 2′,3′-O-(2,4,6-trinitrophenyl)-ATP (TNP-ATP) [42], then exposed to ATP. Preincubation with 50 μM TNP-ATP partially reduced Ca^{2+} responses mediated by 10 μM ATP, but did not significantly inhibit Ca^{2+} responses evoked by 1 μM or 100 μM ATP (Figure 3A–C). This was supported by a significant reduction in net Ca^{2+} movement at 10 μM ATP, but not at other ATP concentrations (Figure 3E). Despite this, no significant change in peak Ca^{2+} response or shift in decay time was observed (Figure 3D,F). Preincubation with TNP-ATP did not result in a significant shift in the peak Ca^{2+} response EC_{50} for ATP, compared to cells preincubated in the absence of TNP-ATP (Figure 3D; pEC_{50} 5.38 ± 0.08 vs. 5.57 ± 0.13, respectively; $p = 0.133$ Student's t-test). In contrast, preincubation with TNP-ATP did result in a significant shift in the net Ca^{2+} movement EC_{50} compared to cells preincubated in absence of TNP-ATP (Figure 2E; pEC_{50} 5.04 ± 0.14 vs. 5.69 ± 0.13, respectively; $p = 0.007$ Student's t-test).

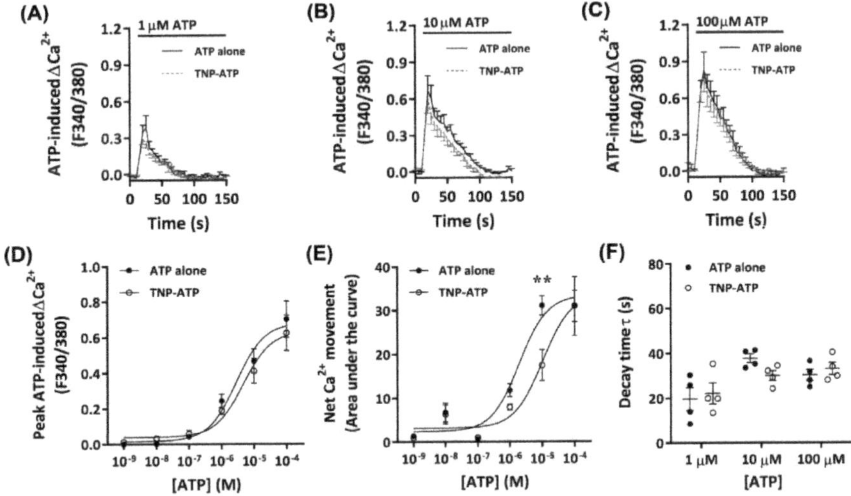

Figure 3. ATP-induced Ca^{2+} responses in DH82 cells in the absence or presence of 2′,3′-O-(2,4,6-trinitrophenyl)-ATP (TNP-ATP). (**A–F**) DH82 cells in ECS were loaded with Fura-2 and preincubated in the absence (ATP alone) or presence of 50 μM TNP-ATP (in ECS) for 5 min. Cells were exposed to increasing concentrations of ATP and Fura-2 fluorescence was recorded. (**A–C**) ATP-induced Ca^{2+} traces (F340/380) were plotted and the (**D**) peak Ca^{2+} response and (**E**) net Ca^{2+} movement were fit to the Hill equation to produce concentration-response curves. (**F**) One phase decay time (τ) calculated from the peak of each Ca^{2+} trace in (**A–C**) (**A–F**) Data shown are the mean ± SEM from four independent experiments. (**D–F**) ** $p < 0.01$ compared to respective concentration of ATP alone analysed using a two-way ANOVA with Bonferroni post hoc test.

2.3.2. Paroxetine Partially Reduces ATP-Induced Net Ca^{2+} Movement

To further investigate the role of P2X receptors in DH82 cells, ATP-induced Ca^{2+} responses were measured in cells preincubated with paroxetine, a selective serotonin reuptake inhibitor which has been shown to inhibit P2X4 receptors [41,43,44] and human (but not rodent) P2X7 receptors [45,46]. Preincubation with paroxetine partially reduced Ca^{2+} responses mediated by ATP concentrations of 10 μM or greater, with a small, but non-significant inhibitory effect observed at 1 μM or below

(Figure 4A–C). Similar to TNP-ATP, preincubation with paroxetine did not significantly reduce peak Ca^{2+} responses (Figure 4D), however, did significantly reduce net Ca^{2+} movement and decay kinetics (Figure 4E,F). Preincubation with paroxetine did not significantly shift the EC_{50} of ATP compared to cells preincubated in absence of paroxetine for peak Ca^+ response (Figure 4D; pEC_{50} 5.55 ± 0.31 vs. 5.51 ± 0.48, respectively; $p = 0.479$ Student's *t*-test) or net Ca^{2+} movement (Figure 4E; pEC_{50} 5.27 ± 0.35 vs. 5.4 ± 0.28, respectively; $p = 0.387$ Student's *t*-test).

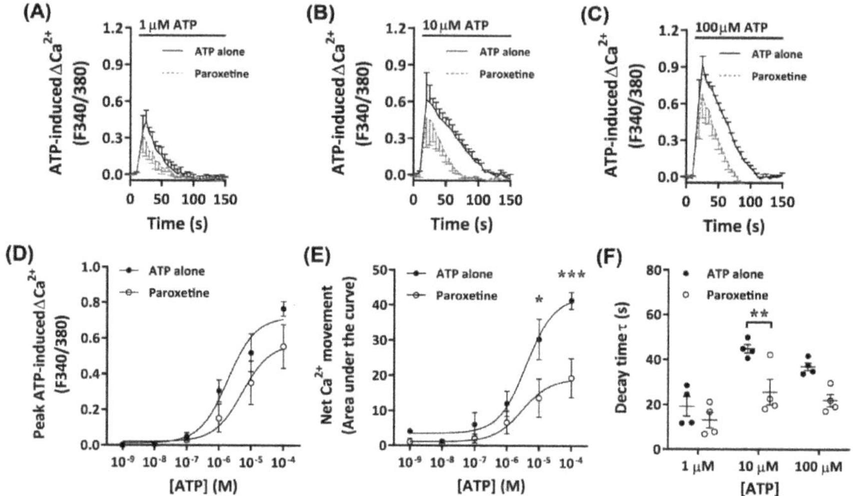

Figure 4. ATP-induced Ca^{2+} responses in DH82 cells in the absence or presence of paroxetine. (**A–F**) DH82 cells in ECS were loaded with Fura-2 and preincubated in the absence (ATP alone) or presence of 100 μM paroxetine (both 0.3% dimethyl sulfoxide; DMSO) for 5 min. Cells were then exposed to increasing concentrations of ATP and Fura-2 fluorescence was recorded. (**A–C**) ATP-induced Ca^{2+} traces (F340/380) were plotted and the (**D**) peak Ca^{2+} response and (**E**) net Ca^{2+} movement were fit to the Hill equation to produce concentration-response curves. (**F**) One phase decay time (τ) calculated from the peak of each Ca^{2+} trace in (**A–C**). (**A–F**) Data shown are the mean ± SEM from four independent experiments. (**D–F**) * $p < 0.05$, ** $p < 0.01$ and *** $p < 0.001$ compared to respective concentration of ATP alone analysed using a two-way ANOVA with Bonferroni post hoc test.

2.3.3. 5-BDBD Partially Reduces ATP-Induced Net Ca^{2+} Movement

To determine further if P2X4 receptors played a role in the observed ATP-induced Ca^{2+} responses, DH82 cells were preincubated with the selective P2X4 receptor antagonist, 5-BDBD, of which was recently demonstrated to inhibit the canine P2X4 receptor [41]. Preincubation with 30 μM 5-BDBD had no significant inhibitory effects on the Ca^{2+} response mediated by ATP (Figure 5A–C). There was no significant difference in the EC_{50} for cells preincubated in the absence or presence of 5-BDBD for ATP-induced peak Ca^{2+} response (Figure 5D; pEC_{50} 5.87 ± 0.12 vs. 5.82 ± 0.10, respectively; $p = 0.372$ Student's *t*-test) or net Ca^{2+} movement (Figure 5E; pEC_{50} 5.73 ± 0.18 vs. 5.64 ± 0.14, respectively; $p = 0.358$ Student's *t*-test), although a trend towards decreased net Ca^{2+} movement was observed at 10 μM ATP (Figure 5E). Of note, there was a significant reduction in decay kinetics at 10 μM ATP for cells preincubated in the presence of 5-BDBD compared to those in absence of the antagonist (Figure 5F).

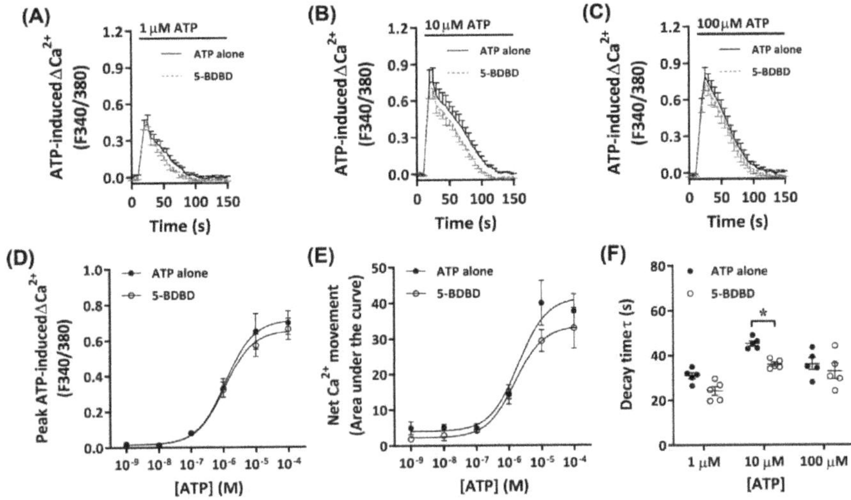

Figure 5. ATP-induced Ca^{2+} responses in DH82 cells in the absence or presence of 5-BDBD. (**A–F**) DH82 cells in ECS were loaded with Fura-2 and preincubated in the absence (ATP alone) or presence of 30 μM 5-BDBD (both 0.3% DMSO) for 5 min. Cells were then exposed to increasing concentrations of ATP and Fura-2 fluorescence was recorded. (**A–C**) ATP-induced Ca^{2+} traces (F340/380) were plotted and the (**D**) peak Ca^{2+} response and (**E**) net Ca^{2+} movement were fit to the Hill equation to produce concentration-response curves. (**F**) One phase decay time (τ) calculated from the peak of each Ca^{2+} trace in (**A–C**). (**A–F**) Data shown are mean $\pm$ SEM from five independent experiments. (**D–F**) * $p < 0.05$ compared to respective concentration of ATP alone analysed using a two-way ANOVA with Bonferroni post hoc test.

2.3.4. Ivermectin Positively Modulates ATP-Induced Net Ca^{2+} Movement

It has recently been demonstrated that ivermectin can effectively potentiate Ca^{2+} responses mediated by canine P2X4 receptors [41]. To further investigate if DH82 cells express functional P2X4 receptors, cells were preincubated with ivermectin prior to activation with increasing concentrations of ATP. Preincubation with 3 μM ivermectin revealed a strong potentiation of ATP-induced Ca^{2+} responses in DH82 cells (Figure 6A–C), with significant increases in ATP-induced net Ca^{2+} movement and peak Ca^{2+} response observed in the presence of ivermectin compared to cells preincubated in absence of ivermectin at ATP concentrations upwards of 10 μM (Figure 6D,E). Despite this, there was no significant difference in the EC_{50} in the absence or presence of ivermectin for ATP-induced peak Ca^{2+} response (Figure 6D; pEC_{50} 5.76 $\pm$ 0.19 vs. 5.37 $\pm$ 0.35; $p = 0.372$ Student's *t*–test) or net Ca^{2+} movement (Figure 6E; pEC_{50} 5.62 $\pm$ 0.01 vs. 5.57 $\pm$ 0.12, respectively; $p = 0.358$ Student's *t*-test). Additionally, there were no significant differences in decay kinetics between cells preincubated in absence or presence of ivermectin (Figure 6F). Collectively this and the above data suggests that DH82 cells express functional P2X4 receptors.

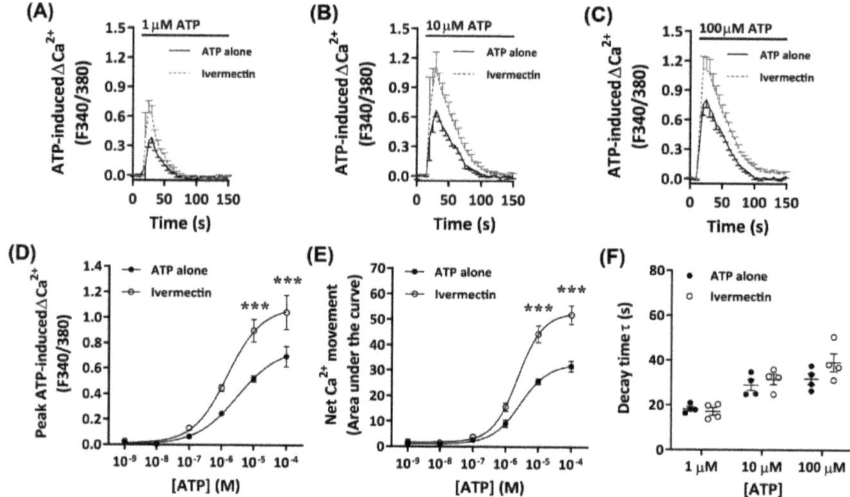

Figure 6. ATP-induced Ca^{2+} responses in DH82 cells in the absence or presence of ivermectin. (**A–F**) DH82 cells in ECS were loaded with Fura-2 and preincubated in the absence (ATP alone) or presence of 3 μM ivermectin (both 0.1% DMSO) for 5 min. Cells were then exposed to increasing concentrations of ATP and Fura-2 fluorescence was recorded. (**A–C**) ATP-induced Ca^{2+} traces (F340/380) were plotted and the (**D**) peak Ca^{2+} response and (**E**) net Ca^{2+} movement were fit to the Hill equation to produce concentration-response curves. (**F**) One phase decay time (τ) calculated from the peak of each Ca^{2+} trace in (**A–C**). (**A–F**) Data shown are mean ± SEM from five independent experiments. (**D–F**) *** $p < 0.001$ compared to respective concentration of ATP alone analysed using a two-way ANOVA with Bonferroni post hoc test.

2.4. ATP and UTP Mediate Both Ca^{2+} Influx and Store-Operated Ca^{2+} Entry in DH82 Cells

Despite the apparent P2X4-mediated effects on Ca^{2+} responses in DH82 cells, a lack of complete inhibition by P2X receptor antagonists, as well as the responsiveness to UTP, suggest the presence of functional P2Y receptors in DH82 cells. To determine if $G_{q/11}$-coupled P2Y receptors, which have demonstrated roles in store-operated Ca^{2+} entry [47], were involved in the observed Ca^{2+} responses in DH82 cells, nucleotide-induced changes in intracellular Ca^{2+} were measured in the presence of extracellular or intracellular Ca^{2+} chelators. Compared to cells in the presence of extracellular Ca^{2+} (Figure 7A,F), cells incubated with ethylene glycol tetraacetic acid (EGTA) demonstrated a partial reduction in ATP- and UTP-induced Ca^{2+} responses (Figure 7B,G). This was supported by significant reductions in decay time (Figure 7K) and net Ca^{2+} movement (Figure 7L,M) for both ATP- and UTP-mediated responses. In contrast, preincubation with the cell-permeant Ca^{2+} chelator, 1,2-*bis*(2-aminophenoxy)ethane-*N,N,N′,N′*-tetraacetic acid tetrakis(acetoxymethyl ester) (BAPTA-AM), near completely reduced ATP- and UTP-induced Ca^{2+} responses (Figure 7C,H) and net Ca^{2+} movement (Figure 7L,M) in DH82 cells. Treatment with thapsigargin, to inhibit sarco/endoplasmic reticulum Ca^{2+} ATPase pumps [48], near completely reduced ATP- and UTP-induced Ca^{2+} responses in the presence of extracellular Ca^{2+} (ECS; Figure 7D,I,L,M) and completely reduced these responses in the absence of extracellular Ca^{2+} (EGTA; Figure 7E,J,L,M). Together, this suggests the involvement of both P2X and P2Y receptors in DH82 cells for mediating changes in intracellular Ca^{2+} in response to activation by nucleotides.

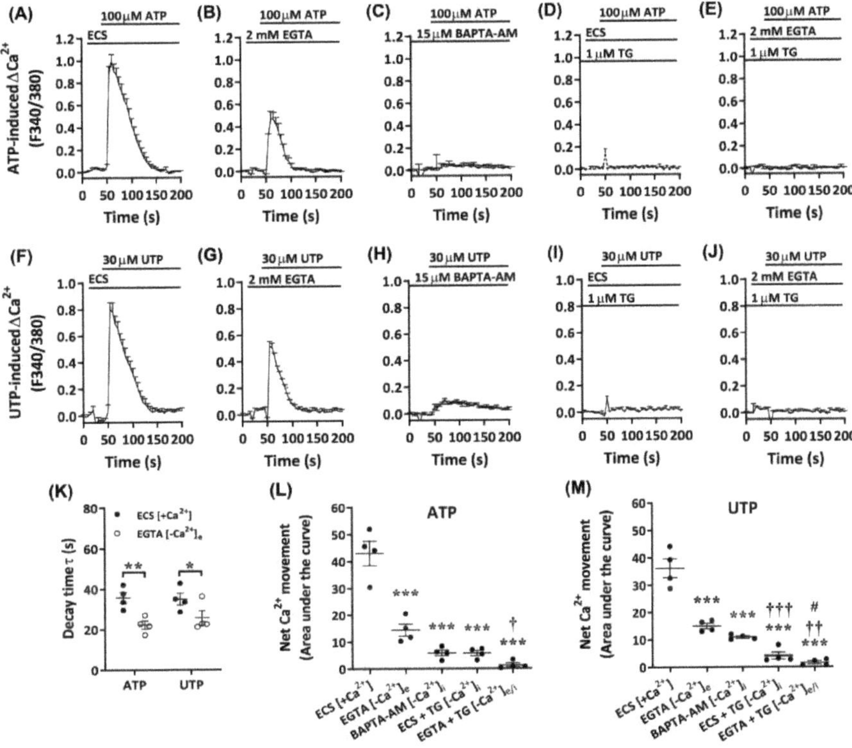

Figure 7. Nucleotide-induced Ca^{2+} responses in the absence or presence of extracellular and/or intracellular Ca^{2+}. (**A–M**) DH82 cells in (**A,D,F,I**) ECS or (**B,C,E,G,H,J**) Ca^{2+}-free solution were loaded with Fura-2 and preincubated in the absence (**A,D,F,I**) (ECS) or presence (**B,E,F,J**) of 2 mM ethylene glycol tetraacetic acid EGTA for 30 s, (**C,H**) 15 µM 1,2-*bis*(2-aminophenoxy)ethane-*N,N,N′,N′*-tetraacetic acid tetrakis(acetoxymethyl ester) (BAPTA-AM) for 30 min or (**D,E,I,J**) 1 µM thapsigargin (TG) for 30 min prior to incubation in the absence (**D,I**) or presence (**E,J**) of 2 mM EGTA. (**A–M**) Cells were then exposed to (**A–E**) 100 µM ATP or (**F–J**) 30 µM UTP and Fura-2 fluorescence was recorded. (**K**) One phase decay time (τ) calculated from the peak of each Ca^{2+} trace in (**A,B,F,G**). (**L,M**) Net Ca^{2+} movement from each trace in **A–J**. (**A–M**) Data shown are the mean ± SEM from four independent experiments. * $p < 0.05$, ** $p < 0.01$ and *** $p < 0.001$ compared to respective ECS alone control; † $p < 0.05$, †† $p < 0.01$ and ††† $p < 0.001$ compared to EGTA; # $p < 0.05$ compared to BAPTA-AM analysed using a (**K**) Student's *t*-test or (**L,M**) one-way ANOVA with Bonferroni post hoc test.

2.5. P2Y₂ Receptor Activation Mediates Ca^{2+} Mobilization in DH82 Cells

2.5.1. Suramin Reduces ATP- and UTP-Induced Ca^{2+} Mobilization

The data presented above suggests a major role for an ATP- and UTP-responsive P2Y receptor in Ca^{2+} mobilization within DH82 cells. Previous studies have demonstrated that the canine P2Y₂ receptor in Madin–Darby canine kidney (MDCK) cells responds to both ATP and UTP with similar potency [49]. Therefore, to determine if nucleotide-induced Ca^{2+} mobilization was mediated by P2Y₂ receptors, DH82 cells were preincubated with increasing concentrations of the non-selective P2 receptor antagonist, suramin [50], which is selective for P2Y₂ over P2Y₄ receptors [51]. Cells were then incubated with ATP or UTP at their respective EC_{80} to determine the optimal concentration for P2Y receptor inhibition. Preincubation of DH82 cells with 1 mM suramin inhibited Ca^{2+} responses evoked by 3 μM ATP, however lower concentrations of suramin (<100 μM) had little to no inhibitory effect (Figure 8A). In contrast, preincubation with 100 μM and 1 mM, but not 10 μM suramin or less inhibited Ca^{2+} responses evoked by 1 μM UTP (Figure 8B). Inhibitory effects observed in the presence of 100 μM suramin resulted in significant shifts in the IC_{50} of suramin between ATP- and UTP-induced peak Ca^{2+} responses (Figure 8C; pIC_{50} 3.02 ± 0.06 and 3.70 ± 0.05, respectively; $p < 0.001$ Student's *t*-test) and net Ca^{2+} movement (Figure 8D; pIC_{50} 3.03 ± 0.12 and 3.54 ± 0.14, respectively; $p = 0.025$ Student's *t*-test).

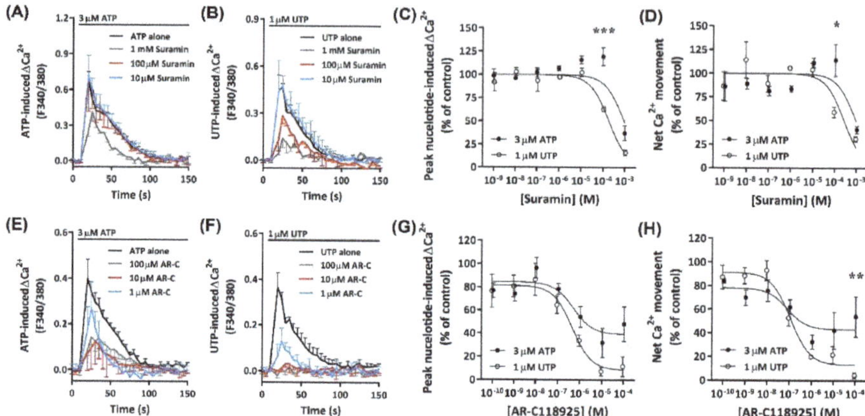

Figure 8. ATP- and UTP-induced Ca^{2+} responses in DH82 cells in the absence or presence of suramin or ARC-118925. (**A–H**) DH82 cells in ECS were loaded with Fura-2 and preincubated in the absence or presence of increasing concentrations of (**A–D**) suramin (in ECS) or (**E–H**) AR-C118925 (AR-C; 0.3% DMSO) for 30 min. Cells were then exposed to 3 μM ATP or 1 μM UTP (respective EC_{80} values) and Fura-2 fluorescence was recorded. Nucleotide-induced (**C,G**) peak Ca^{2+} response and (**D,H**) net Ca^{2+} movement were normalised to 3 μM ATP or 1 μM UTP alone and expressed as a percentage of the response in absence of inhibitor (% of control). Data were then to fit data to the Hill equation to produce concentration-response curves and calculate the IC_{50}. (**A–H**) Data shown are mean ± SEM from three independent experiments. * $p < 0.05$, ** $p < 0.01$ and *** $p < 0.001$ compared to respective concentration of antagonist with ATP or UTP analysed using a two-way ANOVA with Bonferroni post hoc test.

2.5.2. AR-C118925 Reduces ATP- and UTP-Induced Ca^{2+} Mobilization

To determine if the $P2Y_2$ receptor was mediating nucleotide-induced Ca^{2+} mobilization in DH82 cells, the selective $P2Y_2$ receptor antagonist AR-C118925 [52] was preincubated with cells prior to incubation with ATP or UTP at their respective EC_{80} concentrations. AR-C118925 at concentrations of 1 μM or greater could only partially inhibit Ca^{2+} responses evoked by 3 μM ATP (Figure 8E). In contrast, preincubation with AR-C118925 at concentrations of 10 μM or greater near completely inhibited Ca^{2+} responses evoked by 1 μM UTP (Figure 8F). A significant shift was observed in the IC_{50} of AR-C118925 in response to activation by ATP and UTP calculated using peak Ca^{2+} responses (Figure 8G; pIC_{50} 6.18 ± 0.16 and 6.61 ± 0.09, respectively; $p = 0.033$ Student's *t*-test), but not net Ca^{2+} movement (Figure 8H; pIC_{50} 6.67 ± 0.19 and 6.87 ± 0.12, respectively; $p = 0.198$ Student's *t*-test). The inhibition of nucleotide-induced Ca^{2+} responses by AR-C118925 supports the presence of $P2Y_2$ receptors in DH82 cells. Moreover, the differing effect of this antagonist on ATP- and UTP-induced responses indicates the presence of other P2 receptors in this cell line. Additionally, preincubation of DH82 cells together with 5-BDBD and AR-C118925 resulted in a complete inhibition of both ATP- and UTP-induced net Ca^{2+} movement (Figure S1), further suggesting a role for both P2X4 and $P2Y_2$ receptors in nucleotide-mediated Ca^{2+} responses in DH82 cells.

2.6. DH82 Canine Macrophages Predominantly Express Cell Surface P2Y₂ Receptors

To confirm the presence of P2X4 and P2Y$_2$ receptors, DH82 cells were analysed by immunocytochemistry and confocal microscopy using anti-P2X4 or anti-P2Y$_2$ receptor antibodies. Confocal microscopy revealed the presence of both P2X4 and P2Y$_2$ receptors in fixed and permeabilised DH82 cells (Figure 9A,B). The expression of P2X4 receptors was relatively low and largely intracellular (Figure 9A). The expression of P2Y$_2$ receptors on DH82 cells was considerably higher and predominantly localised to the cell surface (Figure 9B), consistent with its reported expression in the membrane of MDCK cells [53,54]. No fluorescence was detected in DH82 cells stained with secondary antibodies alone (Figure S2).

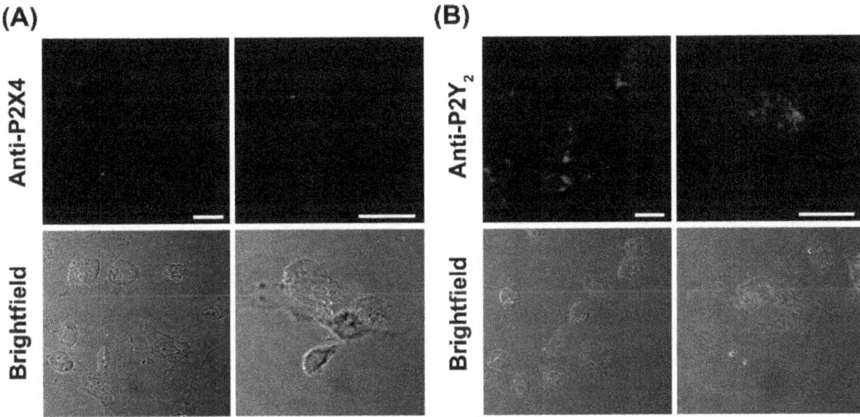

Figure 9. Expression of P2X4 and P2Y$_2$ receptors in DH82 cells. DH82 cells were fixed, permeabilized and labelled with (**A**) anti-P2X4 or (**B**) anti-P2Y$_2$ receptor primary antibodies, then with anti-goat[594] or anti-rabbit[594] secondary antibodies, respectively. Cells were imaged by confocal microscopy. Scale bar = 20 μm. Results are representative of three independent experiments.

2.7. P2Y₂ Receptor Activation Downstream Ca²⁺ Mobilization Is Coupled to the Phospholipase C/Inositol Triphosphate Signal Transduction Pathway in DH82 Cells

Functional P2Y$_2$ receptors have been reported in human and rodent macrophages [9,10], where they can activate phospholipase C (PLC) and inositol trisphosphate (IP$_3$) receptors, leading to Ca^{2+} mobilization from endoplasmic reticulum stores [8,12]. To determine if activation of canine P2Y$_2$ receptors in DH82 macrophage cells results in a similar downstream signalling pathway, cells were preincubated with antagonists of PLC (U-73122) and IP$_3$ receptors (2-aminoethoxydiphenyl borate; 2-APB), and nucleotide-induced intracellular Ca^{2+} mobilization was recorded in absence of extracellular Ca^{2+}. The presence of AR-C118925 under these conditions was also examined. Preincubation of DH82 cells with AR-C118925 near completely inhibited intracellular Ca^{2+} mobilization mediated by ATP (Figure 10A; 90.5 ± 3.3% inhibition) or UTP (Figure 10B; 87.3 ± 7.4% inhibition). Similarly, preincubation of DH82 cells with 75 μM 2-APB near completely inhibited intracellular Ca^{2+} mobilization mediated by ATP (Figure 10A; 90.7 ± 4.4% inhibition) or UTP (Figure 10B; 91.2 ± 3.1% inhibition). In contrast, preincubation with 5 μM U-73122 only partially reduced intracellular Ca^{2+} mobilization mediated by ATP (Figure 10A; 64.9 ± 5.9% inhibition) or UTP (Figure 10B; 55.5 ± 8.7% inhibition). The combination of two or three of these antagonists completely impaired ATP- and UTP-induced Ca^{2+} responses (Figure 10A,B). Thus, preincubation with AR-C118925, 2-APB, U-73122 or any combination of these antagonists resulted in a significant reduction of intracellular Ca^{2+} mobilization ($p < 0.001$, one-way ANOVA) compared to that mediated by ATP (Figure 10A) or UTP (Figure 10B) in the absence of antagonists.

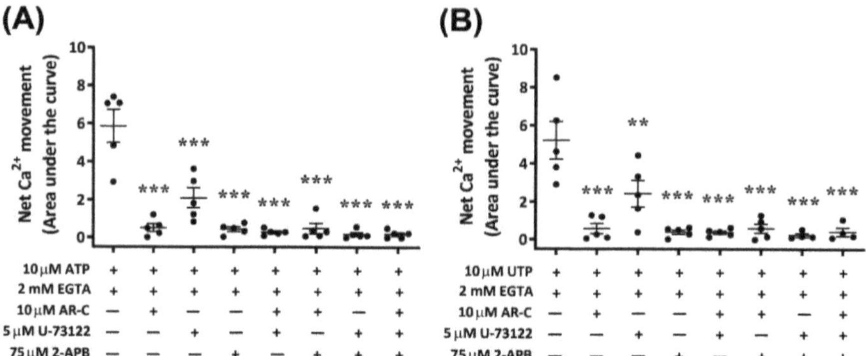

Figure 10. Nucleotide-induced Ca^{2+} mobilization in DH82 cells depleted of extracellular Ca^{2+} in the absence or presence of AR-C118925, U-73122 and 2-APB. DH82 cells were loaded with Fura-2 and preincubated in Ca^{2+}-free solution containing 2 mM EGTA in the absence or presence of 10 μM AR-C118925 (AR-C; 0.03% DMSO), 5 μM U-73122 (0.05% DMSO) or 75 μM 2-APB (0.15% DMSO) for 5 min. Cells were then exposed to (**A**) 10 μM ATP or (**B**) 10 μM UTP and Fura-2 fluorescence was recorded. Data shown are the mean ± SEM from five independent experiments. ** $p < 0.01$ and *** $p < 0.001$ compared to respective nucleotide alone, analysed using a one-way ANOVA with Bonferroni post hoc test.

3. Discussion

To date, the DH82 canine macrophage cell line has primarily been utilized as a model of viral and protozoan infection [23–25], such as for the study of canine distemper and its oncolytic potential [55–57]. Although studies have described the expression of functional P2 receptors in human or rodent macrophages and macrophage cell lines [8,10,11,34], studies directly investigating P2 receptors in canine macrophages have been absent. The current study described, for the first time, the expression and function of P2 receptors in DH82 cells, demonstrating a primary role for cell surface P2Y$_2$ receptors in nucleotide-mediated Ca^{2+} mobilization through PLC/IP$_3$ signal transduction. This study also demonstrates a minor functional role for P2X4 receptors in DH82 cells, suggesting this cell line may present as a suitable model for studying P2 receptor-mediated inflammation and pain signalling in dogs.

The agonist profile of ATP on DH82 cells demonstrated pharmacological similarities to the recombinant canine P2X4 receptor [41], as well as studies of endogenous P2X4 receptors in a human macrophage cell model [34]. BzATP induced partial Ca^{2+} responses in DH82 cells with significantly lower potency compared to ATP, consistent with the recent report that BzATP is a partial agonist of recombinant canine P2X4 receptors [41]. In addition, the increased decay kinetics of Ca^{2+} responses evoked by BzATP, compared to ATP, further supports a role for P2X4 receptors in the observed Ca^{2+} responses [41]. TNP-ATP and paroxetine, two non-selective antagonists of P2X4 receptors [44], as well as 5-BDBD, a selective P2X4 receptor antagonist [58], had minor inhibitory effects on ATP-induced Ca^{2+} responses in DH82 cells. Although it has recently been demonstrated that these antagonists can inhibit recombinant canine P2X4 receptors, the minor inhibition observed with these antagonists in DH82 cells were potentially in part due a lack of potency towards the canine P2X4 receptor [41], as well as the relatively low expression of P2X4 receptors in DH82 cells observed by immunocytochemistry. In contrast, ivermectin, the positive allosteric modulator which is routinely used to investigate P2X4 receptor activity [59], demonstrated strong potentiation of ATP-induced Ca^{2+} responses and efficacy of ATP, with little effect on decay time. This data further supports the expression of functional P2X4 receptors in canine macrophages, however, it suggests that potentiation or upregulation of P2X4 receptors may first be required to observe notable responses.

Consistent with the pharmacological profiles reported for the canine $P2Y_2$ receptor cloned from MDCK cells [49], both ATP and UTP induced robust Ca^{2+} responses in DH82 cells with similar EC_{50} values. These responses were observed even in the absence of extracellular Ca^{2+}, consistent with the P2Y-mediated mobilization of intracellular Ca^{2+} [60]. Similar to the current study with DH82 cells, other studies have also demonstrated that ADP is a low-potency agonist of the canine $P2Y_2$ receptor cloned from MDCK cells [49,61]. In addition, incubation of DH82 cells with BzATP revealed pharmacological similarities to that observed with human $P2Y_2$ receptors, where BzATP is ineffective at concentrations below 100 μM [33,62]. Collectively, this suggests that functional $P2Y_2$ receptors were responsible for nucleotide-induced Ca^{2+} mobilization in DH82 cells. This was supported by inhibition observed in the presence of the $P2Y_2$ receptor antagonists, suramin and AR-C118925. Despite suramin lacking potency and selectivity, it remains a valuable tool in characterising P2Y receptor responses, as it is considered a low-potency antagonist of $P2Y_2$ receptors, but is relatively insensitive to $P2Y_4$ receptors [51,63]. This suggests that the ATP/UTP-sensitive $P2Y_2$, but not $P2Y_4$ receptor, is responsible for the observed Ca^{2+} mobilization, consistent with the expression of $P2Y_2$ receptors in DH82 cells determined by confocal microscopy. Notably, $P2Y_2$ receptor protein expression was greater than that of P2X4 receptor protein expression, with an opposite pattern observed for mRNA expression of these receptors. Reasons for this discrepancy remain unknown, but a lack of correlation between mRNA and protein expression is well documented and attributed to various contributing factors related to post-transcriptional and post-translational regulation of mRNA and protein expression [64].

A number of other canine P2Y receptors from MDCK cells have been cloned and characterised, including the $P2Y_1$ and $P2Y_{11}$ receptors [49,65,66], of which mRNA of both these receptors were detected in DH82 cells. The nucleotide agonist profile of canine $P2Y_{11}$ receptors differs markedly from the human $P2Y_{11}$ receptor, where ATP is a potent agonist of human, but not canine $P2Y_{11}$ receptors, and ADP and its analogue 2MeSADP are potent agonists of canine, but not human $P2Y_{11}$ receptors [66]. In the current study, it was revealed that ATP, but not ADP, was a moderately potent mediator of Ca^{2+} responses in DH82 cells, while no such responses were observed with 2MeSADP. BzATP is also a full agonist of human $P2Y_{11}$ receptors [67], further suggesting that DH82 cells likely do not express functional $P2Y_{11}$ receptors. ADP and 2MeSADP have also been reported as potent agonists of the canine $P2Y_1$ receptor [68]. However, it was demonstrated in the current study that only ADP, but not 2MeSADP, induced a small Ca^{2+} response in DH82 cells. Although this could suggest that DH82 cells express low amounts of functional $P2Y_1$ receptors, the complete lack of response to 2MeSADP suggests that $P2Y_1$ receptors are unlikely to be responsible for P2Y receptor-mediated Ca^{2+} mobilisation in DH82 cells. Additionally, a complete lack of Ca^{2+} response in DH82 cells incubated with the $P2Y_6$ receptor agonist, UDP [69,70], strongly suggests that DH82 cells do not express functional $P2Y_6$ receptors.

The current study demonstrated that nucleotide-mediated Ca^{2+} mobilization in DH82 cells was also inhibited by antagonists of PLC and IP_3 receptors, U-73122 and 2-APB, respectively. This was consistent with previous studies that demonstrate coupling of $P2Y_2$ receptors to $G_{q/11}$ and downstream signalling pathways in MDCK cells [49,65,71]. While 2-APB near completely inhibited Ca^{2+} mobilisation, U-73122 only resulted in partial inhibition, although higher concentrations (>10 μM) have been shown to completely block $P2Y_2$ receptor-mediated Ca^{2+} responses [72]. Notably, pre-incubation with $P2Y_2$ receptor antagonists resulted in approximately two-fold greater inhibition of UTP-induced Ca^{2+} responses compared to ATP-induced responses, suggesting that ATP remained active at other receptors involved in mediating changes in intracellular Ca^{2+}, such as P2X4 receptors, which are also relatively insensitive to suramin [73]. In addition, both ATP- and UTP-induced Ca^{2+} responses could be completely inhibited by co-incubation with 5-BDBD and AR-C118925, supporting a role for both P2X4 and $P2Y_2$ receptors in DH82 cells.

Pro-monocytic and macrophage-like cell lines, such as human THP-1 cells, have recently proven useful models for studying endogenous purinergic signalling via P2X4 and $P2Y_2$ receptors [10,13,34]. However, studies have demonstrated that these cell lines can be polarised towards a more specialised macrophage phenotype, in which the expression of P2 receptors, such as P2X4 and P2X7 receptors, are commonly upregulated [34,38,74]. A study has recently demonstrated that DH82 canine macrophage cells could be polarised towards the M1 or M2a subtype through cytokine stimulation [20]. However, the DH82 cells utilised throughout this study remained unpolarised (M0) and, thus, it remains to be determined if cytokine stimulation influences P2 receptor expression or function. To this end, future studies could determine the purinergic signalling landscape of polarised DH82 cells. Future studies could also analyse P2 receptor expression and signalling in native canine macrophages. Given the known expression of P2X7 receptors on canine monocytes [29,30], canine monocyte-derived macrophages may provide a suitable candidate for the study of other purinergic receptors in native canine macrophages.

The upregulation of P2X4 receptors in macrophages and microglia has been highlighted as a key component in the signalling of inflammatory conditions, including chronic inflammatory and neuropathic pain [75,76], and remyelination of damaged nerves in the CNS [77]. P2X4 receptors, which reside primarily within lysosomes of macrophages, can be upregulated at the cell surface through lysosomal exocytosis [78]. This process plays a key role in Ca^{2+} homeostasis, ATP release and local activation of cell surface purinergic receptors [79]. Notably, activation of the C-C chemokine receptor 2 (CCR2) by C-C chemokine ligand 2 (CCL2) is known to mediate lysosomal exocytosis [80], while in rat alveolar macrophages and human THP-1 cells it has been demonstrated that the activation of cell surface $P2Y_2$ receptors induces the upregulation and secretion of CCL2 [8,10]. Thus, it could be suggested that activation of macrophage $P2Y_2$ receptors results in a P2 receptor signalling feedback mechanism which results in an increase in Ca^{2+} flux through upregulation of lysosomal exocytosis and trafficking of P2X4 receptors to the cell surface, leading to the release of prostaglandin E_2 and subsequent chronic inflammatory pain signalling [2]. These activated macrophages may also modulate microglial P2X4 receptors to control neuroinflammatory signalling following injury to the central nervous system [81]. Given the expression profile of these receptors in DH82 canine macrophages, this cell line may provide a suitable model for studying inflammatory pain signalling mechanisms of dogs in vitro, as well as for the pre-clinical testing of novel therapeutics targeting chronic pain.

Finally, although the sequence of P2X4 and $P2Y_2$ receptors in DH82 cells is yet to be determined, it remains of interest to identify novel single nucleotide polymorphisms should they exist in the genes encoding these receptors in dogs. A recent whole genome study of 582 dogs has revealed at least one missense variant (Ala9Asp) within the canine *P2RX4* gene and two missense variants (Gly193Ser and Val375Ile) within the canine *P2RY2* gene [82] (data accessed from the European Variation Archive; https://www.ebi.ac.uk/eva/). Whilst the effects of single nucleotide polymorphisms in the canine *P2RX4* and *P2RY2* genes are largely unknown, it has been demonstrated that single nucleotide polymorphisms of the genes encoding the human P2X4 and $P2Y_2$ receptors can alter receptor function [83,84]. Notably, in human macrophages, a 312Ser polymorphism of the $P2Y_2$ receptor has been demonstrated to alter secretion of CCL_2 following activation by UTP [10], suggesting a potential association with macrophage-mediated chronic inflammatory pain signalling. Despite this however, studies by our group have demonstrated that the canine *P2RX4* and *P2RX7* genes are much more conserved than their human counterparts [41,85,86] and, as such, naturally-occurring polymorphisms in the canine *P2RY2* gene may also be rare or limited to uncommon breeds not frequently sampled in canine whole genome studies.

In conclusion, the current study demonstrates for the first time, that DH82 canine macrophages primarily express functional $P2Y_2$ receptors and low levels of functional P2X4 receptors. As such, DH82 cells provide the first canine macrophage cell line for the study of endogenous P2X4 and $P2Y_2$ receptors. The data presented here provides indirect evidence that P2X4 and $P2Y_2$ receptors play a role in mediating changes in intracellular Ca^{2+} in canine macrophages in vivo. This mimics events observed in human and rodent macrophages and macrophage cell lines, where these P2 receptors have been suggested to play a key role in inflammation and chronic pain. Thus, DH82 cells may aid in the study of P2 receptor-mediated inflammation, including neuroinflammatory signalling processes, as well as preclinical screening of novel P2 receptor-targeting compounds for potential use in the treatment of inflammatory conditions, such as chronic pain in dogs.

4. Materials and Methods

4.1. Compounds and Reagents

BSA, EGTA and reagents for producing Ca^{2+} solutions were from Amresco (Solon, OH, USA). Fetal bovine serum (FBS) was purchased from Bovogen Biologicals (East Keilor, Melbourne, Australia) and heat inactivated at 56 °C for 30 min before use. 2-APB, U-73122 and UDP were from Cayman Chemical (Ann Arbor, MI, USA). Primers for RT-PCR were from Integrated DNA Technologies (Coralville, IA, USA). 5-BDBD, ADP (pre-treated with hexokinase as per [41]), ATP, BAPTA-AM, BzATP, hexokinase from *Saccharomyces cerevisae*, ivermectin, MEM non-essential amino acid solution, paraformaldehyde, paroxetine, phosphate buffered saline (PBS), poly-ᴅ-lysine hydrobromide (5 µg·mL^{-1} working stock), pluronic F-127, saponin, suramin and UTP were from Sigma-Aldrich (St. Louis, MO, USA). DMEM/F12 medium, ExoSAP-IT, Fura-2 AM, GlutaMAX, penicillin-streptomycin and 0.05% trypsin-EDTA were from ThermoFisher Scientific (Melbourne, Australia). 2MeSADP, AR-C118925, thapsigargin and TNP-ATP were from Tocris Bioscience (Bristol, UK).

4.2. Cells

DH82 cells were obtained from the European Collection of Authenticated Cell Cultures (ECACC cat. no. 94062922, RRID: CVCL_2018). DH82 cells were cultured in DMEM/F12 medium supplemented with 10% FBS, 2 mM GlutaMAX, 100 U/mL penicillin, 100 µg/mL streptomycin and 1% non-essential amino acids at 37 °C/5% CO_2. Cells were routinely found to be negative for mycoplasma contamination using the MycoAlert Mycoplasma Detection Kit (Lonza, Waverley, Australia).

4.3. RNA Isolation, cDNA Synthesis and RT-PCR

Total RNA was extracted from DH82 cells using the ISOLATE II RNA Mini Kit (Bioline, London, UK) according to manufacturer's instructions. cDNA was synthesised from RNA using the qScript cDNA SuperMix Kit (Quanta Biosciences, Gaithersburg, MD, USA) according to manufacturer's instructions. RT-PCR amplification of cDNA was carried out using the primer pairs and conditions listed in Table S1, the MangoTaq DNA polymerase kit (Bioline) and a Mastercycler Pro S (Eppendorf, Hamburg, Germany). PCR cycling consisted of initial denaturation at 95 °C for 2 min, followed by 35 cycles of denaturation at 95 °C for 30 s, annealing at 49–61 °C for 30 s and extension at 72 °C for 1 min. Amplicons were treated with ExoSAP-IT and loaded onto a 1% agarose gel and imaged using GelRed Nucleic Acid Gel Stain (Biotium, Fremont, CA, USA) and a Bio-Rad Molecular Imager Gel Doc XR+ (Hercules, CA, USA). Densitometry quantification was carried out using ImageJ [87] analysis software.

4.4. Measurement of Intracellular Ca^{2+}

Measurements of intracellular Ca^{2+} were determined using Fura-2 AM as previously described [41]. Recordings were performed in extracellular Ca^{2+} solution (ECS; 145 mM NaCl, 2 mM $CaCl_2$, 1 mM $MgCl_2$, 5 mM KCl, 13 mM glucose and 10 mM HEPES, pH 7.4) or in Ca^{2+} free solution (145 mM NaCl, 1 mM $MgCl_2$, 5 mM KCl, 13 mM glucose and 10 mM HEPES, pH 7.4) for recordings in absence of Ca^{2+}. Cells were plated at 6×10^4 cells/well in poly-ᴅ-lysine-coated black-walled µClear bottom 96-well plates (Greiner Bio-One, Frickenheisen, Germany) and incubated at 37 °C/5% CO_2 for 18–24 h. Cells were washed in ECS then preincubated with Fura-2 AM loading buffer (2.5 µM Fura-2 AM/0.2% pluronic acid in ECS) in the dark for 30 min at 37 °C. Prior to recording fluorescence, excess Fura-2 was removed and cells were washed with ECS (or Ca^{2+} free solution for recordings in absence of Ca^{2+}), then incubated for a further 20 min to allow for complete de-esterification. Fura-2 fluorescence emission at 510 nm was recorded every 5 s at 37 °C using a Flexstation3 (Molecular Devices, Sunnyvale, CA, USA) following excitation at 340 and 380 nm. Recordings were taken for 15 s prior to addition of compounds to establish baseline fluorescence then for 3–5 min after addition of agonists. Where indicated, cells were preincubated with antagonists for up to 30 min prior to addition of nucleotides. The relative change in intracellular Ca^{2+} (ΔCa^{2+}) was calculated as ratio of Fura-2 fluorescence following excitation at 340 nm and 380 nm ($F_{340/380}$) was determined and normalized to the mean basal fluorescence according to the formula (1):

$$\Delta Ca^{2+} \; = \; \frac{\Delta F}{F} \; = \; \frac{F - F_{rest}}{F_{rest}} \tag{1}$$

where F is the $F_{340/380}$ at any given time and F_{rest} is the mean fluorescence of the given well prior to the addition of nucleotides [88]. To investigate endogenous P2X and P2Y receptor-mediated Ca^{2+} responses in DH82 canine macrophages, both the peak Ca^{2+} response ($F_{340/380}$) and the net Ca^{2+} movement (calculated as area under the curve using GraphPad Prism) were calculated and, where indicated, used for fitting concentration-response curves fit to the Hill equation using the least squares method. Decay time (τ, time constant) was calculated from the peak $F_{340/380}$ using the nonlinear regression one phase decay model for GraphPad Prism. Decay times were not calculated where no response to agonists was recorded. Where antagonist IC_{50} was calculated against the approximate EC_{80} of ATP or UTP, responses were normalized to the response in absence of antagonist to allow data to be fit to a curve.

4.5. Immunocytochemistry and Confocal Microscopy

Cells were plated at 1×10^5 cells/18 mm glass coverslip in 24-well plates (Greiner Bio-One) and incubated at 37 °C/5% CO_2 overnight prior to use. Cells were fixed with 3% (*w/v*) paraformaldehyde at 4 °C for 15 min then washed three times with PBS. Cells were permeabilized with 0.1% (*w/v*) saponin resuspended in a blocking buffer (2% BSA (*w/v*) in PBS) at room temperature for 15 min and then incubated with anti-P2Y$_2$ (1:250; Alomone, cat no. APR-102, RRID: AB_2756769) or anti-P2X4 (1:250; Sigma-Aldrich cat no. SAB2500734, RRID:AB_10604119) primary antibody in 2% BSA/PBS at room temperature for 2 h. Cells were washed three times with PBS and then incubated with Alexa Fluor594-conjugated anti-rabbit (1:200; Abcam cat no. ab150080, RRID: AB_2650602), or Alexa Fluor594-conjugated anti-goat (1:200; Abcam cat no. ab150136, RRID: AB_2782994) secondary antibody in 2% BSA/PBS at room temperature for 60 min. Cells were washed three times with PBS and then incubated with secondary antibody in 2% BSA/PBS at room temperature for 60 min. Washed coverslips were mounted onto a glass slide using 50% glycerol in PBS and sealed with nail polish. Cells were visualized on a Leica (Mannheim, Germany) SP5 confocal microscope.

4.6. Data and Statistical Analysis

All data were analysed using GraphPad Prism 5. Half-maximal effective and inhibitory concentrations (EC_{50} and IC_{50}, respectively) are expressed as their negative logarithm (pEC_{50}/pIC_{50}) $\pm$ SEM. Data were compared using a two-tailed Student's t test or one-way ANOVA with Bonferroni post hoc test for single or multiple comparisons, respectively. Multiple comparisons involving two interdependent variables were analysed using a two-way ANOVA with Bonferroni post hoc test. Throughout this study $p < 0.05$ was considered statistically significant.

4.7. Nomenclature of Targets and Ligands

All targets and ligands used throughout this manuscript conform with the guidelines outlined by the International Union of Basic and Clinical Pharmacology and British Pharmacological Society (IUPHAR/BPS) Guide to Pharmacology [40,89].

Supplementary Materials: Supplementary materials can be found at http://www.mdpi.com/1422-0067/21/22/8572/s1.

Author Contributions: Conceptualization: R.A.S., L.O., R.S.; methodology: R.A.S.; formal analysis: R.A.S.; investigation: R.A.S., N.A.M.; resources: L.O., R.S.; data curation: R.A.S.; writing—original draft preparation: R.A.S.; writing—review and editing: R.A.S., N.A.M., L.O., R.S.; visualization: R.A.S.; supervision: L.O., R.S.; project administration: R.S.; funding acquisition: R.S. All authors have read and agreed to the published version of the manuscript.

Funding: This research was funded by the American Kennel Club Canine Health Foundation, grant number 01985. R.A.S. was supported through an Australian Government Research Training Program Scholarship. L.O. is supported by a National Health and Medical Research Council (NHMRC) of Australia Boosting Dementia Research Leadership Fellowship (APP1135720). R.S. is supported by Molecular Horizons (University of Wollongong, Wollongong).

Acknowledgments: We thank the staff of the Technical Services Unit of the Illawarra Health and Medical Research Institute for technical assistance.

Conflicts of Interest: The authors declare no conflict of interest. The funders had no role in the design of the study; in the collection, analyses, or interpretation of data; in the writing of the manuscript; or in the decision to publish the results.

Abbreviations

2-APB	2-Aminoethoxydiphenyl borate
2MeSADP	2-Methylthio-ADP
ADP	Adenosine 5′-diphosphate
ATP	Adenosine 5′-triphosphate
BAPTA-AM	1,2-*Bis*(2-aminophenoxy)ethane-*N*,*N*,*N*′,*N*′-tetraacetic acid tetrakis(acetoxymethyl ester)
BzATP	3′-*O*-(4-Benzoyl)benzoyl-ATP
CNS	Central nervous system
CCL2	C-C chemokine ligand 2
CCR2	C-C chemokine receptor 2
DMSO	Dimethyl sulfoxide
EC_{50}	Half maximal effective concentration
EGTA	Ethylene glycol tetraacetic acid
FBS	Fetal Bovine Serum
IC_{50}	Half-maximal inhibitory concentration
IL	Interleukin
IP_3	Inositol triphosphate
LPS	Lipopolysaccharide
MDCK	Madin Darby canine kidney
PLC	Phospholipase C
TNF	Tumor necrosis factor
TNP	2′,3′-*O*-(2,4,6-Trinitrophenyl)
UDP	Uridine 5′-diphosphate
UTP	Uridine 5′-triphosphate

References

1. Ulmann, L.; Hatcher, J.P.; Hughes, J.P.; Chaumont, S.; Green, P.J.; Conquet, F.; Buell, G.N.; Reeve, A.J.; Chessell, I.P.; Rassendren, F. Up-Regulation of P2X4 Receptors in Spinal Microglia after Peripheral Nerve Injury Mediates BDNF Release and Neuropathic Pain. *J. Neurosci.* **2008**, *28*, 11263–11268. [CrossRef] [PubMed]

2. Ulmann, L.; Hirbec, H.; Rassendren, F. P2X4 receptors mediate PGE_2 release by tissue-resident macrophages and initiate inflammatory pain. *EMBO J.* **2010**, *29*, 2290–2300. [CrossRef] [PubMed]

3. Zhu, H.; Yu, Y.; Zheng, L.; Wang, L.; Li, C.; Yu, J.; Wei, J.; Wang, C.; Zhang, J.; Xu, S.; et al. Chronic inflammatory pain upregulates expression of P2Y2 receptor in small-diameter sensory neurons. *Metab. Brain Dis.* **2015**, *30*, 1349–1358. [CrossRef] [PubMed]

4. Li, N.; Lu, Z.Y.; Yu, L.H.; Burnstock, G.; Deng, X.M.; Ma, B. Inhibition of G protein-coupled P2Y2 receptor induced analgesia in a rat model of trigeminal neuropathic pain. *Mol. Pain* **2014**, *10*, 21. [CrossRef]

5. Su, W.F.; Wu, F.; Jin, Z.H.; Gu, Y.; Chen, Y.T.; Fei, Y.; Chen, H.; Wang, Y.X.; Xing, L.Y.; Zhao, Y.Y.; et al. Overexpression of P2X4 receptor in Schwann cells promotes motor and sensory functional recovery and remyelination via BDNF secretion after nerve injury. *Glia* **2019**, *67*, 78–90. [CrossRef]

6. Zabala, A.; Vazquez-Villoldo, N.; Rissiek, B.; Gejo, J.; Martin, A.; Palomino, A.; Perez-Samartín, A.; Pulagam, K.R.; Lukowiak, M.; Capetillo-Zarate, E.; et al. P2X4 receptor controls microglia activation and favors remyelination in autoimmune encephalitis. *EMBO Mol. Med.* **2018**, *10*, e8743. [CrossRef]

7.   Burnstock, G. Pathophysiology and therapeutic potential of purinergic signaling. *Pharmacol. Rev.* **2006**, *58*, 58–86. [CrossRef]

8.   Stokes, L.; Surprenant, A. Purinergic $P2Y_2$ receptors induce increased MCP-1/CCL2 synthesis and release from rat alveolar and peritoneal macrophages. *J. Immunol.* **2007**, *179*, 6016–6023. [CrossRef]

9.   Layhadi, J.A.; Fountain, S.J. ATP-Evoked Intracellular Ca^{2+} Responses in M-CSF Differentiated Human Monocyte-Derived Macrophage are Mediated by P2X4 and $P2Y_{11}$ Receptor Activation. *Int. J. Mol. Sci.* **2019**, *20*, 5113. [CrossRef]

10. Higgins, K.R.; Kovacevic, W.; Stokes, L. Nucleotides regulate secretion of the inflammatory chemokine CCL2 from human macrophages and monocytes. *Mediators Inflamm.* **2014**, *2014*. [CrossRef]

11. Layhadi, J.A.; Turner, J.; Crossman, D.; Fountain, S.J. ATP Evokes Ca^{2+} Responses and CXCL5 Secretion via P2X4 Receptor Activation in Human Monocyte-Derived Macrophages. *J. Immunol.* **2018**, *200*, 1159–1168. [CrossRef] [PubMed]

12. Bowler, J.W.; Bailey, R.J.; North, R.A.; Surprenant, A. P2X4, $P2Y_1$ and $P2Y_2$ receptors on rat alveolar macrophages. *Br. J. Pharmacol.* **2003**, *140*, 567–575. [CrossRef] [PubMed]

13. Stokes, L.; Surprenant, A. Dynamic regulation of the P2X4 receptor in alveolar macrophages by phagocytosis and classical activation. *Eur. J. Immunol.* **2009**, *39*, 986–995. [CrossRef] [PubMed]

14. Vargas-Martínez, E.M.; Gómez-Coronado, K.S.; Espinosa-Luna, R.; Valdez-Morales, E.E.; Barrios-García, T.; Barajas-Espinosa, A.; Ochoa-Cortes, F.; Montaño, L.M.; Barajas-López, C.; Guerrero-Alba, R. Functional expression of P2X1, P2X4 and P2X7 purinergic receptors in human monocyte-derived macrophages. *Eur. J. Pharmacol.* **2020**, *888*, 173460. [CrossRef] [PubMed]

15. Chessell, I.P.; Hatcher, J.P.; Bountra, C.; Michel, A.D.; Hughes, J.P.; Green, P.; Egerton, J.; Murfin, M.; Richardson, J.; Peck, W.L.; et al. Disruption of the P2X7 purinoceptor gene abolishes chronic inflammatory and neuropathic pain. *Pain* **2005**, *114*, 386–396. [CrossRef] [PubMed]

16. Tsuda, M.; Kuboyama, K.; Inoue, T.; Nagata, K.; Tozaki-Saitoh, H.; Inoue, K. Behavioral phenotypes of mice lacking purinergic P2X4 receptors in acute and chronic pain assays. *Mol. Pain* **2009**, *5*, 28. [CrossRef]

17. Trang, T.; Beggs, S.; Wan, X.; Salter, M.W. P2X4-receptor-mediated synthesis and release of brain-derived neurotrophic factor in microglia is dependent on calcium and p38-mitogen-activated protein kinase activation. *J. Neurosci.* **2009**, *29*. [CrossRef] [PubMed]

18. Tsuda, M.; Shigemoto-Mogami, Y.; Koizumi, S.; Mizokoshi, A.; Kohsaka, S.; Salter, M.W.; Inoue, K. P2X4 receptors induced in spinal microglia gate tactile allodynia after nerve injury. *Nature* **2003**, *424*. [CrossRef]

19. Wellman, M.L.; Krakowka, S.; Jacobs, R.M.; Kociba, G.J. A macrophage-monocyte cell line from a dog with malignant histiocytosis. *In Vitro Cell. Dev. Biol.* **1988**, *24*, 223–229. [CrossRef]

20. Herrmann, I.; Gotovina, J.; Fazekas-Singer, J.; Fischer, M.B.; Hufnagl, K.; Bianchini, R.; Jensen-Jarolim, E. Canine macrophages can like human macrophages be in vitro activated toward the M2a subtype relevant in allergy. *Dev. Comp. Immunol.* **2018**, *82*, 118–127. [CrossRef]

21. Heinrich, F.; Contioso, V.B.; Stein, V.M.; Carlson, R.; Tipold, A.; Ulrich, R.; Puff, C.; Baumgärtner, W.; Spitzbarth, I. Passage-dependent morphological and phenotypical changes of a canine histiocytic sarcoma cell line (DH82 cells). *Vet. Immunol. Immunopathol.* **2015**, *163*, 86–92. [CrossRef] [PubMed]

22. Barnes, A.; Bee, A.; Bell, S.; Gilmore, W.; Mee, A.; Morris, R.; Carter, S.D. Immunological and inflammatory characterisation of three canine cell lines: K1, K6 and DH82. *Vet. Immunol Immunopathol.* **2000**, *75*, 9–25. [CrossRef]

23. Mendonça, P.H.B.; da Rocha, R.F.D.B.; Moraes, J.B.d.B.; LaRocque-de-Freitas, I.F.; Logullo, J.; Morrot, A.; Nunes, M.P.; Freire-de-Lima, C.G.; Decote-Ricardo, D. Canine Macrophage DH82 Cell Line As a Model to Study Susceptibility to Trypanosoma cruzi Infection. *Front. Immunol.* **2017**, *8*, 604. [CrossRef] [PubMed]

24. Armando, F.; Gambini, M.; Corradi, A.; Giudice, C.; Pfankuche, V.M.; Brogden, G.; Attig, F.; von Kockritz-Blickwede, M.; Baumgartner, W.; Puff, C. Oxidative Stress in Canine Histiocytic Sarcoma Cells Induced by an Infection with Canine Distemper Virus Led to a Dysregulation of HIF-1alpha Downstream Pathway Resulting in a Reduced Expression of VEGF-B in vitro. *Viruses* **2020**, *12*, 200. [CrossRef]

25. Nadaes, N.R.; Silva da Costa, L.; Santana, R.C.; LaRocque-de-Freitas, I.F.; Vivarini, Á.C.; Soares, D.C.; Wardini, A.B.; Gazos Lopes, U.; Saraiva, E.M.; Freire-de-Lima, C.G.; et al. DH82 Canine and RAW264.7 Murine Macrophage Cell Lines Display Distinct Activation Profiles Upon Interaction with Leishmania infantum and Leishmania amazonensis. *Front. Cell Infect. Microbiol.* **2020**, *10*, 247. [CrossRef]

26. Fujimoto, Y.; Nakatani, N.; Kubo, T.; Semi, Y.; Yoshida, N.; Nakajima, H.; Iseri, T.; Azuma, Y.T.; Takeuchi, T. Adenosine and ATP affect LPS-induced cytokine production in canine macrophage cell line DH82 cells. *J. Vet. Med. Sci.* **2012**, *74*, 27–34. [CrossRef]

27. Zheng, X.; Zhu, Y.; Zhao, Z.; Yan, L.; Xu, T.; Wang, X.; He, H.; Xia, X.; Zheng, W.; Xue, X. RNA sequencing analyses of gene expressions in a canine macrophages cell line DH82 infected with canine distemper virus. *Infect. Genet. Evol.* **2020**, *80*, 104206. [CrossRef]

28. Sluyter, R. The P2X7 Receptor. *Adv. Exp. Med. Biol.* **2017**, *1051*, 17–53.

29. Stevenson, R.O.; Taylor, R.M.; Wiley, J.S.; Sluyter, R. The P2X7 receptor mediates the uptake of organic cations in canine erythrocytes and mononuclear leukocytes: Comparison to equivalent human cell types. *Purinergic Signal.* **2009**, *5*, 385–394. [CrossRef]

30. Jalilian, I.; Peranec, M.; Curtis, B.L.; Seavers, A.; Spildrejorde, M.; Sluyter, V.; Sluyter, R. Activation of the damage-associated molecular pattern receptor P2X7 induces interleukin-1 beta release from canine monocytes. *Vet. Immunol. Immunopathol.* **2012**, *149*, 86–91. [CrossRef]

31. Ase, A.R.; Honson, N.S.; Zaghdane, H.; Pfeifer, T.A.; Seguela, P. Identification and characterization of a selective allosteric antagonist of human P2X4 receptor channels. *Mol. Pharmacol.* **2015**, *87*, 606–616. [CrossRef] [PubMed]

32. Honore, P.; Donnelly-Roberts, D.; Namovic, M.; Zhong, C.M.; Wade, C.; Chandran, P.; Zhu, C.; Carroll, W.; Perez-Medrano, A.; Iwakura, Y.; et al. The antihyperalgesic activity of a selective P2X7 receptor antagonist, A-839977, is lost in IL-1 alpha beta knockout mice. *Behav. Brain Res.* **2009**, *204*, 77–81. [CrossRef] [PubMed]

33. Humphreys, B.D.; Virginio, C.; Surprenant, A.; Rice, J.; Dubyak, G.R. Isoquinolines as antagonists of the P2X7 nucleotide receptor: High selectivity for the human versus rat receptor homologues. *Mol. Pharmacol.* **1998**, *54*, 22–32. [CrossRef] [PubMed]

34. Layhadi, J.A.; Fountain, S.J. P2X4 Receptor-Dependent Ca^{2+} Influx in Model Human Monocytes and Macrophages. *Int. J. Mol. Sci.* **2017**, *18*, 2261. [CrossRef] [PubMed]

35. Micklewright, J.J.; Layhadi, J.A.; Fountain, S.J. P2Y$_{12}$ receptor modulation of ADP-evoked intracellular Ca^{2+} signalling in THP-1 human monocytic cells. *Br. J. Pharmacol.* **2018**, *175*, 2483–2491. [CrossRef] [PubMed]

36. Zhang, Z.; Hao, K.; Li, H.; Lu, R.; Liu, C.; Zhou, M.; Li, B.; Meng, Z.; Hu, Q.; Jiang, C. Design, synthesis and anti-inflammatory evaluation of 3-amide benzoic acid derivatives as novel P2Y$_{14}$ receptor antagonists. *Eur. J. Med. Chem.* **2019**, *181*, 111564. [CrossRef]

37. Tu, Y.M.; Gong, C.X.; Ding, L.; Liu, X.Z.; Li, T.; Hu, F.F.; Wang, S.; Xiong, C.P.; Liang, S.D.; Xu, H. A high concentration of fatty acids induces TNF-alpha as well as NO release mediated by the P2X4 receptor, and the protective effects of puerarin in RAW264.7 cells. *Food Funct.* **2017**, *8*, 4336–4346. [CrossRef]

38. Gadeock, S.; Tran, J.; Georgiou, J.G.; Jalilian, I.; Taylor, R.M.; Wiley, J.S.; Sluyter, R. TGF-beta 1 prevents up-regulation of the P2X7 receptor by IFN-gamma and LPS in leukemic THP-1 monocytes. *Biochim. Biophys. Acta* **2010**, *1798*, 2058–2066. [CrossRef]

39. Jacobson, K.A.; Delicado, E.G.; Gachet, C.; Kennedy, C.; von Kügelgen, I.; Li, B.; Miras-Portugal, M.T.; Novak, I.; Schöneberg, T.; Perez-Sen, R.; et al. Update of P2Y Receptor Pharmacology: IUPHAR Review:27. *Br. J. Pharmacol.* **2020**, *177*, 2413–2433. [CrossRef]

40. Alexander, S.P.H.; Mathie, A.; Peters, J.A.; Veale, E.L.; Striessnig, J.; Kelly, E.; Armstrong, J.F.; Faccenda, E.; Harding, S.D.; Pawson, A.J.; et al. The Concise Guide to Pharmacology 2019/20: Ion channels. *Br. J. Pharmacol* **2019**, *176* (Suppl. 1), S142–S228. [CrossRef]

41. Sophocleous, R.A.; Berg, T.; Finol-Urdaneta, R.K.; Sluyter, V.; Keshiya, S.; Bell, L.; Curtis, S.J.; Curtis, B.L.; Seavers, A.; Bartlett, R.; et al. Pharmacological and genetic characterisation of the canine P2X4 receptor. *Br. J. Pharmacol.* **2020**, *177*, 2812–2829. [CrossRef] [PubMed]

42. Virginio, C.; Robertson, G.; Surprenant, A.; North, R.A. Trinitrophenyl-Substituted Nucleotides Are Potent Antagonists Selective for P2X1, P2X3, and Heteromeric P2X2/3 Receptors. *Mol. Pharmacol.* **1998**, *53*, 969. [PubMed]

43. Nagata, K.; Imai, T.; Yamashita, T.; Tsuda, M.; Tozaki-Saitoh, H.; Inoue, K. Antidepressants inhibit P2X4 Receptor function: A possible involvement in neuropathic pain relief. *Mol. Pain* **2009**, *5*, 20. [CrossRef]

44. Abdelrahman, A.; Namasivayam, V.; Hinz, S.; Schiedel, A.C.; Kose, M.; Burton, M.; El-Tayeb, A.; Gillard, M.; Bajorath, J.; de Ryck, M.; et al. Characterization of P2X4 receptor agonists and antagonists by calcium influx and radioligand binding studies. *Biochem. Pharmacol.* **2017**, *125*, 41–54. [CrossRef] [PubMed]

45. Wang, W.; Xiang, Z.-H.; Jiang, C.-L.; Liu, W.-Z.; Shang, Z.-L. Effects of antidepressants on P2X7 receptors. *Psychiatry Res.* **2016**, *242*, 281–287. [CrossRef]

46. Dao-Ung, P.; Skarratt, K.K.; Fuller, S.J.; Stokes, L. Paroxetine suppresses recombinant human P2X7 responses. *Purinergic Signal.* **2015**, *11*, 481–490. [CrossRef]

47. Abbracchio, M.P.; Burnstock, G.; Boeynaems, J.M.; Barnard, E.A.; Boyer, J.L.; Kennedy, C.; Knight, G.E.; Fumagalli, M.; Gachet, C.; Jacobson, K.A.; et al. International Union of Pharmacology LVIII: Update on the P2Y G protein-coupled nucleotide receptors: From molecular mechanisms and pathophysiology to therapy. *Pharmacol. Rev.* **2006**, *58*, 281–341. [CrossRef]

48. Thastrup, O.; Cullen, P.J.; Drøbak, B.K.; Hanley, M.R.; Dawson, A.P. Thapsigargin, a tumor promoter, discharges intracellular Ca^{2+} stores by specific inhibition of the endoplasmic reticulum Ca^{2+}-ATPase. *Proc. Natl. Acad. Sci. USA* **1990**, *87*, 2466–2470. [CrossRef]

49. Zambon, A.C.; Hughes, R.J.; Meszaros, J.G.; Wu, J.J.; Torres, B.; Brunton, L.L.; Insel, P.A. $P2Y_2$ receptor of MDCK cells: Cloning, expression, and cell-specific signaling. *Am. J. Physiol. Renal Physiol.* **2000**, *279*, F1045–F1052. [CrossRef]

50. Kempson, S.A.; Edwards, J.M.; Osborn, A.; Sturek, M. Acute inhibition of the betaine transporter by ATP and adenosine in renal MDCK cells. *Am. J. Physiol. Renal Physiol.* **2008**, *295*, F108–F117. [CrossRef]

51. Charlton, S.J.; Brown, C.A.; Weisman, G.A.; Turner, J.T.; Erb, L.; Boarder, M.R. Cloned and transfected $P2Y_4$ receptors: Characterization of a suramin and PPADS-insensitive response to UTP. *Br. J. Pharmacol.* **1996**, *119*, 1301–1303. [CrossRef] [PubMed]

52. Rafehi, M.; Burbiel, J.C.; Attah, I.Y.; Abdelrahman, A.; Müller, C.E. Synthesis, characterization, and in vitro evaluation of the selective $P2Y_2$ receptor antagonist AR-C118925. *Purinergic Signal.* **2017**, *13*, 89–103. [CrossRef] [PubMed]

53. Wolff, S.C.; Qi, A.D.; Harden, T.K.; Nicholas, R.A. Polarized expression of human P2Y receptors in epithelial cells from kidney, lung, and colon. *Am. J. Physiol. Cell Physiol.* **2005**, *288*, C624–C632. [CrossRef] [PubMed]

54. Qi, A.D.; Wolff, S.C.; Nicholas, R.A. The apical targeting signal of the $P2Y_2$ receptor is located in its first extracellular loop. *J. Biol. Chem.* **2005**, *280*, 29169–29175. [CrossRef]

55. Xue, X.; Zhu, Y.; Yan, L.; Wong, G.; Sun, P.; Zheng, X.; Xia, X. Antiviral efficacy of favipiravir against canine distemper virus infection in vitro. *BMC Vet. Res.* **2019**, *15*, 316. [CrossRef]

56. Pfankuche, V.M.; Sayed-Ahmed, M.; Contioso, V.B.; Spitzbarth, I.; Rohn, K.; Ulrich, R.; Deschl, U.; Kalkuhl, A.; Baumgärtner, W.; Puff, C. Persistent Morbillivirus Infection Leads to Altered Cortactin Distribution in Histiocytic Sarcoma Cells with Decreased Cellular Migration Capacity. *PLoS ONE* **2016**, *11*, e0167517. [CrossRef]

57. Armando, F.; Gambini, M.; Corradi, A.; Becker, K.; Marek, K.; Pfankuche, V.M.; Mergani, A.E.; Brogden, G.; de Buhr, N.; von Köckritz-Blickwede, M.; et al. Mesenchymal to epithelial transition driven by canine distemper virus infection of canine histiocytic sarcoma cells contributes to a reduced cell motility in vitro. *J. Cell Mol. Med.* **2020**, *24*, 9332–9348. [CrossRef]

58. Balazs, B.; Danko, T.; Kovacs, G.; Koles, L.; Hediger, M.A.; Zsembery, A. Investigation of the inhibitory effects of the benzodiazepine derivative, 5-BDBD on P2X4 purinergic receptors by two complementary methods. *Cell Physiol. Biochem.* **2013**, *32*, 11–24. [CrossRef]

59. Stokes, L.; Bidula, S.; Bibič, L.; Allum, E. To Inhibit or Enhance? Is There a Benefit to Positive Allosteric Modulation of P2X Receptors? *Front. Pharmacol.* **2020**, *11*, 627. [CrossRef]

60. von Kugelgen, I. Pharmacology of P2Y receptors. *Brain Res. Bull.* **2019**. [CrossRef]

61. Jan, C.R.; Ho, C.M.; Wu, S.N.; Tseng, C.J. Mechanisms of rise and decay of ADP-evoked calcium signal in MDCK cells. *Chin. J. Physiol.* **1998**, *41*, 67–73. [PubMed]

62. Katzur, A.C.; Koshimizu, T.-A.; Tomic, M.; Schultze-Mosgau, A.; Ortmann, O.; Stojilkovic, S.S. Expression and Responsiveness of $P2Y_2$ Receptors in Human Endometrial Cancer Cell Lines. *J. Clin. Endocrinol. Metab.* **1999**, *84*, 4085–4091. [CrossRef] [PubMed]

63. Wildman, S.S.; Unwin, R.J.; King, B.F. Extended pharmacological profiles of rat $P2Y_2$ and rat $P2Y_4$ receptors and their sensitivity to extracellular H^+ and Zn^{2+} ions. *Br. J. Pharmacol.* **2003**, *140*, 1177–1186. [CrossRef] [PubMed]

64. Maier, T.; Güell, M.; Serrano, L. Correlation of mRNA and protein in complex biological samples. *FEBS Lett.* **2009**, *583*, 3966–3973. [CrossRef] [PubMed]

65. Post, S.R.; Rump, L.C.; Zambon, A.; Hughes, R.J.; Buda, M.D.; Jacobson, J.P.; Kao, C.C.; Insel, P.A. ATP activates cAMP production via multiple purinergic receptors in MDCK-D1 epithelial cells. Blockade of an autocrine/paracrine pathway to define receptor preference of an agonist. *J. Biol. Chem.* **1998**, *273*, 23093–23097. [CrossRef] [PubMed]

66. Qi, A.D.; Zambon, A.C.; Insel, P.A.; Nicholas, R.A. An arginine/glutamine difference at the juxtaposition of transmembrane domain 6 and the third extracellular loop contributes to the markedly different nucleotide selectivities of human and canine P2Y$_{11}$ receptors. *Mol. Pharmacol.* **2001**, *60*, 1375–1382. [CrossRef]

67. Communi, D.; Robaye, B.; Boeynaems, J.M. Pharmacological characterization of the human P2Y$_{11}$ receptor. *Br. J. Pharmacol* **1999**, *128*, 1199–1206. [CrossRef]

68. Hughes, R.J.; Torres, B.; Zambon, A.; Arthur, D.; Bohmann, C.; Rump, L.C.; Insel, P.A. Expression of multiple P2Y receptors by MDCK-D1 cells: P2Y$_1$ receptor cloning and signaling. *Drug Dev. Res.* **2003**, *59*, 1–7. [CrossRef]

69. Nicholas, R.A.; Watt, W.C.; Lazarowski, E.R.; Li, Q.; Harden, K. Uridine nucleotide selectivity of three phospholipase C-activating P2 receptors: Identification of a UDP-selective, a UTP-selective, and an ATP- and UTP-specific receptor. *Mol. Pharmacol.* **1996**, *50*, 224–229.

70. Communi, D.; Parmentier, M.; Boeynaems, J.M. Cloning, functional expression and tissue distribution of the human P2Y$_6$ receptor. *Biochem. Biophys. Res. Commun.* **1996**, *222*, 303–308. [CrossRef]

71. Insel, P.A.; Ostrom, R.S.; Zambon, A.C.; Hughes, R.J.; Balboa, M.A.; Shehnaz, D.; Gregorian, C.; Torres, B.; Firestein, B.L.; Xing, M.; et al. P2Y receptors of MDCK cells: Epithelial cell regulation by extracellular nucleotides. *Clin. Exp. Pharmacol. Physiol.* **2001**, *28*, 351–354. [CrossRef] [PubMed]

72. Liu, B.; Cao, W.; Li, J.; Liu, J. Lysosomal exocytosis of ATP is coupled to P2Y$_2$ receptor in marginal cells in the stria vascular in neonatal rats. *Cell Calcium* **2018**, *76*, 62–71. [CrossRef] [PubMed]

73. Jones, C.A.; Chessell, I.P.; Simon, J.; Barnard, E.A.; Miller, K.J.; Michel, A.D.; Humphrey, P.P.A. Functional characterization of the P2X4 receptor orthologues. *Br. J. Pharmacol.* **2000**, *129*, 388–394. [CrossRef] [PubMed]

74. Muzzachi, S.; Blasi, A.; Ciani, E.; Favia, M.; Cardone, R.A.; Marzulli, D.; Reshkin, S.J.; Merizzi, G.; Casavola, V.; Soleti, A.; et al. MED1101: A new dialdehydic compound regulating P2x7 receptor cell surface expression in U937 cells. *Biol. Cell* **2013**, *105*, 399–413. [CrossRef] [PubMed]

75. Bernier, L.P.; Ase, A.R.; Seguela, P. P2X receptor channels in chronic pain pathways. *Br. J. Pharmacol.* **2018**, *175*, 2219–2230. [CrossRef] [PubMed]

76. Inoue, K. Role of the P2X4 receptor in neuropathic pain. *Curr. Opin. Pharmacol.* **2019**, *47*, 33–39. [CrossRef]

77. Domercq, M.; Matute, C. Targeting P2X4 and P2X7 receptors in multiple sclerosis. *Curr. Opin. Pharmacol.* **2019**, *47*, 119–125. [CrossRef]

78. Qureshi, O.S.; Paramasivam, A.; Yu, J.C.; Murrell-Lagnado, R.D. Regulation of P2X4 receptors by lysosomal targeting, glycan protection and exocytosis. *J. Cell Sci.* **2007**, *120*, 3838–3849. [CrossRef]

79. Sivaramakrishnan, V.; Bidula, S.; Campwala, H.; Katikaneni, D.; Fountain, S.J. Constitutive lysosome exocytosis releases ATP and engages P2Y receptors in human monocytes. *J. Cell Sci.* **2012**, *125*, 4567–4575. [CrossRef]

80. Toyomitsu, E.; Tsuda, M.; Yamashita, T.; Tozaki-Saitoh, H.; Tanaka, Y.; Inoue, K. CCL2 promotes P2X4 receptor trafficking to the cell surface of microglia. *Purinergic Signal.* **2012**, *8*, 301–310. [CrossRef]

81. Greenhalgh, A.D.; Zarruk, J.G.; Healy, L.M.; Baskar Jesudasan, S.J.; Jhelum, P.; Salmon, C.K.; Formanek, A.; Russo, M.V.; Antel, J.P.; McGavern, D.B.; et al. Peripherally derived macrophages modulate microglial function to reduce inflammation after CNS injury. *PLoS Biol.* **2018**, *16*, e2005264. [CrossRef] [PubMed]

82. Jagannathan, V.; Drogemuller, C.; Leeb, T. A comprehensive biomedical variant catalogue based on whole genome sequences of 582 dogs and eight wolves. *Anim. Genet.* **2019**, *50*, 695–704. [CrossRef] [PubMed]

83. Büscher, R.; Hoerning, A.; Patel, H.H.; Zhang, S.; Arthur, D.B.; Grasemann, H.; Ratjen, F.; Insel, P.A. P2Y$_2$ receptor polymorphisms and haplotypes in cystic fibrosis and their impact on Ca^{2+} influx. *Pharmacogenet. Genom.* **2006**, *16*, 199–205. [CrossRef]

84. Stokes, L.; Scurrah, K.; Ellis, J.A.; Cromer, B.A.; Skarratt, K.K.; Gu, B.J.; Harrap, S.B.; Wiley, J.S. A loss-of-function polymorphism in the human P2X4 receptor is associated with increased pulse pressure. *Hypertension* **2011**, *58*, 1086–1092. [CrossRef]

85. Sophocleous, R.A.; Sluyter, V.; Curtis, B.L.; Curtis, S.J.; Jurak, L.M.; Faulks, M.; Spildrejorde, M.; Gates, S.; Proctor, E.J.; Seavers, A.; et al. Association of a P2RX7 gene missense variant with brachycephalic dog breeds. *Anim. Genet.* **2020**, *51*, 127–131. [CrossRef]

86. Spildrejorde, M.; Bartlett, R.; Stokes, L.; Jalilian, I.; Peranec, M.; Sluyter, V.; Curtis, B.L.; Skarratt, K.K.; Skora, A.; Bakhsh, T.; et al. R270C polymorphism leads to loss of function of the canine P2X7 receptor. *Physiol. Genom.* **2014**, *46*, 512–522. [CrossRef]
87. Rueden, C.T.; Schindelin, J.; Hiner, M.C.; DeZonia, B.E.; Walter, A.E.; Arena, E.T.; Eliceiri, K.W. ImageJ2: ImageJ for the next generation of scientific image data. *BMC Bioinform.* **2017**, *18*, 529. [CrossRef]
88. Paredes, R.M.; Etzler, J.C.; Watts, L.T.; Lechleiter, J.D. Chemical Calcium Indicators. *Methods (San Diego, Calif.)* **2008**, *46*, 143–151. [CrossRef]
89. Alexander, S.P.H.; Christopoulos, A.; Davenport, A.P.; Kelly, E.; Mathie, A.; Peters, J.A.; Veale, E.L.; Armstrong, J.F.; Faccenda, E.; Harding, S.D.; et al. The Concise Guide to Pharmacology 2019/20: G protein-coupled receptors. *Br. J. Pharmacol.* **2019**, *176*, S21–S141. [CrossRef]

International Journal of
Molecular Sciences

Review

Purinergic Signaling in Endometriosis-Associated Pain

Carla Trapero [1,2] and Mireia Martín-Satué [1,2,*]

[1] Departament de Patologia i Terapèutica Experimental, Facultat de Medicina i Ciències de la Salut, Campus Bellvitge, Universitat de Barcelona, 08907 Barcelona, Spain; ctrapero@ub.edu
[2] Institut d'Investigació Biomèdica de Bellvitge (IDIBELL), Oncobell Program, CIBERONC, 08908 Barcelona, Spain
*   Correspondence: martinsatue@ub.edu

Received: 9 October 2020; Accepted: 10 November 2020; Published: 12 November 2020

Abstract: Endometriosis is an estrogen-dependent gynecological disease, with an associated chronic inflammatory component, characterized by the presence of endometrial tissue outside the uterine cavity. Its predominant symptom is pain, a condition notably altering the quality of life of women with the disease. This review is intended to exhaustively gather current knowledge on purinergic signaling in endometriosis-associated pain. Altered extracellular ATP hydrolysis, due to changes in ectonucleotidase activity, has been reported in endometriosis; the resulting accumulation of ATP in the endometriotic microenvironment points to sustained activation of nucleotide receptors (P2 receptors) capable of generating a persistent pain message. P2X3 receptor, expressed in sensory neurons, mediates nociceptive, neuropathic, and inflammatory pain, and is enrolled in endometriosis-related pain. Pharmacological inhibition of P2X3 receptor is under evaluation as a pain relief treatment for women with endometriosis. The role of other ATP receptors is also discussed here, e.g., P2X4 and P2X7 receptors, which are involved in inflammatory cell–nerve and microglia–nerve crosstalk, and therefore in inflammatory and neuropathic pain. Adenosine receptors (P1 receptors), by contrast, mainly play antinociceptive and anti-inflammatory roles. Purinome-targeted drugs, including nucleotide receptors and metabolizing enzymes, are potential non-hormonal therapeutic tools for the pharmacological management of endometriosis-related pain.

Keywords: endometriosis; ATP; adenosine; P2Y; P2X; ectonucleotidases; pain; inflammation; endometrium; CD73; CD39

---

1. Introduction

Endometriosis is an estrogen-dependent gynecological disease characterized by the presence of endometrial tissue, glands, and stroma outside the uterine cavity, including the ovaries, pelvic peritoneum, and gastrointestinal tract, among other locations. In fact, depending on the location and characteristics, there are three different subtypes of endometriosis: peritoneal, ovarian, and deep infiltrating endometriosis [1]. It is estimated that this debilitating disease affects around 10% of women of reproductive age, but the true prevalence is difficult to quantify due to its unspecific symptoms and the lack of non-invasive diagnostic techniques that complicates diagnosis and can sometimes lead to misdiagnoses. Moreover, around 20–25% of women remain asymptomatic [2–4]. Even though pathogenesis of endometriosis is uncertain, the most widely accepted theory is the retrograde menstruation, described by Sampson, that includes three events that have to occur for endometriosis to develop: (i) menstrual blood flow backwards into the pelvic cavity (retrograde menstruation); (ii) the presence of available cells in menstrual debris; and (iii) the ectopic establishment and growth of endometrial cells [5]. More research on endometriosis is needed to fill the gaps in our understanding of its pathogenesis so as to help develop new diagnostic and therapeutic tools.

Associated with endometriosis is a chronic inflammatory component necessary for the establishment and progression of endometriotic lesions and which is related with its main symptoms and signs [6–9]. The main clinical features defining this disorder are pain and infertility, with the most common symptoms being cyclic pelvic pain, dysmenorrhea (painful menstrual periods), dyspareunia (painful sexual relations), dysuria (painful urination), and dyschezia (painful defecation) [10]. Different types of pain have been described in endometriosis: nociceptive, inflammatory, neuropathic, and a mixture of all of them. Interestingly, the predominant type of pain, its intensity, and its cyclicity vary between patients without a clear correlation between the scope of endometriotic lesions and pain experience [11]. This turns endometriosis-associated pain into a complex symptom that is difficult to manage.

Current treatments, both pharmacological and surgical, are addressed to providing symptom relief and are mainly focused on complex pain management, without effective results for all patients [11]. Several studies are attempting to overcome these obstacles by identifying molecular targets to develop new therapeutic approaches to improve the quality of life of affected women.

In recent decades, a large body of data has been published on the role of purinergic signaling in different inflammatory pathologies and its possible use as therapeutic target. The purinergic system is the extracellular signaling with biological effects mediated by nucleotides, such as adenosine triphosphate (ATP), and nucleosides, such as adenosine, involved in a wide range of physiological and pathological inflammatory conditions and in pain generation and transmission [12]. The purinergic signaling complex of a cell is sometimes referred to as the purinome [13]. In recent years, several studies have shown the involvement of purinome elements in endometriosis, but few studies have assessed their role as therapeutic targets for the endometriosis-derived chronic pain.

In this review, we aim to highlight the role of purinergic signaling in the pathogenesis and pathophysiology of endometriosis-associated pain. This information may be useful in presenting the molecular mechanisms underlying endometriosis-associated pain and toward the development of novel pharmacological approaches for endometriosis treatment.

2. Overview of Purinergic Signaling

Extracellular ATP and adenosine are the main purinergic mediators, with multiple roles in physiology and pathophysiology. The release of endogenous nucleotides and nucleosides into extracellular space by different cells in response to cell injury, necrosis, apoptosis, or various mechanical and chemical stimuli represents the beginning of the purinergic signaling cascade, which eventually induces an inflammatory response [12,14].

Under physiological conditions, extracellular ATP concentrations are low (submicromolar levels). However, with the release of endogenous ATP under situations such as inflammation, there is a marked increase of these levels [15].

To avoid sustained ATP signaling and the adverse effects of increased extracellular ATP levels, ectonucleotidases act extracellularly by degrading ATP into adenosine. Ectonucleotidases are specialized nucleotide-hydrolyzing enzymes, broadly expressed at the cell surface of many tissues, which, acting alone or sequentially, control nucleotide and nucleosides levels in the extracellular milieu (Figure 1). Four families of ectonucleotidase have been described: (i) the ectonucleoside triphosphate diphosphohydrolase (ENTPDase) family (also known as CD39 family), with NTPDase1 (CD39), -2, -3, and -8 as plasma membrane-bound members, which hydrolyze extracellular ATP to adenosine diphosphate (ADP), and ADP to adenosine monophosphate (AMP); (ii) the ectonucleotide pyrophosphatase/phosphodiesterase (ENPP) family, which converts ATP to AMP and inorganic pyrophosphate (PPi); (iii) the 5′-nucleotidase family, with only one membrane-bound member, the ecto-5′-nucleotidase, also known as CD73, which hydrolyzes AMP to adenosine; and (iv) the alkaline phosphatase (ALP) family, able to hydrolyze adenine nucleotides and pyrophosphate, releasing inorganic phosphate (Pi) [12,16].

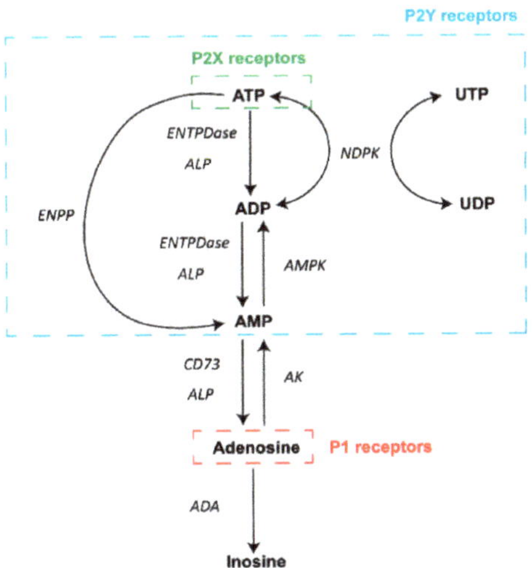

Figure 1. Diagram of the main elements of purinergic signaling. In the extracellular milieu, adenosine and nucleotides can activate P1 and P2 receptors on the surface of a wide variety of cell types. A number of different nucleotides can activate P2Y receptors (blue box), and ATP can also activate P2X receptors (green box). In contrast, adenosine actions involve the activation of P1 receptors (red box). Different enzymes are involved in the metabolism of adenosine and ATP in the process of achieving transient signaling. Abbreviations: adenosine triphosphate, ATP; adenosine diphosphate, ADP; adenosine monophosphate, AMP; uridine triphosphate, UTP; uridine diphosphate, UDP; ectonucleoside triphosphate diphosphohydrolase family, ENTPDase; alkaline phosphatase family, ALP; ectonucleotide pyrophosphatase/phosphodiesterase family, ENPP; ecto-5′-nucleotidase, CD73; nucleoside diphosphate kinase, NDPK; AMP-activated protein kinase, AMPK; adenosine kinase, AK; adenosine deaminase, ADA.

Adenosine also has important biological functions. In general, adenosine has been linked mainly to an anti-inflammatory effect. There are two enzymes responsible for adenosine metabolism: adenosine deaminase (ADA) and adenosine kinase (AK). ADA is a cytoplasmic enzyme but also an ectoenzyme that regulates intra- and extracellular adenosine levels, catalyzing the adenosine deamination yielding inosine [12]. As ectoenzyme, ADA is expressed as a soluble form or as membrane-associated enzyme-forming complexes with CD26/dipeptidyl peptidase IV in humans. AK is a cytosolic enzyme that catalyzes the phosphorylation of intracellular adenosine to AMP [12].

Once in the extracellular microenvironment, adenosine and nucleotides can activate two different purinergic receptor families, P1 and P2, respectively, which are widely and differentially expressed in the surface of most cells. Adenosine binds to P1 receptors, which are G protein coupled receptors classified into four subtypes: A_1, A_{2A}, A_{2B}, and A_3 [17]. These receptors mainly act via adenylate cyclase (AC) activity, modulating cyclic AMP (cAMP) production. The stimulation of the A_{2A} and A_{2B} receptors induces activation of AC, increasing second-messenger cAMP levels. In contrast, the activation of the A_1 and A_3 receptors inhibits AC, causing a decrease in cAMP production [17]. Furthermore, A_3 and A_{2B} receptors lead to the activation of phospholipase C (PLC) and an increase in intracellular calcium levels [18,19]. Moreover, A_1 receptor is involved in the opening of K^+ channels [18,19]. Cation mobilization triggered by adenosine receptors leads to transmission and modulation of pain. In addition, adenosine receptors stimulate mitogen-activated protein kinases (MAPK), regulating growth and proliferation, apoptosis, necrosis, and inflammation, essential for the development of endometriosis [20].

In addition, P2 receptors are nucleotide-selective and include P2X and P2Y receptor subtypes. The ionotropic P2X receptors, comprising seven subtypes (P2X1–P2X7), are ligand-gate ion channels strictly activated by extracellular ATP to mediate K^+ efflux, and Na^+ and Ca^{2+} influx, which mainly mediate short-term (acute) purinergic signaling, and which play an important role as mediators of fast excitatory neurotransmission in the central and peripheral nervous system [17,21]. The metabotropic P2Y receptors are G protein-coupled receptors that activate PLC-β ($P2Y_1$, $P2Y_2$, $P2Y_4$, $P2Y_6$, and $P2Y_{11}$) or inhibit AC ($P2Y_{11-14}$). The P2Y receptors have distinctive nucleotide preferences and, based on this, they can be divided in three groups: (i) adenine nucleotide-preferring receptors, mainly responding to ATP ($P2Y_1$, $P2Y_2$, $P2Y_4$, and $P2Y_{11}$) and ADP ($P2Y_1$, $P2Y_{12}$, $P2Y_{13}$); (ii) uracil nucleotide-preferring receptors, activated by UTP ($P2Y_2$, $P2Y_4$, and $P2Y_6$) and/or UDP ($P2Y_6$ and $P2Y_{14}$); and (iii) the $P2Y_{14}$ nucleotide sugar-preferring receptor, responding to UDP sugars, such as UDP-glucose and UDP-galactose [17,22,23]. P1 and P2Y G protein-coupled receptors are predominantly involved in long-term (trophic) purinergic signaling, such as that found in remodeling, repair, and regeneration events in response to injury. P1 and P2Y receptors are involved in the regulation of proliferation, differentiation, motility, migration, and cell death [21], which are essential processes for the establishment of endometriotic lesions and the progression of endometriosis.

Correct expression and function of the purinome are essential to maintaining tissue homeostasis. Changes in ATP/adenosine balance and in the purinergic receptor activation state can alter the behavior of a wide range of cell types. These changes can trigger a pathological state or promote the progression of a disease, as may occur in endometriosis.

3. Purinergic Signaling in Eutopic and Ectopic Endometrial Tissue

Extracellular purines and pyrimidines play multiples roles in fertilization and embryo development [24,25]. This is made possible by different elements of the purinome, such as purinergic receptors and ectonucleotidases, present in both male and female reproductive organs [26–33].

Endometrium, the innermost layer of the uterus, contains the surface epithelium, the glandular epithelium, and a vascularized stroma. It is a dynamic tissue undergoing repetitive cycles of regeneration and degeneration in which a certain degree of inflammation is physiological. Purinergic signaling plays a role in both inflammation and reproduction and contributes to correct endometrial function. Therefore, it is not surprising that any change in purinergic signaling can alter uterine function and fertility in women, as happens in endometriosis. In addition, purinergic signaling is involved in the control of a number of cell events, such as cell proliferation, migration, and survival, but it is also seen in phenomena such as angiogenesis and fibrosis (reviewed in [34–36]), thus becoming a candidate pathway for playing a causative role in the pathogenesis and development of this disease.

Therefore, the characterization of the purinome elements of eutopic endometrium and endometriotic lesions is essential in order not only to elucidate the role of purinergic signaling in the pathogenesis, but also to identify the main symptoms and signs of endometriosis and new therapeutic targets. Moreover, knowledge of the microenvironment of ectopic lesions can improve understanding of the generation and neurotransmission of pain signals in endometriosis. We present below current knowledge on the expression of ectonucleotidases, the main elements of the purinome regulating the levels of nucleotides and nucleosides in eutopic and ectopic microenvironments.

Ectonucleotidases in the Eutopic and Ectopic Endometria of Women with Endometriosis

Ectonucleotidases are hormone-sensitive enzymes that vary their expression in the endometrium throughout the menstrual cycle [26]. NTPDase2, NTPDase3, NPP1, NPP3, ALP, CD26, and CD73 are expressed by endometrial epithelial cells, while NTPDase1, NTPDase2, and CD73 have been detected in endometrial stromal cells [26]. Although the ectonucleotidases studied to date are mainly present in the endometrial epithelium, most of the changes detected in endometriosis occur in the stroma [37].

Changes in eutopic endometrium of women with endometriosis in comparison with the endometrium of women without the pathology have been detected [26,37]. Moreover, differences between the different types of endometriotic lesions have been described [37]. One notable change in endometriosis is the expression of NPP3. Several studies of NPP3 expression in endometrial tissue have localized the protein in epithelial cells, with changes along the cycle [26,38,39]. In endometriosis, however, NPP3 is found to be expressed by the stroma, as well as the epithelium, in both eutopic and ectopic endometrial tissues [37]. This de novo expression of NPP3 points to its use as a putative histopathological marker of the disease. It has to be noted that although there is no protein expression of NPP3 in the endometrial stroma of women without endometrial pathology, Boggavarapu et al. detected more than double the levels of NPP3 mRNA in the stroma compared to glandular compartment [40]. All of this points to post-translational regulation of protein levels that needs to be further studied. This increased expression of NPP3 in endometrial tissue coincides with the increased ectonucleotidase activity detected in the fluid content of ovarian endometriomas [41]. These results suggest an increase in ATP metabolism, with a concomitant increase in extracellular adenosine levels, as observed in some cancers, in which high levels of adenosine in the tumor microenvironment induce suppression of the local immune response [42,43]. Conversely, the detection of one hundred times higher levels of ADA in the contents of endometriomas refutes the idea of adenosine accumulation [44]. Moreover, the great decrease in, or even the total loss, of the CD39–CD73 axis in endometrial tissue in endometriosis further suggests that extracellular adenosine synthesis is rather limited. [37]. In fact, these data suggest a relation between extracellular ATP accumulation and the severity and progression of endometriosis, since the loss of CD39–CD73 is related to deep infiltrating endometriosis, the most severe form of the illness with high recurrence rates and a high level of associated pain [45,46].

The consequences of the increase in extracellular ATP levels and the subsequent activation of purinergic signaling through P2 receptors in endometriosis-associated pain are discussed throughout this review.

4. Involvement of Purinergic Signaling in Endometriosis-Associated Pain

4.1. Endometriosis-Associated Pain

Pain is recognized as the most common symptom and the primary reason for medical assistance in women with endometriosis. In fact, up to 80% of patients present chronic pain, the most common forms being dysmenorrhea, non-cyclical pelvic pain (chronic pelvic pain), dyspareunia, dysuria, and dyschezia [2]. Endometriosis-associated pain has, in turn, negative effects on women's mental health, including anxiety and depression, thereby altering their quality of life and that of their loved ones [47,48]. For this reason, there is an urgent need to define the molecular mechanisms underlying endometriosis-associated pain to uncover therapeutic targets to minimize the suffering and raise the quality of life of affected women.

Endometriosis-associated pain is complex, and the underlying mechanisms seem to be related to the activation of the peripheral nervous system, involved in nociceptive (a response to a noxious stimulus), inflammatory (due to tissue damage and inflammatory response), and neuropathic (due to a lesion in somatosensory nervous system) pain, and central nervous systems, related with sensitization and hyperalgesia processes [49,50]. The stage of endometriosis, as classified by the American Society of Reproductive Medicine (ASRM), poorly correlates with the degree of pain or symptoms severity, thus hampering clinical management [11].

In recent years, many articles concerning endometriosis-associated pain and treatments for pain-relief have been published (reviewed in [10,11,49,51–63]); independently, the purinergic mechanisms involved in pain (reviewed in [18,64–70]) are also being studied. However, purinergic signaling in relation to endometriosis-associated pain has yet to be fully explored. In the following section, we look at the involvement of the purinergic signaling in pain, especially in peripheral, but also in central processes, in the context of endometriosis.

4.1.1. The Pain Pathway

The neural process of pain starts with a peripheral noxious stimulus detected by the nociceptors on small diameter sensory afferent nerves (fibers Aδ and C) and its transduction into an electric signal. These neurons, which innervate viscera, have the cell bodies in the dorsal root ganglia (DRG) and reach the lamina I-II of the dorsal spinal cord. This information is transmitted along the spinal cord to the brain, where the unpleasant experience called pain is generated. Although multiple painful conditions have their origin in the sensitization and excitation of neurons, immune and glial cells also play key roles in the generation and maintenance of pain signaling [71,72]. Indeed, the pain perceived can be altered, amplified, or reduced by many molecules, including ATP, released from these non-neuronal cells through different mechanisms required for the transition from acute to chronic pain [72,73]. ATP is a peripheral mediator of pain involved in the initiation of this pain perception.

There is evidence that endometriotic lesions are innervated. This innervation is mainly sympathetic, and sensory nerve fibers play a pivotal role in endometriosis-associated pain [74–78]. The inflammation, concomitant to the initial establishment of an endometriotic focus outside the uterus, activates sensory afferent neurons innervating adjacent visceral structures, transmitting the noxious stimulus to the spinal dorsal horn and causing pain. Moreover, local inflammatory cells release neurotrophic factors encouraging the new innervation and cytokines that lead to the implantation of endometrial ectopic cells. Neuronal and non-neuronal cells of endometriotic lesions can release ATP that in turn regulates the action of these cells. For example, the release of ATP and neurotransmitters by afferent neurons also activates spinal glial cells, contributing to central sensitization and overstated pain.

Endometriotic foci present cyclic proliferative and destructive phases similar to the endometrium. During the breakdown of a part of a lesion, high levels of ATP are released in the lesion microenvironment, acting as an acute danger signal on sensory nerve endings of the lesion. Moreover, this internal bleeding in the ectopic locations often leads to local inflammatory reactions that promote the inflammatory state and the release of molecules involved in the pain signaling pathway. In addition, ATP has the potential to modify the pain signaling by activation of pre- and post-synaptic P2 receptors. The two classes of P2 receptors are involved in pain: P2X receptors in short-term neurotransmission responses and P2Y receptors in the slow and continuous pain signaling.

The high levels of extracellular ATP in the ectopic milieu, together with the loss of the CD39–CD73 axis, turns purinergic signaling into a precious source of possible therapeutic targets for endometriosis-associated pain treatment. The role of ATP and adenosine in the pain signaling pathway in the context of endometriosis is reviewed in detail below.

4.1.2. P2X Receptors in Primary Sensory Neuron: The Outset of Nociception

P2X3 and P2X2/3 receptors. P2X3 and P2X2/3 receptors are expressed in terminals of nociceptive fibers and in the sensory neurons of the central nervous system. Homomeric P2X3 receptors mediate transient nociceptive responses through rapidly desensitizing current, and heteromeric P2X2/3 receptors mediate sustained nociceptive responses through a slowly desensitizing current [79]. They mediate neuropathic pain, including inflammatory pain, in acute and chronic processes, and are involved in hyperalgesia and allodynia [80]. In fact, sensory neurons which express transient receptor potential vanilloid-1 (TRPV1) channels and/or P2X3 receptors are essential for the initiation and transduction of nociception and the signaling of pain. Both TRPV1 and P2X3 induce Ca^{2+} influx which activates a cascade of changes, including the phosphorylation of ion channels with the consequent increase in the excitability of sensory neurons.

Estrogens upregulate the expression of both the cation channel TRPV1 and P2X3 receptors of nerve fiber terminals in endometriosis [81]. The implication of TRPV1 in endometriosis-associated pain is clear, but the precise role of P2X3 is still under study, although there is increasing evidence of its central role in the onset of pain sensation in endometriosis. Ding et al. detected the expression of P2X3 in endometriotic epithelial and stromal cells but also in the sensory nerve fibers within endometriotic lesions [82]. Moreover, they positively correlated the levels of P2X3 receptor in endometriotic lesions with the severity of pain [82].

The activation of P2X3 receptor in nerve fibers leads to the release of endogenous ATP via pannexin-1 hemichannels, triggering the activation of P2 receptors of sensory nerve fibers [83]. In addition, this ATP efflux can be enhanced by nerve growth factors [83]. Persistent extracellular ATP with the concomitant P2X3 and P2X2/3 receptor activation has been linked to the induction and the early maintenance phases of allodynia, a clinical manifestation of chronic pain present in endometriosis [84,85].

In rats, the induction of endometriosis produces thermal and mechanical hyperalgesia. Moreover, endometriosis causes the elevation of endogenous ATP content and P2X3 receptor expression in endometriotic and DRG tissues, which correlate with the severity of hyperalgesia in these animals [86]. In fact, the implication of ATP and P2X3 in endometriosis-associated pain was also confirmed, since administration of A-317491, a selective P2X3 receptor antagonist, attenuated endometriosis-associated pain in rats [87].

Study of the upregulation of P2X3 receptor in endometriosis-related pain showed ATP and ADP, but not UTP, as the effector molecules of this process [86]. In accordance with P2 receptor affinities, $P2Y_1$, $P2Y_{12}$, and/or $P2Y_{13}$ receptors could be responsible for this upregulation. Ding et al. determined that P2X3 upregulation by ADP is mediated by the activation of the transcription factors ATF3 and AP-1 [86]. Moreover, based on the literature, they discussed the possible role of $P2Y_1$ as the promoter of this pathway, but functional studies are needed to confirm this point [86].

Hence, extracellular ATP seems to have a close relationship with initiation, amplification, and maintenance of endometriosis-related pain [87]. In a persistent damaging inflammatory microenvironment, such as the one generally found in the peritoneum of women with endometriosis, high levels of inflammatory mediators are detected. Endometrial cells and immune cells in the ectopic lesions secrete different inflammatory mediators such as interleukin (IL)-1β, IL-6, and tumor necrosis factor alpha (TNF-α) [71,88], as well as growth factors such as vascular endothelial growth factor (VEGF), neurotropin nerve growth factor (NGF), and brain-derived neurotrophic factor (BDNF), often promoted by the presence of high estrogen levels [71,89–91]. These factors not only promote the inflammatory state and the aberrant neuroangiogenic milieu, but are also involved in processes that reduce the threshold of ion channels of sensory fibers, increasing their membrane excitability and the expression of receptors involved in nociception, thus favoring the activation of pain signaling pathway. These molecules promote the release of endogenous ATP during inflammation with the subsequent activation of P2X3 and P2X2/3 receptors, allowing Ca^{2+} influx and depolarization in nearby nociceptive fibers of the endometriotic foci and leading to the sensitization of sensory neurons, which send the pain message to the central nervous system. Therefore, ATP triggers the purinergic signaling cascade in nervous cells involved in pain symptoms. However, P2X3 activation by ATP as a potential action generator is only the beginning of ATP participation in the pain signaling pathway, as shown below.

The activation of the receptors tyrosine kinase A (TrkA), p75, and VEGFR2 in primary sensory neurons mediated by NGF and VEGF, secreted by endometriotic and immune cells, triggers the upregulation of P2X3, causing repeated neuronal sensitization by increased P2X3 receptor signaling and provoking a persistent sensation of pain [92,93]. Moreover, NGF plays a role in the production of the neuropeptide substance P (SP) and calcitonin gene-related peptide (CGRP), which sensitize the sensory nerve fibers, stimulate the immune system, cause the degranulation of mast cells, and promote fibrosis of the lesion through activation of their receptors neurokinin-1 receptor (NK1R) and calcitonin gene-related peptide receptor (CGRPR), respectively [94,95]. Moreover, the high levels of estrogens detected in the microenvironment of endometriotic lesions trigger the degranulation and secretion of NGF by mast cells [96]. Mast cell granules contain inflammatory mediators and neuro-sensitizing molecules including IL-1β, IL-6, TNF-α, and histamine, which, once released into the endometriotic milieu, encourage nerve sensitization and the inflammatory state [97].

4.1.3. P2Y Receptors in Primary Sensory Neuron: The Modulation of Nociception

Primary sensory neurons also express P2Y receptors and their role is mainly pain modulation. P2Y receptors potentiate pain induced by chemical or physical stimuli via capsaicin-sensitive TRPV1 channels and facilitate the P2X receptor-mediated currents [98,99].

P2Y$_1$ and P2Y$_2$ receptors. Activation of P2Y$_1$ and P2Y$_2$ receptors is involved in the activation of TRPV1 channels of nociceptors. Although in humans it is not clear, in rats, upregulation of TRPV1 channel expression by P2Y$_1$ receptors is mediated via p38/MAPK [100,101]. Therefore, it is probably the case that ATP and ADP are involved in short- and long-term effects in nociceptors. On the one hand, they can play a role in the modulation of Ca^{2+} influx that potentiates the sensitization of sensory neurons, while, on the other hand, they may be involved in the long-term nociceptor changes that produce hyperalgesia through the upregulation of TRPV1 channels. Moreover, as stated above, P2Y$_1$ may also play a role in the upregulation of P2X3 receptor ion channel in endometriosis-associated pain [86]. In inflammation, P2Y$_2$ receptor upregulation occurs in sensory neurons of inflamed tissue [98,102]. ATP (and UTP) stimulus on P2Y$_2$ receptors activates TRPV1 channels [103]. This points to the contribution of ATP to chronic inflammatory pain, and therefore, endometriosis.

4.1.4. P2X4 and P2X7 in Macrophage–Nerve Interaction: The Base of Inflammatory Pain

As a consequence of the inflammatory process, large numbers of macrophages, mastocytes, and neutrophils are recruited in the endometriotic focus and macrophages infiltrate DRG [71,96,104]. Macrophages are among the most numerous immune cells in endometriotic lesions. They produce pro-inflammatory cytokines such as IL-1β, TNF-α, and IL-6, which intervene in the pain phenomena [105] and have a role in endometriosis-associated pain (reviewed in [71]). The activation of P2 receptors expressed by immune cells, such as P2X7 and P2X4 receptors in macrophages [106], allows the activation of the immune system via ATP and leads to the production of cytokines, thus maintaining the persistent inflammatory state. Moreover, the activation of macrophage P2X4 receptors is involved in the release of COX-dependent release of prostaglandin E2 (PGE2), mediated by cytosolic phospholipase A2 (cPLA2) [107]. PGE2 is involved in the sensitization of primary sensory neurons [107]. The interaction among endometrial cells, inflammatory cells, and peripheral sensory neurons at the ectopic foci, and the ATP-mediated molecular pathways, are represented in Figure 2.

4.1.5. P2 Receptors in Activated Microglia: The Modulation of Pain Transmission

Endometriosis-associated pain is not only inflammatory but also neuropathic [82,108]. Nerve damage and persistent stimulation of peripheral fibers can lead to the secretion of inflammatory neurotransmitters and neuromodulators from nerve fibers, including ATP, which acts on glial cells. In the meantime, glial cells react, becoming the main source of neuroactive substances, including pro-inflammatory cytokines, trophic factors, and neurotransmitters (such as ATP), which regulate neuronal excitability and are fundamental to the transition from acute to chronic pain.

P2 receptors are present in the surface of activated spinal microglia and are involved in the recruitment and activation of microglia and the interaction between neurons and microglia (summarized in Figure 3). All of these play important roles in neuropathic pain, mainly in allodynia and hyperalgesia, and in neuroinflammation.

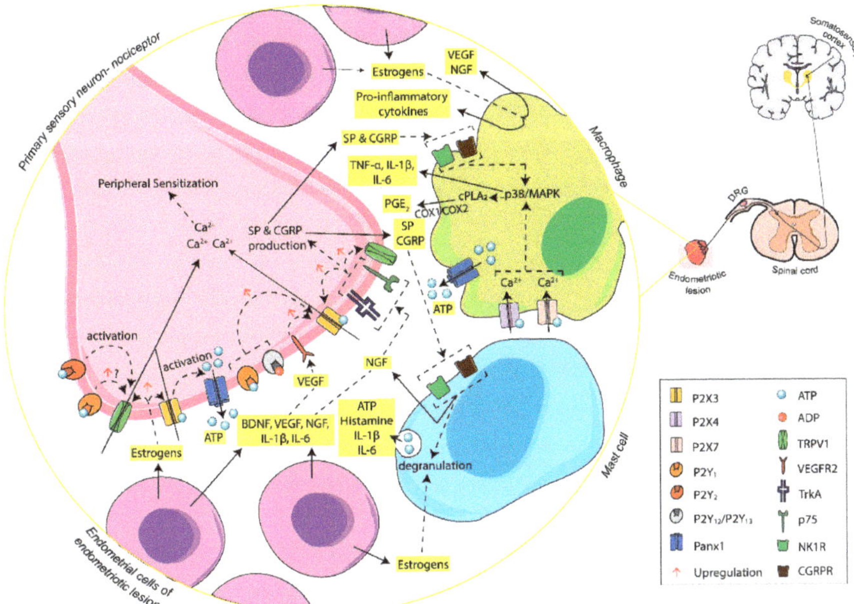

Figure 2. Schematic summary of the involvement of ATP, through activation of P2X and P2Y receptors, in the initiation of endometriosis-associated pain. In the endometriotic lesion, ATP, released by different cell sources, carries out Ca²⁺ influx via P2X3 receptor activation at the endings of primary sensory neurons, triggering a cascade of changes that increase the excitability of afferent sensory neurons. The activation of P2Y receptors potentiates the action of P2X3 receptor and TRPV1, triggering induction of nociception and the maintenance of overstated pain. Moreover, ectopic endometrial cells and inflammatory cells of the lesion release inflammatory mediators that boost nerve sensitization and promote the inflammatory state typical of women with endometriosis. Abbreviations: adenosine triphosphate, ATP; adenosine diphosphate, ADP; pannexin-1, Panx1; brain-derived neurotrophic factor, BDNF; neurotropin nerve growth factor, NGF; tyrosine kinase A receptor, TrkA; p75 neurotrophin receptor, p75; vascular endothelial growth factor, VEGF; vascular endothelial growth factor receptor 2, VEGFR2; interleukin-1 beta, IL-1β; interleukin-6, IL-6; tumor necrosis factor alpha, TNF-α; substance P, SP; neurokinin-1 receptor, NK1R; calcitonin gene-related peptide, CGRP; calcitonin gene-related peptide receptor, CGRPR; transient receptor potential vanilloid-1 channel, TRPV1; p38 mitogen-activated protein kinases, p38/MAPK; cytosolic phospholipase A2, cPLA₂; prostaglandin E2, PGE2; cyclooxygenase-1 and -2, COX-1/COX-2; dorsal root ganglia, DRG.

Apparently, the activation of microglial P2 receptors, particularly P2X4 and P2X7, promotes neuronal excitability. Therefore, blocking microglia–neuron signaling must be considered as a possible therapeutic strategy for treating endometriosis-associated pain.

P2X4 receptor. The role of the P2X4 receptor in activated microglia in endometriosis-associated pain has not yet been studied, but P2X4 receptor involvement in neuropathic pain is clear. Following peripheral nerve injury, P24 receptor is expressed by the microglia of the dorsal horn [109]. The activation of P2X4 receptor causes Ca²⁺ flux and p38/MAPK activation, which promotes the synthesis and release of BDNF, a key molecule for maintaining pain hypersensitivity. ATP-mediated BDNF release from activated microglia, via its receptor TrkB, mediates the downregulation of the K⁺/Cl⁻ cotransporter KCC2 in dorsal horn neurons. KCC2 maintains the anion gradient necessary for the inhibitory actions of gamma-aminobutyric acid (GABA) through gamma-aminobutyric acid A receptor (GABAₐR). KCC2 downregulation increases intracellular chloride levels, allowing the accumulation of anions in dorsal horn neurons. Meanwhile, GABA is released from inhibitory interneurons. GABA activates the GABA$_A$R on dorsal horn neurons with the subsequent chloride outflow, causing the depolarization of the second order neurons. Hence, BDNF alters the chloride gradient of dorsal horn neurons, triggering a decrease in the inhibitory control of the GABAergic interneurons. Consequently, there is an increase in dorsal horn neuron excitability that enables low threshold information to gain access to nociceptive circuits. This evokes pain transmission and neuropathic pain [109–111]. In addition, P2X4 receptor, through p38/MAPK activation, leads to the synthesis and release of pro-inflammatory cytokines and increases the expression of COX, enhancing PGE2 levels, involved in pain-related inflammatory responses and dorsal horn neuronal excitability [112,113]. It is noteworthy, however, that P2X4 is not the only one. The activation of P2X7 (by ATP) and P2Y$_{12}$ and P2Y$_{13}$ receptors (by ADP) also induces microglial pro-inflammatory cytokine release, upregulating excitatory synaptic transmission in the dorsal horn and participating in neuropathic pain [112,114–116]. Moreover, these P2 receptors can be activated by the release of ATP by spinal astrocytes, promoting cytokine production and a pro-inflammatory milieu [117].

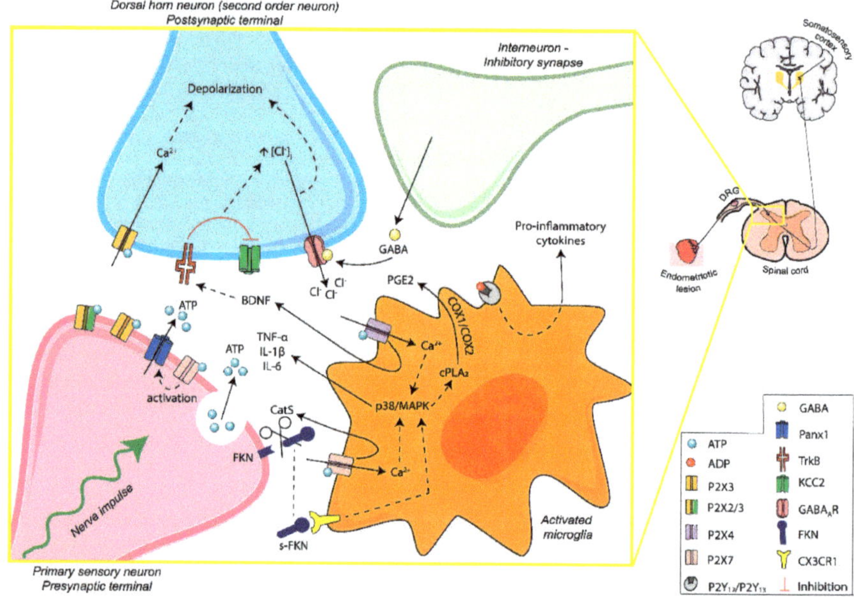

Figure 3. Schematic overview of ATP signaling in microglia–nerve interaction in endometriosis-associated pain. Persistent stimulation of peripheral fibers leads to the release of ATP into the synaptic space. The activation of P2X3 receptors of postsynaptic terminals induces excitability of dorsal horn neurons. Moreover, P2X4 and P2X7 receptors of activated microglia are involved in the secretion of neuroactive substances, contributing to the microglia–nerve interaction needed for central sensitization and modulation of the sensation of pain. ATP-mediated signaling is fundamental in the transition from acute to chronic pain, including endometriosis-related neuropathic and inflammatory pain. Abbreviations: adenosine triphosphate, ATP; adenosine diphosphate, ADP; pannexin-1, Panx1; brain-derived neurotrophic factor, BDNF; interleukin-1 beta, IL-1β; interleukin-6, IL-6; tumor necrosis factor alpha, TNF-α; gamma-aminobutyric acid, GABA; gamma-aminobutyric acid A receptor, GABA$_A$R; p38 mitogen-activated protein kinases, p38/MAPK; cytosolic phospholipase A2, cPLA$_2$; prostaglandin E2, PGE2; cyclooxygenase-1 and -2, COX-1/COX-2; fractalkine, FKN; soluble-fractalkine, s-FKN; cathepsin S, CatS; CX3C chemokine receptor 1, CX3CR1; tyrosine kinase B receptor, TrkB; potassium-chloride cotransporter 2, KCC2; dorsal root ganglia, DRG.

P2X7 receptor. P2X7 receptor functions as an ion channel, but sustained stimulation with large amounts of extracellular ATP induces its conformation as nonselective large pores in cell membrane. It is present in microglia and plays a role in the maintenance of neuropathic pain. Specific inhibition of P2X7 receptor pore formation, without affecting its cation channel activity, reduces chronic pain [118].

In microglia, the activation of P2X7 receptor causes the release of the enzyme Cathepsin S (CatS). Extracellular CatS splits the chemokine domain of neuronal membrane-bound fractalkine (FKN) present in dorsal horn neurons. Soluble-FKN (s-FKN) interacts with microglia CX3C chemokine receptor 1 (CX3CR1), stimulates p38/MAPK, and promotes the release of cytokines that sensitize the second order neurons of the sensory pathway [119–121]. Liu et al. showed an increase in the expression of FKN/CX3CR1/p38/MAPK and in the amount of microglia in the dorsal horn of the sciatic nerve, in an endometriosis rat model [122]. This microglia–nerve crosstalk plays a role in central sensitization and could explain one of the mechanisms associated with hypersensitivity and allodynia in endometriosis. Moreover, increased levels of s-FKN were found in the peritoneal fluid of women with endometriosis [123]. In the ectopic endometrial lesions of a sciatic endometriosis model in rats, increased levels of FKN and s-FKN as well as CX3CR1 have been detected, with a positive correlation with the severity of hyperalgesia [122]. In ectopic lesions, membrane-bound FKN was found in the macrophage surface, whereas CX3CR1 was present in nerve fibers, specifically in Schwann cells. As previously noted, macrophages express P2X7 receptor, which could be an essential piece in FKN cleavage in the endometriotic foci. s-FKN can act as a CX3CR1 ligand as well as being a potent chemoattractant that would favor the development and maintenance of the inflammatory microenvironment in the ectopic lesion, mainly around nerve fibers. On the other hand, a direct interaction between macrophage membrane-bound FKN and CX3CR1 of Schwann cells has been suggested, which could cause myelin phagocytosis and activation of Schwann cells involved in peripheral sensitization in ectopic lesions. Therefore, P2X7 receptor could contribute to peripheral and central hypersensitivity in endometriosis [122].

Although microglial P2X7 receptor activation by presynaptic neurons is mainly related to the release of pro-inflammatory cytokines, it also causes more ATP release into the dorsal horn microenvironment. This is made possible by the increase in the intracellular calcium resulting from P2X7 action that activates pannexin-1. The increasing extracellular ATP levels potentiate purinergic signaling in pain effectors and induce chronification of pain [124].

4.1.6. Adenosine and Adenosine Receptors (AR): Analgesic and Anti-Inflammatory Effects

Adenosine, the hydrolysis product of purine nucleotides such as ATP and ADP, often plays the opposite role to them, leading to a compensatory system in physiological and pathological conditions. In fact, endogenous adenosine has modulating effects on neuronal and glial cells with implications in pain transmission. For this reason, adenosine has been proposed as a potential analgesic target for nociceptive, inflammatory, and neuropathic pain. Therefore, the adenosine signaling pathway may be an interesting target to treat endometriosis-associated pain. However, determining the effects of adenosine in endometriosis-associated pain is quite complex. It has been experimentally proven that adenosine can produce antinociceptive and anti-inflammatory effects as well as their opposite, depending on the site of action, the receptor activated, the extent of exposure, and the context.

Although the role of P1 receptors in endometriosis-associated pain has not yet been studied, we briefly outline the different effects that have been described in the literature and their possible usefulness for the treatment of endometriosis-associated pain.

A_1 adenosine receptor (AR). A_1AR is the main adenosine receptor associated with inhibitory neuromodulation of pain. A_1AR regulates neurotransmitter release, neuronal excitability, and pain reduction. This high adenosine affinity receptor is expressed at peripheral sensory nerve endings [125], in dorsal horn neurons of the superficial layers of spinal cord [126], and in microglia [127].

A_1AR is G_i-coupled and its activation inhibits AC activity (inhibition of cAMP production), which leads to hyperpolarization by increasing potassium conductance, blocks transient calcium channel opening, and stimulates PLC, inducing an increase in inositol 1,4,5-triphosphate (IP_3) and intracellular Ca^{2+} levels and stimulation of calcium-binding proteins such as protein kinase C (PKC) [128].

Furthermore, A_1AR presynaptically inhibits primary sensory neuron transmission onto dorsal spinal neurons by blocking neurotransmitter release [126,129]. Moreover, when primary sensory afferent neurons depolarize, afferent nerve terminals release glutamate and SP, but also adenosine, in the dorsal spinal cord [129]. In this situation, adenosine can act as a negative modulator by activating A_1AR at postsynaptic sites.

As mentioned, A_1AR is also present in spinal cord microglia where it is upregulated by high levels of extracellular ATP. The activation of A_1ARs curbs activation of microglia and blocks their role in neuronal sensitization [127].

Several clinical studies have demonstrated that specific A_1AR stimulation results in analgesic effects sufficient to ameliorate nociceptive, neuropathic, and inflammatory pain (reviewed in [18,130]). This antinociception, and specifically the reduction of hyperalgesia in several pain models, suggests A_1AR agonists as possible tools for treating endometriosis-associated chronic pain.

$A_{2A}AR$. $A_{2A}AR$ is present in immune cells, neurons, and glial cells, among other cell types. In fact, $A_{2A}AR$ is considered a potential therapeutic target in treating chronic pain of neuroinflammatory origin. $A_{2A}AR$ action is, however, controversial due to an apparent dichotomy between peripheral and central signaling.

At the molecular level, $A_{2A}ARs$ are directly coupled to G_s intracellular proteins, leading to the activation of AC (activation of cAMP production) with the consequent activation of the protein kinase A (PKA) and PKC signaling cascade. Increased cAMP levels in primary afferent neurons contribute to peripheral sensory nerve stimulation and increased excitatory neurotransmitter release in the spinal cord, enhancing nociception [131].

In addition, $A_{2A}AR$ is considered the main mediator of anti-inflammatory responses through its expression in a wide range of immune cells. $A_{2A}AR$ activation increases intracellular immunosuppressive cAMP in peripheral immune cells. This may contribute to reducing inflammatory pain by modulating these cells and consequently decreasing the levels of sensitizing substances (e.g., inflammatory cytokines from macrophages or histamine from mast cells) in the endometriotic microenvironment [130].

In addition, $A_{2A}AR$ modulates immunoresponses of microglia. As in peripheral immune cells, $A_{2A}AR$ agonists produce a cAMP signaling cascade that attenuates pro-inflammatory cytokine production and increases anti-inflammatory cytokine release, mainly IL-10. The reduction of the inflammatory state relieves neurophatic pain [132–135].

Conversely, under pathologic conditions, $A_{2A}AR$ activation is also related to a pro-inflammatory role [136,137]. During neuropathic pain, persistent activation of spinal microglia occurs. The sustained release of stimulating molecules such as BDNF in spinal cord exacerbates neuronal hypersensitivity and chronic neuroinflammation. As noted above, ATP mediates BDNF release via P2X4 receptor from activated microglia, as well as by adenosine via $A_{2A}AR$. Constitutive release of BDNF is under the control of PKC, but, in some pathological conditions, upregulation of $A_{2A}AR$ in microglia has been detected. The high levels of intracellular cAMP and the subsequent activation of PKA prevails over PKC actions inducing increased BDNF release [137]. However, the increase in $A_{2A}AR$ levels in microglia is mainly described in brain with chronic neuroinflammation in cases with neurodegenerative diseases, where $A_{2A}AR$ interaction with several neurotransmitters plays an important role in pathology progression [138]. Further studies are needed to determine if $A_{2A}AR$ upregulation in activated microglia also occurs in endometriosis-associated pain. If that is the case, the therapeutic strategies involving $A_{2A}AR$ activation might not be good for the evolution of chronic pain related with endometriosis.

In conclusion, therapeutic tools that potentiate A_1AR-mediated antinociceptive and $A_{2A}AR$-mediated anti-inflammatory actions of adenosine could have analgesic effects that would be useful for the management of endometriosis-associated pain.

$A_{2B}AR$. $A_{2B}AR$ is mainly present in immune cells, as well as, at low levels, in spinal cord and CNS, and prominently in astrocytes [18,19]. Its ubiquitous expression in inflammatory cells suggests a role in inflammatory pain through the release of various inflammatory mediators.

A$_{2B}$AR has a low affinity for adenosine, and it is therefore significantly activated under pathophysiological conditions, when adenosine concentration is high. Similar to A$_{2A}$AR, A$_{2B}$AR activates AC, leading to a PKA phosphorylation cascade. This cellular signaling mainly promotes an anti-inflammatory response in immune cells and plays a part in the attenuation of acute inflammation [128].

While it may appear that A$_{2B}$AR plays the same role as A$_{2A}$AR, its effect is sometimes opposed. This is possible because, as with A$_1$ and A$_3$AR, A$_{2B}$AR is coupled to PLC, leading to the accumulation of intracellular Ca^{2+}, the generation of IP$_3$ and diacylglycerol (DAG), and the activation of PKC, involved in the release of pro-inflammatory mediators such as IL-6 from macrophages and IL-1β and VEGF from mast cells, with indirect effects on inflammatory pain [19,130,139]. In fact, Hu et al. described a mechanism underlying A$_{2B}$AR-mediated chronic pain in three independent models of chronic pain [140]. The prolonged elevated adenosine levels activate A$_{2B}$ARs of myeloid cells. This activation increases the production and release of IL-6 and soluble IL-6 receptor (sIL-6R). The complex IL-6/sIL-6R trans-activates gp130 on primary sensory fibers, which in turn stimulates the transcription factor pSTAT3, inducing TRPV1 gene transcription. The increase of TRPV1 in nociceptors produces sensory neuron hyperexcitability and can produce chronic inflammatory pain.

These opposing effects of A$_{2B}$AR hamper the use of its agonists or antagonists as potential therapeutic agents for inflammatory disorders and pain. Evaluating the possible use of A$_{2B}$AR as therapeutic target of endometriosis-associated pain requires a better understanding of the function of A$_{2B}$AR in this disease. Moreover, the loss of expression of some ectonucleotidases in endometriotic tissue and the large amount of ADA in the fluid contents of endometriomas suggests that the adenosine levels in endometriotic milieu are low, rendering activation of A$_{2B}$ARs unlikely. Consequently, A$_{2B}$AR does not seem to be a suitable target for endometriosis treatment.

A$_3$AR. A$_3$ARs are described in peripheral and central neurons, glial cells, and immune cells [141,142]. Over the last decades, experimental studies with A$_3$AR agonists showed the attenuation of nociception and neuropathic pain [130,141–143]. This has made A$_3$AR a focus of development of new therapeutic strategies for pain.

Similar to A$_1$AR, A$_3$AR is coupled to G$_i$, which inhibits AC, and to G$_q$, which activates PKC. On primary sensory neurons, A$_3$AR inhibits the Ca^{2+}-dependent K$^+$ currents, reducing their excitability and neurotransmitter release [142]. This mechanism exerts an antinociceptive effect, but it is not the only one.

Agonists of A$_3$AR enhance and recover inhibitory GABA signaling in spinal cord neurons. In neuropathic pain states, reduced levels of spinal GABA and GAD65 [144,145], its synthesis enzyme, are detected, whereas the expression of GABA transporter GAT-1, which uptakes the neurotransmitter from the synapse, is increased [145]. Additionally, there is a reduction in the activity and expression of KCC2. As noted above, P2X4 receptor activation promotes the production and release of BDNF, which, via TrkB, downregulates KCC2 and disrupts Cl$^-$ homeostasis, altering GABA$_A$ receptor postsynaptic inhibitory control and leading to neuronal hyperexcitability [109–111]. A$_3$AR activation reverses these situations. The use of A$_3$AR agonists maintains the phosphorylation status of GAD65 and GAT-1, which stabilizes and activates GAD65 and leads to the internalization and inactivation of GAT-1 [146]. In addition, A$_3$AR agonists enhance KCC2 phosphorylation, increasing its activity [146]. Therefore, the activation of A$_3$AR allows the increase of GABA levels and the restoration of inhibitory actions of GABA in the spinal cord, alleviating neuropathic pain. Moreover, A$_3$AR activation also diminishes the production of pro-inflammatory cytokines (TNF-α and IL-1β) and enhances the formation of anti-inflammatory cytokines (IL-10) in the spinal cord by inhibiting the p38/MAPK and NF-$_{kB}$ signaling pathways [143,147,148]. This enhances the reduction of glial activation, decreasing neuroinflammation and pain hypersensitivity.

In recent years, antinociceptive properties of A$_1$AR and A$_{2A}$AR agonists have been the most studied targets of adenosine signaling to relieve acute and chronic pain. However, A$_3$AR agonists have increased in importance in pain treatment since they avoid the undesirable cardiovascular side effects of A_1AR and $A_{2A}AR$ agonists [149,150]. For this reason, the use of selective A_3AR agonists seems promising as a safe and effective analgesic treatment for chronic pain, and it is also being considered to treat endometriosis-associated pain.

5. Perspectives of ATP and Adenosine Signaling Modulation: Possible Tools to Treat Endometriosis-Associated Pain

Most medical treatments for endometriosis are aimed at relieving the chronic pain associated with the disease, but in many cases endometriosis-associated pain symptoms are not improved, and, if they are, they return with treatment cessation (reviewed in [53]).

As we try to document in this review, there is clear involvement of purinergic signaling in the generation and modulation of the sensation of pain. ATP participates in the hyperexcitability of sensory neurons and the development and maintenance of different types of pain (nociceptive, inflammatory, and neuropathic). Evidence mostly suggests ATP is responsible for triggering nociceptive pain, as well as inflammatory responses in the body, in contrast to adenosine. ATP also influences fecundation, endometrial receptivity, and embryo implantation [2].

In endometriosis, the adhesion of viable endometrial cells to establish an ectopic endometriotic focus triggers signals of injury and promotes the activation of an immune system that is inefficient in clearing ectopic cells. This situation is reflected in a significant increase in extracellular ATP levels in the endometriotic lesion. The ATP-rich microenvironment of the endometriotic lesion contributes to the two main symptoms of endometriosis: pain and infertility. The increase in ATP levels is enhanced by the loss of expression of certain ectonucleotidases in endometrial ectopic cells, prolonging signaling and activating pain. It seems, then, reasonable that development of therapeutic approaches for endometriosis-associated pain be based on achieving a decrease in ATP levels and their signaling cascade, and/or increased adenosine levels and activation of their receptors. Despite the potential of purinergic-based drugs and their analgesic effects in various pain models, there is no medication presently available for clinical use.

Currently, most pharmacological therapies for endometriosis are hormonal therapies. Neither these nor surgical treatment are compatible with the pregnancy desire of women with endometriosis, and it is usually a matter of choosing between alleviating symptoms and getting pregnant. A non-hormonal alternative to alleviate endometriosis-associated pain may lie in purinergic-based drugs. The development of clinical drugs targeting purinergic receptors is not free of difficulties, such as the ubiquitous expression and wide action of purinergic receptors throughout the body, and the large number of receptor subtypes combined with a lack of complete knowledge of their physiological and pathophysiological functions.

We present below a selection of potential purinergic signaling targets for pharmacological treatment of endometriosis-associated pain, without ruling out their utility in the improvement of the fertility of these women.

(I)   P2 Receptor Antagonists

ATP, through P2X and P2Y receptors, induces cell signaling directly involved in hyperexcitability of sensory neurons and sustained glial cell reactivity in neuropathic pain. P2 receptor antagonists and other molecules that alter their function have been used to describe the involvement of ATP in pain signaling. Moreover, the attenuation of nociception, hyperalgesia, and allodynia by blocking some P2 receptors suggests that P2 receptor-related drugs are potential candidates for the treatment of chronic pain conditions. In fact, P2X3 receptors of sensory neurons seem to be the main receptors involved in pain, and P2X4 and P2X7 receptors appear to be key elements in neuropathic and inflammatory pain for their function in glial and immune cells.

P2X3 receptor is involved in the development and progression of endometriosis-associated pain [81,82,86,87], and there is a growing interest in the development of substances to interfere or inhibit its function. A P2X3 antagonist, MK-7264/AF-219, known as gefapixant, is currently being tested in various advanced clinical trials with subjects with pulmonary disease and chronic cough (NCT01432730, NCT02502097, and NCT02477709). Recently, a clinical trial has been begun to evaluate the efficacy and safety of gefapixant in women with endometriosis-associated pain (NCT03654326). Experimental evidence of the relief of endometriosis-related pain has been obtained with the use of P2X3 receptor antagonists (e.g., A-317491), in endometriosis-induced animal models [87].

The use of P2X4 receptor antagonists (such as CORM-2, 5-BDBD, and NP-1815-PX) and P2X7 receptor antagonists (such as AZD9056 and AZ11645373) has been reported to inhibit microglia activation, significantly reducing inflammation and alleviating pain (reviewed in [151,152]). Although the administration of AZD9056 yielded good results in pain models in animals, clinical trials were not successful in alleviating symptoms in patients with rheumatoid arthritis, a chronic inflammatory disorder [153]. This suggests that new pharmacological strategies with P2X4 and P2X7 as targets are needed to achieve the desired clinical results in the treatment of inflammation and pain in endometriosis.

P2Y receptors are also potential pharmacological targets. For example, P2Y$_{12}$ receptor regulates microglial activation and the neurotransmission of the excitatory signal in spinal cord neurons; the administration of P2Y$_{12}$ antagonists blocked microglia action in nerve injury-induced pain models [115,154]. There are clinical trials with P2Y$_{12}$ antagonists, in cardiovascular pathologies, but not yet in pain. However, the antithrombotic actions of P2Y$_{12}$ antagonist drugs could certainly complicate their use.

(II)   The Control of ATP Release

ATP release can occur via vesicular and non-vesicular mechanisms. Targeting ATP release on the endometriotic lesions, on sensory and central neurons, would represent a fine-tuning regulation of the P2 receptors involved in pain.

Vesicular ATP release involves the mechanism of exocytosis. Vesicular nucleotide transporter (VNUT), encoded by *SLC17A9* gene, is responsible for the vesicular storage and release of ATP in neurons, astrocytes, and microglia [155,156]. It is known that secretion of ATP through VNUT-dependent vesicular release mechanisms is involved in purinergic signaling in pain and inflammation [155].

After peripheral nerve injury, increases in VNUT expression and extracellular ATP levels are detected in spinal cord [156]. Masuda et al. showed that mice lacking VNUT in the dorsal horn neurons reduced the alloydina evoked by peripheral nerve injury. This did not occur with mice lacking VNUT in primary sensory neurons, astrocytes, or microglia. Increased extracellular ATP levels and neuropathic pain were restored in these mice lacking VNUT in dorsal horn neurons when VNUT expression was restored [156].

Clodronate and etidronate are biphosphonates used in osteoporosis therapy that have analgesic properties. In vitro assays demonstrated that they inhibited VNUT, leading to the modulation of ATP release and purinergic transmission [157,158]. In addition, in vivo studies with clodronate showed attenuation of neuropathic and inflammatory pain [158]. As clodronate is approved for clinical use in the treatment of osteoporosis and its safety is proven, it may be a good candidate for the treatment of endometriosis-associated pain.

ATP release in neurons, astrocytes, and microglia also occurs through membrane channels, such as pannexin hemichannels, connexins, and the P2X7 receptor itself. In recent years, a wealth of evidence has pointed to pannexin-1 [124,159,160] and connexin-43 (reviewed in [161–163]) as crucial elements in the induction and maintenance of chronic pain. These ATP-permeable channels are, therefore, pain relief targets for further investigation.

(III) Recombinant Ectonucleotidases

In endometriosis, a loss of ectonucleotidase expression is reported in association with the severity of the disease. This might contribute to a rise in ATP levels. Restoring the ectonucleotidase activity in the endometriotic tissue would consequently reduce ATP-induced pain and enhance the antinociceptive effects of adenosine in the pain pathway.

CD39, CD73, and prostatic acid phosphatase (PAP) have been described as the main ectonucleotidases involved in the production of adenosine in DRG and spinal cord neurons [128,164]. A single intrathecal injection of recombinant soluble CD73 or PAP had long-lasting antinociceptive effects, dependent on A_1AR activation, including antihyperalgesic and antiallodynic effects, in naïve mice and in mouse models of inflammatory and neuropathic pain [165,166].

Recombinant ectonucleotidases are potential tools for endometriosis-associated pain treatment. Additional preclinical and clinical studies are required to confirm their benefit in inflammatory and neuropathic pain.

(IV) P1 Receptor Agonists

Adenosine has a limited use due to its short life in vivo. Alternatively, agonists and positive allosteric modulators of AR have been described as pharmacological tools to treat inflammation and pain (reviewed in [167–169]). Their analgesic and anti-inflammatory effects have been studied with A_1AR, $A_{2A}AR$, and A_3AR agonists. The controversial role of $A_{2B}AR$ limits their therapeutic use.

Several preclinical and clinical trials with agonists of A_1AR (e.g., GW493838 and NCT00376454) and $A_{2A}AR$ (e.g., BVT.115929 and NCT00452777) have been performed. Despite the good results of AR agonists in several pain models [126,132–135], the lack of analgesia in humans together with the evidence of undesirable side effects, such as cardiovascular involvement, jeopardizes their use in any therapy [168,169].

On the other hand, preclinical and human clinical studies with A_3AR agonists did not have significant side effects [170]. Antinociceptive effects have been described with the use of moderately selective agonists of A_3ARs, such as IB-MECA and Cl-IB-MECA, and highly selective agonists, such as MRS5698 and MRS5980. In general, A_3AR agonists tend to restore the altered pain signaling of chronic pain. For example, A_3AR agonists decrease glial activation and the generation of pro-inflammatory cytokines [143,147,148], increase the production of anti-inflammatory cytokines [147,148], and restore the inhibitory action of GABA [146]. Interestingly, A_3AR agonists selectively modify pathological pain but seem not to alter protective pain [171]. Phase II and III clinical trials with IB-MECA have been completed in rheumatoid arthritis (NCT01034306) and psoriasis (NCT01265667), respectively. A phase II clinical trial with Cl-IB-MECA was also completed in chronic hepatitis C (NCT00790673). Moreover, Cl-IB-MECA is currently in phase II for its antitumor effects in hepatocellular carcinoma (NCT02128958). Although these clinical trials do not attend to their effectiveness in chronic pain, the safety profile in chronic inflammatory diseases, liver disease, and cancer envisages an optimistic future in the pharmacological treatment of neurophatic endometriosis-associated pain.

Several studies have demonstrated that acupuncture improves endometriosis-associated pain [54,172,173]. Interestingly, mechanisms underlying acupuncture-induced analgesia involve purinergic signaling [174–176]. Acupuncture produces local release of ATP leading to the activation of purinoreceptors on sensory nerve endings, triggering the neurotransmission of pain signal to brain. Local release of adenosine in certain centers of the brain cortex can modulate and inhibit pain sensation through the activation of AR, mainly A_1AR [175,177]. Thus, acupuncture is a possible complementary treatment to relieve endometriosis-related pain.

In addition, the potential analgesic effect of adenosine receptor antagonists such as caffeine has also been studied. Caffeine, at low doses, and in combination with analgesic drugs, acts as an adjuvant (reviewed in [178]). At dietary levels, caffeine has a high affinity for A_1, $A_{2A,}$ and $A_{2B}ARs$ [179]. Adenosine-based mechanisms involved in caffeine pharmacological antinociception are attributed to the A_{2A} and $A_{2B}AR$ blockade [178]. Despite its auspicious effects, caffeine has not been tested in endometriosis-associated pain. In fact, we predict that the use of caffeine in relieving endometriosis-associated pain is a complex matter due to the role of caffeine in female hormone pathways, in turn influencing the endometriosis outcome [180–182]. Moreover, no consistent association has been found between coffee/caffeine intake and the risk of this hormone-dependent disease [183–185]. Furthermore, the inhibition of A_1AR by caffeine might interfere with the effectiveness of several analgesic agents and treatments, e.g., acupuncture (reviewed in [178]), which should be taken into account in the clinical management of patients.

(V) Inhibitors of Equilibrative Nucleoside Transporters

In contrast to ATP, adenosine is neither stored nor released in synaptic vesicles. However, adenosine can be released by the cell via nucleoside transporters. One way to increase local extracellular adenosine levels and the concomitant antinociceptive signaling is the blocking of the equilibrative nucleoside transporters (ENTs). ENTs regulate facilitated diffusion and bidirectional nucleoside transport across the cell membrane, following the concentration gradient, in a number of tissues, including central nervous system. ENT-1 is highly expressed in superficial dorsal horn laminae and in DRG, colocalizing with A_1 and A_2AR [186,187].

Maes et al. demonstrated that systemic administration of ENT-1 inhibitors can reverse hyperalgesia in guinea pig inflammatory pain models [188]. The mechanism underlying this analgesia seems to be the blocking of adenosine reuptake into cells, allowing greater activation of A_1 and A_2AR [188]. These results show that it is necessary to investigate the potential therapeutic effect of ENT inhibition. Moreover, the use of ENT inhibitors in combination with AR agonists or ADA inhibitors might enhance the antinociceptive effects of these molecules.

Consequently, the use of ENT inhibitors for analgesia in endometriosis is worth studying.

(VI) Inhibitors of AK and ADA

Adenosine metabolism is mainly the responsibility of AK and ADA enzymes; their inhibition would result in an increase in local extracellular adenosine levels, enhancing their antinociceptive signaling.

Several studies have shown that the supply of an orally active non-nucleoside AK inhibitor (ABT-702) produced effective antinociceptive and anti-inflammatory effects both in vitro and in vivo [189–191]. Increased adenosine concentration by AK inhibition seems to produce therapeutic effects through the activation of A_1 and $A_{2A}AR$ [192]. Moreover, its pharmacological action is achieved with lower doses than with AR agonists, thus reducing the probability of producing psychomotor and cardiovascular side effects. Nevertheless, other AK inhibitors have toxic side effects, such as neurotoxicity (reviewed in [192]). The findings about ABT-702 indicate this to be an efficient and safe drug for the treatment of neuropathic pain and inflammatory states. Studies of the effects of ABT-702 in induced-endometriosis animal models are needed to assess its utility in therapies against endometriosis-associated pain.

It is notable that ADA, another metabolic enzyme of adenosine, and AK have different kinetics: with low or moderate inflammation, AK is the one modulating adenosine levels; with greater inflammation, where there is a substantial elevation of adenosine levels, ADA activity gains importance [193]. Hence, the effect of their inhibitors would also vary depending on adenosine levels in the environment. While the antinociceptive effect of AK inhibitors has been demonstrated, inhibition of ADA does not produce an intrinsic antinociceptive effect without very high levels of adenosine [194]. This is why the use of ADA inhibitors is only considered as an enhancer of the antinociceptive effects of AK inhibitors [195]. ADA inhibitors, such as deoxycoformycin, have been shown to have anticancer effects and to be useful in the treatment of infectious diseases [196–198]. Unfortunately, these compounds are usually toxic at effective doses [196,198–200]. The co-administration of AK inhibitors, such as ABT-702, with ADA inhibitors at lower doses as adjuvants, seems to be a suitable strategy for analgesia in endometriosis patients, but exhaustive prior safety studies are required.

6. Conclusions

Pain has a strong impact on the quality of life of women with endometriosis. Current surgical and pharmacological treatments for endometriosis have as their primary goal the relief of pain. Nevertheless, these treatments have a limited success rate and in general hamper pregnancy. Although there is increasing understanding of the essential role of purinergic signaling in the development and progression of nociceptive, inflammatory, and neuropathic pain, its implication in endometriosis-associated pain is still poorly studied. In this review, we examine the role of purinergic signaling, and particularly the role of extracellular ATP as a triggering factor for acute and chronic pain signaling, in the context of endometriosis.

Ectonucleotidases, the enzymes regulating ATP levels in the extracellular milieu, are altered in endometrial tissue in endometriosis. Of note is the decrease in the expression of the CD39–CD73 axis that supports the hypothesis of ATP (rather than adenosine) accumulation. Concomitant sustained activation of P2 receptors, capable of generating a persistent pain message, is compatible with the onset and maintenance of endometriosis-associated pain. It is known that P2X3 receptor, expressed in sensory neurons, mediates nociceptive, neuropathic, and inflammatory pain, and it is enrolled in endometriosis-related pain; therefore, pharmacological P2X3 inhibition is a worthy candidate for testing; in this sense, the use of the P2X3 receptor antagonist gefapixant is under clinical study. Although the P2X3 receptor fulfills the requirements to be a suitable molecule to be targeted, other ATP receptors have to be considered as well, such as the P2X4 and P2X7 receptors that are involved in macrophage–nerve and microglia–nerve interactions, promoting a persistent inflammatory state and the chronification of pain. P2X4 receptor triggers the generation of the inflammatory pain sensitization mediator PGE2 and is involved in the decrease of inhibitory control of GABAergic interneurons in neuropathic pain. In contrast, P2X7 receptor promotes the maintenance of neuropathic pain through the FKN/CX3C/CX3CR1 pathway, which seems to be altered in endometriosis. The use of P2X4 and P2X7 antagonists has yielded good results in reducing inflammation and alleviating pain in animal models. However, a clinical trial with the P2X7 receptor antagonist AZD9056 has been unsuccessful in a chronic inflammatory disorder, rheumatoid arthritis.

$P2Y_1$ and $P2Y_2$ receptors are involved in the activation and regulation of P2X3 and TRPV1 receptor ion channels of nociceptors. Moreover, microglial $P2Y_{12}$ and $P2Y_{13}$ receptors activation triggers the release of pro-inflammatory cytokines, increasing excitatory synaptic transmission in the dorsal horn. Although P2Y receptors are involved in the modulation of pain, there are no clinical trials evaluating the use of P2Y antagonists in pain.

In addition, the antinociceptive and anti-inflammatory actions of the ARs, mainly A_1AR and A_3AR, have encouraged studies using agonists of these receptors for the treatment of pain and chronic inflammatory diseases. With the current knowledge, A_3AR-targeting drugs are potential tools to treat neuropathic pain, as well as endometriosis-associated pain, but clinical studies in this regard are needed.

In summary, purinergic signaling-based strategies need to be further explored in the medical management of endometriosis-associated pain. This will also improve treatment of other symptoms of endometriosis, such as infertility, in which purinergic signaling also plays a role.

Author Contributions: C.T., Writing—original draft preparation; and M.M.-S., Writing—review, and editing. Both authors have read and agreed to the published version of the manuscript.

Funding: This research was funded by the Instituto de Salud Carlos III (grant number: PI18/00541), co-funded by FEDER funds/European Regional Development Fund (ERDF) ("a Way to Build Europe"), FONDOS FEDER ("una manera de hacer Europa"), and a grant from the Fundación Merck Salud (Ayuda Merck de Investigación 2016-Fertilidad).

Acknowledgments: We thank CERCA Programme (Generalitat de Catalunya) for institutional support, and Tom Yohannan for language editing.

References

1.   Nisolle, M.; Donnez, J. Peritoneal endometriosis, ovarian endometriosis, and adenomyotic nodules of the rectovaginal septum are three different entities. *Fertil. Steril.* **1997**, *68*, 585–596. [CrossRef]

2.   Bulletti, C.; Coccia, M.E.; Battistoni, S.; Borini, A. Endometriosis and infertility. *J. Assist. Reprod. Genet.* **2010**, *27*, 441–447. [CrossRef] [PubMed]

3. May, K.E.; Villar, J.; Kirtley, S.; Kennedy, S.H.; Becker, C.M. Endometrial alterations in endometriosis: A systematic review of putative biomarkers. *Hum. Reprod. Update* **2011**, *17*, 637–653. [CrossRef] [PubMed]

4. As-Sanie, S.; Black, R.; Giudice, L.C.; Gray Valbrun, T.; Gupta, J.; Jones, B.; Laufer, M.R.; Milspaw, A.T.; Missmer, S.A.; Norman, A.; et al. Assessing research gaps and unmet needs in endometriosis. *Am. J. Obstet. Gynecol.* **2019**, *221*, 86–94. [CrossRef]

5. Sampson, J. Peritoneal endometriosis due to the menstrual dissemination of endometrial tissue into the peritoneal cavity. *Am. J. Obstet. Gynecol.* **1927**, *14*, 422–469. [CrossRef]

6. Grandi, G.; Mueller, M.D.; Papadia, A.; Kocbek, V.; Bersinger, N.A.; Petraglia, F.; Cagnacci, A.; McKinnon, B. Inflammation influences steroid hormone receptors targeted by progestins in endometrial stromal cells from women with endometriosis. *Am. J. Reprod. Immunol.* **2016**, *117*, 30–38. [CrossRef]

7. Nothnick, W.; Alali, Z. Recent advances in the understanding of endometriosis: The role of inflammatory mediators in disease pathogenesis and treatment. *F1000Research* **2016**, *5*. [CrossRef]

8. Ahn, S.H.; Khalaj, K.; Young, S.L.; Lessey, B.A.; Koti, M.; Tayade, C. Immune-inflammation gene signatures in endometriosis patients. *Fertil. Steril.* **2016**, *106*, 1420–1431.e7. [CrossRef]

9. Zhang, T.; De Carolis, C.; Man, G.C.W.; Wang, C.C. The link between immunity, autoimmunity and endometriosis: A literature update. *Autoimmun. Rev.* **2018**, *17*, 945–955. [CrossRef]

10. Zheng, P.; Zhang, W.; Leng, J.; Lang, J. Research on central sensitization of endometriosis-associated pain: A systematic review of the literature. *J. Pain Res.* **2019**, *12*, 1447–1456. [CrossRef]

11. Morotti, M.; Vincent, K.; Becker, C.M. Mechanisms of pain in endometriosis. *Eur. J. Obstet. Gynecol. Reprod. Biol.* **2017**, *209*, 8–13. [CrossRef] [PubMed]

12. Yegutkin, G.G. Enzymes involved in metabolism of extracellular nucleotides and nucleosides: Functional implications and measurement of activities. *Crit. Rev. Biochem. Mol. Biol.* **2014**, *49*, 473–497. [CrossRef] [PubMed]

13. Volonté, C.; D'Ambrosi, N. Membrane compartments and purinergic signalling: The purinome, a complex interplay among ligands, degrading enzymes, receptors and transporters. *FEBS J.* **2009**, *276*, 318–329. [CrossRef] [PubMed]

14. Faas, M.M.; Saez, T.; de Vos, P. Extracellular ATP and adenosine: The Yin and Yang in immune responses? *Mol. Asp. Med.* **2017**, *55*, 9–19. [CrossRef] [PubMed]

15. Falzoni, S.; Donvito, G.; Di Virgilio, F. Detecting adenosine triphosphate in the pericellular space. *Interface Focus* **2013**, *3*, 20120101. [CrossRef]

16. Zimmermann, H.; Zebisch, M.; Strater, N. Cellular function and molecular structure of ecto-nucleotidases. *Purinergic Signal.* **2012**, *8*, 437–502. [CrossRef]

17. Ralevic, V.; Burnstock, G. Receptors for purines and pyrimidines. *Pharmacol. Rev.* **1998**, *50*, 413–492.

18. Sawynok, J. Adenosine receptor targets for pain. *Neuroscience* **2016**, *338*, 1–18. [CrossRef]

19. Chandrasekaran, B.; Samarneh, S.; Jaber, A.M.Y.; Kassab, G.; Agrawal, N. Therapeutic Potentials of A2B Adenosine Receptor Ligands: Current Status and Perspectives. *Curr. Pharm. Des.* **2019**, *25*, 2741–2771. [CrossRef]

20. Schulte, G.; Fredholm, B.B. Signalling from adenosine receptors to mitogen-activated protein kinases. *Cell. Signal.* **2003**, *15*, 813–827. [CrossRef]

21. Burnstock, G. Short- and long-term (trophic) purinergic signalling. *Philos. Trans. R. Soc. B Biol. Sci.* **2016**, *371*. [CrossRef] [PubMed]

22. Scarfi, S. Purinergic receptors and nucleotide processing ectoenzymes: Their roles in regulating mesenchymal stem cell functions. *World J. Stem Cells* **2014**, *6*, 153–162. [CrossRef] [PubMed]

23. Roszek, K.; Wujak, M. How to influence the mesenchymal stem cells fate? Emerging role of ectoenzymes metabolizing nucleotides. *J. Cell. Physiol.* **2018**, *234*, 320–334. [CrossRef] [PubMed]

24. Burnstock, G. Purinergic signalling in the reproductive system in health and disease. *Purinergic Signal.* **2014**, *10*, 157–187. [CrossRef] [PubMed]

25. Burnstock, G.; Dale, N. Purinergic signalling during development and ageing. *Purinergic Signal.* **2015**, *11*, 277–305. [CrossRef]

26. Aliagas, E.; Vidal, A.; Torrejon-Escribano, B.; Taco Mdel, R.; Ponce, J.; de Aranda, I.G.; Sevigny, J.; Condom, E.; Martin-Satue, M. Ecto-nucleotidases distribution in human cyclic and postmenopausic endometrium. *Purinergic Signal.* **2013**, *9*, 227–237. [CrossRef]

27. Villamonte, M.L.; Torrejon-Escribano, B.; Rodriguez-Martinez, A.; Trapero, C.; Vidal, A.; Gomez de Aranda, I.; Sevigny, J.; Matias-Guiu, X.; Martin-Satue, M. Characterization of ecto-nucleotidases in human oviducts with an improved approach simultaneously identifying protein expression and in situ enzyme activity. *Histochem. Cell Biol.* **2018**, *149*, 269–276. [CrossRef]

28. Battistone, M.A.; Merkulova, M.; Park, Y.J.; Peralta, M.A.; Gombar, F.; Brown, D.; Breton, S. Unravelling purinergic regulation in the epididymis: Activation of V-ATPase-dependent acidification by luminal ATP and adenosine. *J. Physiol.* **2019**, *597*, 1957–1973. [CrossRef]

29. Casali, E.A.; de Souza, L.F.; Gelain, D.P.; Kaiser, G.R.; Battastini, A.M.; Sarkis, J.J. Changes in ectonucleotidase activities in rat Sertoli cells during sexual maturation. *Mol. Cell. Biochem.* **2003**, *247*, 111–119. [CrossRef]

30. Kauffenstein, G.; Pelletier, J.; Lavoie, E.G.; Kukulski, F.; Martin-Satue, M.; Dufresne, S.S.; Frenette, J.; Ribas Furstenau, C.; Sereda, M.J.; Toutain, B.; et al. Nucleoside triphosphate diphosphohydrolase-1 ectonucleotidase is required for normal vas deferens contraction and male fertility through maintaining P2X1 receptor function. *J. Biol. Chem.* **2014**, *289*, 28629–28639. [CrossRef]

31. Martin-Satue, M.; Lavoie, E.G.; Pelletier, J.; Fausther, M.; Csizmadia, E.; Guckelberger, O.; Robson, S.C.; Sevigny, J. Localization of plasma membrane bound NTPDases in the murine reproductive tract. *Histochem. Cell Biol.* **2009**, *131*, 615–628. [CrossRef] [PubMed]

32. Martinez-Ramirez, A.S.; Vazquez-Cuevas, F.G. Purinergic signaling in the ovary. *Mol. Reprod. Dev.* **2015**, *82*, 839–848. [CrossRef] [PubMed]

33. Trapero, C.; Vidal, A.; Rodriguez-Martinez, A.; Sevigny, J.; Ponce, J.; Coroleu, B.; Matias-Guiu, X.; Martin-Satue, M. The ectonucleoside triphosphate diphosphohydrolase-2 (NTPDase2) in human endometrium: A novel marker of basal stroma and mesenchymal stem cells. *Purinergic Signal.* **2019**, *15*, 225–236. [CrossRef] [PubMed]

34. Lu, D.; Insel, P.A. Cellular mechanisms of tissue fibrosis. 6. Purinergic signaling and response in fibroblasts and tissue fibrosis. *Am. J. Physiol. Cell Physiol.* **2014**, *306*, C779–C788. [CrossRef] [PubMed]

35. Burnstock, G. Purinergic Signaling in the Cardiovascular System. *Circ. Res.* **2017**, *120*, 207–228. [CrossRef]

36. Burnstock, G.; Verkhratsky, A. Long-term (trophic) purinergic signalling: Purinoceptors control cell proliferation, differentiation and death. *Cell Death Dis.* **2010**, *1*, e9. [CrossRef]

37. Trapero, C.; Vidal, A.; Fernandez-Montoli, M.E.; Coroleu, B.; Tresserra, F.; Barri, P.; Gomez de Aranda, I.; Sevigny, J.; Ponce, J.; Matias-Guiu, X.; et al. Impaired Expression of Ectonucleotidases in Ectopic and Eutopic Endometrial Tissue Is in Favor of ATP Accumulation in the Tissue Microenvironment in Endometriosis. *Int. J. Mol. Sci.* **2019**, *20*, 5532. [CrossRef]

38. Hood, B.L.; Liu, B.; Alkhas, A.; Shoji, Y.; Challa, R.; Wang, G.; Ferguson, S.; Oliver, J.; Mitchell, D.; Bateman, N.W.; et al. Proteomics of the human endometrial glandular epithelium and stroma from the proliferative and secretory phases of the menstrual cycle. *Biol. Reprod.* **2015**, *92*, 106. [CrossRef]

39. Chen, Q.; Xin, A.; Qu, R.; Zhang, W.; Li, L.; Chen, J.; Lu, X.; Gu, Y.; Li, J.; Sun, X. Expression of ENPP3 in human cyclic endometrium: A novel molecule involved in embryo implantation. *Reprod. Fertil. Dev.* **2018**, *30*, 1277–1285. [CrossRef]

40. Boggavarapu, N.R.; Lalitkumar, S.; Joshua, V.; Kasvandik, S.; Salumets, A.; Lalitkumar, P.G.; Gemzell-Danielsson, K. Compartmentalized gene expression profiling of receptive endometrium reveals progesterone regulated ENPP3 is differentially expressed and secreted in glycosylated form. *Sci. Rep.* **2016**, *6*, 33811. [CrossRef]

41. Texido, L.; Romero, C.; Vidal, A.; Garcia-Valero, J.; Fernandez Montoli, M.E.; Baixeras, N.; Condom, E.; Ponce, J.; Garcia-Tejedor, A.; Martin-Satue, M. Ecto-nucleotidases activities in the contents of ovarian endometriomas: Potential biomarkers of endometriosis. *Mediat. Inflamm.* **2014**, *2014*, 120673. [CrossRef] [PubMed]

42. Leone, R.D.; Emens, L.A. Targeting adenosine for cancer immunotherapy. *J. Immunother. Cancer* **2018**, *6*, 57. [CrossRef] [PubMed]

43. Aliagas, E.; Vidal, A.; Texido, L.; Ponce, J.; Condom, E.; Martin-Satue, M. High expression of ecto-nucleotidases CD39 and CD73 in human endometrial tumors. *Mediat. Inflamm.* **2014**, *2014*, 509027. [CrossRef] [PubMed]

44. Trapero, C.; Jover, L.; Fernandez-Montoli, M.E.; Garcia-Tejedor, A.; Vidal, A.; Gomez de Aranda, I.; Ponce, J.; Matias-Guiu, X.; Martin-Satue, M. Analysis of the ectoenzymes ADA, ALP, ENPP1, and ENPP3, in the contents of ovarian endometriomas as candidate biomarkers of endometriosis. *Am. J. Reprod. Immunol.* **2018**, *79*. [CrossRef]

45. Busacca, M.; Chiaffarino, F.; Candiani, M.; Vignali, M.; Bertulessi, C.; Oggioni, G.; Parazzini, F. Determinants of long-term clinically detected recurrence rates of deep, ovarian, and pelvic endometriosis. *Am. J. Obstet. Gynecol.* **2006**, *195*, 426–432. [CrossRef] [PubMed]

46. Dai, Y.; Leng, J.H.; Lang, J.H.; Li, X.Y.; Zhang, J.J. Anatomical distribution of pelvic deep infiltrating endometriosis and its relationship with pain symptoms. *Chin. Med. J.* **2012**, *125*, 209–213. [PubMed]

47. Rush, G.; Misajon, R. Examining subjective wellbeing and health-related quality of life in women with endometriosis. *Health Care Women Int.* **2018**, *39*, 303–321. [CrossRef]

48. Facchin, F.; Barbara, G.; Dridi, D.; Alberico, D.; Buggio, L.; Somigliana, E.; Saita, E.; Vercellini, P. Mental health in women with endometriosis: Searching for predictors of psychological distress. *Hum. Reprod.* **2017**, *32*, 1855–1861. [CrossRef]

49. Coxon, L.; Horne, A.W.; Vincent, K. Pathophysiology of endometriosis-associated pain: A review of pelvic and central nervous system mechanisms. *Best Pract. Res. Clin. Obstet. Gynaecol.* **2018**, *51*, 53–67. [CrossRef]

50. Vercellini, P.; Vigano, P.; Somigliana, E.; Fedele, L. Endometriosis: Pathogenesis and treatment. *Nat. Rev. Endocrinol.* **2014**, *10*, 261–275. [CrossRef]

51. Aredo, J.V.; Heyrana, K.J.; Karp, B.I.; Shah, J.P.; Stratton, P. Relating Chronic Pelvic Pain and Endometriosis to Signs of Sensitization and Myofascial Pain and Dysfunction. *Semin. Reprod. Med.* **2017**, *35*, 88–97. [CrossRef] [PubMed]

52. Rolla, E. Endometriosis: Advances and controversies in classification, pathogenesis, diagnosis, and treatment. *F1000Research* **2019**, *8*. [CrossRef] [PubMed]

53. Becker, C.M.; Gattrell, W.T.; Gude, K.; Singh, S.S. Reevaluating response and failure of medical treatment of endometriosis: A systematic review. *Fertil. Steril.* **2017**, *108*, 125–136. [CrossRef] [PubMed]

54. Xu, Y.; Zhao, W.; Li, T.; Zhao, Y.; Bu, H.; Song, S. Effects of acupuncture for the treatment of endometriosis-related pain: A systematic review and meta-analysis. *PLoS ONE* **2017**, *12*, e0186616. [CrossRef]

55. Brown, J.; Crawford, T.J.; Allen, C.; Hopewell, S.; Prentice, A. Nonsteroidal anti-inflammatory drugs for pain in women with endometriosis. *Cochrane Database Syst. Rev.* **2017**, *1*, Cd004753. [CrossRef]

56. Singh, S.S.; Suen, M.W. Surgery for endometriosis: Beyond medical therapies. *Fertil. Steril.* **2017**, *107*, 549–554. [CrossRef]

57. Chaichian, S.; Kabir, A.; Mehdizadehkashi, A.; Rahmani, K.; Moghimi, M.; Moazzami, B. Comparing the Efficacy of Surgery and Medical Therapy for Pain Management in Endometriosis: A Systematic Review and Meta-analysis. *Pain Physician* **2017**, *20*, 185–195.

58. Kim, J.H.; Han, E. Endometriosis and Female Pelvic Pain. *Semin. Reprod. Med.* **2018**, *36*, 143–151. [CrossRef]

59. Andres, M.P.; Borrelli, G.M.; Abrao, M.S. Endometriosis classification according to pain symptoms: Can the ASRM classification be improved? *Best Pract. Res. Clin. Obstet. Gynaecol.* **2018**, *51*, 111–118. [CrossRef]

60. Brown, J.; Crawford, T.J.; Datta, S.; Prentice, A. Oral contraceptives for pain associated with endometriosis. *Cochrane Database Syst. Rev.* **2018**, *5*, Cd001019. [CrossRef]

61. Vercellini, P.; Buggio, L.; Frattaruolo, M.P.; Borghi, A.; Dridi, D.; Somigliana, E. Medical treatment of endometriosis-related pain. *Best Pract. Res. Clin. Obstet. Gynaecol.* **2018**, *51*, 68–91. [CrossRef] [PubMed]

62. Flyckt, R.; Kim, S.; Falcone, T. Surgical Management of Endometriosis in Patients with Chronic Pelvic Pain. *Semin. Reprod. Med.* **2017**, *35*, 54–64. [CrossRef] [PubMed]

63. Yan, D.; Liu, X.; Guo, S.W. Nerve fibers and endometriotic lesions: Partners in crime in inflicting pains in women with endometriosis. *Eur. J. Obstet. Gynecol. Reprod. Biol.* **2017**, *209*, 14–24. [CrossRef] [PubMed]

64. Bele, T.; Fabbretti, E. P2X receptors, sensory neurons and pain. *Curr. Med. Chem.* **2015**, *22*, 845–850. [CrossRef]

65. Burnstock, G. Purinergic Mechanisms and Pain. *Adv. Pharmacol* **2016**, *75*, 91–137. [CrossRef]

66. Inoue, K. Purinergic signaling in microglia in the pathogenesis of neuropathic pain. *Proc. Jpn. Acad. Ser. B Phys. Biol. Sci.* **2017**, *93*, 174–182. [CrossRef]

67. Magni, G.; Ceruti, S. The role of adenosine and P2Y receptors expressed by multiple cell types in pain transmission. *Brain Res. Bull.* **2019**. [CrossRef]

68. Magni, G.; Riccio, D.; Ceruti, S. Tackling Chronic Pain and Inflammation through the Purinergic System. *Curr. Med. Chem.* **2018**, *25*, 3830–3865. [CrossRef]

69. Tsuda, M. P2 receptors, microglial cytokines and chemokines, and neuropathic pain. *J. Neurosci. Res.* **2017**, *95*, 1319–1329. [CrossRef]

70. Kuan, Y.H.; Shyu, B.C. Nociceptive transmission and modulation via P2X receptors in central pain syndrome. *Mol. Brain* **2016**, *9*, 58. [CrossRef]

71. Liang, Y.; Xie, H.; Wu, J.; Liu, D.; Yao, S. Villainous role of estrogen in macrophage-nerve interaction in endometriosis. *Reprod. Biol. Endocrinol.* **2018**, *16*, 122. [CrossRef] [PubMed]

72. Inoue, K.; Tsuda, M. Microglia in neuropathic pain: Cellular and molecular mechanisms and therapeutic potential. *Nat. Rev. Neurosci.* **2018**, *19*, 138–152. [CrossRef] [PubMed]

73. Scholz, J.; Woolf, C.J. The neuropathic pain triad: Neurons, immune cells and glia. *Nat. Neurosci.* **2007**, *10*, 1361–1368. [CrossRef]

74. Tokushige, N.; Markham, R.; Russell, P.; Fraser, I.S. Nerve fibres in peritoneal endometriosis. *Hum. Reprod.* **2006**, *21*, 3001–3007. [CrossRef] [PubMed]

75. Tokushige, N.; Russell, P.; Black, K.; Barrera, H.; Dubinovsky, S.; Markham, R.; Fraser, I.S. Nerve fibers in ovarian endometriomas. *Fertil. Steril.* **2010**, *94*, 1944–1947. [CrossRef]

76. Zhang, X.; Yao, H.; Huang, X.; Lu, B.; Xu, H.; Zhou, C. Nerve fibres in ovarian endometriotic lesions in women with ovarian endometriosis. *Hum. Reprod.* **2010**, *25*, 392–397. [CrossRef]

77. Wang, G.; Tokushige, N.; Markham, R.; Fraser, I.S. Rich innervation of deep infiltrating endometriosis. *Hum. Reprod.* **2009**, *24*, 827–834. [CrossRef]

78. Arnold, J.; Barcena de Arellano, M.L.; Ruster, C.; Vercellino, G.F.; Chiantera, V.; Schneider, A.; Mechsner, S. Imbalance between sympathetic and sensory innervation in peritoneal endometriosis. *Brain Behav. Immun.* **2012**, *26*, 132–141. [CrossRef]

79. Ueno, S.; Tsuda, M.; Iwanaga, T.; Inoue, K. Cell type-specific ATP-activated responses in rat dorsal root ganglion neurons. *Br. J. Pharmacol.* **1999**, *126*, 429–436. [CrossRef]

80. Chen, Y.; Li, G.W.; Wang, C.; Gu, Y.; Huang, L.Y. Mechanisms underlying enhanced P2X receptor-mediated responses in the neuropathic pain state. *Pain* **2005**, *119*, 38–48. [CrossRef]

81. Greaves, E.; Grieve, K.; Horne, A.W.; Saunders, P.T. Elevated peritoneal expression and estrogen regulation of nociceptive ion channels in endometriosis. *J. Clin. Endocrinol. Metab.* **2014**, *99*, E1738–E1743. [CrossRef] [PubMed]

82. Ding, S.; Zhu, L.; Tian, Y.; Zhu, T.; Huang, X.; Zhang, X. P2X3 receptor involvement in endometriosis pain via ERK signaling pathway. *PLoS ONE* **2017**, *12*, e0184647. [CrossRef] [PubMed]

83. Bele, T.; Fabbretti, E. The scaffold protein calcium/calmodulin-dependent serine protein kinase controls ATP release in sensory ganglia upon P2X3 receptor activation and is part of an ATP keeper complex. *J. Neurochem.* **2016**, *138*, 587–597. [CrossRef] [PubMed]

84. Stratton, P.; Khachikyan, I.; Sinaii, N.; Ortiz, R.; Shah, J. Association of chronic pelvic pain and endometriosis with signs of sensitization and myofascial pain. *Obstet. Gynecol.* **2015**, *125*, 719–728. [CrossRef] [PubMed]

85. Nakagawa, T.; Wakamatsu, K.; Zhang, N.; Maeda, S.; Minami, M.; Satoh, M.; Kaneko, S. Intrathecal administration of ATP produces long-lasting allodynia in rats: Differential mechanisms in the phase of the induction and maintenance. *Neuroscience* **2007**, *147*, 445–455. [CrossRef]

86. Ding, S.; Yu, Q.; Wang, J.; Zhu, L.; Li, T.; Guo, X.; Zhang, X. Activation of ATF3/AP-1 signaling pathway is required for P2X3-induced endometriosis pain. *Hum. Reprod.* **2020**, *35*, 1130–1144. [CrossRef]

87. Yuan, M.; Ding, S.; Meng, T.; Lu, B.; Shao, S.; Zhang, X.; Yuan, H.; Hu, F. Effect of A-317491 delivered by glycolipid-like polymer micelles on endometriosis pain. *Int. J. Nanomed.* **2017**, *12*, 8171–8183. [CrossRef]

88. Tsudo, T.; Harada, T.; Iwabe, T.; Tanikawa, M.; Nagano, Y.; Ito, M.; Taniguchi, F.; Terakawa, N. Altered gene expression and secretion of interleukin-6 in stromal cells derived from endometriotic tissues. *Fertil. Steril.* **2000**, *73*, 205–211. [CrossRef]

89. Wei, X.; Shao, X. Nobiletin alleviates endometriosis via down-regulating NF-κB activity in endometriosis mouse model. *Biosci. Rep.* **2018**, *38*. [CrossRef]

90. Yu, J.; Francisco, A.M.C.; Patel, B.G.; Cline, J.M.; Zou, E.; Berga, S.L.; Taylor, R.N. IL-1β Stimulates Brain-Derived Neurotrophic Factor Production in Eutopic Endometriosis Stromal Cell Cultures: A Model for Cytokine Regulation of Neuroangiogenesis. *Am. J. Pathol.* **2018**, *188*, 2281–2292. [CrossRef]

91. Makabe, T.; Koga, K.; Miyashita, M.; Takeuchi, A.; Sue, F.; Taguchi, A.; Urata, Y.; Izumi, G.; Takamura, M.; Harada, M.; et al. Drospirenone reduces inflammatory cytokines, vascular endothelial growth factor (VEGF) and nerve growth factor (NGF) expression in human endometriotic stromal cells. *J. Reprod. Immunol.* **2017**, *119*, 44–48. [CrossRef] [PubMed]

92. Lin, J.; Li, G.; Den, X.; Xu, C.; Liu, S.; Gao, Y.; Liu, H.; Zhang, J.; Li, X.; Liang, S. VEGF and its receptor-2 involved in neuropathic pain transmission mediated by P2X$_{2/3}$ receptor of primary sensory neurons. *Brain Res. Bull.* **2010**, *83*, 284–291. [CrossRef] [PubMed]

93. Ye, Y.; Ono, K.; Bernabé, D.G.; Viet, C.T.; Pickering, V.; Dolan, J.C.; Hardt, M.; Ford, A.P.; Schmidt, B.L. Adenosine triphosphate drives head and neck cancer pain through P2X2/3 heterotrimers. *Acta Neuropathol. Commun.* **2014**, *2*, 62. [CrossRef] [PubMed]

94. Yan, D.; Liu, X.; Guo, S.W. Neuropeptides Substance P and Calcitonin Gene Related Peptide Accelerate the Development and Fibrogenesis of Endometriosis. *Sci. Rep.* **2019**, *9*, 2698. [CrossRef] [PubMed]

95. Kritas, S.K.; Caraffa, A.; Antinolfi, P.; Saggini, A.; Pantalone, A.; Rosati, M.; Tei, M.; Speziali, A.; Saggini, R.; Pandolfi, F.; et al. Nerve growth factor interactions with mast cells. *Int. J. Immunopathol. Pharmacol.* **2014**, *27*, 15–19. [CrossRef] [PubMed]

96. Zhu, T.H.; Ding, S.J.; Li, T.T.; Zhu, L.B.; Huang, X.F.; Zhang, X.M. Estrogen is an important mediator of mast cell activation in ovarian endometriomas. *Reproduction* **2018**, *155*, 73–83. [CrossRef]

97. Theoharides, T.C.; Tsilioni, I.; Bawazeer, M. Mast Cells, Neuroinflammation and Pain in Fibromyalgia Syndrome. *Front. Cell. Neurosci.* **2019**, *13*, 353. [CrossRef]

98. Zhu, H.; Yu, Y.; Zheng, L.; Wang, L.; Li, C.; Yu, J.; Wei, J.; Wang, C.; Zhang, J.; Xu, S.; et al. Chronic inflammatory pain upregulates expression of P2Y2 receptor in small-diameter sensory neurons. *Metab. Brain Dis.* **2015**, *30*, 1349–1358. [CrossRef]

99. Chen, X.; Molliver, D.C.; Gebhart, G.F. The P2Y2 receptor sensitizes mouse bladder sensory neurons and facilitates purinergic currents. *J. Neurosci.* **2010**, *30*, 2365–2372. [CrossRef]

100. Kwon, S.G.; Roh, D.H.; Yoon, S.Y.; Moon, J.Y.; Choi, S.R.; Choi, H.S.; Kang, S.Y.; Han, H.J.; Beitz, A.J.; Oh, S.B.; et al. Acid evoked thermal hyperalgesia involves peripheral P2Y1 receptor mediated TRPV1 phosphorylation in a rodent model of thrombus induced ischemic pain. *Mol. Pain* **2014**, *10*, 2. [CrossRef]

101. Kwon, S.G.; Roh, D.H.; Yoon, S.Y.; Moon, J.Y.; Choi, S.R.; Choi, H.S.; Kang, S.Y.; Han, H.J.; Beitz, A.J.; Lee, J.H. Blockade of peripheral P2Y1 receptors prevents the induction of thermal hyperalgesia via modulation of TRPV1 expression in carrageenan-induced inflammatory pain rats: Involvement of p38 MAPK phosphorylation in DRGs. *Neuropharmacology* **2014**, *79*, 368–379. [CrossRef] [PubMed]

102. Luo, Y.; Feng, C.; Wu, J.; Wu, Y.; Liu, D.; Dai, F.; Zhang, J. P2Y1, P2Y2, and TRPV1 Receptors Are Increased in Diarrhea-Predominant Irritable Bowel Syndrome and P2Y2 Correlates with Abdominal Pain. *Dig. Dis. Sci.* **2016**, *61*, 2878–2886. [CrossRef] [PubMed]

103. Lakshmi, S.; Joshi, P.G. Co-activation of P2Y2 receptor and TRPV channel by ATP: Implications for ATP induced pain. *Cell. Mol. Neurobiol.* **2005**, *25*, 819–832. [CrossRef] [PubMed]

104. Munrós, J.; Tàssies, D.; Reverter, J.C.; Martin, L.; Pérez, A.; Carmona, F.; Martínez-Zamora, M. Circulating Neutrophil Extracellular Traps Are Elevated in Patients With Deep Infiltrating Endometriosis. *Reprod. Sci.* **2019**, *26*, 70–76. [CrossRef]

105. Oliveira-Fusaro, M.C.; Gregory, N.S.; Kolker, S.J.; Rasmussen, L.; Allen, L.H.; Sluka, K.A. P2X4 Receptors on Muscle Macrophages Are Required for Development of Hyperalgesia in an Animal Model of Activity-Induced Muscle Pain. *Mol. Neurobiol.* **2020**, *57*, 1917–1929. [CrossRef]

106. Janks, L.; Sprague, R.S.; Egan, T.M. ATP-Gated P2X7 Receptors Require Chloride Channels To Promote Inflammation in Human Macrophages. *J. Immunol.* **2019**, *202*, 883–898. [CrossRef]

107. Ulmann, L.; Hirbec, H.; Rassendren, F. P2X4 receptors mediate PGE2 release by tissue-resident macrophages and initiate inflammatory pain. *EMBO J.* **2010**, *29*, 2290–2300. [CrossRef]

108. Possover, M. Five-Year Follow-Up After Laparoscopic Large Nerve Resection for Deep Infiltrating Sciatic Nerve Endometriosis. *J. Minim. Invasive Gynecol.* **2017**, *24*, 822–826. [CrossRef]

109. Coull, J.A.; Beggs, S.; Boudreau, D.; Boivin, D.; Tsuda, M.; Inoue, K.; Gravel, C.; Salter, M.W.; De Koninck, Y. BDNF from microglia causes the shift in neuronal anion gradient underlying neuropathic pain. *Nature* **2005**, *438*, 1017–1021. [CrossRef]

110. Ulmann, L.; Hatcher, J.P.; Hughes, J.P.; Chaumont, S.; Green, P.J.; Conquet, F.; Buell, G.N.; Reeve, A.J.; Chessell, I.P.; Rassendren, F. Up-regulation of P2X4 receptors in spinal microglia after peripheral nerve injury mediates BDNF release and neuropathic pain. *J. Neurosci.* **2008**, *28*, 11263–11268. [CrossRef]

111. Long, T.; He, W.; Pan, Q.; Zhang, S.; Zhang, D.; Qin, G.; Chen, L.; Zhou, J. Microglia P2X4R-BDNF signalling contributes to central sensitization in a recurrent nitroglycerin-induced chronic migraine model. *J. Headache Pain* **2020**, *21*, 4. [CrossRef]

112. Xu, F.; Yang, J.; Lu, F.; Liu, R.; Zheng, J.; Zhang, J.; Cui, W.; Wang, C.; Zhou, W.; Wang, Q.; et al. Fast Green FCF Alleviates Pain Hypersensitivity and Down-Regulates the Levels of Spinal P2X4 Expression and Pro-inflammatory Cytokines in a Rodent Inflammatory Pain Model. *Front. Pharmacol.* **2018**, *9*, 534. [CrossRef] [PubMed]

113. Ji, R.R.; Chamessian, A.; Zhang, Y.Q. Pain regulation by non-neuronal cells and inflammation. *Science* **2016**, *354*, 572–577. [CrossRef] [PubMed]

114. Liu, P.W.; Yue, M.X.; Zhou, R.; Niu, J.; Huang, D.J.; Xu, T.; Luo, P.; Liu, X.H.; Zeng, J.W. P2Y$_{12}$ and P2Y$_{13}$ receptors involved in ADP β s induced the release of IL-1 β, IL-6 and TNF-α from cultured dorsal horn microglia. *J. Pain Res.* **2017**, *10*, 1755–1767. [CrossRef]

115. Yu, T.; Zhang, X.; Shi, H.; Tian, J.; Sun, L.; Hu, X.; Cui, W.; Du, D. P2Y12 regulates microglia activation and excitatory synaptic transmission in spinal lamina II neurons during neuropathic pain in rodents. *Cell Death Dis.* **2019**, *10*, 165. [CrossRef] [PubMed]

116. Shieh, C.H.; Heinrich, A.; Serchov, T.; van Calker, D.; Biber, K. P2X7-dependent, but differentially regulated release of IL-6, CCL2, and TNF-α in cultured mouse microglia. *Glia* **2014**, *62*, 592–607. [CrossRef] [PubMed]

117. Garré, J.M.; Yang, G.; Bukauskas, F.F.; Bennett, M.V. FGF-1 Triggers Pannexin-1 Hemichannel Opening in Spinal Astrocytes of Rodents and Promotes Inflammatory Responses in Acute Spinal Cord Slices. *J. Neurosci.* **2016**, *36*, 4785–4801. [CrossRef]

118. Sorge, R.E.; Trang, T.; Dorfman, R.; Smith, S.B.; Beggs, S.; Ritchie, J.; Austin, J.S.; Zaykin, D.V.; Vander Meulen, H.; Costigan, M.; et al. Genetically determined P2X7 receptor pore formation regulates variability in chronic pain sensitivity. *Nat. Med.* **2012**, *18*, 595–599. [CrossRef]

119. Clark, A.K.; Yip, P.K.; Grist, J.; Gentry, C.; Staniland, A.A.; Marchand, F.; Dehvari, M.; Wotherspoon, G.; Winter, J.; Ullah, J.; et al. Inhibition of spinal microglial cathepsin S for the reversal of neuropathic pain. *Proc. Natl. Acad. Sci. USA* **2007**, *104*, 10655–10660. [CrossRef]

120. Clark, A.K.; Yip, P.K.; Malcangio, M. The liberation of fractalkine in the dorsal horn requires microglial cathepsin S. *J. Neurosci.* **2009**, *29*, 6945–6954. [CrossRef]

121. Clark, A.K.; Wodarski, R.; Guida, F.; Sasso, O.; Malcangio, M. Cathepsin S release from primary cultured microglia is regulated by the P2X7 receptor. *Glia* **2010**, *58*, 1710–1726. [CrossRef] [PubMed]

122. Liu, Z.; Chen, S.; Qiu, C.; Sun, Y.; Li, W.; Jiang, J.; Zhang, J.M. Fractalkine/CX3CR1 Contributes to Endometriosis-Induced Neuropathic Pain and Mechanical Hypersensitivity in Rats. *Front. Cell. Neurosci.* **2018**, *12*, 495. [CrossRef] [PubMed]

123. Hou, X.X.; Zhou, W.J.; Wang, X.Q.; Li, D.J. Fractalkine/CX3CR1 is involved in the pathogenesis of endometriosis by regulating endometrial stromal cell proliferation and invasion. *Am. J. Reprod. Immunol.* **2016**, *76*, 318–325. [CrossRef] [PubMed]

124. Mousseau, M.; Burma, N.E.; Lee, K.Y.; Leduc-Pessah, H.; Kwok, C.H.T.; Reid, A.R.; O'Brien, M.; Sagalajev, B.; Stratton, J.A.; Patrick, N.; et al. Microglial pannexin-1 channel activation is a spinal determinant of joint pain. *Sci. Adv.* **2018**, *4*, eaas9846. [CrossRef] [PubMed]

125. Katz, N.K.; Ryals, J.M.; Wright, D.E. Central or peripheral delivery of an adenosine A1 receptor agonist improves mechanical allodynia in a mouse model of painful diabetic neuropathy. *Neuroscience* **2015**, *285*, 312–323. [CrossRef]

126. Imlach, W.L.; Bhola, R.F.; May, L.T.; Christopoulos, A.; Christie, M.J. A Positive Allosteric Modulator of the Adenosine A1 Receptor Selectively Inhibits Primary Afferent Synaptic Transmission in a Neuropathic Pain Model. *Mol. Pharmacol.* **2015**, *88*, 460–468. [CrossRef]

127. Luongo, L.; Guida, F.; Imperatore, R.; Napolitano, F.; Gatta, L.; Cristino, L.; Giordano, C.; Siniscalco, D.; Di Marzo, V.; Bellini, G.; et al. The A1 adenosine receptor as a new player in microglia physiology. *Glia* **2014**, *62*, 122–132. [CrossRef]

128. Borea, P.A.; Gessi, S.; Merighi, S.; Vincenzi, F.; Varani, K. Pharmacology of Adenosine Receptors: The State of the Art. *Physiol. Rev.* **2018**, *98*, 1591–1625. [CrossRef]

129. El-Hashim, A.Z.; Mathews, S.; Al-Shamlan, F. Central adenosine A$_1$ receptors inhibit cough *via* suppression of excitatory glutamatergic and tachykininergic neurotransmission. *Br. J. Pharmacol.* **2018**, *175*, 3162–3174. [CrossRef]

130. Effendi, W.I.; Nagano, T.; Kobayashi, K.; Nishimura, Y. Focusing on Adenosine Receptors as a Potential Targeted Therapy in Human Diseases. *Cells* **2020**, *9*, 785. [CrossRef]

131. Li, Z.H.; Cui, D.; Qiu, C.J.; Song, X.J. Cyclic nucleotide signaling in sensory neuron hyperexcitability and chronic pain after nerve injury. *Neurobiol. Pain* **2019**, *6*, 100028. [CrossRef] [PubMed]

132. Loram, L.C.; Taylor, F.R.; Strand, K.A.; Harrison, J.A.; Rzasalynn, R.; Sholar, P.; Rieger, J.; Maier, S.F.; Watkins, L.R. Intrathecal injection of adenosine 2A receptor agonists reversed neuropathic allodynia through protein kinase (PK)A/PKC signaling. *Brain Behav. Immun.* **2013**, *33*, 112–122. [CrossRef] [PubMed]

133. Loram, L.C.; Harrison, J.A.; Sloane, E.M.; Hutchinson, M.R.; Sholar, P.; Taylor, F.R.; Berkelhammer, D.; Coats, B.D.; Poole, S.; Milligan, E.D.; et al. Enduring reversal of neuropathic pain by a single intrathecal injection of adenosine 2A receptor agonists: A novel therapy for neuropathic pain. *J. Neurosci.* **2009**, *29*, 14015–14025. [CrossRef] [PubMed]

134. Kwilasz, A.J.; Green Fulgham, S.M.; Ellis, A.; Patel, H.P.; Duran-Malle, J.C.; Favret, J.; Harvey, L.O.; Rieger, J.; Maier, S.F.; Watkins, L.R. A single peri-sciatic nerve administration of the adenosine 2A receptor agonist ATL313 produces long-lasting anti-allodynia and anti-inflammatory effects in male rats. *Brain Behav. Immun.* **2019**, *76*, 116–125. [CrossRef]

135. Kwilasz, A.J.; Ellis, A.; Wieseler, J.; Loram, L.; Favret, J.; McFadden, A.; Springer, K.; Falci, S.; Rieger, J.; Maier, S.F.; et al. Sustained reversal of central neuropathic pain induced by a single intrathecal injection of adenosine A2A receptor agonists. *Brain Behav. Immun.* **2018**, *69*, 470–479. [CrossRef]

136. Meng, F.; Guo, Z.; Hu, Y.; Mai, W.; Zhang, Z.; Zhang, B.; Ge, Q.; Lou, H.; Guo, F.; Chen, J.; et al. CD73-derived adenosine controls inflammation and neurodegeneration by modulating dopamine signalling. *Brain* **2019**, *142*, 700–718. [CrossRef]

137. Gomes, C.; Ferreira, R.; George, J.; Sanches, R.; Rodrigues, D.I.; Gonçalves, N.; Cunha, R.A. Activation of microglial cells triggers a release of brain-derived neurotrophic factor (BDNF) inducing their proliferation in an adenosine A2A receptor-dependent manner: A2A receptor blockade prevents BDNF release and proliferation of microglia. *J. Neuroinflamm.* **2013**, *10*, 16. [CrossRef]

138. Cunha, R.A. How does adenosine control neuronal dysfunction and neurodegeneration? *J. Neurochem.* **2016**, *139*, 1019–1055. [CrossRef]

139. Feoktistov, I.; Biaggioni, I. Role of adenosine A(2B) receptors in inflammation. *Adv. Pharmacol.* **2011**, *61*, 115–144. [CrossRef]

140. Hu, X.; Adebiyi, M.G.; Luo, J.; Sun, K.; Le, T.T.; Zhang, Y.; Wu, H.; Zhao, S.; Karmouty-Quintana, H.; Liu, H.; et al. Sustained Elevated Adenosine via ADORA2B Promotes Chronic Pain through Neuro-immune Interaction. *Cell Rep.* **2016**, *16*, 106–119. [CrossRef]

141. Janes, K.; Symons-Liguori, A.M.; Jacobson, K.A.; Salvemini, D. Identification of A3 adenosine receptor agonists as novel non-narcotic analgesics. *Br. J. Pharmacol.* **2016**, *173*, 1253–1267. [CrossRef] [PubMed]

142. Coppi, E.; Cherchi, F.; Fusco, I.; Failli, P.; Vona, A.; Dettori, I.; Gaviano, L.; Lucarini, E.; Jacobson, K.A.; Tosh, D.K.; et al. Adenosine A3 receptor activation inhibits pronociceptive N-type Ca2+ currents and cell excitability in dorsal root ganglion neurons. *Pain* **2019**, *160*, 1103–1118. [CrossRef] [PubMed]

143. Terayama, R.; Tabata, M.; Maruhama, K.; Iida, S. A$_3$ adenosine receptor agonist attenuates neuropathic pain by suppressing activation of microglia and convergence of nociceptive inputs in the spinal dorsal horn. *Exp. Brain Res.* **2018**, *236*, 3203–3213. [CrossRef]

144. Lorenzo, L.E.; Magnussen, C.; Bailey, A.L.; St Louis, M.; De Koninck, Y.; Ribeiro-da-Silva, A. Spatial and temporal pattern of changes in the number of GAD65-immunoreactive inhibitory terminals in the rat superficial dorsal horn following peripheral nerve injury. *Mol. Pain* **2014**, *10*, 57. [CrossRef] [PubMed]

145. Ge, M.M.; Chen, S.P.; Zhou, Y.Q.; Li, Z.; Tian, X.B.; Gao, F.; Manyande, A.; Tian, Y.K.; Yang, H. The therapeutic potential of GABA in neuron-glia interactions of cancer-induced bone pain. *Eur. J. Pharmacol.* **2019**, *858*, 172475. [CrossRef]

146. Ford, A.; Castonguay, A.; Cottet, M.; Little, J.W.; Chen, Z.; Symons-Liguori, A.M.; Doyle, T.; Egan, T.M.; Vanderah, T.W.; De Koninck, Y.; et al. Engagement of the GABA to KCC2 signaling pathway contributes to the analgesic effects of A3AR agonists in neuropathic pain. *J. Neurosci.* **2015**, *35*, 6057–6067. [CrossRef] [PubMed]

147. Janes, K.; Esposito, E.; Doyle, T.; Cuzzocrea, S.; Tosh, D.K.; Jacobson, K.A.; Salvemini, D. A3 adenosine receptor agonist prevents the development of paclitaxel-induced neuropathic pain by modulating spinal glial-restricted redox-dependent signaling pathways. *Pain* **2014**, *155*, 2560–2567. [CrossRef]

148. Janes, K.; Wahlman, C.; Little, J.W.; Doyle, T.; Tosh, D.K.; Jacobson, K.A.; Salvemini, D. Spinal neuroimmune activation is independent of T-cell infiltration and attenuated by A3 adenosine receptor agonists in a model of oxaliplatin-induced peripheral neuropathy. *Brain Behav. Immun.* **2015**, *44*, 91–99. [CrossRef]

149. Zylka, M.J. Pain-relieving prospects for adenosine receptors and ectonucleotidases. *Trends Mol. Med.* **2011**, *17*, 188–196. [CrossRef]

150. Guerrero, A. A2A Adenosine Receptor Agonists and their Potential Therapeutic Applications. An Update. *Curr. Med. Chem.* **2018**, *25*, 3597–3612. [CrossRef]

151. Zhang, W.J.; Zhu, Z.M.; Liu, Z.X. The role of P2X4 receptor in neuropathic pain and its pharmacological properties. *Pharmacol. Res.* **2020**, *158*, 104875. [CrossRef] [PubMed]

152. Zhang, W.J.; Zhu, Z.M.; Liu, Z.X. The role and pharmacological properties of the P2X7 receptor in neuropathic pain. *Brain Res. Bull.* **2020**, *155*, 19–28. [CrossRef] [PubMed]

153. Keystone, E.C.; Wang, M.M.; Layton, M.; Hollis, S.; McInnes, I.B.; Team, D.C.S. Clinical evaluation of the efficacy of the P2X7 purinergic receptor antagonist AZD9056 on the signs and symptoms of rheumatoid arthritis in patients with active disease despite treatment with methotrexate or sulphasalazine. *Ann. Rheum. Dis.* **2012**, *71*, 1630–1635. [CrossRef] [PubMed]

154. Horváth, G.; Göloncsér, F.; Csölle, C.; Király, K.; Andó, R.D.; Baranyi, M.; Koványi, B.; Máté, Z.; Hoffmann, K.; Algaier, I.; et al. Central P2Y12 receptor blockade alleviates inflammatory and neuropathic pain and cytokine production in rodents. *Neurobiol. Dis.* **2014**, *70*, 162–178. [CrossRef]

155. Moriyama, Y.; Hiasa, M.; Sakamoto, S.; Omote, H.; Nomura, M. Vesicular nucleotide transporter (VNUT): Appearance of an actress on the stage of purinergic signaling. *Purinergic Signal.* **2017**, *13*, 387–404. [CrossRef]

156. Masuda, T.; Ozono, Y.; Mikuriya, S.; Kohro, Y.; Tozaki-Saitoh, H.; Iwatsuki, K.; Uneyama, H.; Ichikawa, R.; Salter, M.W.; Tsuda, M.; et al. Dorsal horn neurons release extracellular ATP in a VNUT-dependent manner that underlies neuropathic pain. *Nat. Commun.* **2016**, *7*, 12529. [CrossRef]

157. Yamagata, R.; Nemoto, W.; Nakagawasai, O.; Hung, W.Y.; Shima, K.; Endo, Y.; Tan-No, K. Etidronate attenuates tactile allodynia by spinal ATP release inhibition in mice with partial sciatic nerve ligation. *Naunyn Schmiedeberg Arch. Pharmacol.* **2019**, *392*, 349–357. [CrossRef]

158. Kato, Y.; Hiasa, M.; Ichikawa, R.; Hasuzawa, N.; Kadowaki, A.; Iwatsuki, K.; Shima, K.; Endo, Y.; Kitahara, Y.; Inoue, T.; et al. Identification of a vesicular ATP release inhibitor for the treatment of neuropathic and inflammatory pain. *Proc. Natl. Acad. Sci. USA* **2017**, *114*, E6297–E6305. [CrossRef]

159. Yamakita, S.; Horii, Y.; Takemura, H.; Matsuoka, Y.; Yamashita, A.; Yamaguchi, Y.; Matsuda, M.; Sawa, T.; Amaya, F. Synergistic activation of ERK1/2 between A-fiber neurons and glial cells in the DRG contributes to pain hypersensitivity after tissue injury. *Mol. Pain* **2018**, *14*, 1744806918767508. [CrossRef]

160. Zhang, Y.; Laumet, G.; Chen, S.R.; Hittelman, W.N.; Pan, H.L. Pannexin-1 Up-regulation in the Dorsal Root Ganglion Contributes to Neuropathic Pain Development. *J. Biol. Chem.* **2015**, *290*, 14647–14655. [CrossRef]

161. Wang, A.; Xu, C. The role of connexin43 in neuropathic pain induced by spinal cord injury. *Acta Biochim. Biophys. Sin.* **2019**, *51*, 555–561. [CrossRef] [PubMed]

162. Morioka, N.; Nakamura, Y.; Zhang, F.F.; Hisaoka-Nakashima, K.; Nakata, Y. Role of Connexins in Chronic Pain and Their Potential as Therapeutic Targets for Next-Generation Analgesics. *Biol. Pharm. Bull.* **2019**, *42*, 857–866. [CrossRef] [PubMed]

163. Spray, D.C.; Hanani, M. Gap junctions, pannexins and pain. *Neurosci. Lett.* **2019**, *695*, 46–52. [CrossRef] [PubMed]

164. Street, S.E.; Walsh, P.L.; Sowa, N.A.; Taylor-Blake, B.; Guillot, T.S.; Vihko, P.; Wightman, R.M.; Zylka, M.J. PAP and NT5E inhibit nociceptive neurotransmission by rapidly hydrolyzing nucleotides to adenosine. *Mol. Pain* **2011**, *7*, 80. [CrossRef]

165. Sowa, N.A.; Vadakkan, K.I.; Zylka, M.J. Recombinant mouse PAP has pH-dependent ectonucleotidase activity and acts through A(1)-adenosine receptors to mediate antinociception. *PLoS ONE* **2009**, *4*, e4248. [CrossRef]

166. Sowa, N.A.; Voss, M.K.; Zylka, M.J. Recombinant ecto-5′-nucleotidase (CD73) has long lasting antinociceptive effects that are dependent on adenosine A1 receptor activation. *Mol. Pain* **2010**, *6*, 20. [CrossRef]

167. Deb, P.K.; Deka, S.; Borah, P.; Abed, S.N.; Klotz, K.N. Medicinal Chemistry and Therapeutic Potential of Agonists, Antagonists and Allosteric Modulators of A1 Adenosine Receptor: Current Status and Perspectives. *Curr. Pharm. Des.* **2019**, *25*, 2697–2715. [CrossRef]

168. Jacobson, K.A.; Tosh, D.K.; Jain, S.; Gao, Z.G. Historical and Current Adenosine Receptor Agonists in Preclinical and Clinical Development. *Front. Cell. Neurosci.* **2019**, *13*, 124. [CrossRef]

169. Jacobson, K.A.; Giancotti, L.A.; Lauro, F.; Mufti, F.; Salvemini, D. Treatment of chronic neuropathic pain: Purine receptor modulation. *Pain* **2020**, *161*, 1425–1441. [CrossRef]

170. Jacobson, K.A.; Merighi, S.; Varani, K.; Borea, P.A.; Baraldi, S.; Aghazadeh Tabrizi, M.; Romagnoli, R.; Baraldi, P.G.; Ciancetta, A.; Tosh, D.K.; et al. A$_3$ Adenosine Receptors as Modulators of Inflammation: From Medicinal Chemistry to Therapy. *Med. Res. Rev.* **2018**, *38*, 1031–1072. [CrossRef]

171. Little, J.W.; Ford, A.; Symons-Liguori, A.M.; Chen, Z.; Janes, K.; Doyle, T.; Xie, J.; Luongo, L.; Tosh, D.K.; Maione, S.; et al. Endogenous adenosine A3 receptor activation selectively alleviates persistent pain states. *Brain* **2015**, *138*, 28–35. [CrossRef] [PubMed]

172. Mira, T.A.A.; Buen, M.M.; Borges, M.G.; Yela, D.A.; Benetti-Pinto, C.L. Systematic review and meta-analysis of complementary treatments for women with symptomatic endometriosis. *Int. J. Gynecol. Obstet.* **2018**, *143*, 2–9. [CrossRef] [PubMed]

173. Liang, R.; Li, P.; Peng, X.; Xu, L.; Fan, P.; Peng, J.; Zhou, X.; Xiao, C.; Jiang, M. Efficacy of acupuncture on pelvic pain in patients with endometriosis: Study protocol for a randomized, single-blind, multi-center, placebo-controlled trial. *Trials* **2018**, *19*, 314. [CrossRef] [PubMed]

174. Burnstock, G. Acupuncture: A novel hypothesis for the involvement of purinergic signalling. *Med. Hypotheses* **2009**, *73*, 470–472. [CrossRef]

175. Tang, Y.; Yin, H.Y.; Rubini, P.; Illes, P. Acupuncture-Induced Analgesia: A Neurobiological Basis in Purinergic Signaling. *Neuroscientist* **2016**, *22*, 563–578. [CrossRef]

176. Zhang, Y.; Huang, L.; Kozlov, S.A.; Rubini, P.; Tang, Y.; Illes, P. Acupuncture alleviates acid- and purine-induced pain in rodents. *Br. J. Pharmacol.* **2020**, *177*, 77–92. [CrossRef]

177. Goldman, N.; Chen, M.; Fujita, T.; Xu, Q.; Peng, W.; Liu, W.; Jensen, T.K.; Pei, Y.; Wang, F.; Han, X.; et al. Adenosine A1 receptors mediate local anti-nociceptive effects of acupuncture. *Nat. Neurosci.* **2010**, *13*, 883–888. [CrossRef]

178. Sawynok, J. Caffeine and pain. *Pain* **2011**, *152*, 726–729. [CrossRef]

179. Fredholm, B.B.; Bättig, K.; Holmén, J.; Nehlig, A.; Zvartau, E.E. Actions of caffeine in the brain with special reference to factors that contribute to its widespread use. *Pharmacol. Rev.* **1999**, *51*, 83–133.

180. Kwak, Y.; Choi, H.; Bae, J.; Choi, Y.Y.; Roh, J. Peri-pubertal high caffeine exposure increases ovarian estradiol production in immature rats. *Reprod. Toxicol.* **2017**, *69*, 43–52. [CrossRef]

181. Klonoff-Cohen, H.; Bleha, J.; Lam-Kruglick, P. A prospective study of the effects of female and male caffeine consumption on the reproductive endpoints of IVF and gamete intra-Fallopian transfer. *Hum. Reprod.* **2002**, *17*, 1746–1754. [CrossRef] [PubMed]

182. Lucero, J.; Harlow, B.L.; Barbieri, R.L.; Sluss, P.; Cramer, D.W. Early follicular phase hormone levels in relation to patterns of alcohol, tobacco, and coffee use. *Fertil. Steril.* **2001**, *76*, 723–729. [CrossRef]

183. Hemmert, R.; Schliep, K.C.; Willis, S.; Peterson, C.M.; Louis, G.B.; Allen-Brady, K.; Simonsen, S.E.; Stanford, J.B.; Byun, J.; Smith, K.R. Modifiable life style factors and risk for incident endometriosis. *Paediatr. Perinat. Epidemiol.* **2019**, *33*, 19–25. [CrossRef] [PubMed]

184. Chiaffarino, F.; Bravi, F.; Cipriani, S.; Parazzini, F.; Ricci, E.; Viganò, P.; La Vecchia, C. Coffee and caffeine intake and risk of endometriosis: A meta-analysis. *Eur. J. Nutr.* **2014**, *53*, 1573–1579. [CrossRef] [PubMed]

185. Parazzini, F.; Viganò, P.; Candiani, M.; Fedele, L. Diet and endometriosis risk: A literature review. *Reprod. Biomed. Online* **2013**, *26*, 323–336. [CrossRef]

186. Governo, R.J.; Deuchars, J.; Baldwin, S.A.; King, A.E. Localization of the NBMPR-sensitive equilibrative nucleoside transporter, ENT1, in the rat dorsal root ganglion and lumbar spinal cord. *Brain Res.* **2005**, *1059*, 129–138. [CrossRef]

187. Choca, J.I.; Proudfit, H.K.; Green, R.D. Identification of A1 and A2 adenosine receptors in the rat spinal cord. *J. Pharmacol. Exp. Ther.* **1987**, *242*, 905–910.

188. Maes, S.S.; Pype, S.; Hoffmann, V.L.; Biermans, M.; Meert, T.F. Antihyperalgesic activity of nucleoside transport inhibitors in models of inflammatory pain in guinea pigs. *J. Pain Res.* **2012**, *5*, 391–400. [CrossRef]

189. Suzuki, R.; Stanfa, L.C.; Kowaluk, E.A.; Williams, M.; Jarvis, M.F.; Dickenson, A.H. The effect of ABT-702, a novel adenosine kinase inhibitor, on the responses of spinal neurones following carrageenan inflammation and peripheral nerve injury. *Br. J. Pharmacol.* **2001**, *132*, 1615–1623. [CrossRef]

190. Jarvis, M.F.; Yu, H.; Kohlhaas, K.; Alexander, K.; Lee, C.H.; Jiang, M.; Bhagwat, S.S.; Williams, M.; Kowaluk, E.A. ABT-702 (4-amino-5-(3-bromophenyl)-7-(6-morpholinopyridin-3-yl)pyrido[2, 3-d]pyrimidine), a novel orally effective adenosine kinase inhibitor with analgesic and anti-inflammatory properties: I. In vitro characterization and acute antinociceptive effects in the mouse. *J. Pharmacol. Exp. Ther.* **2000**, *295*, 1156–1164.

191. Kowaluk, E.A.; Mikusa, J.; Wismer, C.T.; Zhu, C.Z.; Schweitzer, E.; Lynch, J.J.; Lee, C.H.; Jiang, M.; Bhagwat, S.S.; Gomtsyan, A.; et al. ABT-702 (4-amino-5-(3-bromophenyl)-7-(6-morpholino-pyridin-3-yl)pyrido[2,3-d]pyrimidine), a novel orally effective adenosine kinase inhibitor with analgesic and anti-inflammatory properties. II. In vivo characterization in the rat. *J. Pharmacol. Exp. Ther.* **2000**, *295*, 1165–1174. [PubMed]

192. Jarvis, M.F. Therapeutic potential of adenosine kinase inhibition-Revisited. *Pharmacol. Res. Perspect.* **2019**, *7*, e00506. [CrossRef] [PubMed]

193. Liu, X.J.; White, T.D.; Sawynok, J. Potentiation of formalin-evoked adenosine release by an adenosine kinase inhibitor and an adenosine deaminase inhibitor in the rat hind paw: A microdialysis study. *Eur. J. Pharmacol.* **2000**, *408*, 143–152. [CrossRef]

194. Keil, G.J.; DeLander, G.E. Adenosine kinase and adenosine deaminase inhibition modulate spinal adenosine- and opioid agonist-induced antinociception in mice. *Eur. J. Pharmacol.* **1994**, *271*, 37–46. [CrossRef]

195. Poon, A.; Sawynok, J. Antinociceptive and anti-inflammatory properties of an adenosine kinase inhibitor and an adenosine deaminase inhibitor. *Eur. J. Pharmacol.* **1999**, *384*, 123–138. [CrossRef]

196. Yoshikawa, N.; Nakamura, K.; Yamaguchi, Y.; Kagota, S.; Shinozuka, K.; Kunitomo, M. Reinforcement of antitumor effect of Cordyceps sinensis by 2′-deoxycoformycin, an adenosine deaminase inhibitor. *In Vivo* **2007**, *21*, 291–295.

197. Bennett, M.; Matutes, E.; Gaulard, P. Hepatosplenic T cell lymphoma responsive to 2′-deoxycoformycin therapy. *Am. J. Hematol.* **2010**, *85*, 727–729. [CrossRef]

198. Spiers, A.S. Deoxycoformycin (pentostatin): Clinical pharmacology, role in the chemotherapy of cancer, and use in other diseases. *Haematologia* **1996**, *27*, 55–84.

199. Todd, S.A.; Morris, T.C.; Alexander, H.D. Myelodysplasia terminating in acute myeloid leukemia in a hairy cell leukemia patient treated with 2-deoxycoformycin. *Leuk. Lymphoma* **2002**, *43*, 1343–1344. [CrossRef]

200. Dalla Rosa, L.; Da Silva, A.S.; Oliveira, C.B.; Gressler, L.T.; Arnold, C.B.; Baldissera, M.D.; Sagrillo, M.; Sangoi, M.; Moresco, R.; Mendes, R.E.; et al. Dose finding of 3′deoxyadenosine and deoxycoformycin for the treatment of Trypanosoma evansi infection: An effective and nontoxic dose. *Microb. Pathog.* **2015**, *85*, 21–28. [CrossRef]

Publisher's Note: MDPI stays neutral with regard to jurisdictional claims in published maps and institutional affiliations.

International Journal of
Molecular Sciences

Review

Neonatal Seizures and Purinergic Signalling

Aida Menéndez Méndez [1,†], Jonathon Smith [1,2,†] and Tobias Engel [1,2,*]

[1] Department of Physiology and Medical Physics, RCSI University of Medicine and Health Sciences, Dublin D02 YN77, Ireland; aidammendez@rcsi.ie (A.M.M.); jonathonsmith@rcsi.ie (J.S.)
[2] FutureNeuro, Science Foundation Ireland Research Centre for Chronic and Rare Neurological Diseases, RCSI University of Medicine and Health Sciences, Dublin D02 YN77, Ireland
* Correspondence: tengel@rcsi.ie; Tel.: +35-314-025-199
† These authors have contributed equally to this work.

Received: 30 September 2020; Accepted: 20 October 2020; Published: 22 October 2020

Abstract: Neonatal seizures are one of the most common comorbidities of neonatal encephalopathy, with seizures aggravating acute injury and clinical outcomes. Current treatment can control early life seizures; however, a high level of pharmacoresistance remains among infants, with increasing evidence suggesting current anti-seizure medication potentiating brain damage. This emphasises the need to develop safer therapeutic strategies with a different mechanism of action. The purinergic system, characterised by the use of adenosine triphosphate and its metabolites as signalling molecules, consists of the membrane-bound P1 and P2 purinoreceptors and proteins to modulate extracellular purine nucleotides and nucleoside levels. Targeting this system is proving successful at treating many disorders and diseases of the central nervous system, including epilepsy. Mounting evidence demonstrates that drugs targeting the purinergic system provide both convulsive and anticonvulsive effects. With components of the purinergic signalling system being widely expressed during brain development, emerging evidence suggests that purinergic signalling contributes to neonatal seizures. In this review, we first provide an overview on neonatal seizure pathology and purinergic signalling during brain development. We then describe in detail recent evidence demonstrating a role for purinergic signalling during neonatal seizures and discuss possible purine-based avenues for seizure suppression in neonates.

Keywords: neonatal seizures; development; ATP; purinergic signalling; P2X7 receptor

1. Introduction

Neonatal seizures are a clinical emergency affecting 3–5 out every 1000 live births and are one of the most common comorbidities of neonatal encephalopathy, with seizures aggravating acute injury and clinical outcomes [1,2]. Neonatal seizures result in a mortality rate up to 20% and contribute to long-term outcomes including epilepsy, cerebral palsy, developmental delay and psychomotor deficits [3,4]. Current treatment strategies aim to reduce the hyperexcitability of brain tissue via the use of anti-seizure drugs (ASDs), with phenobarbital being the first-line drug for neonatal seizures. ASDs, however, fail to resolve seizures in 50% of infants and may exacerbate symptoms and later life neurological deficits [2,5]. Therefore, there is a pressing need to identify novel treatment options with higher response rates and without affecting normal development of the brain.

Purinergic signalling refers to the extracellular communication between cells mediated via purine nucleotides and nucleosides, such as adenosine triphosphate (ATP) and adenosine. The purinergic system involves a complex regulatory machinery including regulatory proteins of purine release and uptake, cell membrane receptors and metabolizing enzymes to remove purines from the extracellular space [6,7]. Research over the past decades has demonstrated purinergic signalling to be involved in literally all human pathological conditions ranging from bone diseases, cancer and diabetes to diseases of the central nervous system (CNS) [8]. In the CNS, targeting different components of the purinergic signalling cascade has been proposed as a potential treatment strategy for a range of different diseases including chronic neurodegenerative diseases (e.g., Alzheimer's disease), psychiatric diseases (e.g., depression), neurological disease epilepsy and acute insults to the brain such as a stroke or traumatic brain injury [9–11]. Emerging evidence also suggests a role for purinergic signalling during early developmental disorders such as schizophrenia and autism spectrum [8,12]. Early brain development comprises a sequence of specific events including proliferation (neurogenesis/gliogenesis), differentiation, migration of neuronal precursors, neuronal network formation and synaptogenesis. Critically, purinergic signalling has been shown to be involved in all of these processes [13]. More recent data now also suggests purinergic signalling to be involved during acute insults to the immature brain including neonatal seizures [14].

In this review, we first provide a summary of neonatal seizures, including current treatments and animal models for its study. We then summarize the different elements of the purinergic system and its role during CNS development. Finally, we discuss current knowledge regarding the role of purinergic signalling during neonatal seizures and provide potential directions for future research.

2. Neonatal Seizures

Seizures are a period of excessive and highly synchronous neuronal brain activity and are one of the most common neurological disorders in newborns admitted to the intensive care unit [2]. Normally, seizures are indicative of an underlying dysfunction in the brain. Early life seizures are widely described as a neurological emergency due to a mortality rate as high as 23% and are well documented to cause later life comorbidities such as postnatal epilepsy and global neurodevelopmental delay [3,15]. A seizure is presented when the physiology of the brain abnormally favours excitatory neurotransmission, i.e., promotion of glutamatergic and disinhibition of γ-aminobutyric acid (GABA)ergic transmission. The neonatal brain is in a hyperexcitable state, essential for normal brain development including processes such as synaptogenesis, dendritic spine density development, glial proliferation, myelination and axon guidance [16,17]. Unfortunately, this hyperexcitable state renders the neonatal population at a greater risk to develop seizures particularly within the first two days of life [18,19]. In fact, the incidence rate of seizures in neonates is between 1.8–3.5 per 1000 live births and 10-fold higher in pre-terms [20,21]. Furthermore, any interference, such as a seizure, during these critical neurodevelopmental mechanisms may produce serious consequences persisting into adulthood. For example, early elevated inflammation is associated with network reorganisation with the potential for epileptogenic circuits and psychiatric disorders, as seen in animal models [16,22–24]. Depending on the study, 20–50% of seizure survivors will express some form of neurodevelopment disability in later life [3,25]. In fact, a comprehensive review of studies which evaluated an overall population of 4538 newborns with neonatal seizures observed that 17.9% developed postneonatal epilepsy [26].

2.1. Aetiologies of Neonatal Seizures

Presentation of neonatal seizures is most commonly symptomatic of an underlying aetiology rather than idiopathic. Many risk factors associated with neonatal seizures are related to a metabolic imbalance during pregnancy or immediately postdelivery, including perinatal infection, hypoglycaemia and intracranial haemorrhage [4,15,27]. Moreover, rare cases of an inborn genetic component of neonatal seizures exist, with the majority altering metabolic pathways, including KCNQ2 mutations, infantile hypophosphatasia (mutations in the tissue nonspecific alkaline phosphatase (TNAP)) and propionic acidaemia (deficiency of propionyl-CoA carboxylase) [27]. However, the most common aetiologies of neonatal seizures are acute neurological insults to the brain that limit oxygen and glucose delivery. This includes ischemic stroke; intracranial haemorrhage; and the most common cause, accounting for 40–60% of neonatal seizure cases, hypoxic-ischemic encephalopathy (HIE) [28–30]. Birth asphyxia, that precedes HIE, is the third most common cause of neonatal mortality (23%), behind infection (36%) and preterm births (28%) [31]. HIE is caused by events that limit efficient oxygen delivery to the preterm or neonatal brain tissue, such as foetal distress or placental pathology. However, neonatal seizures are only presented in moderate or severe HIE [32,33]. Neonatal seizure aetiology can be difficult to determine, with the timing of the first seizure normally a good indicator. In line with this, HIE-induced seizures usually present within the first 48 h of life with the other aetiologies having a later seizure onset [34].

2.2. Animal Models of Neonatal Seizures

Clinical investigation can provide information on aetiologies and consequences of neonatal seizures; however, animal studies are a requirement to elucidate pathogenic mechanisms and possible novel treatments. Many animal models of neonatal seizures are derivatives of adult seizure models. This is typical of models where a chemoconvulsant (e.g., kainic acid (KA), pentylenetetrazole (PTZ) or flurothyl) is used to trigger seizures [35–38]. Direct delivery into the brain of KA, a glutamate receptor agonist, to illicit seizures was first achieved by Ben Ari et al. in 1978 [39]. This model can be translated for use in neonatal rats (P10), in which Mesuret et al. microinjected KA into the amygdala to illicit electrographic nonterminating seizures that persist for at least 1 h, with hippocampal neuronal damage observed 72 h later [40]. An intraperitoneal injection of PTZ, a $GABA_A$ receptor antagonist, can also induce neonatal seizures at any age; however, the pattern of seizures and dose required is age-dependent [41]. It is also possible to illicit seizures with multiple low doses of PTZ [42]. These models are widely used to screen preclinical and currently available drugs at various ages ranging from neonatal to adult [43]. PTZ-induced neonatal seizures at P10 in rats produced neuronal damage yet not neuronal death, a feature common among neonatal seizure models [44]. Despite these models not encompassing a translatable seizure induction to the clinic, they are extremely useful in investigating seizure pathophysiology. However, before stark conclusions can be made, results must be validated in a model more similar to the human condition.

With limited oxygen and glucose delivery predominately responsible for most neonatal seizure cases, many experimental models are built to recapitulate clinical features of HIE and subsequent seizures. The Rice–Vannucci model, first published in 1981, was first to encapsulate features of neonatal ischemia and is the basis of current animal models in which hypoxia-ischemia induces neonatal seizures [45]. This model involves ligating the common carotid artery unilaterally (MCAO (medial carotid artery occlusion)), followed by a brief period of hypoxia in neonatal rats (P7). This was developed from a previous model of hypoxia ischemia in adult rats [46]. Further reiterations of the Rice–Vannucci model have been utilised that vary in the degree of the hypoxia insult (8% O_2, for 30 min–2.5 h), the species of rodent and the age of the rodent used (P2—adulthood). This model is primarily used to study HIE. However, using video-electroencephalogram (EEG), Cuaycong et al. validated this model for use in neonatal seizures in which a period of 90 min hypoxic (8% O_2) insult is required to illicit acute seizures in P10–12 rats [47]. Kadam et al. also observed epileptogenesis in this model (P7 rats, MCAO and 8% O_2 for 2 h), with 56% of rats developing spontaneous seizures in later life [48]. The age of the rodent is an important consideration to make due to the neurodevelopment of the rodent occurring rapidly. P7 age is widely used as it relates to the same brain maturation state as 36-week gestation in a human infant, the final week of gestation, with P10 representing a term infant [49].

More recently, mice have been utilized for neonatal seizure studies by using either a combination of MCAO and hypoxia [50] or hypoxia alone [51,52]. In 2015, Rodriguez et al., building upon these studies, developed a noninvasive model of global hypoxia in mice [53]. Briefly, P7 mice were subjected to 15 min of hypoxic conditions (5% O_2) and presented with symptomatic seizures during at least 1 h post-hypoxia. When assaying other ages, mice either had high mortality or did not present with seizures, highlighting how the age of mice must be carefully considered. This model also encapsulates post-seizure morbidity, with mice who underwent infantile hypoxia showing an increased seizure susceptibility and development of multiple behavioural deficits in later life. These studies invite the use of transgenic mouse lines in neonatal seizure studies. This could add great power by dissecting the complex network of pathophysiological systems following a neonatal insult.

2.3. Current Treatment for Neonatal Seizures

Treatment for neonatal seizures with a known genetic or metabolic component can be relatively simple. For example, seizures attributed to a mutation in the TNAP gene resulting in hypophosphatasia, a deficiency in vitamin B6 metabolism, can be controlled with pyridoxine, the phosphorylated form of vitamin B6 [54,55]. Acute symptomatic seizures, such as those following HIE in infants, have proven much harder to treat. The current standard of care for HIE is to initiate therapeutic hypothermia, and if neonatal seizures are present, a course of at least one anti-seizure medication as well [56,57]. Therapeutic hypothermia has proved very successful to reduce the acute seizure burden and mortality following HIE [58–60]. Unfortunately, therapeutic hypothermia is only effective to reduce the seizure burden in moderate HIE cases and not in severe cases [59]. Therapeutic hypothermia's ability to prevent the later life comorbidities remains inconclusive due to the limited number of studies investigating this. Rates of developing cerebral palsy and neurodevelopmental delay were reduced following therapeutic hypothermia when examined at 18–22 months [60]; however, no significant conclusion regarding therapeutic hypothermia's ability to prevent disability could be made when followed up in later life [61].

For acquired seizures not initiated by HIE, currently, the only treatment strategy is anti-seizure medications, which act to inhibit excitatory glutamatergic or promote GABAergic neurotransmission. These medications are useful and certainly are effective in many cases, yet a level of pharmacoresistance remains, particularly in symptomatic neonatal seizures [62]. Also, concerns have been raised with safety of anti-seizure medications in the developing brain. The three most popular anti-seizure drugs, phenobarbital, valproate and phenytoin, that all act upon different neurotransmitter systems, have all been shown to induce apoptotic neurodegeneration in the developing rodent brain [63]. This can be attributed to the developmental expression levels of these neurotransmitter systems, and hence, the onset of certain drug administration needs careful consideration. The first-line anti-seizure medication is phenobarbital, acting as a positive allosteric modulator of the $GABA_A$ receptor. However, phenobarbital remains ineffective in around 50% of neonates to manage seizures [64]. In the immature brain, $GABA_A$ activation leads to an efflux of chloride ions to promote excitatory neurotransmission needed for natural brain development [65,66]. Nevertheless, this makes the immature brain more susceptible to seizures and therapies targeting GABA may even potentiate seizures and excitotoxicity. There are multiple studies raising concern with phenobarbital's safety due to potentiation of neuronal damage and behavioural deficits observed in rodent models [5,67,68] and various reports of patients developing behavioural abnormalities in later life [69]. In fact, Torolina et al. observed that phenobarbital and midazolam exacerbate neonatal seizure damage even at subclinical doses [68]. Due to the damage seen with current medications, careful consideration is needed to outweigh the risks of seizure management with the possibility to potentiate neuronal damage. Therefore, there is an urgent need to develop new treatments that act upon nonclassical mechanisms of seizure prevention with minimal impact on neurodevelopment. Furthermore, there is limited evidence of therapies to protect against long-term consequences of neonatal seizures, and as such, current clinical focus is targeting initial neonatal seizure [4,70]. In recent years, with better standard of care and earlier diagnosis for neonates, mortality rates have decreased, yet the levels of later life neurological sequelae remain unchanged [71,72], suggesting that current medications are not tackling this aspect effectively.

3. The Purinergic System

Purinergic signalling represents probably one of the most ancient cellular signalling systems. Accordingly, purinergic signalling is an essential signalling system employed by the majority of cells across species with key roles during health and disease [73]. Purinergic signalling comprises a complex regulatory system including nucleoside and nucleotide channels and transporters, purinergic receptors, ectonucleotide-metabolizing enzymes and ectonucleoside transporters [10] (Figure 1). The particular high expression of different components of the purinergic system within the CNS highlights its importance in normal brain function. As such, purinergic signalling is involved in a plethora of different cellular pathways including synaptic transmission, in which purine nucleotides and nucleosides act as neuro- and gliotransmitters or modulators [74–76]; cell proliferation and differentiation [77,78]; mediation of communication between astrocytes and reciprocal communication between neurons and glia [79–81]; and inflammatory processes [82–86]. The following section will briefly introduce the different components of the purinergic system and highlight their relevance to normal brain function.

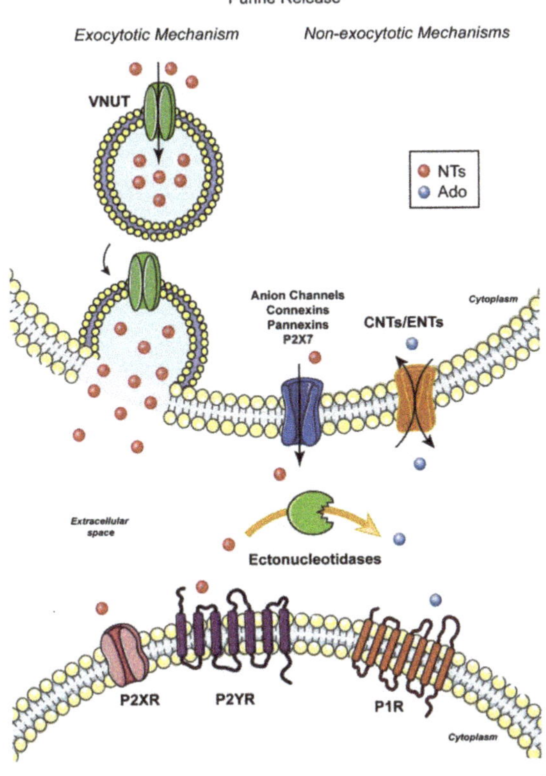

Figure 1. Purine release mechanisms: purines such as ATP and adenosine can be actively released from neurons and glial cells including microglia and astrocytes or passively from damaged or dying cells. Schematic showing the different release mechanisms including exocytotic and non-exocytotic mechanisms. Exocytotic mechanisms require previous storage of nucleotides via the vesicular nucleotide transporter (VNUT) in secretory/synaptic vesicles. Non-exocytotic mechanisms include the release of nucleotides by different types of channels, such as anion channels, pannexins and connexins. In contrast to ATP, adenosine can also be released into the extracellular space via Concentrative Nucleoside Transporters (CNTs) and Equilibrative Nucleoside Transporters (ENTS). Released nucleotides activate P2X and P2Y receptors localized on neuronal or glial membranes. Simultaneously, the hydrolysis of nucleotides by ectonucleotidases produces adenosine which, in turn, activates P1 receptors. Abbreviations: NTs, nucleotides; Ado, adenosine; VNUT, vesicular nucleotide transporter; CNTs, Concentrative Nucleoside Transporters; ENTs, Equilibrative Nucleoside Transporters.

3.1. Purine Release

The release of ATP and other nucleotides and nucleoside including adenosine into the extracellular space occurs via different mechanisms depending on cell type and physiological context. Non-exocytotic mechanisms include anion channels, such as plasmalemma voltage-dependent anion channels [87]; ATP-binding cassette transporters, such as the cystic fibrosis transmembrane conductance regulator Cl^- channel [88]; the purinergic P2X7 receptor [89,90]; and hemichannels, including connexin-43 [91] and pannexins [92,93]. The pannexin family comprises three members: Pannexin 1 (Panx1), Panx2 and Panx3 [94]. Among members of this family, Panx1 is the only one which forms functional channels and is expressed in both neuronal and glial cells in the brain [95,96]. Panx1 can be activated by different mechanisms including depolarization, mechanical stress or elevated intracellular Ca^{2+} concentrations [97–100]. Moreover, Panx1 may also contribute to ATP release after P2X7 activation, suggesting a direct connection between P2X7 and Panx1 [93,101]. Release of ATP and other nucleotides via exocytosis in the CNS has been reported from several cell types including neurons [102,103], astrocytes [104] and microglia [105]. Finally, the Cl^--dependent vesicular nucleotide transporter (VNUT) has been described to mediate the storage of ATP and other nucleotides in secretory and synaptic vesicles [106]. This transporter is highly expressed in different brain regions including the olfactory bulb, hippocampus and cerebellum [103] and has been shown to be functional in different types of neurons [102,107–109] and populations of glial cells [104,105,109].

3.2. The Purinergic Receptor Family

Nucleotides and nucleosides activate a large number of different cell-surface receptors divided into two major families termed purinergic P1 and P2 receptors. Whereas P1 receptors respond to adenosine and adenosine mono-phosphate (AMP), P2 receptors can be activated by ATP, adenosine diphospate (ADP), uridine triphospate (UTP), uridine monophosphate (UDP), nucleotide sugars, dinucleoside polyphosphates and NAD^+ [75].

3.2.1. P1 Receptor Family

P1 receptors are G protein-coupled and include four isoforms: A1, $A2_A$, $A2_B$ and A3 receptors. While in general $A2_A$ and $A2_B$ receptors induce the production of cyclic AMP (cAMP) via the Gs family, A1 and A3 receptors are usually coupled to Gi/o proteins, thereby inhibiting the production of cAMP. Other G protein combinations have, however, been described [110,111]. In the CNS, adenosine plays several roles, such as the modulation of neural and glial functions, neuron-glia signalling, neural development and the control of neurotransmitter release [112–116], with adenosine receptors expressed in both neurons and glia (astrocytes and microglia). Among the different P1 receptor subtypes, A1 and $A2_B$ receptors are usually associated with physiological neuronal processes (e.g., control of neurotransmitters release [117,118]) whereas $A2_A$ and A3 receptors are thought to be mostly activated under pathological conditions (e.g., epilepsy, neuropathy, neurodegenerative disorders or psychiatric conditions [119–122]). Because the dysregulation of the adenosinergic system is implicated in different pathologies, several studies have focused on this system as an avenue for new treatments. While $A2_A$ inhibition has shown neuroprotective properties during clinical trials in patients with Parkinson's [123], the activation of A1 receptors has been shown to reduce chronic pain [124] and to protect against epilepsy [125] and cerebral ischemia [126].

3.2.2. P2 Receptor Family

Among the P1 receptors, the A1 receptor subtype presents the most extensive distribution in the CNS, including limbic and neocortical brain regions, basal ganglia, brainstem, diencephalon and cerebellum. Regarding its cell type-specific expression, A1 receptor subtypes have been shown to be expressed in neurons, astrocytes, oligodendrocytes and microglia. A1 receptors, via Gi and Go interaction, inhibit adenyl cyclase with the subsequent decrease of cAMP levels, reduction of protein kinase A activation and inhibition of GABA uptake into astrocytes [127,128]. A1 receptor activation has been linked to several pathological conditions including neurodegeneration, pain and seizures [129–133]. Counteracting increased hyperexcitability states in the brain, A1 receptors mediate the inhibition of N-type calcium channels and the activation of G protein-coupled inwardly rectifying potassium channels [134,135], block presynaptic glutamate release and decrease the activation of the postsynaptic glutamate receptor N-methyl-D-aspartate receptor (NMDA), resulting in the suppression of neuronal activity [136,137]. The expression of $A2_A$ receptors is mainly located at the postsynaptic region of the encephalin-containing medium spiny neurons of the indirect pathway of the basal ganglia. $A2_A$ receptors are related to the activation of adenylate cyclase [138] through its coupling with Gs proteins. $A2_A$ receptor activation promotes the increase of NMDA receptor function and glutamate release at glutamatergic axon terminals [139–142]. Similar to $A2_A$ receptors, $A2_B$ receptors also activate adenylate cyclase and are ubiquitously expressed in the brain. However, there is not a clear link between these receptors and physiological or behavioural responses most likely due to the lack of specific agonists or antagonists. Finally, A3 receptors can uncouple A1 receptors and decrease thereby their inhibitory effects [143] via a protein kinase C-dependent mechanism. Although the presence of A3 receptors is low in the brain [144], its expression has been detected in hippocampal neurons [143], astrocytes [145] and microglial cells [146].

P2 receptors are subdivided into two subfamilies according to mechanism of action, pharmacology and molecular cloning, including the fast-acting P2X ligand-gated ion channels [147,148] and the slower-acting G-protein coupled P2Y receptors [149–151]. P2 receptors diverge in their molecular properties, amino acid sequences and relative sensitivities to ATP (e.g., nanomolar (P2Y receptors), low micromolar (most P2X receptors) and high micromolar (P2X7 receptor)). The structure of P2X receptors consists of two transmembrane domains: an intracellular C- and N- terminus and a large extracellular loop [147]. Most of the conserved regions are located in the extracellular loop, whereas transmembrane domains are less conserved between P2X receptors [152–154]. To date, seven mammalian subunits have been cloned (P2X1-7) [147] which form either functional homo- or heterotrimers exhibiting a high diversity due to the assembly of different individual subunits [155–158]. Among the P2X receptors, the P2X7 receptor has unique characteristics including the lowest affinity for ATP (approximately 100 μM) and a slower desensitization [159]. Functional expression of all P2X subunits has been shown within the brain on both neurons and glia [160]. Ionotropic P2X receptors, via the binding of extracellular ATP, open a permeable pore to the cations Na^+, K^+ and Ca^{2+}. In the brain, P2X receptor activation is involved in the regulation of synaptic plasticity in different brain circuits and fast synaptic transmission [161,162]. Synaptic currents induced by P2X activation contribute only 5–15% to fast excitatory transmission, possibly due to their high Ca^{2+} permeability at hyperpolarized membrane potentials [151,163]. However, the contribution of P2X-mediated currents might be higher under pathological conditions, such as a seizure, by increasing the influx of Ca^{2+} and by elevating the release of neurotransmitters such as glutamate [164,165]. P2X receptors are involved in a multitude of Ca^{2+}-sensitive processes including cellular proliferation, differentiation, maturation and survival, cell communication, migration and inflammation [166].

The metabotropic P2Y receptor family comprises eight G-protein coupled receptors: $P2Y_1$, $P2Y_2$, $P2Y_4$, $P2Y_6$, $P2Y_{11}$, $P2Y_{12}$, $P2Y_{13}$ and $P2Y_{14}$. All P2Y receptors share the topology of G-protein coupled receptors, which is characterised by seven transmembrane-domains, an extracellular amino and an intracellular carboxyl terminus. Moreover, P2Y receptors form homo- or heterodimers with other P2Y subunits [167] or with other receptors such as adenosine receptors [149]. Several P2Y receptors, including $P2Y_1$, $P2Y_2$, $P2Y_4$, $P2Y_6$ and $P2Y_{11}$, are coupled to Gq/G_{11}, which promotes endoplasmic reticulum Ca^{2+} release through the phospholipase C/inositol triphosphate pathway. $P2Y_{12}$, $P2Y_{13}$ and $P2Y_{14}$ receptors are coupled to Gi/Go proteins which inhibit adenylyl cyclase, resulting in a decrease of cAMP production. The $P2Y_{11}$ receptor is an exception as it is also able to couple to Gs, which stimulates adenylyl cyclase, thereby increasing cAMP production [150,168]. Depending on the P2Y receptor subtype, P2Y receptors can be activated by different nucleotides including ATP, ADP, UDP and sugar nucleotides [150,169–172]. Similar to P2X receptors, P2Y receptors are present at very early stages of embryonic CNS development [158] and are expressed on both neurons and glia, being involved in different processes such as modulation of neurotransmitter release [173,174], cell survival and neuroinflammation [175,176].

The purinergic system and neuroinflammation are tightly linked, with purinergic signalling described as fundamental for microglia's physiological roles and proconvulsive cytokine release [177]. A diverse number of purinergic receptors is expressed on microglial cells, where they exert different effects. Activation of microglial $A1_A$ receptors potentially removes microglia from a pro-inflammatory phenotype [178], with $A2_A$ receptors critical for microglia process retraction [86]. Among P2X receptors, P2X7 is often portrayed as a key driver of pathological inflammation. P2X7 is widely expressed in microglia [179,180] and has been described as essential for the NLRP3 inflammasome activation and subsequent release of Interleukin-1β (IL-1β) [181]. The P2X4 receptor has a described role in microglia chemotaxis and activation [182,183]. Likewise, the activation of several P2Y receptors in microglia, such as $P2Y_1$ and $P2Y_{12}$, promotes its phagocytic activity, migration towards damaged region and the release of IL-1β [184–187]. Alves et al. showed the context-dependent role of the $P2Y_1$ receptor to seizure pathology involving its expression in microglia [188]. Also, astrocytic $P2Y_1$ has been described as responsible for the spread of neuronal hyperexcitability throughout the brain via mediating glutamate gliotransmission [189]. $P2Y_{12}$ is expressed in microglia throughout its life cycle and again has prominent roles in microglia upregulation and migration [190]. With inflammation becoming increasing associated with seizure pathology [191], purinergic signalling will have major roles in mediating this.

3.3. Ectonucleotidases

Ectonucleotidases are enzymes with an extracellularly oriented catalytic site which rapidly hydrolyses ATP and other nucleotides after their release. These enzymes, operating in concert or consecutively, control the lifetime of extracellular released nucleotides by degrading or interconverting the originally released nucleotide generating ligands for additional P2 or P1 receptors. Ectonucleotidases comprise several families of enzymes divided by their functional and molecular properties including substrate specificity, product formation, optimal catalytic pH and cationic dependence [192].

All ectonucleotidase families are expressed in the brain including ectonucleoside triphosphate diphosphohydrolases (E-NTPDases/CD39), ectonucleotide pyrophosphatase and/or phosphodiesterases (E-NPPs), alkaline phosphatases and ecto-5′-nucleotidase [192]. The E-NTPDase/CD39 family comprises four surface-located members (E-NTPDase 1, 2, 3 and 8) which hydrolyse ATP into ADP or AMP, and ADP to AMP, exhibiting a different affinity for each nucleotide. E-NTPDase1 (also called CD39) presents equal affinity for ATP and ADP, whereas E-NTPDases 2, 3 and 8 are more selective for ATP [193]. The E-NPP consists of 7 enzymes (NPP1–7) which are able to cleave ATP directly into AMP [194]. Moreover, E-NPPs also hydrolyse dinucleoside polyphosphates and UDP sugars. AMP produced by E-NTPDases and E-NPPs is in turn metabolized to adenosine by ecto-5′-nucleotidase/CD73 [195]. Nucleoside tri, di and monophosphates are equally hydrolysed by alkaline phosphatases including TNAP, which is highly expressed in the CNS [196,197]. In the case of adenosine, this metabolite is generally the product of the ectoenzymatic breakdown of ATP; however, certain neurons and astrocytes are able to release adenosine also directly [198,199]. Adenosine can be removed from the extracellular space by different mechanisms such as its phosphorylation to AMP mediated by adenosine kinase (ADK) or deamination to inosine via the action of adenosine deaminase [200].

4. Purinergic Signalling during CNS Development

The early and predominant expression of purinergic receptors and ectonucleotidases in the developing CNS and the capacity of different cells to release ATP gives a cue of the many roles purinergic signalling carries out at the different neurodevelopmental stages. Numerous studies have demonstrated the involvement of purinergic signalling in proliferation, migration and differentiation of neural precursor cells [201–204]. Likewise, purinergic signalling is also involved in neuronal migration and the subsequent establishment of synaptic contacts as well as synaptogenesis [205,206] processes known to be dysregulated following neonatal seizures [205–209].

4.1. Expression and Function of Proteins Involved in Purine Release during Development

During CNS development, several proteins involved in purine release have been described including VNUT, which is expressed by granule cell precursors of the mouse cerebellum [107] and hemichannels such as connexins and pannexins. Regarding connexins, nine members of this family are expressed differentially throughout development [210–215], with their expression linked to cell proliferation and migration [211]. Connexins are involved in the regulation of the migration of the neural precursor cells by modulating cell–cell adhesion such as connexin-43, which is located in radial glial fibers [211]. Likewise, the expression pattern of pannexin changes throughout brain development, corresponding these changes with neurogenic and gliogenic processes of embryonic and early postnatal development [95]. Postnatally, Panx1 is expressed by neural and progenitor cells, playing a role in cell proliferation [216]. During CNS development, Panx1 transcripts have been found in the periventricular postnatal neural stem cells (NSCs) and neural progenitor cells (NPCs) [216]. In vitro studies with ventricular zone (VZ)-derived neurospheres have demonstrated that Panx1 is involved in cell proliferation. In line with this, blocking of Panx1 activity with the specific blocker probenecid reduced the proliferative capacity of VZ neurosphere cultures [216]. Moreover, Panx1 mediates the release of ATP, which in turn activates P2 receptors and increases proliferation of NSCs and NPCs [216]. Panx1 has also been linked to cell migration and the control of neurite outgrowth [217]. Panx2, another member of the pannexin family, is expressed in different subsets of neural progenitor cells of the postnatal hippocampus. However, when these cells differentiate into a neuronal lineage, Panx2 expression is downregulated [218].

4.1.1. P1 Receptor Expression and Function during CNS Development

Purinergic receptors are differentially expressed at different stages of embryonic and postnatal neurodevelopment. The expression of P1 purinergic receptors is already detected during embryonic neurodevelopmental stages. The A1 receptor is expressed from E14 and presents a similar allocation to adulthood at E21, being found in the cerebral cortex, hippocampus, thalamus, midbrain and cerebellum of the rat brain [219,220]. Expression of the A2 receptor has been detected from E13 onwards, increasing its expression levels after birth [220,221]. During CNS development, A1 and A_{2A} receptors were involved in processes regulating cell migration, neuronal connectivity and synaptogenesis. Tangential migration of medial ganglionic eminence (MGE)-derived GABAergic interneurons was delayed during pregnancy and lactation periods due to exposure to caffeine, an antagonist of A1 and A_{2A} receptors [221]. The same effect has been observed by using a specific A_{2A} receptor antagonist or A_{2A} receptor knockout mouse pups, demonstrating the involvement of the A_{2A} receptor in the migration of MGE-interneurons [221]. Regarding neuronal connectivity, adenosine receptors may contribute to neurite growth counteractively. In vitro studies have described that activation of the A1 receptor inhibits neurite outgrowth via the Rho-kinase pathway [222]. In contrast, A_{2A} receptor activation promotes the outgrowth of dendrites [223,224] and axonal elongation [224] through different signalling pathways. Finally, the A1 receptor modulates immature neuronal activity in different regions of the brain, including the hippocampus and cortex. In immature CA1 neurons, adenosine inhibits GABA release from the presynaptic nerve terminals through activation of the A1 receptor [225]. Since previous studies have described that A1 receptor activation inhibits glutamatergic release in adult hippocampal neurons [226–229], these results might confer an additional role to the A1 receptor during development. Moreover, activation of presynaptic A1 receptors inhibits excitatory GABAergic transmission from Cajal–Retzius cells, the early born neurons in layer I of the cortex, to pyramidal neurons in lower cortical layers [230]. Likewise, adenosine can regulate oligodendrogenesis in a bidirectional manner via A1 and A_{2A} receptors. A1 receptor stimulation contributes to maturation and prevents proliferation of the oligodendrocyte precursor cells (OPCs) [231,232]. Conversely, $A2_A$ receptor activation inhibits maturation and induces proliferation of OPCs [77].

4.1.2. P2 Receptor Expression and Function during CNS Development

P2X5 is the earliest expressed P2X receptor during development, with P2X5 being detected in mouse neural tubes from E8 and being upregulated to E13. The expression of P2X3 has been detected in mouse neuroectodermal cells [233] and rat brain from E11 onwards [234,235] and its activation induces the proliferation of embryonic stem cells [236]. From E14 onwards, both P2X2 and P2X7 are expressed [235]. Previous data has shown that silencing of the P2X2 receptor promotes proliferation, suggesting that P2X2 regulates this process negatively [237], whereas the P2X7 receptor is expressed in mouse embryonic stem cells and modulates processes involved in proliferation and neural differentiation [238]. The remaining P2X receptors, P2X1, 4 and 6 appear at postnatal stages of rat brain development [235]. P2X1 and P2X3 expression within the brain remains consistent from birth to adulthood, whereas P2X2 expression is downregulated with age. Conversely, neocortical P2X4 and P2X7 expression is upregulated incrementally with age, reaching its peak in adulthood [14]. At P7 age, P2X7 expression is predominately found in microglia and is also expressed in Bergmann glia of the cerebellum [179]. Unfortunately, a definitive answer on neuronal and astrocytic P2X7 expression, not just in infants, is under debate [239]. Multiple groups have observed that neuronal P2X7 localised to presynaptic terminals [240,241]. P2X7 is also expressed in primary neuronal and astrocytic in vitro cultures [242]. However, when using a transgenic P2X7 reporter mouse, in which the green fluorescent protein is fused to the P2X7 receptor, thus allowing visualization of P2X7 expression, neuronal and astrocytic P2X7 is not observed [179,243]. Neuronal and astrocytic P2X7 immunoreactivity was also absent when using a P2X7-specific nanobody in both the immature and adult mouse brain [179]. However, one could hypothesis that neuronal P2X7 expression is below the detection limit with immunoreactivity techniques, is localised to intracellular compartments or is upregulated only in pathology [179,244].

P2Y expression has been mostly located in proliferative regions and at early stages of neurodevelopment. $P2Y_1$ expression has been detected from E11 onwards [245], being expressed by radial glial and intermediate precursor cells located in proliferative regions of the developing cortex [201,203]. Neurospheres cultured from the adult subventricular zone (SVZ) exhibited an increase of cell proliferation after $P2Y_1$ activation by using several agonists (2-MeSATP, ADPβS, 2-ClATP and 2-MeSADP) [246] suggesting that the $P2Y_1$ receptor may be involved in cell proliferation. On the contrary, blocking of $P2Y_1$ by the antagonist MRS2179 reduced cell proliferation, and the same effect was observed in $P2Y_1$ receptor knockout mice [246]. $P2Y_4$ expression has also been detected at E11 stage, being located together with $P2Y_2$ in ventricular regions of the E14-telencephalon. Similar to P1 receptors, P2 purinergic receptors present different profiles of location and temporal expression during postnatal stages, suggesting that they also play a role in various stages of brain development [160,205,234,235,245,247–250].

4.1.3. The Dual Role of the P2X7 Receptor during CNS Development

During CNS development, the P2X7 receptor plays a dual role promoting opposing processes such as cell death and cell proliferation. These opposing effects driven by the same receptor may depend on the cell type that expresses it, the extracellular concentration of ATP or the duration of P2X7 receptor activation. However, the involvement of P2X7 in neuronal cell death is still unclear since there is still controversy about the expression of this receptor in the different cell types of the CNS [11,239], as explained previously.

In mouse embryonic stem cells, the P2X7 receptor promotes its proliferation and maintenance in an undifferentiated state, while for its neural differentiation, P2X7 receptor expression needs to be suppressed [238]. Likewise, the P2X7 receptor is able to induce necrosis of NPCs when activated with high concentrations of ATP or the agonist Bz-ATP [251]. In contrast, stimulation of P2X7 with low concentrations of Bz-ATP leads to neuronal differentiation of NPCs [252]. Moreover, depending on the duration of P2X7 receptor activation, this receptor can mediate pro-survival or pro-death signalling [253]. Additionally, P2X7 might regulate the population of NPCs through innate phagocytosis of dead cells throughout development. In this regard, neuroblasts isolated from human foetal telencephalons are able to phagocytose apoptotic cells in the absence of P2X7 receptor activation [254].

P2X7 has also been identified on microglial cells of the rat brain from late E16 onwards, exhibiting a wide distribution in the forebrain at P30 stage [255]. In line with its known role driving microglia proliferation [181,256], P2X7 has been shown to control microglial proliferation in the embryonic spinal cord of mice at E13.5 stage [257]. Conversely, prolonged P2X7 stimulation with high concentrations of Bz-ATP induces microglia cell death in the cortex of newborn mice [180]. Thus, similar to NPCs, the outcome of P2X7 activation in microglia cells might depend on the amount of available extracellular ATP and the duration of stimulation of the receptor. As such, it can be concluded that P2X7 may act to regulate itself to prevent excessive microglia proliferation during neurodevelopment. Finally, the P2X7 receptor is also expressed on oligodendrocyte progenitors contributing to stimulation of migration and driving oligodendrocyte differentiation [258].

4.2. Extracellular Purine Metabolism during Development

The temporal expression of purinergic receptors during brain development is accompanied by modifications in the expression of ectonucleotidases. Individual ectonucleotidase expression varies according to developmental stage and brain region. NTPDase 2, which is the dominant ectonucleotidase expressed by progenitors in the late embryonic and adult mouse brain, has been identified from E18 in neurogenic regions [259,260], whereas NTPDases 1, 3, 5 and 6 are detected in later stages of brain development (P7-21) [261]. Concerning ecto-5′-ectonucleotidase, its expression increases during postnatal stages (it has been identified in migrating neuroblasts of the cerebellum and is related with synaptogenesis processes [192,262–265]). Certain ectonucleotidases of the E-NNPs family are also expressed at early stages of neural development, such as E-NNP-2, for which the splice variant autotaxin was identified in the floor plate of the neural tube at E9.5 [266]. Postnatally, the expression of E-NNP 1–3 is detected in several regions of the rat brain [267].

Finally, TNAP expression begins at very early stages of neural development, being highly expressed by neuroepithelial stem cells of the neural tube from E8.5 and a migrating subpopulation of neuroectodermal cells [268–270]. Moreover, a strong activity of TNAP has been identified in ventricular and subventricular zones, which are high cell proliferative regions, at the E14 stage [260], and postnatally, its activity is related to synaptogenesis in the cerebral cortex [271]. Therefore, TNAP might contribute to cell proliferation or cell differentiation in the neurogenic niche. In NSCs cultured from adult mice, downregulation of TNAP causes a strong decrease in progenitor cell proliferation [272]. In addition, TNAP might be involved in the control of axonal growth during development. Studies with cultured hippocampal neurons have shown that TNAP expressed by outgrowing axons promotes axonal elongation through the hydrolysis of extracellular ATP [273]. As a result of this, extracellular ATP levels are drastically reduced, indirectly modulating activation of purine receptors. Interestingly, TNAP and P2X7 are tightly linked, with the addition of exogenous TNAP increasing P2X7 receptor expression, whereas TNAP expression is downregulated when P2X7 is inhibited. Importantly, TNAP knockout mice exhibit perinatal lethality, with P9 being the maximum reached age [274,275], and present with a decrease in the number of matured cortical synapses and an absence of myelinated cortical axons [276].

In summary, the fundamental role of the purinergic system during neurodevelopment is clear, and with its diverse expression and functionality, there are many avenues to explore that could be effective treatments to early life disorders.

5. Purinergic Signalling and Neonatal Seizures

As stated earlier, purinergic signalling is widespread in the immature brain and many studies have targeted this system effectively to modulate seizures in the adult scenario [9]. This section will discuss our current knowledge of how purinergic signalling modulates neonatal seizures and future potential therapeutic avenues to explore (Figure 2). This encompasses studies on both P1 and P2 receptors. An overview of studies investigating neonatal seizures and the purinergic system is displayed in Table 1.

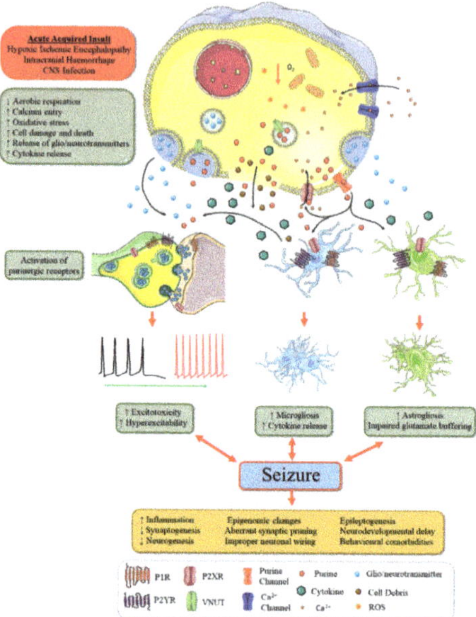

Figure 2. Cellular mechanisms of acute symptomatic neonatal seizure ictogenesis and the potential role of purinergic signalling: following an acute insult to the neonatal brain, cells are placed under high cellular stress, leading to increases in calcium entry and cell death pathways. In the case of hypoxic-ischemic encephalopathy (HIE)-induced seizures, the lack of oxygen and glucoses limits aerobic respiration, forming radical oxygen species (ROS) causing further oxidative stress on cells. Increases in intracellular calcium and cell death can trigger the release of glio/neurotransmitters (e.g., glutamate) into the extracellular space that increases neurotransmission. Cell debris can trigger microgliosis, astrogliosis and release of proconvulsive cytokines. Purines (e.g., ATP and adenosine) are also hypothesised to be released into the extracellular space following cell death and through a combination of exocytotic and non-exocytotic mechanisms under cellular stress. ATP acts upon P2X7 to further increase intracellular calcium, contributing to cell death mechanisms and to increasing neurotransmission and, in turn, seizure severity. P2X7 activation is known to potentiate proconvulsive cytokine release following neonatal seizures, which in turn can lower seizure thresholds. Other P2 receptors are known to modulate many mechanisms of seizure ictogenesis, such as direct modulation of neurotransmission and inflammatory signalling cascades. A2$_A$ receptors may also contribute to neonatal seizures via similar mechanism to P2X7. Conversely, A1 receptor activation is anticonvulsive in neonatal seizures, acting as an endogenous compensatory mechanism. Once these outlined mechanisms create a system that favours excitatory neurotransmission, seizures are elicited. A seizure can also create further cellular stress and neuroinflammation, increasing the likelihood of recurrent seizures. Elevated neuroinflammation and hyperexcitability alter many mechanisms critical for brain development, leading to long-lasting changes of the brain. Purinergic signalling can be hypothesised to modulate this and may be targeted in the future to prevent comorbidities following neonatal seizures.

5.1. Targeting of P1 Receptors

As early as 1988, when the purinergic signalling field was in its infancy, the nucleoside adenosine was proposed as an endogenous anticonvulsant [277]. When adenosine is applied to resected epileptic hippocampal slices, it was shown to arrest epileptiform activity [278]. In fact, adenosine is released into the brain following seizures of temporal lobe epilepsy patients, where it may act as an endogenous mechanism to arrest seizures [278], whereas caffeine, a nonspecific adenosine receptor antagonist, acts as a convulsant compound, potentiating the seizure phenotype following PTZ injection [279,280]. Various case reports also show caffeine to induce seizures in non-epileptic persons [281].

Currently, there is only evidence of A1 and A2$_A$ receptors modulating seizure phenotypes. Pometlova et al., 2010, showed the potential of targeting the P1 receptors, with the nonspecific adenosine receptor agonist, 2-chloradenosine, having an anticonvulsive effect in immature rats following cortical stimulation [282]. These effects were not model-specific, with PTZ-induced seizures in immature rats being suppressed by 2-chloradenosine administration [283]. Building upon this, using specific agonists and antagonists of A1 and A2$_A$ receptors, Mares observed that the anticonvulsive effects seen was primarily due to action upon A1 receptors, with pharmacological targeting of A2$_A$ having little effect in P12 rats, the age that relates most to a human neonate [283]. Anticonvulsant action of A1 receptors was reinforced with agonistic action reducing the magnitude of elicited cortical discharges [284]. Again, this effect was more pronounced in P12 rats rather than P25, suggesting a possible developmental shift in the sensitivity of adenosine receptors [284]. Interestingly, in this model, both agonistic and antagonistic action of A2$_A$ receptors had an anticonvulsive effect in P12 rats, yet blocking A2$_A$ receptors in P25 produces a proconvulsive effect [284]. These results were replicated, even when a different area of the brain (hippocampus) was stimulated to induce seizures [285], with the A1 receptor agonist 2-chloro-$N6$-cyclopentyladenosine having an anticonvulsive effect at all ages except P25. These studies highlight the developmental regulation of P1 receptors and the possible age-dependent modulation of seizure phenotypes. Altered expression levels of adenosine receptors has been observed 48 h following induced febrile seizures in neonatal rats, with the A1 receptor increasing and the A2$_A$ receptor decreasing [286]. This suggests that the adenosine system may act as endogenous compensatory mechanism for seizures. Currently, there are limited studies investigating the anti-epileptogenic capacity of targeting P1 receptors. Possible adverse effects of modulation of the adenosine system have been unexplored in these neonatal seizures studies, yet the authors acknowledge the necessity for this. Likewise, only the effect of caffeine, an A1 antagonist, on neurodevelopment has been studied. Caffeine is shown to ameliorate phenobarbital impairment of neurogenesis in neonatal rats [287], possibly due to caffeine's ability to suppress GABAergic action [288]. However, when it is given in isolation, caffeine reduces the proliferative capacity of the brain [287]. In the adult mouse, caffeine can reduce long-term potentiation and can alter synaptic plasticity, which could be detrimental in the immature brain [289]. One such rodent study shows that early life exposure to caffeine can increase the seizure susceptibility in adulthood [290]. Conversely, at low doses, caffeine may act to reduce acute seizures, particularly in the infant brain, with neonatal rats having an increased seizure threshold to chemoconvulsants following a low dose of caffeine [291]. Interestingly, many studies have shown a neuroprotective effect of a low dose of caffeine in the setting of HIE, reducing white matter injury and protecting against memory impairment [292] and motor deficits in later life [293]. These studies highlight the complex nature of early life seizures and how mechanisms of seizure ictogenesis may differ from epileptogenesis.

Despite the presence of adenosine signalling in the majority of biological systems, little is known about the adverse effects of adenosine receptors in the CNS and concerns for unwanted side effects are well warranted. With P1 receptors having a large role in cardiovascular and respiratory function via action upon the brainstem, the sudden rise of endogenous adenosine following seizures is hypothesised as one contributing factor to Sudden Unexpected Death in Epilepsy (SUDEP) [294]. Also, despite the documented use of many P1 receptor ligands reported in the literature, only adenosine and regadenoson (A2$_A$ receptor antagonist) are approved for use in the clinic [295].

Table 1. Overview of studies investigating purinergic signalling modulating neonatal seizures.

Target Receptor	Compound	Seizure Model	Species, Age and Gender	Effect	Reference
P1					
Nonspecific P1	2-chloroadenosine (1, 4 and 10 mg/kg, i.p) (agonist)	Cortical epileptic after discharges (drug administered 5 min first after discharge)	Rats (P12, P18 and P25); sex not specified	Behavioural and EEG-detected seizures were only reduced at P18.	[282]
Nonspecific P1	2-chloroadenosine (1, 5, 10 and 15 mg/kg, i.p.) (agonist)	PTZ 100 mg/kg s.c. (90 mg/kg in P18). (drugs were administered 30 min before seizure induction)	Rats (P7, P12, P18, P25 and P90); males	Anticonvulsive effect was seen at all ages. Suppression of tonic seizures was only at P12 and younger. Suppression of generalised seizures was at P18 and above.	[283]
A1	2-chloro-*N6*-cyclopentyladenosine (0.2, 0.5 and 1 mg/kg to 12-day-old rats and 0.5, 1 and 2 mg/kg to 25-day-old rats, i.p) (agonist)		Rats (P12 and P25); males	2-chloro-*N6*-cyclopentyladenosine led to marked anticonvulsant effects in P12. Minimal effects were seen in P25. No effect was seen with DPCPX.	
	DPCPX (1 and 2 mg/kg i.p.) (Antagonist)				
A2A	CGS 21680 (0.1, 0.2, 0.5, 1, 2 and 5 mg/kg, i.p.) (agonist)			Highest dose of CGS 21680 (5mg/kg) reduced seizure severity only at P25. No effect was observed in P12 at any dose. No effect was observed with ZM 241385.	
	ZM 241385 (1, 2 and 5 mg/kg, i.p.) (antagonist)				
A1	2-chloro-*N6*-cyclopentyladenosine (0.5) and 1 mg/kg i.p.) (agonist)	Cortical epileptic after discharges (drugs were administered 5 min after first stimulation)	Rats (P12, P18 and P25); males	Duration reduced after discharges with agonist and proconvulsant action of antagonist at P12 and P18. At P25, both agonistic and antagonistic action are proconvulsive.	[284]
	DPCPX (1 and 2 mg/kg, i.p.) (antagonist)				
A2A	CGS 21680 (0.5 and 5 mg/kg i.p.) (Agonist)			CGS 21680 is anticonvulsive at all ages. While ZM 241385 action is anticonvulsive at P12 and P18, it is proconvulsive at P25 at the highest dose.	
	ZM 241385 (1 and 5 mg/kg i.p.) (antagonist)				
Nonspecific P1	2-chloro-*N6*-cyclopentyladenosine (0.5 and 1 mg/kg, i.p) (agonist)	Hippocampal epileptic after discharges. (drug administered 10 min prior to the stimulation procedure)	Rats (P12–P60); males	Anticonvulsive effect was seen in all ages bar P25. Hippocampal A1 protein expression peaks at P10 and decreases with age.	[285]

Table 1. *Cont.*

Target Receptor	Compound	Seizure Model	Species, Age and Gender	Effect	Reference
		P2			
P2X7	A-438079 (5 and 15 mg/kg, i.p) (antagonist)	Intra-amygdala KA (2 µg in 0.2 µL PBS) (drug administered 1 h post-KA injection)	Rats (P10); mixed sex group	A-438079 reduced seizure severity, subsequent neuronal damage and inflammation.	[40]
P2X7	A-438079 0.5, 5, 15, 25 and 50 mg/kg, i.p.) (antagonist) JNJ-47965567 (10 and 30 mg/kg, i.p.) (antagonist)	Global hypoxia (5% O_2 15 min) (drugs administered 5 min prior to hypoxia)	Mice (P7); mixed sex group	P2X7 expression is increased 24 h following hypoxia-induced seizures in the hippocampus. P2X7 expression increased in tissue from patients who experienced HIE and seizures. Both compounds reduced seizure severity. A-438079 reduced post-seizure inflammation.	[14]

Abbreviations: DPCPX, 8-Cyclopentyl-1,3-dipropylxanthine; EEG, electroencephalogram; HIE, hypoxia-ischemia encephalopathy, i.p., intraperitoneal; KA, kainic acid; s.c. subcutaneous; PTZ, Pentylenetetrazole.

5.2. Targeting of P2 Receptors

Of the P2 receptors, targeting of the P2X7 receptor has shown the most promise in neonatal seizures. A role for the P2X7 receptor in seizures was first examined in adult seizures, where using transgenic and pharmacological tools showed it to have a proconvulsive or anticonvulsive action depending on experimental model used [244].

Mesuret et al., 2014, first investigated the P2X7 receptor in the neonatal seizure scenario. P2X7 receptor expression was upregulated as early as one hour in the hippocampus following seizures induced via intra-amygdala injection of KA in P10 rats. P2X7 expression increased to a maximum at 72 h post-KA that was also accompanied by elevated levels of the cytokine IL-β [40]. Interestingly, treatment with the P2X7 antagonist A-438079 reduced the acute electrographic seizures by over 50%, was neuroprotective and reduced levels of seizure-induced neuroinflammation. Importantly, treatment with A-438079 had greater neuroprotective effects than treatment with current clinical used drugs, phenobarbital and bumetanide. In fact, phenobarbital and bumetanide failed to show any neuroprotective effects [40]. These results have been translated in a model more clinically relevant. Using global hypoxia to induce seizures (5% O_2, 15 min), P2X7 receptor expression was increased 24 h post-seizure. Interestingly, P2X4 receptor expression was also increased 24 h post-seizures suggesting a new avenue to explore. More importantly, P2X7 receptor protein levels were elevated in human infant brain tissue 3 months after a HIE/seizure event [14]. Two different P2X7 antagonists, A-438079 and JNJ-47965567, were able to reduce hypoxia-induced electrographic seizures in neonatal mouse pups. Antagonistic action was also able to reduce levels of pro-inflammatory markers (e.g., IL-1β) 24 h post-seizures. The limitation of these two studies is that P2X7 receptor antagonists were given before seizures ictogenesis, which is not clinically viable. In addition, as P2X7 receptor antagonism reduced the acute insult, it cannot be concluded that P2X7 receptor antagonism alone is able to reduce post-seizure inflammation. However, this is the most likely case due to the major role of P2X7 in pathological inflammation. With inflammation heavily involved in pathology following neonatal seizures, the P2X7 receptor might have a role in epileptogenesis following an insult to the infant brain. Further investigation with post-seizure treatment to investigate the ability of P2X7 to prevent neonatal seizure comorbidities is much anticipated. As aforementioned, more P2 receptors are currently being explored in adult seizures, whereas now, only P2X7 has been targeted in neonatal seizures. A further therapeutic target requiring further investigation is P2X4, with its expression upregulated following neonatal seizures [14]. Under hypoxic conditions (5% O_2, 3.5 h, P0), P2X4 was again upregulated in immature rats. Furthermore, this upregulation was greater than that observed with P2X7 and $P2Y_{12}$ [296]. P2X4 is described to mediate ATP-gated microglia activation and release of proconvulsive inflammatory cytokines in these hypoxic conditions [296]. Again, with inflammation heavily involved in seizure ictogenesis and epileptogenesis, one could hypothesis targeting P2X4 to be advantageous for the treatment of neonatal seizures.

5.3. Potential Purinergic Targets to Explore

Finally, apart from direct action upon membrane-bound receptors, another strategy to explore would be to regulate concentrations in the extracellular space of purine nucleotides and nucleosides. This can be achieved via inhibition of enzymes, such as ADK, to reduce the clearance of adenosine. Pharmacological and genetic evidence shows that ADK has a role in adult epilepsy development and seizure generation [297–299]. Hypophosphatasia, in which neonatal seizures are a major component, is heavily associated with mutations in the TNAP gene. In fact, mice deficient in TNAP show spontaneous seizures by P6 [274,275]. Interestingly, TNAP-related seizures are mediated via P2X7. Mice double deficient in TNAP and P2X7, as well as TNAP knock-out mice treated with a P2X7 receptor antagonist, did not present with spontaneous seizures [274]. Furthermore, antagonistic action against TNAP increased the seizure duration in adults, and thus, it would be interesting to investigate TNAP in neonatal seizure models [274]. With TNAP's roles in regulating synaptic function during neuronal development [300], targeting TNAP in the neonatal seizure scenario could aid in preventing the comorbidities seen in this condition.

Apart from regulating the metabolism of purines, one potential strategy would be to prevent the release of ATP into the extracellular space. VNUT is a relatively unexplored target in relation to seizure modulation yet, with its prominent role in ATP release, is an exciting avenue to explore. With its prominent expression in the immature brain, targeting Panx1 may also show promise in modulating neonatal seizures. Panx1 is shown to be active in KA-induced seizures in juvenile mice (P13–14) which corresponds with a doubling in extracellular ATP levels [301]. In Panx1 null mice and when Panx1 was blocked with pharmacological tools, behavioural seizure manifestations were reduced [301]. Interestingly, Panx1 seems to not be involved in seizure ictogenesis yet is involved in maintaining seizure activity. It would be interesting to see if this result can be translated to an age more appropriate to neonatal seizures.

As stated earlier, there are many components of the purinergic system that are present early on during development, with the majority unexplored in the role of seizure generation and epileptogenesis. It would be advantageous to investigate these in further detail to uncover the full picture of purinergic system in neonatal seizures, to maximise the efficiency of future pharmacological drugs and to minimise adverse effects. Currently, no study examines purinergic signalling away from the initial neonatal seizure event. In the clinic, it may prove difficult to prevent the initial neonatal seizure, and further investigation into preventing further recurrent seizures is needed. Purinergic signalling is involved in many processes known to contribute to epileptogenesis and to potentiate damage. Targeting inflammation following neonatal seizures, a process in which purinergic signalling is heavily involved, has shown promise to reduce the development of epilepsy and behavioural deficits [302].

6. Conclusions

Current therapies for neonatal seizures seemed to be limited to direct modulation of ion channels on neurones. As we have progressed in understanding seizure pathology, we now know that many mechanisms, such as chronic neuroinflammation, blood–brain barrier dysfunction and aberrant neurogenesis, can influence seizure ictogenesis. This allows us to use many more potential mechanisms to target greater efficacy. As outlined in this review, the purinergic system is widely expressed within the CNS and has a multitude of physiological and pathological functions. We are still lacking knowledge in many aspects of what role the purinergic system has in contributing to neonatal seizure pathology, but studies have shown great promise in targeting this biological system, particularly in targeting the P2X7 receptor. Further studies are needed not only in uncovering mechanisms of how purinergic signalling may influence neonatal seizures and subsequent pathologies but also in investigating the fundamental mechanisms of neonatal seizure pathology itself.

Author Contributions: A.M.M. wrote the manuscript and designed figures; J.S., wrote the manuscript and designed figures; T.E., wrote and edited the manuscript. All authors have read and agreed to the published version of the manuscript.

Funding: This work was supported by funding from Science Foundation Ireland (17/CDA/4708 (approval date 1 April 2018) and cofunded under the European Regional Development Fund and by FutureNeuro industry partners 16/RC/3948) and from the H2020 Marie Skłodowska-Curie Actions Individual Fellowship (No 884956, approval date 18 September 2017).

Conflicts of Interest: The authors declare no conflict of interest.

Abbreviations

ADK	Adenosine kinase
Ado	Adenosine
ADP	Adenosine diphosphate
AMP	Adenosine monophosphate
ASD	Antiseizure drugs
ATP	Adenosine triphosphate

cAMP	Cyclic AMP
CNS	Central nervous system
CNTs	Concentrative nucleoside transporters
EEG	Electroencephalogram
E-NPPs	Ectonucleotide pyrophosphatase and/or phosphodiesterases
E-NTPDases	Ectonucleoside triphosphate diphosphohydrolases
ENTs	Equilibrative nucleoside transporters
GABA	γ-aminobutyric acid
HIE	Hypoxic-ischemic encephalopathy
IL-1β	Interleukin-1β
KA	Kainic acid
MCAO	Medial carotid artery occlusion
MGE	Medial ganglionic eminence
NAD$^+$	Nicotinamide adenine dinucleotide
NMDA	*N*-methyl-D-aspartate receptor
NPCs	Neural progenitor cells
NSCs	Neural stem cells
NTs	Nucleotides
OPCs	Oligodendrocyte precursor cells
Panx	Pannexin
PTZ	Pentylenetetrazole
SVZ	Subventricular zone
TNAP	Tissue nonspecific alkaline phosphatase
UDP	Uridine monophosphate
UTP	Uridine triphosphate
VNUT	Vesicular nucleotide transporter
VZ	Ventricular zone

References

1. Shetty, J. Neonatal seizures in hypoxic-ischaemic encephalopathy–risks and benefits of anticonvulsant therapy. *Dev. Med. Child. Neurol.* **2015**, *57*, 40–43. [CrossRef] [PubMed]
2. Rennie, J.M.; de Vries, L.S.; Blennow, M.; Foran, A.; Shah, D.K.; Livingstone, V.; van Huffelen, A.C.; Mathieson, S.R.; Pavlidis, E.; Weeke, L.C.; et al. Characterisation of neonatal seizures and their treatment using continuous EEG monitoring: A multicentre experience. *Arch. Dis. Child. Fetal. Neonatal Ed.* **2019**, *104*, F493–F501. [CrossRef] [PubMed]
3. Glass, H.C.; Numis, A.L.; Gano, D.; Bali, V.; Rogers, E.E. Outcomes After Acute Symptomatic Seizures in Children Admitted to a Neonatal Neurocritical Care Service. *Pediatr. Neurol.* **2018**, *84*, 39–45. [CrossRef] [PubMed]
4. Kang, S.K.; Kadam, S.D. Neonatal Seizures: Impact on Neurodevelopmental Outcomes. *Front. Pediatr.* **2015**, *3*. [CrossRef] [PubMed]
5. Quinlan, S.M.M.; Rodriguez-Alvarez, N.; Molloy, E.J.; Madden, S.F.; Boylan, G.B.; Henshall, D.C.; Jimenez-Mateos, E.M. Complex spectrum of phenobarbital effects in a mouse model of neonatal hypoxia-induced seizures. *Sci. Rep.* **2018**, *8*, 9986. [CrossRef]
6. Burnstock, G. Introduction to purinergic signalling in the brain. *Adv. Exp. Med. Biol.* **2013**, *986*, 1–12. [CrossRef]
7. Burnstock, G. Historical review: ATP as a neurotransmitter. *Trends Pharm. Sci.* **2006**, *27*, 166–176. [CrossRef]
8. Burnstock, G. Purinergic Signalling: Therapeutic Developments. *Front. Pharm.* **2017**, *8*, 661. [CrossRef]
9. Engel, T.; Alves, M.; Sheedy, C.; Henshall, D.C. ATPergic signalling during seizures and epilepsy. *Neuropharmacology* **2016**, *104*, 140–153. [CrossRef]
10. Burnstock, G. An introduction to the roles of purinergic signalling in neurodegeneration, neuroprotection and neuroregeneration. *Neuropharmacology* **2016**, *104*, 4–17. [CrossRef]
11. Miras-Portugal, M.T.; Sebastián-Serrano, Á.; de Diego García, L.; Díaz-Hernández, M. Neuronal P2X7 Receptor: Involvement in Neuronal Physiology and Pathology. *J. Neurosci.* **2017**, *37*, 7063–7072. [CrossRef]

12. Huang, L.; Otrokocsi, L.; Sperlágh, B. Role of P2 receptors in normal brain development and in neurodevelopmental psychiatric disorders. *Brain Res. Bull.* **2019**, *151*, 55–64. [CrossRef] [PubMed]

13. Rodrigues, R.J.; Marques, J.M.; Cunha, R.A. Purinergic signalling and brain development. *Semin. Cell Dev. Biol.* **2019**, *95*, 34–41. [CrossRef]

14. Rodriguez-Alvarez, N.; Jimenez-Mateos, E.M.; Engel, T.; Quinlan, S.; Reschke, C.R.; Conroy, R.M.; Bhattacharya, A.; Boylan, G.B.; Henshall, D.C. Effects of P2X7 receptor antagonists on hypoxia-induced neonatal seizures in mice. *Neuropharmacology* **2017**, *116*, 351–363. [CrossRef] [PubMed]

15. Heljic, S.; Uzicanin, S.; Catibusic, F.; Zubcevic, S. Predictors of Mortality in Neonates with Seizures; A Prospective Cohort Study. *Med. Arch.* **2016**, *70*, 182–185. [CrossRef] [PubMed]

16. Mottahedin, A. Effect of Neuroinflammation on Synaptic Organization and Function in the Developing Brain: Implications for Neurodevelopmental and Neurodegenerative Disorders. *Front. Cell Neurosci.* **2017**, *11*, 190. [CrossRef]

17. Rakhade, S.N.; Jensen, F.E. Epileptogenesis in the immature brain: Emerging mechanisms. *Nat. Rev. Neurol.* **2009**, *5*, 380–391. [CrossRef] [PubMed]

18. Liu, S.; Yu, W.; Lu, Y. The causes of new-onset epilepsy and seizures in the elderly. *Neuropsychiatr. Dis. Treat.* **2016**, *12*, 1425–1434. [CrossRef] [PubMed]

19. Hauser, W.A.; Annegers, J.F.; Kurland, L.T. Incidence of epilepsy and unprovoked seizures in Rochester, Minnesota: 1935–1984. *Epilepsia* **1993**, *34*, 453–468. [CrossRef]

20. Wirrell, E.C. Neonatal seizures: To treat or not to treat? *Semin. Pediatr. Neurol.* **2005**, *12*, 97–105. [CrossRef]

21. Vasudevan, C.; Levene, M. Epidemiology and aetiology of neonatal seizures. *Semin. Fetal. Neonatal Med.* **2013**, *18*, 185–191. [CrossRef] [PubMed]

22. Numis, A.L.; Foster-Barber, A.; Deng, X.; Rogers, E.E.; Barkovich, A.J.; Ferriero, D.M.; Glass, H.C. Early changes in pro-inflammatory cytokine levels in neonates with encephalopathy are associated with remote epilepsy. *Pediatr. Res.* **2019**. [CrossRef] [PubMed]

23. Müller, N.; Weidinger, E.; Leitner, B.; Schwarz, M.J. The role of inflammation in schizophrenia. *Front. Neurosci.* **2015**, *9*, 372. [CrossRef] [PubMed]

24. Yin, P.; Li, Z.; Wang, Y.Y.; Qiao, N.N.; Huang, S.Y.; Sun, R.P.; Wang, J.W. Neonatal immune challenge exacerbates seizure-induced hippocampus-dependent memory impairment in adult rats. *Epilepsy Behav.* **2013**, *27*, 9–17. [CrossRef] [PubMed]

25. Kharoshankaya, L.; Stevenson, N.J.; Livingstone, V.; Murray, D.M.; Murphy, B.P.; Ahearne, C.E.; Boylan, G.B. Seizure burden and neurodevelopmental outcome in neonates with hypoxic-ischemic encephalopathy. *Dev. Med. Child. Neurol.* **2016**, *58*, 1242–1248. [CrossRef] [PubMed]

26. Pisani, F.; Facini, C.; Pavlidis, E.; Spagnoli, C.; Boylan, G. Epilepsy after neonatal seizures: Literature review. *Eur. J. Paediatr. Neurol.* **2015**, *19*, 6–14. [CrossRef]

27. Soul, J.S. Acute symptomatic seizures in term neonates: Etiologies and treatments. *Semin. Fetal. Neonatal Med.* **2018**, *23*, 183–190. [CrossRef]

28. Levene, M.I.; Trounce, J.Q. Cause of neonatal convulsions. Towards more precise diagnosis. *Arch. Dis. Child.* **1986**, *61*, 78–79. [CrossRef]

29. Han, J.Y.; Moon, C.J.; Youn, Y.A.; Sung, I.K.; Lee, I.G. Efficacy of levetiracetam for neonatal seizures in preterm infants. *BMC Pediatr.* **2018**, *18*, 131. [CrossRef]

30. Jensen, F.E. Neonatal seizures: An update on mechanisms and management. *Clin. Perinatol.* **2009**, *36*, 881–vii. [CrossRef]

31. Jehan, I.; Harris, H.; Salat, S.; Zeb, A.; Mobeen, N.; Pasha, O.; McClure, E.M.; Moore, J.; Wright, L.L.; Goldenberg, R.L. Neonatal mortality, risk factors and causes: A prospective population-based cohort study in urban Pakistan. *Bull. World Health Organ.* **2009**, *87*, 130–138. [CrossRef] [PubMed]

32. Sarnat, H.B.; Sarnat, M.S. Neonatal encephalopathy following fetal distress. A clinical and electroencephalographic study. *Arch. Neurol.* **1976**, *33*, 696–705. [CrossRef]

33. Levene, M.L.; Kornberg, J.; Williams, T.H. The incidence and severity of post-asphyxial encephalopathy in full-term infants. *Early Hum. Dev.* **1985**, *11*, 21–26. [CrossRef]

34. Glass, H.C.; Shellhaas, R.A.; Wusthoff, C.J.; Chang, T.; Abend, N.S.; Chu, C.J.; Cilio, M.R.; Glidden, D.V.; Bonifacio, S.L.; Massey, S.; et al. Contemporary Profile of Seizures in Neonates: A Prospective Cohort Study. *J. Pediatr.* **2016**, *174*, 98.e101–103.e101. [CrossRef] [PubMed]

35. Huang, L.; Cilio, M.R.; Silveira, D.C.; McCabe, B.K.; Sogawa, Y.; Stafstrom, C.E.; Holmes, G.L. Long-term effects of neonatal seizures: A behavioral, electrophysiological, and histological study. *Brain Res. Dev. Brain Res.* **1999**, *118*, 99–107. [CrossRef]
36. Parker, A.K.; Le, M.M.; Smith, T.S.; Hoang-Minh, L.B.; Atkinson, E.W.; Ugartemendia, G.; Semple-Rowland, S.; Coleman, J.E.; Sarkisian, M.R. Neonatal seizures induced by pentylenetetrazol or kainic acid disrupt primary cilia growth on developing mouse cortical neurons. *Exp. Neurol.* **2016**, *282*, 119–127. [CrossRef] [PubMed]
37. Holmes, G.L.; Sarkisian, M.; Ben-Ari, Y.; Chevassus-Au-Louis, N. Mossy fiber sprouting after recurrent seizures during early development in rats. *J. Comp. Neurol.* **1999**, *404*, 537–553. [CrossRef]
38. Stafstrom, C.E.; Thompson, J.L.; Holmes, G.L. Kainic acid seizures in the developing brain: Status epilepticus and spontaneous recurrent seizures. *Brain Res. Dev. Brain Res.* **1992**, *65*, 227–236. [CrossRef]
39. Ben-Ari, Y.; Lagowska, J. Epileptogenic action of intra-amygdaloid injection of kainic acid. *C R Acad. Hebd. Seances Acad. Sci. D* **1978**, *287*, 813–816.
40. Mesuret, G.; Engel, T.; Hessel, E.V.; Sanz-Rodriguez, A.; Jimenez-Pacheco, A.; Miras-Portugal, M.T.; Diaz-Hernandez, M.; Henshall, D.C. P2X7 receptor inhibition interrupts the progression of seizures in immature rats and reduces hippocampal damage. *CNS Neurosci. Ther.* **2014**, *20*, 556–564. [CrossRef]
41. Klioueva, I.A.; van Luijtelaar, E.L.; Chepurnova, N.E.; Chepurnov, S.A. PTZ-induced seizures in rats: Effects of age and strain. *Physiol. Behav.* **2001**, *72*, 421–426. [CrossRef]
42. Velísek, L.; Mares, P. Influence of clonazepam on electrocorticographic changes induced by metrazol in rats during ontogenesis. *Arch. Int. Pharm. Ther.* **1987**, *288*, 256–269.
43. Auvin, S.; Nehlig, A. Chapter 38 - Models of Seizures and Status Epilepticus Early in Life. In *Models of Seizures and Epilepsy*, 2nd ed.; Pitkänen, A., Buckmaster, P.S., Galanopoulou, A.S., Moshé, S.L., Eds.; Academic Press: Cambridge, MA, USA, 2017; pp. 569–586. [CrossRef]
44. Pineau, N.; Charriaut-Marlangue, C.; Motte, J.; Nehlig, A. Pentylenetetrazol seizures induce cell suffering but not death in the immature rat brain. *Brain Res. Dev. Brain Res.* **1999**, *112*, 139–144. [CrossRef]
45. Rice, J.E., 3rd; Vannucci, R.C.; Brierley, J.B. The influence of immaturity on hypoxic-ischemic brain damage in the rat. *Ann. Neurol.* **1981**, *9*, 131–141. [CrossRef] [PubMed]
46. Levine, S. Anoxic-ischemic encephalopathy in rats. *Am. J. Pathol.* **1960**, *36*, 1–17.
47. Cuaycong, M.; Engel, M.; Weinstein, S.L.; Salmon, E.; Perlman, J.M.; Sunderam, S.; Vannucci, S.J. A novel approach to the study of hypoxia-ischemia-induced clinical and subclinical seizures in the neonatal rat. *Dev. Neurosci.* **2011**, *33*, 241–250. [CrossRef] [PubMed]
48. Kadam, S.D.; White, A.M.; Staley, K.J.; Dudek, F.E. Continuous electroencephalographic monitoring with radio-telemetry in a rat model of perinatal hypoxia-ischemia reveals progressive post-stroke epilepsy. *J. Neurosci.* **2010**, *30*, 404–415. [CrossRef]
49. Romijn, H.J.; Hofman, M.A.; Gramsbergen, A. At what age is the developing cerebral cortex of the rat comparable to that of the full-term newborn human baby? *Early Hum. Dev.* **1991**, *26*, 61–67. [CrossRef]
50. Peng, J.; Li, R.; Arora, N.; Lau, M.; Lim, S.; Wu, C.; Eubanks, J.H.; Zhang, L. Effects of neonatal hypoxic-ischemic episodes on late seizure outcomes in C57 black mice. *Epilepsy Res.* **2015**, *111*, 142–149. [CrossRef]
51. Zanelli, S.; Goodkin, H.P.; Kowalski, S.; Kapur, J. Impact of transient acute hypoxia on the developing mouse EEG. *Neurobiol. Dis.* **2014**, *68*, 37–46. [CrossRef]
52. Leonard, A.S.; Hyder, S.N.; Kolls, B.J.; Arehart, E.; Ng, K.C.; Veerapandiyan, A.; Mikati, M.A. Seizure predisposition after perinatal hypoxia: Effects of subsequent age and of an epilepsy predisposing gene mutation. *Epilepsia* **2013**, *54*, 1789–1800. [CrossRef] [PubMed]
53. Rodriguez-Alvarez, N.; Jimenez-Mateos, E.M.; Dunleavy, M.; Waddington, J.L.; Boylan, G.B.; Henshall, D.C. Effects of hypoxia-induced neonatal seizures on acute hippocampal injury and later-life seizure susceptibility and anxiety-related behavior in mice. *Neurobiol. Dis.* **2015**, *83*, 100–114. [CrossRef] [PubMed]
54. Demirbilek, H.; Alanay, Y.; Alikaşifoğlu, A.; Topçu, M.; Mornet, E.; Gönç, N.; Özön, A.; Kandemir, N. Hypophosphatasia presenting with pyridoxine-responsive seizures, hypercalcemia, and pseudotumor cerebri: Case report. *J. Clin. Res. Pediatr. Endocrinol.* **2012**, *4*, 34–38. [CrossRef] [PubMed]
55. Belachew, D.; Kazmerski, T.; Libman, I.; Goldstein, A.C.; Stevens, S.T.; Deward, S.; Vockley, J.; Sperling, M.A.; Balest, A.L. Infantile hypophosphatasia secondary to a novel compound heterozygous mutation presenting with pyridoxine-responsive seizures. *JIMD Rep.* **2013**, *11*, 17–24. [CrossRef]

56. Kwon, J.M.; Guillet, R.; Shankaran, S.; Laptook, A.R.; McDonald, S.A.; Ehrenkranz, R.A.; Tyson, J.E.; O'Shea, T.M.; Goldberg, R.N.; Donovan, E.F.; et al. Clinical seizures in neonatal hypoxic-ischemic encephalopathy have no independent impact on neurodevelopmental outcome: Secondary analyses of data from the neonatal research network hypothermia trial. *J. Child. Neurol.* **2011**, *26*, 322–328. [CrossRef]

57. Dizon, M.L.V.; Rao, R.; Hamrick, S.E.; Zaniletti, I.; DiGeronimo, R.; Natarajan, G.; Kaiser, J.R.; Flibotte, J.; Lee, K.S.; Smith, D.; et al. Practice variation in anti-epileptic drug use for neonatal hypoxic-ischemic encephalopathy among regional NICUs. *BMC Pediatr.* **2019**, *19*, 67. [CrossRef]

58. Orbach, S.A.; Bonifacio, S.L.; Kuzniewicz, M.W.; Glass, H.C. Lower incidence of seizure among neonates treated with therapeutic hypothermia. *J. Child. Neurol.* **2014**, *29*, 1502–1507. [CrossRef]

59. Low, E.; Boylan, G.B.; Mathieson, S.R.; Murray, D.M.; Korotchikova, I.; Stevenson, N.J.; Livingstone, V.; Rennie, J.M. Cooling and seizure burden in term neonates: An observational study. *Arch. Dis. Child. Fetal. Neonatal Ed.* **2012**, *97*, F267-272. [CrossRef]

60. Shankaran, S.; Laptook, A.R.; Ehrenkranz, R.A.; Tyson, J.E.; McDonald, S.A.; Donovan, E.F.; Fanaroff, A.A.; Poole, W.K.; Wright, L.L.; Higgins, R.D.; et al. Whole-body hypothermia for neonates with hypoxic-ischemic encephalopathy. *N. Engl. J. Med.* **2005**, *353*, 1574–1584. [CrossRef]

61. Shankaran, S.; Pappas, A.; McDonald, S.A.; Vohr, B.R.; Hintz, S.R.; Yolton, K.; Gustafson, K.E.; Leach, T.M.; Green, C.; Bara, R.; et al. Childhood outcomes after hypothermia for neonatal encephalopathy. *N. Engl. J. Med.* **2012**, *366*, 2085–2092. [CrossRef]

62. Slaughter, L.A.; Patel, A.D.; Slaughter, J.L. Pharmacological Treatment of Neonatal Seizures: A Systematic Review. *J. Child. Neurol.* **2013**, *28*, 351–364. [CrossRef] [PubMed]

63. Bittigau, P.; Sifringer, M.; Ikonomidou, C. Antiepileptic drugs and apoptosis in the developing brain. *Ann. N. Y. Acad. Sci.* **2003**, *993*, 103–114, discussion 123-104. [CrossRef] [PubMed]

64. Painter, M.J.; Scher, M.S.; Stein, A.D.; Armatti, S.; Wang, Z.; Gardiner, J.C.; Paneth, N.; Minnigh, B.; Alvin, J. Phenobarbital compared with phenytoin for the treatment of neonatal seizures. *N. Engl. J. Med.* **1999**, *341*, 485–489. [CrossRef]

65. Ben-Ari, Y. Excitatory actions of gaba during development: The nature of the nurture. *Nat. Rev. Neurosci.* **2002**, *3*, 728–739. [CrossRef] [PubMed]

66. Owens, D.F.; Kriegstein, A.R. Is there more to GABA than synaptic inhibition? *Nat. Rev. Neurosci.* **2002**, *3*, 715–727. [CrossRef]

67. Gutherz, S.B.; Kulick, C.V.; Soper, C.; Kondratyev, A.; Gale, K.; Forcelli, P.A. Brief postnatal exposure to phenobarbital impairs passive avoidance learning and sensorimotor gating in rats. *Epilepsy Behav.* **2014**, *37*, 265–269. [CrossRef]

68. Torolira, D.; Suchomelova, L.; Wasterlain, C.G.; Niquet, J. Phenobarbital and midazolam increase neonatal seizure-associated neuronal injury. *Ann. Neurol.* **2017**, *82*, 115–120. [CrossRef]

69. Farwell, J.R.; Lee, Y.J.; Hirtz, D.G.; Sulzbacher, S.I.; Ellenberg, J.H.; Nelson, K.B. Phenobarbital for febrile seizures–effects on intelligence and on seizure recurrence. *N. Engl. J. Med.* **1990**, *322*, 364–369. [CrossRef]

70. Jehi, L.; Wyllie, E.; Devinsky, O. Epileptic encephalopathies: Optimizing seizure control and developmental outcome. *Epilepsia* **2015**, *56*, 1486–1489. [CrossRef]

71. Ramantani, G.; Schmitt, B.; Plecko, B.; Pressler, R.M.; Wohlrab, G.; Klebermass-Schrehof, K.; Hagmann, C.; Pisani, F.; Boylan, G.B. Neonatal Seizures-Are We there Yet? *Neuropediatrics* **2019**, *50*, 280–293. [CrossRef]

72. Ramantani, G. Neonatal epilepsy and underlying aetiology: To what extent do seizures and EEG abnormalities influence outcome? *Epileptic Disord.* **2013**, *15*, 365–375. [CrossRef] [PubMed]

73. Verkhratsky, A.; Burnstock, G. Biology of purinergic signalling: Its ancient evolutionary roots, its omnipresence and its multiple functional significance. *Bioessays* **2014**, *36*, 697–705. [CrossRef] [PubMed]

74. Halassa, M.M.; Fellin, T.; Haydon, P.G. Tripartite synapses: Roles for astrocytic purines in the control of synaptic physiology and behavior. *Neuropharmacology* **2009**, *57*, 343–346. [CrossRef] [PubMed]

75. Burnstock, G.; Krügel, U.; Abbracchio, M.P.; Illes, P. Purinergic signalling: From normal behaviour to pathological brain function. *Prog. Neurobiol.* **2011**, *95*, 229–274. [CrossRef] [PubMed]

76. Khakh, B.S.; North, R.A. Neuromodulation by extracellular ATP and P2X receptors in the CNS. *Neuron* **2012**, *76*, 51–69. [CrossRef]

77. Coppi, E.; Cellai, L.; Maraula, G.; Pugliese, A.M.; Pedata, F. Adenosine A_2A receptors inhibit delayed rectifier potassium currents and cell differentiation in primary purified oligodendrocyte cultures. *Neuropharmacology* **2013**, *73*, 301–310. [CrossRef]

78. Fumagalli, M.; Lecca, D.; Abbracchio, M.P. CNS remyelination as a novel reparative approach to neurodegenerative diseases: The roles of purinergic signaling and the P2Y-like receptor GPR17. *Neuropharmacology* **2016**, *104*, 82–93. [CrossRef]

79. Cotrina, M.L.; Lin, J.H.; Nedergaard, M. Cytoskeletal assembly and ATP release regulate astrocytic calcium signaling. *J. Neurosci.* **1998**, *18*, 8794–8804. [CrossRef]

80. Koizumi, S.; Fujishita, K.; Tsuda, M.; Shigemoto-Mogami, Y.; Inoue, K. Dynamic inhibition of excitatory synaptic transmission by astrocyte-derived ATP in hippocampal cultures. *Proc. Natl. Acad. Sci. USA* **2003**, *100*, 11023–11028. [CrossRef]

81. Koizumi, S.; Fujishita, K.; Inoue, K. Regulation of cell-to-cell communication mediated by astrocytic ATP in the CNS. *Purinergic Signal.* **2005**, *1*, 211–217. [CrossRef]

82. Ferrari, D.; Chiozzi, P.; Falzoni, S.; Hanau, S.; Di Virgilio, F. Purinergic modulation of interleukin-1 beta release from microglial cells stimulated with bacterial endotoxin. *J. Exp. Med.* **1997**, *185*, 579–582. [CrossRef] [PubMed]

83. Koizumi, S.; Shigemoto-Mogami, Y.; Nasu-Tada, K.; Shinozaki, Y.; Ohsawa, K.; Tsuda, M.; Joshi, B.V.; Jacobson, K.A.; Kohsaka, S.; Inoue, K. UDP acting at P2Y6 receptors is a mediator of microglial phagocytosis. *Nature* **2007**, *446*, 1091–1095. [CrossRef] [PubMed]

84. Irino, Y.; Nakamura, Y.; Inoue, K.; Kohsaka, S.; Ohsawa, K. Akt activation is involved in P2Y12 receptor-mediated chemotaxis of microglia. *J. Neurosci. Res.* **2008**, *86*, 1511–1519. [CrossRef] [PubMed]

85. Neligan, A.; Shorvon, S.D. Prognostic factors, morbidity and mortality in tonic-clonic status epilepticus: A review. *Epilepsy Res.* **2011**, *93*, 1–10. [CrossRef]

86. Orr, A.G.; Orr, A.L.; Li, X.J.; Gross, R.E.; Traynelis, S.F. Adenosine A(2A) receptor mediates microglial process retraction. *Nat. Neurosci.* **2009**, *12*, 872–878. [CrossRef]

87. Murphy, R.; Cherny, V.V.; Morgan, D.; DeCoursey, T.E. Voltage-gated proton channels help regulate pHi in rat alveolar epithelium. *Am. J. Physiol. Lung Cell Mol. Physiol.* **2005**, *288*, 398–408. [CrossRef]

88. Johnson, M.D.; Bao, H.F.; Helms, M.N.; Chen, X.J.; Tigue, Z.; Jain, L.; Dobbs, L.G.; Eaton, D.C. Functional ion channels in pulmonary alveolar type I cells support a role for type I cells in lung ion transport. *Proc. Natl. Acad. Sci. USA* **2006**, *103*, 4964–4969. [CrossRef]

89. Pellegatti, P.; Falzoni, S.; Pinton, P.; Rizzuto, R.; Di Virgilio, F. A novel recombinant plasma membrane-targeted luciferase reveals a new pathway for ATP secretion. *Mol. Biol. Cell* **2005**, *16*, 3659–3665. [CrossRef]

90. Suadicani, S.O.; Brosnan, C.F.; Scemes, E. P2X7 receptors mediate ATP release and amplification of astrocytic intercellular Ca2+ signaling. *J. Neurosci.* **2006**, *26*, 1378–1385. [CrossRef]

91. Kang, J.; Kang, N.; Lovatt, D.; Torres, A.; Zhao, Z.; Lin, J.; Nedergaard, M. Connexin 43 hemichannels are permeable to ATP. *J. Neurosci.* **2008**, *28*, 4702–4711. [CrossRef]

92. Bao, L.; Locovei, S.; Dahl, G. Pannexin membrane channels are mechanosensitive conduits for ATP. *FEBS Lett.* **2004**, *572*, 65–68. [CrossRef] [PubMed]

93. Locovei, S.; Scemes, E.; Qiu, F.; Spray, D.C.; Dahl, G. Pannexin1 is part of the pore forming unit of the P2X(7) receptor death complex. *FEBS Lett.* **2007**, *581*, 483–488. [CrossRef] [PubMed]

94. Baranova, A.; Ivanov, D.; Petrash, N.; Pestova, A.; Skoblov, M.; Kelmanson, I.; Shagin, D.; Nazarenko, S.; Geraymovych, E.; Litvin, O.; et al. The mammalian pannexin family is homologous to the invertebrate innexin gap junction proteins. *Genomics* **2004**, *83*, 706–716. [CrossRef]

95. Vogt, A.; Hormuzdi, S.G.; Monyer, H. Pannexin1 and Pannexin2 expression in the developing and mature rat brain. *Brain Res. Mol. Brain Res.* **2005**, *141*, 113–120. [CrossRef]

96. Zappalà, A.; Cicero, D.; Serapide, M.F.; Paz, C.; Catania, M.V.; Falchi, M.; Parenti, R.; Pantò, M.R.; La Delia, F.; Cicirata, F. Expression of pannexin1 in the CNS of adult mouse: Cellular localization and effect of 4-aminopyridine-induced seizures. *Neuroscience* **2006**, *141*, 167–178. [CrossRef]

97. D'Hondt, C.; Ponsaerts, R.; De Smedt, H.; Bultynck, G.; Himpens, B. Pannexins, distant relatives of the connexin family with specific cellular functions? *Bioessays* **2009**, *31*, 953–974. [CrossRef]

98. Giaume, C.; Leybaert, L.; Naus, C.C.; Sáez, J.C. Connexin and pannexin hemichannels in brain glial cells: Properties, pharmacology, and roles. *Front. Pharm.* **2013**, *4*, 88. [CrossRef]

99. Wang, N.; De Bock, M.; Decrock, E.; Bol, M.; Gadicherla, A.; Vinken, M.; Rogiers, V.; Bukauskas, F.F.; Bultynck, G.; Leybaert, L. Paracrine signaling through plasma membrane hemichannels. *Biochim. Biophys. Acta* **2013**, *1828*, 35–50. [CrossRef]

100. MacVicar, B.A.; Thompson, R.J. Non-junction functions of pannexin-1 channels. *Trends Neurosci.* **2010**, *33*, 93–102. [CrossRef]
101. Pelegrin, P.; Surprenant, A. Pannexin-1 mediates large pore formation and interleukin-1beta release by the ATP-gated P2X7 receptor. *EMBO J.* **2006**, *25*, 5071–5082. [CrossRef] [PubMed]
102. Ho, T.; Jobling, A.I.; Greferath, U.; Chuang, T.; Ramesh, A.; Fletcher, E.L.; Vessey, K.A. Vesicular expression and release of ATP from dopaminergic neurons of the mouse retina and midbrain. *Front. Cell. Neurosci.* **2015**, *9*, 389. [CrossRef] [PubMed]
103. Larsson, M.; Sawada, K.; Morland, C.; Hiasa, M.; Ormel, L.; Moriyama, Y.; Gundersen, V. Functional and anatomical identification of a vesicular transporter mediating neuronal ATP release. *Cereb. Cortex* **2012**, *22*, 1203–1214. [CrossRef] [PubMed]
104. Oya, M.; Kitaguchi, T.; Yanagihara, Y.; Numano, R.; Kakeyama, M.; Ikematsu, K.; Tsuboi, T. Vesicular nucleotide transporter is involved in ATP storage of secretory lysosomes in astrocytes. *Biochem. Biophys. Res. Commun.* **2013**, *438*, 145–151. [CrossRef] [PubMed]
105. Imura, Y.; Morizawa, Y.; Komatsu, R.; Shibata, K.; Shinozaki, Y.; Kasai, H.; Moriishi, K.; Moriyama, Y.; Koizumi, S. Microglia release ATP by exocytosis. *Glia* **2013**, *61*, 1320–1330. [CrossRef] [PubMed]
106. Sawada, K.; Echigo, N.; Juge, N.; Miyaji, T.; Otsuka, M.; Omote, H.; Yamamoto, A.; Moriyama, Y. Identification of a vesicular nucleotide transporter. *Proc. Natl. Acad. Sci. USA* **2008**, *105*, 5683–5686. [CrossRef] [PubMed]
107. Menéndez-Méndez, A.; Díaz-Hernández, J.I.; Ortega, F.; Gualix, J.; Gómez-Villafuertes, R.; Miras-Portugal, M.T. Specific Temporal Distribution and Subcellular Localization of a Functional Vesicular Nucleotide Transporter (VNUT) in Cerebellar Granule Neurons. *Front. Pharm.* **2017**, *8*, 951. [CrossRef] [PubMed]
108. Vessey, K.A.; Fletcher, E.L. Rod and cone pathway signalling is altered in the P2X7 receptor knock out mouse. *PLoS ONE* **2012**, *7*, e29990. [CrossRef]
109. Moriyama, S.; Hiasa, M. Expression of Vesicular Nucleotide Transporter in the Mouse Retina. *Biol. Pharm. Bull.* **2016**, *39*, 564–569. [CrossRef]
110. Olah, M.E.; Stiles, G.L. The role of receptor structure in determining adenosine receptor activity. *Pharm. Ther.* **2000**, *85*, 55–75. [CrossRef]
111. Yaar, R.; Jones, M.R.; Chen, J.F.; Ravid, K. Animal models for the study of adenosine receptor function. *J. Cell Physiol.* **2005**, *202*, 9–20. [CrossRef]
112. Ribeiro, J.A.; Sebastião, A.M.; de Mendonça, A. Adenosine receptors in the nervous system: Pathophysiological implications. *Prog. Neurobiol.* **2002**, *68*, 377–392. [CrossRef]
113. Oliveira, L.; Timóteo, M.A.; Correia-de-Sá, P. Modulation by adenosine of both muscarinic M1-facilitation and M2-inhibition of [3H]-acetylcholine release from the rat motor nerve terminals. *Eur. J. Neurosci.* **2002**, *15*, 1728–1736. [CrossRef] [PubMed]
114. Sebastião, A.M.; Ribeiro, J.A. Fine-tuning neuromodulation by adenosine. *Trends Pharm. Sci.* **2000**, *21*, 341–346. [CrossRef]
115. Daré, E.; Schulte, G.; Karovic, O.; Hammarberg, C.; Fredholm, B.B. Modulation of glial cell functions by adenosine receptors. *Physiol. Behav.* **2007**, *92*, 15–20. [CrossRef] [PubMed]
116. Fellin, T.; Pascual, O.; Haydon, P.G. Astrocytes coordinate synaptic networks: Balanced excitation and inhibition. *Physiology (Bethesda)* **2006**, *21*, 208–215. [CrossRef] [PubMed]
117. Rebola, N.; Coelho, J.E.; Costenla, A.R.; Lopes, L.V.; Parada, A.; Oliveira, C.R.; Soares-da-Silva, P.; de Mendonça, A.; Cunha, R.A. Decrease of adenosine A1 receptor density and of adenosine neuromodulation in the hippocampus of kindled rats. *Eur. J. Neurosci.* **2003**, *18*, 820–828. [CrossRef]
118. Popoli, P.; Betto, P.; Reggio, R.; Ricciarello, G. Adenosine A2A receptor stimulation enhances striatal extracellular glutamate levels in rats. *Eur. J. Pharm.* **1995**, *287*, 215–217. [CrossRef]
119. Benarroch, E.E. Adenosine and its receptors: Multiple modulatory functions and potential therapeutic targets for neurologic disease. *Neurology* **2008**, *70*, 231–236. [CrossRef]
120. Cunha, R.A. Neuroprotection by adenosine in the brain: From A(1) receptor activation to A (2A) receptor blockade. *Purinergic Signal.* **2005**, *1*, 111–134. [CrossRef]
121. Fredholm, B.B. Adenosine receptors as targets for drug development. *Drug News Perspect.* **2003**, *16*, 283–289. [CrossRef]
122. Salvemini, D.; Jacobson, K.A. Highly selective A3 adenosine receptor agonists relieve chronic neuropathic pain. *Expert Opin. Ther. Pat.* **2017**, *27*, 967. [CrossRef] [PubMed]

123. Schwarzschild, M.A.; Agnati, L.; Fuxe, K.; Chen, J.F.; Morelli, M. Targeting adenosine A2A receptors in Parkinson's disease. *Trends Neurosci.* **2006**, *29*, 647–654. [CrossRef] [PubMed]

124. McGaraughty, S.; Jarvis, M.F. Purinergic control of neuropathic pain. *Drug Dev. Res.* **2006**, *67*, 376–388. [CrossRef]

125. Boison, D. Cell and gene therapies for refractory epilepsy. *Curr. Neuropharmacol.* **2007**, *5*, 115–125. [CrossRef] [PubMed]

126. Trendelenburg, G.; Dirnagl, U. Neuroprotective role of astrocytes in cerebral ischemia: Focus on ischemic preconditioning. *Glia* **2005**, *50*, 307–320. [CrossRef] [PubMed]

127. Fredholm, B.B.; AP, I.J.; Jacobson, K.A.; Linden, J.; Müller, C.E. International Union of Basic and Clinical Pharmacology. LXXXI. Nomenclature and classification of adenosine receptors–an update. *Pharm. Rev.* **2011**, *63*, 1–34. [CrossRef]

128. Cristóvão-Ferreira, S.; Navarro, G.; Brugarolas, M.; Pérez-Capote, K.; Vaz, S.H.; Fattorini, G.; Conti, F.; Lluis, C.; Ribeiro, J.A.; McCormick, P.J.; et al. A1R-A2AR heteromers coupled to Gs and G i/0 proteins modulate GABA transport into astrocytes. *Purinergic Signal.* **2013**, *9*, 433–449. [CrossRef]

129. Stenberg, D.; Litonius, E.; Halldner, L.; Johansson, B.; Fredholm, B.B.; Porkka-Heiskanen, T. Sleep and its homeostatic regulation in mice lacking the adenosine A1 receptor. *J. Sleep Res.* **2003**, *12*, 283–290. [CrossRef]

130. Ciruela, F.; Casadó, V.; Rodrigues, R.J.; Luján, R.; Burgueño, J.; Canals, M.; Borycz, J.; Rebola, N.; Goldberg, S.R.; Mallol, J.; et al. Presynaptic control of striatal glutamatergic neurotransmission by adenosine A1-A2A receptor heteromers. *J. Neurosci.* **2006**, *26*, 2080–2087. [CrossRef]

131. Gessi, S.; Merighi, S.; Fazzi, D.; Stefanelli, A.; Varani, K.; Borea, P.A. Adenosine receptor targeting in health and disease. *Expert Opin. Investig. Drugs* **2011**, *20*, 1591–1609. [CrossRef]

132. Cunha, R.A. How does adenosine control neuronal dysfunction and neurodegeneration? *J. Neurochem.* **2016**, *139*, 1019–1055. [CrossRef] [PubMed]

133. Sawynok, J. Adenosine receptor targets for pain. *Neuroscience* **2016**, *338*, 1–18. [CrossRef] [PubMed]

134. Wu, L.G.; Saggau, P. Adenosine inhibits evoked synaptic transmission primarily by reducing presynaptic calcium influx in area CA1 of hippocampus. *Neuron* **1994**, *12*, 1139–1148. [CrossRef]

135. Gundlfinger, A.; Bischofberger, J.; Johenning, F.W.; Torvinen, M.; Schmitz, D.; Breustedt, J. Adenosine modulates transmission at the hippocampal mossy fibre synapse via direct inhibition of presynaptic calcium channels. *J. Physiol.* **2007**, *582*, 263–277. [CrossRef]

136. Yoon, K.W.; Rothman, S.M. Adenosine inhibits excitatory but not inhibitory synaptic transmission in the hippocampus. *J. Neurosci.* **1991**, *11*, 1375–1380. [CrossRef]

137. Von Lubitz, D.K.; Paul, I.A.; Ji, X.D.; Carter, M.; Jacobson, K.A. Chronic adenosine A1 receptor agonist and antagonist: Effect on receptor density and N-methyl-D-aspartate induced seizures in mice. *Eur. J. Pharm.* **1994**, *253*, 95–99. [CrossRef]

138. Kull, B.; Svenningsson, P.; Fredholm, B.B. Adenosine A(2A) receptors are colocalized with and activate g(olf) in rat striatum. *Mol. Pharm.* **2000**, *58*, 771–777. [CrossRef]

139. Lopes, L.V.; Cunha, R.A.; Kull, B.; Fredholm, B.B.; Ribeiro, J.A. Adenosine A(2A) receptor facilitation of hippocampal synaptic transmission is dependent on tonic A(1) receptor inhibition. *Neuroscience* **2002**, *112*, 319–329. [CrossRef]

140. Marchi, M.; Raiteri, L.; Risso, F.; Vallarino, A.; Bonfanti, A.; Monopoli, A.; Ongini, E.; Raiteri, M. Effects of adenosine A1 and A2A receptor activation on the evoked release of glutamate from rat cerebrocortical synaptosomes. *Br. J. Pharm.* **2002**, *136*, 434–440. [CrossRef]

141. Rebola, N.; Lujan, R.; Cunha, R.A.; Mulle, C. Adenosine A2A receptors are essential for long-term potentiation of NMDA-EPSCs at hippocampal mossy fiber synapses. *Neuron* **2008**, *57*, 121–134. [CrossRef]

142. Rebola, N.; Simões, A.P.; Canas, P.M.; Tomé, A.R.; Andrade, G.M.; Barry, C.E.; Agostinho, P.M.; Lynch, M.A.; Cunha, R.A. Adenosine A2A receptors control neuroinflammation and consequent hippocampal neuronal dysfunction. *J. Neurochem.* **2011**, *117*, 100–111. [CrossRef] [PubMed]

143. Dunwiddie, T.V.; Diao, L.; Kim, H.O.; Jiang, J.L.; Jacobson, K.A. Activation of hippocampal adenosine A3 receptors produces a desensitization of A1 receptor-mediated responses in rat hippocampus. *J. Neurosci.* **1997**, *17*, 607–614. [CrossRef] [PubMed]

144. Fredholm, B.B.; Arslan, G.; Halldner, L.; Kull, B.; Schulte, G.; Wasserman, W. Structure and function of adenosine receptors and their genes. *Naunyn Schmiedebergs Arch. Pharm.* **2000**, *362*, 364–374. [CrossRef]

145. Abbracchio, M.P.; Rainaldi, G.; Giammarioli, A.M.; Ceruti, S.; Brambilla, R.; Cattabeni, F.; Barbieri, D.; Franceschi, C.; Jacobson, K.A.; Malorni, W. The A3 adenosine receptor mediates cell spreading, reorganization of actin cytoskeleton, and distribution of Bcl-XL: Studies in human astroglioma cells. *Biochem. Biophys. Res. Commun.* **1997**, *241*, 297–304. [CrossRef]

146. Hammarberg, C.; Schulte, G.; Fredholm, B.B. Evidence for functional adenosine A3 receptors in microglia cells. *J. Neurochem.* **2003**, *86*, 1051–1054. [CrossRef] [PubMed]

147. Khakh, B.S.; North, R.A. P2X receptors as cell-surface ATP sensors in health and disease. *Nature* **2006**, *442*, 527–532. [CrossRef]

148. Köles, L.; Fürst, S.; Illes, P. Purine ionotropic (P2X) receptors. *Curr. Pharm. Des.* **2007**, *13*, 2368–2384. [CrossRef]

149. Fischer, W.; Krügel, U. P2Y receptors: Focus on structural, pharmacological and functional aspects in the brain. *Curr. Med. Chem.* **2007**, *14*, 2429–2455. [CrossRef]

150. Abbracchio, M.P.; Burnstock, G.; Boeynaems, J.M.; Barnard, E.A.; Boyer, J.L.; Kennedy, C.; Knight, G.E.; Fumagalli, M.; Gachet, C.; Jacobson, K.A.; et al. International Union of Pharmacology LVIII: Update on the P2Y G protein-coupled nucleotide receptors: From molecular mechanisms and pathophysiology to therapy. *Pharm. Rev.* **2006**, *58*, 281–341. [CrossRef]

151. Brunschweiger, A.; Müller, C.E. P2 receptors activated by uracil nucleotides–an update. *Curr. Med. Chem.* **2006**, *13*, 289–312. [CrossRef]

152. Ennion, S.J.; Evans, R.J. Conserved cysteine residues in the extracellular loop of the human P2X(1) receptor form disulfide bonds and are involved in receptor trafficking to the cell surface. *Mol. Pharm.* **2002**, *61*, 303–311. [CrossRef] [PubMed]

153. Clyne, J.D.; Wang, L.F.; Hume, R.I. Mutational analysis of the conserved cysteines of the rat P2X2 purinoceptor. *J. Neurosci.* **2002**, *22*, 3873–3880. [CrossRef] [PubMed]

154. Young, M.T. P2X receptors: Dawn of the post-structure era. *Trends Biochem. Sci.* **2010**, *35*, 83–90. [CrossRef] [PubMed]

155. Nicke, A.; Bäumert, H.G.; Rettinger, J.; Eichele, A.; Lambrecht, G.; Mutschler, E.; Schmalzing, G. P2X1 and P2X3 receptors form stable trimers: A novel structural motif of ligand-gated ion channels. *EMBO J.* **1998**, *17*, 3016–3028. [CrossRef] [PubMed]

156. Roberts, J.A.; Vial, C.; Digby, H.R.; Agboh, K.C.; Wen, H.; Atterbury-Thomas, A.; Evans, R.J. Molecular properties of P2X receptors. *Pflugers Arch.* **2006**, *452*, 486–500. [CrossRef] [PubMed]

157. North, R.A. Molecular physiology of P2X receptors. *Physiol. Rev.* **2002**, *82*, 1013–1067. [CrossRef]

158. Guo, C.; Masin, M.; Qureshi, O.S.; Murrell-Lagnado, R.D. Evidence for functional P2X4/P2X7 heteromeric receptors. *Mol. Pharm.* **2007**, *72*, 1447–1456. [CrossRef]

159. Sperlagh, B.; Vizi, E.S.; Wirkner, K.; Illes, P. P2X7 receptors in the nervous system. *Prog. Neurobiol.* **2006**, *78*, 327–346. [CrossRef]

160. Burnstock, G. Physiology and pathophysiology of purinergic neurotransmission. *Physiol. Rev.* **2007**, *87*, 659–797. [CrossRef]

161. Pankratov, Y.; Lalo, U.; Krishtal, O.A.; Verkhratsky, A. P2X receptors and synaptic plasticity. *Neuroscience* **2009**, *158*, 137–148. [CrossRef]

162. Lalo, U.; Palygin, O.; Rasooli-Nejad, S.; Andrew, J.; Haydon, P.G.; Pankratov, Y. Exocytosis of ATP from astrocytes modulates phasic and tonic inhibition in the neocortex. *PLoS Biol.* **2014**, *12*, e1001747. [CrossRef] [PubMed]

163. North, R.A.; Verkhratsky, A. Purinergic transmission in the central nervous system. *Pflugers Arch.* **2006**, *452*, 479–485. [CrossRef] [PubMed]

164. Sperlágh, B.; Zsilla, G.; Baranyi, M.; Illes, P.; Vizi, E.S. Purinergic modulation of glutamate release under ischemic-like conditions in the hippocampus. *Neuroscience* **2007**, *149*, 99–111. [CrossRef] [PubMed]

165. Andó, R.D.; Sperlágh, B. The role of glutamate release mediated by extrasynaptic P2X7 receptors in animal models of neuropathic pain. *Brain Res. Bull.* **2013**, *93*, 80–85. [CrossRef]

166. Burnstock, G. P2X ion channel receptors and inflammation. *Purinergic Signal.* **2016**, *12*, 59–67. [CrossRef]

167. Ecke, D.; Hanck, T.; Tulapurkar, M.E.; Schäfer, R.; Kassack, M.; Stricker, R.; Reiser, G. Hetero-oligomerization of the P2Y11 receptor with the P2Y1 receptor controls the internalization and ligand selectivity of the P2Y11 receptor. *Biochem. J.* **2008**, *409*, 107–116. [CrossRef]

168. Verkhratsky, A. Physiology and pathophysiology of the calcium store in the endoplasmic reticulum of neurons. *Physiol. Rev.* **2005**, *85*, 201–279. [CrossRef]
169. Von Kugelgen, I. Pharmacological profiles of cloned mammalian P2Y-receptor subtypes. *Pharm. Ther.* **2006**, *110*, 415–432. [CrossRef]
170. Burnstock, G. Purine and pyrimidine receptors. *Cell Mol. Life Sci.* **2007**, *64*, 1471–1483. [CrossRef]
171. Von Kugelgen, I.; Harden, T.K. Molecular pharmacology, physiology, and structure of the P2Y receptors. *Adv. Pharm.* **2011**, *61*, 373–415. [CrossRef]
172. Weisman, G.A.; Woods, L.T.; Erb, L.; Seye, C.I. P2Y receptors in the mammalian nervous system: Pharmacology, ligands and therapeutic potential. *CNS Neurol. Disord. Drug Targets* **2012**, *11*, 722–738. [CrossRef]
173. Krügel, U.; Kittner, H.; Franke, H.; Illes, P. Stimulation of P2 receptors in the ventral tegmental area enhances dopaminergic mechanisms in vivo. *Neuropharmacology* **2001**, *40*, 1084–1093. [CrossRef]
174. Koch, H.; Bespalov, A.; Drescher, K.; Franke, H.; Krügel, U. Impaired cognition after stimulation of P2Y1 receptors in the rat medial prefrontal cortex. *Neuropsychopharmacology* **2015**, *40*, 305–314. [CrossRef] [PubMed]
175. Jacobson, K.A.; Boeynaems, J.M. P2Y nucleotide receptors: Promise of therapeutic applications. *Drug Discov. Today* **2010**, *15*, 570–578. [CrossRef] [PubMed]
176. Guzman, S.J.; Gerevich, Z. P2Y Receptors in Synaptic Transmission and Plasticity: Therapeutic Potential in Cognitive Dysfunction. *Neural Plast.* **2016**, *2016*, 1207393. [CrossRef]
177. Calovi, S.; Mut-Arbona, P.; Sperlágh, B. Microglia and the Purinergic Signaling System. *Neuroscience* **2019**, *405*, 137–147. [CrossRef]
178. Luongo, L.; Guida, F.; Imperatore, R.; Napolitano, F.; Gatta, L.; Cristino, L.; Giordano, C.; Siniscalco, D.; Di Marzo, V.; Bellini, G.; et al. The A1 adenosine receptor as a new player in microglia physiology. *Glia* **2014**, *62*, 122–132. [CrossRef] [PubMed]
179. Kaczmarek-Hajek, K.; Zhang, J.; Kopp, R.; Grosche, A.; Rissiek, B.; Saul, A.; Bruzzone, S.; Engel, T.; Jooss, T.; Krautloher, A.; et al. Re-evaluation of neuronal P2X7 expression using novel mouse models and a P2X7-specific nanobody. *Elife* **2018**, *7*. [CrossRef]
180. He, Y.; Taylor, N.; Fourgeaud, L.; Bhattacharya, A. The role of microglial P2X7: Modulation of cell death and cytokine release. *J. Neuroinflammation* **2017**, *14*, 135. [CrossRef]
181. Monif, M.; Reid, C.A.; Powell, K.L.; Smart, M.L.; Williams, D.A. The P2X7 receptor drives microglial activation and proliferation: A trophic role for P2X7R pore. *J. Neurosci.* **2009**, *29*, 3781–3791. [CrossRef]
182. Ohsawa, K.; Irino, Y.; Nakamura, Y.; Akazawa, C.; Inoue, K.; Kohsaka, S. Involvement of P2X4 and P2Y12 receptors in ATP-induced microglial chemotaxis. *Glia* **2007**, *55*, 604–616. [CrossRef]
183. Zabala, A.; Vazquez-Villoldo, N.; Rissiek, B.; Gejo, J.; Martin, A.; Palomino, A.; Perez-Samartín, A.; Pulagam, K.R.; Lukowiak, M.; Capetillo-Zarate, E.; et al. P2X4 receptor controls microglia activation and favors remyelination in autoimmune encephalitis. *EMBO Mol. Med.* **2018**, *10*. [CrossRef]
184. Davalos, D.; Grutzendler, J.; Yang, G.; Kim, J.V.; Zuo, Y.; Jung, S.; Littman, D.R.; Dustin, M.L.; Gan, W.B. ATP mediates rapid microglial response to local brain injury in vivo. *Nat. Neurosci.* **2005**, *8*, 752–758. [CrossRef]
185. Engel, T.; Gomez-Villafuertes, R.; Tanaka, K.; Mesuret, G.; Sanz-Rodriguez, A.; Garcia-Huerta, P.; Miras-Portugal, M.T.; Henshall, D.C.; Diaz-Hernandez, M. Seizure suppression and neuroprotection by targeting the purinergic P2X7 receptor during status epilepticus in mice. *FASEB J.* **2012**, *26*, 1616–1628. [CrossRef]
186. Swiatkowski, P.; Murugan, M.; Eyo, U.B.; Wang, Y.; Rangaraju, S.; Oh, S.B.; Wu, L.J. Activation of microglial P2Y12 receptor is required for outward potassium currents in response to neuronal injury. *Neuroscience* **2016**, *318*, 22–33. [CrossRef]
187. Inoue, K. UDP facilitates microglial phagocytosis through P2Y6 receptors. *Cell Adh. Migr.* **2007**, *1*, 131–132. [CrossRef]
188. Alves, M.; De Diego Garcia, L.; Conte, G.; Jimenez-Mateos, E.M.; D'Orsi, B.; Sanz-Rodriguez, A.; Prehn, J.H.M.; Henshall, D.C.; Engel, T. Context-Specific Switch from Anti- to Pro-epileptogenic Function of the P2Y(1) Receptor in Experimental Epilepsy. *J. Neurosci.* **2019**, *39*, 5377–5392. [CrossRef] [PubMed]
189. Wellmann, M.; Álvarez-Ferradas, C.; Maturana, C.J.; Sáez, J.C.; Bonansco, C. Astroglial Ca(2+)-Dependent Hyperexcitability Requires P2Y(1) Purinergic Receptors and Pannexin-1 Channel Activation in a Chronic Model of Epilepsy. *Front. Cell Neurosci.* **2018**, *12*, 446. [CrossRef] [PubMed]

190. Moore, C.S.; Ase, A.R.; Kinsara, A.; Rao, V.T.; Michell-Robinson, M.; Leong, S.Y.; Butovsky, O.; Ludwin, S.K.; Séguéla, P.; Bar-Or, A.; et al. P2Y12 expression and function in alternatively activated human microglia. *Neurol. Neuroimmunol. Neuroinflamm.* **2015**, *2*, e80. [CrossRef]
191. Vezzani, A.; French, J.; Bartfai, T.; Baram, T.Z. The role of inflammation in epilepsy. *Nat. Rev. Neurol.* **2011**, *7*, 31–40. [CrossRef] [PubMed]
192. Zimmermann, H. Nucleotide signaling in nervous system development. *Pflugers Arch.* **2006**, *452*, 573–588. [CrossRef] [PubMed]
193. Robson, S.C.; Sévigny, J.; Zimmermann, H. The E-NTPDase family of ectonucleotidases: Structure function relationships and pathophysiological significance. *Purinergic Signal.* **2006**, *2*, 409–430. [CrossRef]
194. Massé, K.; Bhamra, S.; Allsop, G.; Dale, N.; Jones, E.A. Ectophosphodiesterase/nucleotide phosphohydrolase (Enpp) nucleotidases: Cloning, conservation and developmental restriction. *Int. J. Dev. Biol.* **2010**, *54*, 181–193. [CrossRef] [PubMed]
195. Sträter, N. Ecto-5′-nucleotidase: Structure function relationships. *Purinergic Signal.* **2006**, *2*, 343–350. [CrossRef] [PubMed]
196. Fonta, C.; Negyessy, L.; Renaud, L.; Barone, P. Areal and subcellular localization of the ubiquitous alkaline phosphatase in the primate cerebral cortex: Evidence for a role in neurotransmission. *Cereb. Cortex* **2004**, *14*, 595–609. [CrossRef] [PubMed]
197. Langer, D.; Hammer, K.; Koszalka, P.; Schrader, J.; Robson, S.; Zimmermann, H. Distribution of ectonucleotidases in the rodent brain revisited. *Cell Tissue Res.* **2008**, *334*, 199–217. [CrossRef] [PubMed]
198. Lovatt, D.; Xu, Q.; Liu, W.; Takano, T.; Smith, N.A.; Schnermann, J.; Tieu, K.; Nedergaard, M. Neuronal adenosine release, and not astrocytic ATP release, mediates feedback inhibition of excitatory activity. *Proc. Natl. Acad. Sci. USA* **2012**, *109*, 6265–6270. [CrossRef] [PubMed]
199. Martín, E.D.; Fernández, M.; Perea, G.; Pascual, O.; Haydon, P.G.; Araque, A.; Ceña, V. Adenosine released by astrocytes contributes to hypoxia-induced modulation of synaptic transmission. *Glia* **2007**, *55*, 36–45. [CrossRef] [PubMed]
200. Lloyd, H.G.; Fredholm, B.B. Involvement of adenosine deaminase and adenosine kinase in regulating extracellular adenosine concentration in rat hippocampal slices. *Neurochem. Int.* **1995**, *26*, 387–395. [CrossRef]
201. Weissman, T.A.; Riquelme, P.A.; Ivic, L.; Flint, A.C.; Kriegstein, A.R. Calcium waves propagate through radial glial cells and modulate proliferation in the developing neocortex. *Neuron* **2004**, *43*, 647–661. [CrossRef]
202. Scemes, E.; Duval, N.; Meda, P. Reduced expression of P2Y1 receptors in connexin43-null mice alters calcium signaling and migration of neural progenitor cells. *J. Neurosci.* **2003**, *23*, 11444–11452. [CrossRef]
203. Liu, X.; Hashimoto-Torii, K.; Torii, M.; Haydar, T.F.; Rakic, P. The role of ATP signaling in the migration of intermediate neuronal progenitors to the neocortical subventricular zone. *Proc. Natl. Acad. Sci. USA* **2008**, *105*, 11802–11807. [CrossRef]
204. Noctor, S.C.; Martínez-Cerdeño, V.; Ivic, L.; Kriegstein, A.R. Cortical neurons arise in symmetric and asymmetric division zones and migrate through specific phases. *Nat. Neurosci.* **2004**, *7*, 136–144. [CrossRef] [PubMed]
205. Heine, C.; Sygnecka, K.; Franke, H. Purines in neurite growth and astroglia activation. *Neuropharmacology* **2016**, *104*, 255–271. [CrossRef] [PubMed]
206. Peterson, T.S.; Camden, J.M.; Wang, Y.; Seye, C.I.; Wood, W.G.; Sun, G.Y.; Erb, L.; Petris, M.J.; Weisman, G.A. P2Y2 nucleotide receptor-mediated responses in brain cells. *Mol. Neurobiol.* **2010**, *41*, 356–366. [CrossRef] [PubMed]
207. Miller, S.M.; Goasdoue, K.; Björkman, S.T. Neonatal seizures and disruption to neurotransmitter systems. *Neural Regen Res.* **2017**, *12*, 216–217. [CrossRef] [PubMed]
208. Karnam, H.B.; Zhao, Q.; Shatskikh, T.; Holmes, G.L. Effect of age on cognitive sequelae following early life seizures in rats. *Epilepsy Res.* **2009**, *85*, 221–230. [CrossRef] [PubMed]
209. Isaeva, E.; Isaev, D.; Holmes, G.L. Alteration of synaptic plasticity by neonatal seizures in rat somatosensory cortex. *Epilepsy Res.* **2013**, *106*, 280–283. [CrossRef]
210. Gulisano, M.; Parenti, R.; Spinella, F.; Cicirata, F. Cx36 is dynamically expressed during early development of mouse brain and nervous system. *Neuroreport* **2000**, *11*, 3823–3828. [CrossRef]
211. Nadarajah, B.; Jones, A.M.; Evans, W.H.; Parnavelas, J.G. Differential expression of connexins during neocortical development and neuronal circuit formation. *J. Neurosci.* **1997**, *17*, 3096–3111. [CrossRef]

212. Söhl, G.; Eiberger, J.; Jung, Y.T.; Kozak, C.A.; Willecke, K. The mouse gap junction gene connexin29 is highly expressed in sciatic nerve and regulated during brain development. *Biol. Chem.* **2001**, *382*, 973–978. [CrossRef] [PubMed]

213. Söhl, G.; Theis, M.; Hallas, G.; Brambach, S.; Dahl, E.; Kidder, G.; Willecke, K. A new alternatively spliced transcript of the mouse connexin32 gene is expressed in embryonic stem cells, oocytes, and liver. *Exp. Cell Res.* **2001**, *266*, 177–186. [CrossRef] [PubMed]

214. Rouach, N.; Avignone, E.; Même, W.; Koulakoff, A.; Venance, L.; Blomstrand, F.; Giaume, C. Gap junctions and connexin expression in the normal and pathological central nervous system. *Biol. Cell.* **2002**, *94*, 457–475. [CrossRef]

215. Lo, C.W. The role of gap junction membrane channels in development. *J. Bioenerg. Biomembr.* **1996**, *28*, 379–385. [CrossRef]

216. Wicki-Stordeur, L.E.; Dzugalo, A.D.; Swansburg, R.M.; Suits, J.M.; Swayne, L.A. Pannexin 1 regulates postnatal neural stem and progenitor cell proliferation. *Neural Dev.* **2012**, *7*, 11. [CrossRef]

217. Wicki-Stordeur, L.E.; Swayne, L.A. Panx1 regulates neural stem and progenitor cell behaviours associated with cytoskeletal dynamics and interacts with multiple cytoskeletal elements. *Cell. Commun. Signal.* **2013**, *11*, 62. [CrossRef]

218. Swayne, L.A.; Sorbara, C.D.; Bennett, S.A. Pannexin 2 is expressed by postnatal hippocampal neural progenitors and modulates neuronal commitment. *J. Biol. Chem.* **2010**, *285*, 24977–24986. [CrossRef]

219. Rivkees, S.A. The ontogeny of cardiac and neural A1 adenosine receptor expression in rats. *Brain Res. Dev. Brain Res.* **1995**, *89*, 202–213. [CrossRef]

220. Weaver, D.R. A1-adenosine receptor gene expression in fetal rat brain. *Brain Res. Dev. Brain Res.* **1996**, *94*, 205–223. [CrossRef]

221. Silva, C.G.; Metin, C.; Fazeli, W.; Machado, N.J.; Darmopil, S.; Launay, P.S.; Ghestem, A.; Nesa, M.P.; Bassot, E.; Szabo, E.; et al. Adenosine receptor antagonists including caffeine alter fetal brain development in mice. *Sci. Transl. Med.* **2013**, *5*, 197. [CrossRef]

222. Thevananther, S.; Rivera, A.; Rivkees, S.A. A1 adenosine receptor activation inhibits neurite process formation by Rho kinase-mediated pathways. *Neuroreport* **2001**, *12*, 3057–3063. [CrossRef]

223. Jeon, S.J.; Rhee, S.Y.; Ryu, J.H.; Cheong, J.H.; Kwon, K.; Yang, S.I.; Park, S.H.; Lee, J.; Kim, H.Y.; Han, S.H.; et al. Activation of adenosine A2A receptor up-regulates BDNF expression in rat primary cortical neurons. *Neurochem. Res.* **2011**, *36*, 2259–2269. [CrossRef]

224. Ribeiro, F.F.; Neves-Tome, R.; Assaife-Lopes, N.; Santos, T.E.; Silva, R.F.; Brites, D.; Ribeiro, J.A.; Sousa, M.M.; Sebastiao, A.M. Axonal elongation and dendritic branching is enhanced by adenosine A2A receptors activation in cerebral cortical neurons. *Brain Struct. Funct.* **2016**, *221*, 2777–2799. [CrossRef]

225. Jeong, H.J.; Jang, I.S.; Nabekura, J.; Akaike, N. Adenosine A1 receptor-mediated presynaptic inhibition of GABAergic transmission in immature rat hippocampal CA1 neurons. *J. Neurophysiol.* **2003**, *89*, 1214–1222. [CrossRef]

226. Dolphin, A.C.; Archer, E.R. An adenosine agonist inhibits and a cyclic AMP analogue enhances the release of glutamate but not GABA from slices of rat dentate gyrus. *Neurosci. Lett.* **1983**, *43*, 49–54. [CrossRef]

227. Lambert, N.A.; Teyler, T.J. Adenosine depresses excitatory but not fast inhibitory synaptic transmission in area CA1 of the rat hippocampus. *Neurosci. Lett.* **1991**, *122*, 50–52. [CrossRef]

228. Thompson, S.M.; Haas, H.L.; Gahwiler, B.H. Comparison of the actions of adenosine at pre- and postsynaptic receptors in the rat hippocampus in vitro. *J. Physiol.* **1992**, *451*, 347–363. [CrossRef]

229. Scanziani, M.; Capogna, M.; Gahwiler, B.H.; Thompson, S.M. Presynaptic inhibition of miniature excitatory synaptic currents by baclofen and adenosine in the hippocampus. *Neuron* **1992**, *9*, 919–927. [CrossRef]

230. Kirmse, K.; Dvorzhak, A.; Grantyn, R.; Kirischuk, S. Developmental downregulation of excitatory GABAergic transmission in neocortical layer I via presynaptic adenosine A(1) receptors. *Cereb. Cortex* **2008**, *18*, 424–432. [CrossRef] [PubMed]

231. Stevens, B.; Porta, S.; Haak, L.L.; Gallo, V.; Fields, R.D. Adenosine: A neuron-glial transmitter promoting myelination in the CNS in response to action potentials. *Neuron* **2002**, *36*, 855–868. [CrossRef]

232. Othman, T.; Yan, H.; Rivkees, S.A. Oligodendrocytes express functional A1 adenosine receptors that stimulate cellular migration. *Glia* **2003**, *44*, 166–172. [CrossRef]

233. Boldogkoi, Z.; Schutz, B.; Sallach, J.; Zimmer, A. P2X(3) receptor expression at early stage of mouse embryogenesis. *Mech. Dev.* **2002**, *118*, 255–260. [CrossRef]

234. Cheung, K.K.; Burnstock, G. Localization of P2X3 receptors and coexpression with P2X2 receptors during rat embryonic neurogenesis. *J. Comp. Neurol.* **2002**, *443*, 368–382. [CrossRef]
235. Cheung, K.K.; Chan, W.Y.; Burnstock, G. Expression of P2X purinoceptors during rat brain development and their inhibitory role on motor axon outgrowth in neural tube explant cultures. *Neuroscience* **2005**, *133*, 937–945. [CrossRef] [PubMed]
236. Heo, J.S.; Han, H.J. ATP stimulates mouse embryonic stem cell proliferation via protein kinase C, phosphatidylinositol 3-kinase/Akt, and mitogen-activated protein kinase signaling pathways. *Stem Cells* **2006**, *24*, 2637–2648. [CrossRef]
237. Yuahasi, K.K.; Demasi, M.A.; Tamajusuku, A.S.; Lenz, G.; Sogayar, M.C.; Fornazari, M.; Lameu, C.; Nascimento, I.C.; Glaser, T.; Schwindt, T.T.; et al. Regulation of neurogenesis and gliogenesis of retinoic acid-induced P19 embryonal carcinoma cells by P2X2 and P2X7 receptors studied by RNA interference. *Int. J. Dev. Neurosci.* **2012**, *30*, 91–97. [CrossRef] [PubMed]
238. Glaser, T.; de Oliveira, S.L.; Cheffer, A.; Beco, R.; Martins, P.; Fornazari, M.; Lameu, C.; Junior, H.M.; Coutinho-Silva, R.; Ulrich, H. Modulation of mouse embryonic stem cell proliferation and neural differentiation by the P2X7 receptor. *PLoS ONE* **2014**, *9*, e96281. [CrossRef]
239. Illes, P.; Khan, T.M.; Rubini, P. Neuronal P2X7 Receptors Revisited: Do They Really Exist? *J. Neurosci.* **2017**, *37*, 7049–7062. [CrossRef]
240. Deuchars, S.A.; Atkinson, L.; Brooke, R.E.; Musa, H.; Milligan, C.J.; Batten, T.F.; Buckley, N.J.; Parson, S.H.; Deuchars, J. Neuronal P2X7 receptors are targeted to presynaptic terminals in the central and peripheral nervous systems. *J. Neurosci.* **2001**, *21*, 7143–7152. [CrossRef]
241. Miras-Portugal, M.T.; Díaz-Hernández, M.; Giráldez, L.; Hervás, C.; Gómez-Villafuertes, R.; Sen, R.P.; Gualix, J.; Pintor, J. P2X7 receptors in rat brain: Presence in synaptic terminals and granule cells. *Neurochem. Res.* **2003**, *28*, 1597–1605. [CrossRef]
242. Metzger, M.W.; Walser, S.M.; Aprile-Garcia, F.; Dedic, N.; Chen, A.; Holsboer, F.; Arzt, E.; Wurst, W.; Deussing, J.M. Genetically dissecting P2rx7 expression within the central nervous system using conditional humanized mice. *Purinergic Signal.* **2017**, *13*, 153–170. [CrossRef]
243. Morgan, J.; Alves, M.; Conte, G.; Menéndez-Méndez, A.; de Diego-Garcia, L.; de Leo, G.; Beamer, E.; Smith, J.; Nicke, A.; Engel, T. Characterization of the Expression of the ATP-Gated P2X7 Receptor Following Status Epilepticus and during Epilepsy Using a P2X7-EGFP Reporter Mouse. *Neurosci. Bull.* **2020**. [CrossRef]
244. Beamer, E.; Fischer, W.; Engel, T. The ATP-Gated P2X7 Receptor As a Target for the Treatment of Drug-Resistant Epilepsy. *Front. Neurosci.* **2017**, *11*, 21. [CrossRef] [PubMed]
245. Cheung, K.K.; Ryten, M.; Burnstock, G. Abundant and dynamic expression of G protein-coupled P2Y receptors in mammalian development. *Dev. Dyn.* **2003**, *228*, 254–266. [CrossRef] [PubMed]
246. Mishra, S.K.; Braun, N.; Shukla, V.; Füllgrabe, M.; Schomerus, C.; Korf, H.W.; Gachet, C.; Ikehara, Y.; Sévigny, J.; Robson, S.C.; et al. Extracellular nucleotide signaling in adult neural stem cells: Synergism with growth factor-mediated cellular proliferation. *Development* **2006**, *133*, 675–684. [CrossRef] [PubMed]
247. Xiang, Z.; Burnstock, G. Changes in expression of P2X purinoceptors in rat cerebellum during postnatal development. *Brain Res. Dev. Brain Res.* **2005**, *156*, 147–157. [CrossRef]
248. Safiulina, V.F.; Kasyanov, A.M.; Sokolova, E.; Cherubini, E.; Giniatullin, R. ATP contributes to the generation of network-driven giant depolarizing potentials in the neonatal rat hippocampus. *J. Physiol.* **2005**, *565*, 981–992. [CrossRef] [PubMed]
249. Heine, C.; Sygnecka, K.; Scherf, N.; Grohmann, M.; Brasigk, A.; Franke, H. P2Y(1) receptor mediated neuronal fibre outgrowth in organotypic brain slice co-cultures. *Neuropharmacology* **2015**, *93*, 252–266. [CrossRef]
250. Beamer, E.; Kovacs, G.; Sperlagh, B. ATP released from astrocytes modulates action potential threshold and spontaneous excitatory postsynaptic currents in the neonatal rat prefrontal cortex. *Brain Res. Bull.* **2017**, *135*, 129–142. [CrossRef]
251. Delarasse, C.; Gonnord, P.; Galante, M.; Auger, R.; Daniel, H.; Motta, I.; Kanellopoulos, J.M. Neural progenitor cell death is induced by extracellular ATP via ligation of P2X7 receptor. *J. Neurochem.* **2009**, *109*, 846–857. [CrossRef]
252. Tsao, H.K.; Chiu, P.H.; Sun, S.H. PKC-dependent ERK phosphorylation is essential for P2X7 receptor-mediated neuronal differentiation of neural progenitor cells. *Cell Death Dis.* **2013**, *4*, e751. [CrossRef] [PubMed]
253. Thompson, B.A.; Storm, M.P.; Hewinson, J.; Hogg, S.; Welham, M.J.; MacKenzie, A.B. A novel role for P2X7 receptor signalling in the survival of mouse embryonic stem cells. *Cell. Signal.* **2012**, *24*, 770–778. [CrossRef]

254. Lovelace, M.D.; Gu, B.J.; Eamegdool, S.S.; Weible, M.W., 2nd; Wiley, J.S.; Allen, D.G.; Chan-Ling, T. P2X7 receptors mediate innate phagocytosis by human neural precursor cells and neuroblasts. *Stem Cells* **2015**, *33*, 526–541. [CrossRef] [PubMed]

255. Xiang, Z.; Burnstock, G. Expression of P2X receptors on rat microglial cells during early development. *Glia* **2005**, *52*, 119–126. [CrossRef]

256. Bianco, F.; Ceruti, S.; Colombo, A.; Fumagalli, M.; Ferrari, D.; Pizzirani, C.; Matteoli, M.; Di Virgilio, F.; Abbracchio, M.P.; Verderio, C. A role for P2X7 in microglial proliferation. *J. Neurochem.* **2006**, *99*, 745–758. [CrossRef] [PubMed]

257. Rigato, C.; Swinnen, N.; Buckinx, R.; Couillin, I.; Mangin, J.M.; Rigo, J.M.; Legendre, P.; Le Corronc, H. Microglia proliferation is controlled by P2X7 receptors in a Pannexin-1-independent manner during early embryonic spinal cord invasion. *J. Neurosci.* **2012**, *32*, 11559–11573. [CrossRef]

258. Matute, C. P2X7 receptors in oligodendrocytes: A novel target for neuroprotection. *Mol. Neurobiol.* **2008**, *38*, 123–128. [CrossRef]

259. Braun, N.; Sevigny, J.; Mishra, S.K.; Robson, S.C.; Barth, S.W.; Gerstberger, R.; Hammer, K.; Zimmermann, H. Expression of the ecto-ATPase NTPDase2 in the germinal zones of the developing and adult rat brain. *Eur. J. Neurosci.* **2003**, *17*, 1355–1364. [CrossRef]

260. Langer, D.; Ikehara, Y.; Takebayashi, H.; Hawkes, R.; Zimmermann, H. The ectonucleotidases alkaline phosphatase and nucleoside triphosphate diphosphohydrolase 2 are associated with subsets of progenitor cell populations in the mouse embryonic, postnatal and adult neurogenic zones. *Neuroscience* **2007**, *150*, 863–879. [CrossRef]

261. Da Silva, R.S.; Richetti, S.K.; Tonial, E.M.; Bogo, M.R.; Bonan, C.D. Profile of nucleotide catabolism and ectonucleotidase expression from the hippocampi of neonatal rats after caffeine exposure. *Neurochem. Res.* **2012**, *37*, 23–30. [CrossRef]

262. Fenoglio, C.; Scherini, E.; Vaccarone, R.; Bernocchi, G. A re-evaluation of the ultrastructural localization of 5'-nucleotidase activity in the developing rat cerebellum, with a cerium-based method. *J. Neurosci. Methods* **1995**, *59*, 253–263. [CrossRef]

263. Schoen, S.W.; Leutenecker, B.; Kreutzberg, G.W.; Singer, W. Ocular dominance plasticity and developmental changes of 5'-nucleotidase distributions in the kitten visual cortex. *J. Comp. Neurol.* **1990**, *296*, 379–392. [CrossRef]

264. Schoen, S.W.; Kreutzberg, G.W.; Singer, W. Cytochemical redistribution of 5'-nucleotidase in the developing cat visual cortex. *Eur. J. Neurosci.* **1993**, *5*, 210–222. [CrossRef]

265. Grkovic, I.; Bjelobaba, I.; Nedeljkovic, N.; Mitrovic, N.; Drakulic, D.; Stanojlovic, M.; Horvat, A. Developmental increase in ecto-5'-nucleotidase activity overlaps with appearance of two immunologically distinct enzyme isoforms in rat hippocampal synaptic plasma membranes. *J. Mol. Neurosci.* **2014**, *54*, 109–118. [CrossRef] [PubMed]

266. Bachner, D.; Ahrens, M.; Betat, N.; Schroder, D.; Gross, G. Developmental expression analysis of murine autotaxin (ATX). *Mech. Dev.* **1999**, *84*, 121–125. [CrossRef]

267. Cognato, G.P.; Czepielewski, R.S.; Sarkis, J.J.; Bogo, M.R.; Bonan, C.D. Expression mapping of ectonucleotide pyrophosphatase/phosphodiesterase 1-3 (E-NPP1-3) in different brain structures during rat development. *Int. J. Dev. Neurosci.* **2008**, *26*, 593–598. [CrossRef] [PubMed]

268. Foster, B.L.; Nagatomo, K.J.; Nociti, F.H., Jr.; Fong, H.; Dunn, D.; Tran, A.B.; Wang, W.; Narisawa, S.; Millan, J.L.; Somerman, M.J. Central role of pyrophosphate in acellular cementum formation. *PLoS ONE* **2012**, *7*, e38393. [CrossRef]

269. Narisawa, S.; Hasegawa, H.; Watanabe, K.; Millan, J.L. Stage-specific expression of alkaline phosphatase during neural development in the mouse. *Dev. Dyn.* **1994**, *201*, 227–235. [CrossRef]

270. Chiquoine, A.D. The identification, origin, and migration of the primordial germ cells in the mouse embryo. *Anat. Rec.* **1954**, *118*, 135–146. [CrossRef] [PubMed]

271. Fonta, C.; Negyessy, L.; Renaud, L.; Barone, P. Postnatal development of alkaline phosphatase activity correlates with the maturation of neurotransmission in the cerebral cortex. *J. Comp. Neurol.* **2005**, *486*, 179–196. [CrossRef]

272. Kermer, V.; Ritter, M.; Albuquerque, B.; Leib, C.; Stanke, M.; Zimmermann, H. Knockdown of tissue nonspecific alkaline phosphatase impairs neural stem cell proliferation and differentiation. *Neurosci. Lett.* **2010**, *485*, 208–211. [CrossRef]

273. Diez-Zaera, M.; Diaz-Hernandez, J.I.; Hernandez-Alvarez, E.; Zimmermann, H.; Diaz-Hernandez, M.; Miras-Portugal, M.T. Tissue-nonspecific alkaline phosphatase promotes axonal growth of hippocampal neurons. *Mol. Biol. Cell.* **2011**, *22*, 1014–1024. [CrossRef]

274. Sebastian-Serrano, A.; Engel, T.; de Diego-Garcia, L.; Olivos-Ore, L.A.; Arribas-Blazquez, M.; Martinez-Frailes, C.; Perez-Diaz, C.; Millan, J.L.; Artalejo, A.R.; Miras-Portugal, M.T.; et al. Neurodevelopmental alterations and seizures developed by mouse model of infantile hypophosphatasia are associated with purinergic signalling deregulation. *Hum. Mol. Genet.* **2016**, *25*, 4143–4156. [CrossRef]

275. Waymire, K.G.; Mahuren, J.D.; Jaje, J.M.; Guilarte, T.R.; Coburn, S.P.; MacGregor, G.R. Mice lacking tissue non-specific alkaline phosphatase die from seizures due to defective metabolism of vitamin B-6. *Nat. Genet.* **1995**, *11*, 45–51. [CrossRef]

276. Hanics, J.; Barna, J.; Xiao, J.; Millan, J.L.; Fonta, C.; Negyessy, L. Ablation of TNAP function compromises myelination and synaptogenesis in the mouse brain. *Cell Tissue Res.* **2012**, *349*, 459–471. [CrossRef]

277. Dragunow, M.; Goddard, G.V.; Laverty, R. Is adenosine an endogenous anticonvulsant? *Epilepsia* **1985**, *26*, 480–487. [CrossRef]

278. During, M.J.; Spencer, D.D. Adenosine: A potential mediator of seizure arrest and postictal refractoriness. *Ann. Neurol.* **1992**, *32*, 618–624. [CrossRef] [PubMed]

279. Cutrufo, C.; Bortot, L.; Giachetti, A.; Manzini, S. Differential effects of various xanthines on pentylenetetrazole-induced seizures in rats: An EEG and behavioural study. *Eur. J. Pharm.* **1992**, *222*, 1–6. [CrossRef]

280. Jailani, M.; Mubarak, M.; Sarkhouh, M.; Al Mahrezi, A.; Abdulnabi, H.; Naiser, M.; Alaradi, H.; Alabbad, A.; Hassan, M.; Kamal, A. The Effect of Low-Doses of Caffeine and Taurine on Convulsive Seizure Parameters in Rats. *Behav. Sci.* **2020**, *10*. [CrossRef]

281. Van Koert, R.R.; Bauer, P.R.; Schuitema, I.; Sander, J.W.; Visser, G.H. Caffeine and seizures: A systematic review and quantitative analysis. *Epilepsy Behav.* **2018**, *80*, 37–47. [CrossRef]

282. Pometlová, M.; Kubová, H.; Mares, P. Effects of 2-chloroadenosine on cortical epileptic afterdischarges in immature rats. *Pharm. Rep.* **2010**, *62*, 62–67. [CrossRef]

283. Mareš, P. Anticonvulsant action of 2-chloroadenosine against pentetrazol-induced seizures in immature rats is due to activation of A1 adenosine receptors. *J. Neural Transm. (Vienna)* **2010**, *117*, 1269–1277. [CrossRef] [PubMed]

284. Mareš, P. A1 not A2A adenosine receptors play a role in cortical epileptic afterdischarges in immature rats. *J. Neural Transm. (Vienna)* **2014**, *121*, 1329–1336. [CrossRef] [PubMed]

285. Fabera, P.; Parizkova, M.; Uttl, L.; Vondrakova, K.; Kubova, H.; Tsenov, G.; Mares, P. Adenosine A1 Receptor Agonist 2-chloro-N6-cyclopentyladenosine and Hippocampal Excitability During Brain Development in Rats. *Front. Pharm.* **2019**, *10*, 656. [CrossRef]

286. León-Navarro, D.A.; Albasanz, J.L.; Martín, M. Hyperthermia-induced seizures alter adenosine A1 and A2A receptors and 5′-nucleotidase activity in rat cerebral cortex. *J. Neurochem.* **2015**, *134*, 395–404. [CrossRef]

287. Endesfelder, S.; Weichelt, U.; Schiller, C.; Winter, K.; von Haefen, C.; Bührer, C. Caffeine Protects Against Anticonvulsant-Induced Impaired Neurogenesis in the Developing Rat Brain. *Neurotox. Res.* **2018**, *34*, 173–187. [CrossRef]

288. Isokawa, M. Caffeine-Induced Suppression of GABAergic Inhibition and Calcium-Independent Metaplasticity. *Neural Plast.* **2016**, *2016*, 1239629. [CrossRef]

289. Blaise, J.H.; Park, J.E.; Bellas, N.J.; Gitchell, T.M.; Phan, V. Caffeine consumption disrupts hippocampal long-term potentiation in freely behaving rats. *Physiol. Rep.* **2018**, *6*. [CrossRef]

290. Guillet, R.; Dunham, L. Neonatal caffeine exposure and seizure susceptibility in adult rats. *Epilepsia* **1995**, *36*, 743–749. [CrossRef]

291. Guillet, R. Neonatal caffeine exposure alters seizure susceptibility in rats in an age-related manner. *Brain Res. Dev. Brain Res.* **1995**, *89*, 124–128. [CrossRef]

292. Alexander, M.; Smith, A.L.; Rosenkrantz, T.S.; Fitch, R.H. Therapeutic effect of caffeine treatment immediately following neonatal hypoxic-ischemic injury on spatial memory in male rats. *Brain Sci.* **2013**, *3*, 177–190. [CrossRef]

293. Winerdal, M.; Urmaliya, V.; Winerdal, M.E.; Fredholm, B.B.; Winqvist, O.; Ådén, U. Single Dose Caffeine Protects the Neonatal Mouse Brain against Hypoxia Ischemia. *PLoS ONE* **2017**, *12*, e0170545. [CrossRef] [PubMed]

294. Richerson, G.B.; Boison, D.; Faingold, C.L.; Ryvlin, P. From unwitnessed fatality to witnessed rescue: Pharmacologic intervention in sudden unexpected death in epilepsy. *Epilepsia* **2016**, *57*, 35–45. [CrossRef] [PubMed]

295. Borah, P.; Deka, S.; Mailavaram, R.P.; Deb, P.K. P1 Receptor Agonists/Antagonists in Clinical Trials—Potential Drug Candidates of the Future. *Curr. Pharm. Des.* **2019**, *25*, 2792–2807. [CrossRef]

296. Li, F.; Wang, L.; Li, J.W.; Gong, M.; He, L.; Feng, R.; Dai, Z.; Li, S.Q. Hypoxia induced amoeboid microglial cell activation in postnatal rat brain is mediated by ATP receptor P2X4. *BMC Neurosci.* **2011**, *12*, 111. [CrossRef]

297. Boison, D. The adenosine kinase hypothesis of epileptogenesis. *Prog. Neurobiol.* **2008**, *84*, 249–262. [CrossRef] [PubMed]

298. Fedele, D.E.; Gouder, N.; Güttinger, M.; Gabernet, L.; Scheurer, L.; Rülicke, T.; Crestani, F.; Boison, D. Astrogliosis in epilepsy leads to overexpression of adenosine kinase, resulting in seizure aggravation. *Brain* **2005**, *128*, 2383–2395. [CrossRef] [PubMed]

299. Gouder, N.; Scheurer, L.; Fritschy, J.M.; Boison, D. Overexpression of adenosine kinase in epileptic hippocampus contributes to epileptogenesis. *J. Neurosci.* **2004**, *24*, 692–701. [CrossRef]

300. Sebastián-Serrano, Á.; de Diego-García, L.; Martínez-Frailes, C.; Ávila, J.; Zimmermann, H.; Millán, J.L.; Miras-Portugal, M.T.; Díaz-Hernández, M. Tissue-nonspecific Alkaline Phosphatase Regulates Purinergic Transmission in the Central Nervous System During Development and Disease. *Comput. Struct. Biotechnol. J.* **2015**, *13*, 95–100. [CrossRef]

301. Santiago, M.F.; Veliskova, J.; Patel, N.K.; Lutz, S.E.; Caille, D.; Charollais, A.; Meda, P.; Scemes, E. Targeting pannexin1 improves seizure outcome. *PLoS ONE* **2011**, *6*, e25178. [CrossRef]

302. Quinlan, S.; Merino-Serrais, P.; Di Grande, A.; Dussmann, H.; Prehn, J.H.M.; Ní Chonghaile, T.; Henshall, D.C.; Jimenez-Mateos, E.M. The Anti-inflammatory Compound Candesartan Cilexetil Improves Neurological Outcomes in a Mouse Model of Neonatal Hypoxia. *Front. Immunol.* **2019**, *10*. [CrossRef] [PubMed]

Publisher's Note: MDPI stays neutral with regard to jurisdictional claims in published maps and institutional affiliations.

International Journal of
Molecular Sciences

Review

Bone Marrow and Adipose Tissue Adenosine Receptors Effect on Osteogenesis and Adipogenesis

Anna Eisenstein [1,*], Shlok V. Chitalia [2] and Katya Ravid [2]

[1]   Department of Dermatology, Yale School of Medicine, New Haven, CT 06520, USA
[2]   Department of Medicine and Whitaker Cardiovascular Institute, Boston University School of Medicine, Boston, MA 02118, USA; shlokvipulc@gmail.com (S.V.C.); kravid@bu.edu (K.R.)
*   Correspondence: anna.eisenstein@yale.edu

Received: 2 September 2020; Accepted: 6 October 2020; Published: 10 October 2020

Abstract: Adenosine is an extracellular signaling molecule that is particularly relevant in times of cellular stress, inflammation and metabolic disturbances when the levels of the purine increase. Adenosine acts on two G-protein-coupled stimulatory and on two G-protein-coupled inhibitory receptors, which have varying expression profiles in different tissues and conditions, and have different affinities for the endogenous ligand. Studies point to significant roles of adenosine and its receptors in metabolic disease and bone health, implicating the receptors as potential therapeutic targets. This review will highlight our current understanding of the dichotomous effects of adenosine and its receptors on adipogenesis versus osteogenesis within the bone marrow to maintain bone health, as well as its relationship to obesity. Therapeutic implications will also be reviewed.

Keywords: adenosine receptors; adipogenesis; osteogenesis; adipose tissue; bone marrow; obesity

---

1. Introduction

Adenosine is a nucleoside that is released by all cells, including adipocytes in vitro and the subcutaneous adipose tissue in vivo [1,2]. Adenosine is formed by the degradation of adenosine triphosphate (ATP), adenosine diphosphate (ADP) and adenosine monophosphate (AMP). Levels of adenosine rise in response to metabolic stress, tissue injury, hypoxia and inflammation [3]. Adenosine acts broadly via four receptors. The A1 adenosine receptor (A1AR) and the A3 adenosine receptor (A3AR) are coupled to Gi, which inhibits adenylyl cyclase and decreases cytosolic levels of cyclic AMP (cAMP). The A2a and the A2b adenosine receptors (A2aAR and A2bAR, respectively) are coupled to the stimulatory G alpha protein (Gs), activate adenylyl cyclase and increase cytosolic levels of cAMP. The adenosine receptors may also act on other signaling cascades (Figure 1). The expression of these receptors and the receptor affinity for adenosine allow for modulation of the effect of adenosine in different physiologic states and in different tissues. As such, the action of adenosine may be contradictory depending on these various factors [3]

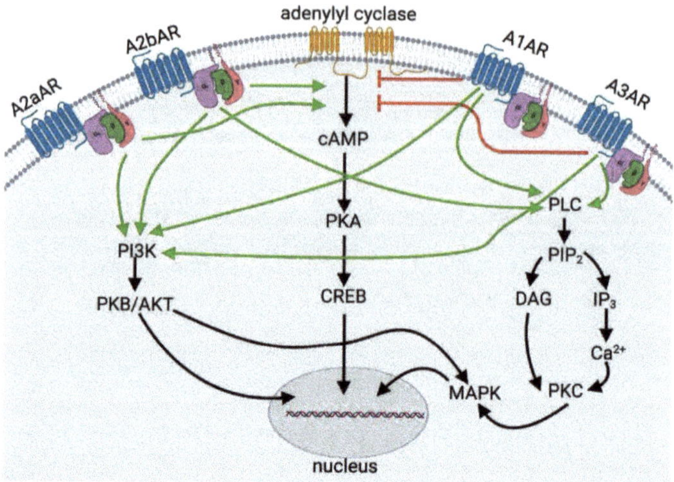

Figure 1. Adenosine Receptor Signaling. The A2aAR and the A2bAR are adenylyl cyclase stimulating receptors through Gαs subunit. The A1AR and the A3AR are adenylyl cyclase inhibitory receptors through Gαi subunit. The A1AR, A3AR and A2bAR also act through phospholipase C (PLC) to increase cytosolic calcium and diacylglycerol (DAG) to activate protein kinase C (PKC) and MAP kinase (MAPK) signalling. All adenosine receptors are capable of activating phosphatidylinositol 3-kinase (PI3K).

2. Adenosine Receptors Are Expressed on Bone Marrow-Derived Mesenchymal Stem Cells and Preadipocytes

Adenosine is found extracellularly under pathologic situations including hypoxia, ischemia or cell damage. Cells release adenosine into the extracellular space through equilibrative nucleoside transporters (ENTs). Adenosine can also be formed by the conversion of extracellular ATP, released from damaged cells, by ectonucleoside triphosphate diphosphorylase (CD39) and ecto-5′-nucleotidase (CD73) enzymes. Adenosine can also be metabolized to inosine by adenosine deaminase or phosphorylated to AMP by adenosine kinase. As one can imagine, the coordination of adenosine release is well regulated at baseline and, during times of cellular stress, can become a signal of disequilibrium [4]. Similarly, while adenosine receptors are present on many different cell types and found generally throughout the body, the level of expression varies in different cell types, pathologic states and developmental stages [4,5], and hence the response to adenosine can be, at times, contradictory. The A1AR and A3AR are more ubiquitously expressed in most tissues, while the A2aAR and the A2bAR have more selective expression [6]. The A1AR is expressed at high levels in the brain, spinal cord, adrenal glands and the heart, with slightly lower expression in the liver, bladder, adipose tissue and testis [7,8]. Highest expression of the A3AR occurs in the lung and liver but is also expressed in brain, testis, spleen, thyroid, kidney, bladder, heart, eosinophils and mast cells [7,8]. The A2aAR is most highly expressed in the brain, spleen, thymus, leukocytes and platelets; this receptor is also expressed in moderate levels in the heart, lung and vasculature [7,8]. The A2bAR is expressed in the vasculature of large intestine, ovaries, testis, liver, spleen, adipose tissue, muscle, pancreas, brain, lung, heart, kidneys, eye and lung [6].

The expression of adenosine receptors is increased in various pathologic conditions (review in [9,10]). The A1AR is induced by oxidative stress [11] and hypoxia [12]. The A3AR was found to be upregulated in rheumatoid arthritis and during methotrexate treatment (which increases adenosine levels) as well as in breast and colon cancer [13–16]. The A2aAR is upregulated in inflammatory states [17,18], hypoxia [19,20] and cellular stress such as during food restriction [21]. The A2aAR was also upregulated in individuals receiving methotrexate therapy [16]. Similarly, the A2bAR is upregulated by inflammation, hypoxia and extracellular adenosine [22–25]. Expression of the A2bAR is upregulated in conditions of inflammation or hypoxia in the vasculature, intestine, kidneys, heart and lung [9,19,26,27]. Not only does the variable and inducible expression of adenosine receptors indicate the complexity of adenosine receptor signaling, it also indicates the consideration that must be taken when designing therapeutics.

Bone-marrow-derived mesenchymal stem cell (MSCs), precursors to adipocytes and osteoblasts, express the A2bAR and the A2aAR [28]. In testing the activity of adenosine receptors in vitro and in vivo, numerous agonists and antagonists have been developed (Table 1). In Ob1771 preadipocytes, the nonspecific adenosine receptor agonist, 5′-N-Ethylcarboxamidoadenosine (NECA), increased cAMP levels, while the A2aAR specific agonist, CGS 21680 (3-(4-(2-((6-amino-9-((2R,3R,4S,5S)-5-(ethylcarbomoyl)-3,4-dihydroxytetrahydrofuran-2-yl)-9H-purin-2-yl)amino)ethyl)phenyl)propanoic acid), did not increase cAMP levels, which would suggest that the A2bAR is the predominant adenosine receptor in these preadipocytes [29]. All adenosine receptors and adenosine metabolic and catabolic enzymes (adenosine deaminase, adenosine kinase and CD73) are expressed on MSCs (although no A3AR protein was identified by Western blotting), though at different levels and activities [30]. The A2bAR was shown to be the predominant adenosine receptor subtype, based on a cAMP assay [30].

Table 1. Experimental agonists and antagonists of the adenosine receptors (see the abreviation section for drugs full names).

Receptor	Agonist (+)/Antagonist (−)	Drug Acronym
A1AR	+	CPA, PIA, CCPA
A1AR	−	DPCPX
A2aAR	+	CGS 21680
A2aAR	−	SCH442416
A2bAR	+	BAY 60-6583
A2bAR	−	MRS1706, MRS-1754, ATL-801
A3AR	+	IB-MECA, C1-IB-MECA, MRS5698
A3AR	−	MRS1523, MRS1220
All ARs	+	NECA

The A1AR is the most highly expressed adenosine receptor in white adipose tissue, followed by the A3AR, A2aAR and A2bAR [31–33]. Adipose tissue can be divided into the mature adipocytes and the stromal vascular fraction, which contains preadipocytes as well as endothelial cells and fibroblasts. The stromal vascular fraction from mice and humans expresses the A2bAR [34]. In vitro, adipocyte precursors (preadipocytes) have a higher expression of the A2aAR and A2bAR, but, following differentiation to a mature adipocyte, there is greater expression of A1AR [29,34–36]. Adenosine receptors are also expressed in adipose tissue. The A2bAR is notably also expressed in other metabolic tissues such as the pancreas, liver and muscle [6,37,38]. Moreover, mice fed a high-fat diet (HFD) have increased expression of the A2bAR in the liver, gastrocnemius muscle and the epididymal visceral adipose tissue [37]. Interestingly, the A2bAR mRNA expression is significantly elevated in the subcutaneous adipose tissue of obese individuals as compared to lean individuals [37]. Of interest, adenosine is released from adipocytes. The adenosine content of abdominal subcutaneous adipose tissue has been measured in obese and lean individuals. The levels show a trend towards increased adenosine levels in the adipose tissue of obese (0.67 pmol/g wet weight) versus lean (0.42 pmol/g wet weight) individuals [39]. Within the bone, osteoclasts express all four adenosine receptors, and mature osteoblasts express A1 and A2bAR [40].

3. Regulation of Adipogenesis and Osteogenesis by Bone Marrow Cell Adenosine Receptors

Adipogenesis results from the differentiation of adipocyte precursors into mature adipocytes capable of storing lipid in lipid droplets and secreting adipokines as well as cytokines either within the bone marrow or within adipose tissue depots. Adipocyte precursors are found in the bone marrow and within the stromal vascular fraction of peripheral adipose tissue depots. In the bone marrow, adipocytes differentiate from the MSCs, which are common progenitors that are also capable of differentiating into osteoblasts and chondrocytes. The process of adipogenesis requires a cascade of transcription factors. The master regulator of adipogenesis is the nuclear hormone receptor, peroxisome proliferator-activated receptor gamma (PPARγ). PPARγ is necessary for adipogenesis; mice lacking PPARγ and humans with dominant negative mutations in PPARγ have severe lipodystrophy and are insulin-resistant [41,42]. Moreover, PPARγ expression is sufficient for adipogenesis, as the retroviral expression of PPARγ in cultured fibroblasts resulted in adipogenesis [43]. It is generally thought that in the early stage of adipogenesis, CCAAT enhancer binding protein-beta and -delta (C/EBP-β and C/EBP-δ) induce the expression of PPARγ and C/EBP-α [44–47].

One of the first studies investigating the role of adenosine receptors in adipogenesis found that the treatment of rat primary preadipocytes with the adenosine analogue, NECA, promoted differentiation [36]. In bone-marrow-derived MSCs, activation of the A2aAR promoted the expression of the above-described adipogenic transcription factors and enhanced adipocyte differentiation, while activation of the A1AR promoted lipogenesis and lipid accumulation in the mature adipocyte [30]. Transfection of the A1AR into a murine osteoblast precursor cell line promoted adipogenesis. In the same model, forced expression of the A2bAR inhibited adipogenesis and promoted osteogenesis [48]. Activation of the A2bAR with BAY 60-6583, an A2bAR-specific agonist, inhibits adipogenesis in ex vivo stromal vascular cells isolated from subcutaneous inguinal adipose tissue [34]. This effect was found to be dependent on the expression of the transcription and differentiation regulator, Krüppel-like factor 4 (KLF4), which was also induced by A2bAR agonism [34]. Notably, while BAY 60-6583 is the most selective A2bAR agonist available, in certain cell lines BAY 60-6583 was reported to act as a partial agonist or even as an antagonist of the A2bAR [49].

Within the bone marrow, there is thought to be a dichotomous relationship between adipogenesis and osteogenesis, whereby the MSC is directed to one lineage or the other. The control of this fate decision is still being investigated. Some research points to adenosine receptors playing a role in dictating the formation of adipocytes or osteoblasts. Clinically, these decisions may manifest during aging, when the bone marrow contains more fat than younger marrow, or in osteoporosis, where there is loss of bone and an accumulation of fat. Transcription factors important for bone development include runt-related gene family members, such as Runx2. Osteoblast differentiation is known to be stimulated by fibroblast growth factor, insulin-like growth factor, bone morphogenic proteins and the Wnt/β-catenin pathway.

There has been interest in the role of adenosine and adenosine receptors in bone metabolism. Using pharmacologic inhibitors and A2aAR knock out (KO) mice, the A2aAR was shown to promote proliferation of bone-marrow-derived MSCs [50]. During osteoblast differentiation, the A2bAR is upregulated in the first few days of differentiation, and then is downregulated [30]. Osteoblast differentiation was greater with the nonspecific adenosine receptor agonist, NECA, and this effect was inhibited by the A2bAR antagonist, MRS1706, but not by the A2aAR antagonist, SCH442416. Additionally, the A2aAR agonist, CGS21680, had no effect on bone differentiation [30]. This all suggests that the A2bAR is the predominant adenosine receptor in augmenting osteoblast differentiation. These findings were supported and advanced in the study by Carroll et al. Using an A2bAR-specific agonist and A2bAR KO mice, the authors showed that activation of the A2bAR promotes osteoblast differentiation [28]. Furthermore, microcomputed tomography analysis of adult femurs from A2bAR KO and wild type (WT) mice showed lower bone density in the KO mice compared to the WT mice [28]. In addition, the A2bAR KO mice had delayed fracture repair as compared to the WT mice [28]. In a study of A2bAR KO from youth to adult age, alterations in bone homeostasis were observed in the KO mice. Most notably, there was a decrease in bone mineral density and trabecular volume as well as reduced formation of new bone in A2bAR KO mice as compared to WT mice [51]. In contrast, the application of an A2bAR agonist reduced bone loss in ovariectomized mice [52], underscoring the need to carefully evaluate the results based on the usage of ligands versus genetic ablation. Regarding the A1AR, pharmacologic inhibition of or genetic knockout of this receptor in mice impairs osteoclast differentiation [53,54]. Furthermore, A1AR KO mice that were ovariectomized, as a model of post-menopausal bone loss, maintained bone volume and had decreased osteoclast bone resorption compared to wild type [55]. Importantly, treatment with an A1AR antagonist prevented ovariectomy-induced bone loss [55]. Conversely, A2aAR agonists have been shown to inhibit osteoclast differentiation and function, and femurs from A2aAR KO mice had decreased bone volume to trabecular bone volume ratio and increased osteoclasts [56]. Thus, the A2bAR and the A2aAR both regulate bone turnover by affecting the osteoblast and osteoclast lineages, respectively. In a study utilizing human bone marrow cells, similar results were found: notably, that inhibition of the A1AR and activation of the A2aAR and A2bAR inhibits osteoclast formation, while A2bAR and A2aAR promote osteoblast formation [57].

The ectonucleotidase CD73, which regulates levels of extracellular adenosine through the metabolism AMP to adenosine, plays an important role in bone metabolism. CD73 KO mice have impaired osteoblast differentiation and decreased bone formation with the development of osteopenia [58]. In vitro, this was found to be mediated by the A2bAR, and this is more fully reviewed in the above section [58]. CD73 was also found to be important for bone repair following injury in aged mice [59]. The expression of CD39, which hydrolyzes ATP and ADP, and CD73 as well as adenosine levels in the bone marrow, were found to be decreased in a mouse model of osteoporosis. This same group also found that an A2bAR agonist reduced bone loss in ovariectomized mice [52]. A more recent study reported CD39 on gingiva-derived mesenchymal stem cells as a promoter of osteogenesis through the Wnt/β-catenin signaling pathway [60].

While the above studies focus on the cellular mechanism of osteogenesis, further direct investigation is needed on the effect of adenosine and its receptors on bone density in humans. Due to the beneficial effect of activation of adenosine receptors on bone density in various mice models [61–63], Zarebska et al. investigated the influence of exercise endurance on plasma ATP levels with the hypothesis that ATP concentration, a predecessor molecule of adenosine, is likely to correlate with exercise level [63]. This was a one-year study that compared 18 to 34-year old athletes, including futsal players, sprinters and endurance athletes, to a control group. A treadmill test was taken until the participants reached exhaustion and ATP levels were measured. There were higher ATP levels in athletes who had better exercise endurance, postulating a resulting increase in adenosine and adenosine receptor activation. Further studies are warranted to elucidate the contribution of adenosine receptor activation under different physiological conditions, including exercise and with varying bone density.

4. Adenosine Receptors in Glucose Homeostasis and Obesity

The function of the A1AR in inhibiting lipolysis has been extensively studied. It is thought that endogenously released adenosine maintains a constant inhibitory signal to adipocyte lipolysis. When adenosine deaminase, which removes endogenous adenosine by degrading it to inosine, is added to the culture of adipocytes from fasted rats, lipolysis reaches maximal levels [64,65]. It has also been shown that adipocytes isolated from humans have a similar response to A1AR agonism, such that treatment with PIA, an A1AR agonist, inhibits isoprenaline-stimulated lipolysis [39]. Conversely, when an A1AR antagonist, 8-Cyclopentyl-1,3-dipropylxanthine (DPCPX), was applied to rat adipocytes, lipolysis increased [66]. A different A1AR antagonist, 8-phenyltheophylline, also increased lipolysis in adipocytes from obese Zucker (fa/fa) rats [67]. When an A1AR agonist was given to Wistar and Zucker fatty rats, levels of free fatty acids, glycerol and triglycerides were reduced [33,68]. These findings were also verified in A1AR KO mice. In these conditions, when an adenosine analog, 2-chloroadenosine, was added to A1AR KO adipocytes, there was no inhibition of lipolysis, as occurred in control WT mice [69]. Furthermore, systemic administration of N-Cyclopentyladenosine (CPA), an A1AR agonist, reduced plasma levels of free fatty acids, glycerol and triglycerides (reducing lipolysis) in WT but not A1AR KO mice. Importantly, A1AR KO mice at baseline had elevated free fatty acids as compared to WT mice, suggesting a tonic-suppressive effect on lipolysis by adenosine through the A1AR in vivo [69]. Furthermore, the body weight of A1AR KO mice was significantly increased by 7–8.5% in older (>5 months) male as compared to controls [69]. Conversely, in aged A1AR KO mice (14–16 months old), the KO mice weighed slightly less than WT age-matched controls, with more lean mass and less abdominal fat mass [70]. Correspondingly, aged A1AR KO mice had better glucose tolerance and insulin sensitivity systemically and within visceral adipose tissue [70]. In an HFD model of obesity, A1AR KO mice were heavier than WT mice, had elevated fasting plasma and insulin levels and had impaired glucose clearance, as determined in hyperinsulinemic-euglycemic clamp studies [71]. Overexpression of the A1AR in the adipose tissue of mice resulted in decreased plasma free fatty acid levels as compared to controls [72]. There was no effect on body weight on a normal chow diet or with HFD challenge. Interestingly, overexpression of the A1AR was protective against insulin resistance [72]. Adipocytes from lean individuals as compared to obese individuals were more responsive to the effect of A1AR agonism on reducing lipolysis. The mechanism for this effect was shown to be a result of a greater ability of A1AR agonist to decrease adenylate cyclase activity in adipocytes from lean individuals. The role of the A1AR on leptin secretion by adipocytes has also been studied [73]. Sprague-Dawley rats were treated with the A1AR-selective agonist, CPA (N6-Cyclopentyladenosine), and serum leptin levels were measured. Interestingly, stimulation of the A1AR increased leptin concentration in the serum, a finding that was not seen in the vehicle-treated rats. The authors contend that this finding was a result of increased secretion of pre-formed leptin, as the mRNA levels of leptin did not change with A1AR activation. This finding would fit with the idea that leptin secretion is negatively regulated by cAMP. A1AR agonism of adipocytes from lean and obese individuals revealed enhanced glucose uptake with A1AR agonism by PIA in adipocytes from lean individuals as compared to obese individuals [39]. During glucose challenge, freely fed A1AR KO mice have increased insulin and glucagon secretion, though no difference was seen in plasma glucose as compared to WT mice [32]. Clearly, A1ARs play an important role in lipid homeostasis, a consideration that should be made when developing therapeutic targets of the A1AR.

The A2bAR has also been studied in the context of lipid and glucose homeostasis. Koupenova et al. showed that lack of the A2bAR in ApoE-deficient mice fed a HFD resulted in elevated liver and plasma cholesterol and triglycerides, hepatic steatosis and atherosclerotic plaques [38]. Furthermore, administration of the A2bAR agonist, BAY 60-6583 (2-[[6-Amino-3,5-dicyano-4-[4-(cyclopropylmethoxy)penyl]-2-pyridinyl]thio]-acetamide) to ApoE-deficient mice on HFD reduced atherosclerotic plaque formation and circulating plasma triglycerides and cholesterol [38]. In a type I diabetes model, the adenosine receptor agonist, NECA, reduced plasma glucose levels and ameliorated the diabetes-induced decrease in pancreatic insulin [74]. In an attempt to determine which adenosine receptor was responsible for this effect of NECA, the authors found that the A1AR agonist, 2-Chloro-N6-cyclopentyladenosine (CCPA), decreased hyperglycemia, but to a lesser extent than that of NECA. The A2aAR agonist, CGS21680, did not alter glucose levels and NECA was able to suppress hyperglycemia in A2aAR KO, suggesting that this receptor was not playing a significant role. When NECA and an A2bAR antagonist, MRS 1754, were injected together, the suppressive effect of NECA was abolished, suggesting that the NECA action through the A2bAR played a dominant role in reducing plasma glucose in this model [74]. In one study, KKAy mice (named after Kyoji Kondo who developed a diabetic mouse strain) were used to model type 2 diabetes mellitus. In this case, administration of the A2bAR antagonist, ATL-801, for two days reduced hepatic glucose production and increased glucose uptake in skeletal muscle and brown adipose tissue [75]. Acute challenge with the adenosine receptor agonist, NECA, delayed glucose disposal and increased fasting glucose levels [75]. This effect was eliminated in A2bAR KO mice, which suggests that NECA activation of the A2bAR impaired glucose uptake and utilization [75]. On the other hand, A2bAR KO mice fed an HFD for 16 weeks have increased adiposity, delayed glucose clearance and impaired insulin sensitivity [37]. Intraperitoneal injection of the A2bAR-specific agonist, BAY 60-6583, during HFD feeding lowered fasting glucose levels and improved glucose and insulin tolerance [37]. Further work showed that restoration of macrophage expression of A2bAR in an otherwise full-body A2bAR KO mouse restored insulin signaling and glucose homeostasis to the levels of WT mice [76]. Furthermore, glucose homeostasis and systemic and tissue inflammatory markers were ameliorated with restoration of A2bAR signaling in macrophages [76]. In a third model, A2bAR KO mice fed a chow diet, but not an HFD, gained more weight and had impaired glucose and lipid homeostasis as compared to WT mice [77]. This effect was shown to be a result of the ability of A2bAR to promote alternative macrophage activation and prevent adipose tissue inflammation [77]. The role of the A2bAR in insulin and glucose sensitivity has been extensively studied in mouse models of obesity and diabetes and appears to play an active role in maintaining glucose homeostasis [37,78].

The A2aAR has also been implicated in adipose tissue dynamics, obesity and inflammation. Mice fed a HFD for 12 weeks had increased expression of the A2aAR in macrophages within the adipose tissue as compared to mice on a low-fat diet [79]. These macrophages were also more proinflammatory (M1) as compared to WT mice [79]. A2aAR KO mice fed an HFD gained more weight than WT mice, despite similar food consumption and energy expenditure [79]. A2aAR KO mice had greater abdominal fat mass, increased adipocyte size than WT mice, greater insulin resistance and glucose intolerance [79,80]. Furthermore, in another study, A2aAR KO mice fed an HFD for 16 weeks had reduced glucose tolerance due to a decrease in insulin secretion by the pancreas rather than a difference in insulin sensitivity [81]. In an HFD model of obesity in Swiss mice, the administration of CGS21680, an A2aAR agonist, improved glucose homeostasis, as determined by insulin tolerance test, and reduced inflammatory markers such as tumor necrosis factor alpha (TNFα) systemically and in the visceral adipose tissue [82]. Overall, activation of the A2aAR may also be a potential therapeutic target in metabolic disease.

An exciting study explored the effect of the A2aAR on brown adipose tissue. The authors found that the A2aAR is the most highly expressed adenosine receptor in brown adipose tissue [31]. A2aAR KO mice or pharmacologic inhibition of the A2aAR decreased thermogenesis [31]. Conversely, the administration of A2aAR agonists increased energy expenditure [31]. This finding was explained by browning of the white adipose tissue which was induced by activation of the A2aAR [31]. Finally, in mice fed an HFD, administration of an A2aAR agonist decreased weight gain and improved glucose tolerance [31]. Ruan et al. showed that A2aAR expression in brown adipose tissue was important in cardiac remodeling during hypertension [83]. A2aAR KO mice had worsened hypertensive cardiac remodeling compared to WT mice in a mouse model of hypertension [83]. The A2aAR was highly induced in the brown adipose tissue in this model, and lack of the A2aAR in KO mice impaired brown adipose tissue thermogenesis induced by the hypertensive model in WT mice [83]. Furthermore, A2aAR-induced thermogenesis in brown adipose tissue was important for secretion of fibroblast growth factor 21 (FGF21), which was necessary to protect against hypertension-induced cardiac remodeling [83]. In a study in lean men using Positron Emission Tomography—Computed Tomography (PET/CT), adenosine was found to increase perfusion in brown adipose tissue to a greater extent than cold exposure [84]. Interestingly, using the PET radioligand, [^{11}C]TMSX, which binds specifically to the A2aAR, lean men were found to have reduced brown adipose tissue radioligand binding (a marker of A2aAR receptor density) following cold exposure [84]. However, in mice lacking all four ARs, termed quadruple knockout mice (QKO), body temperature did not differ from WT mice [85], suggesting that ARs are not necessary for baseline body temperature or that the four adenosine receptors may have reciprocal and opposing effects on body temperature such that deletion of one adenosine receptor is balanced by the deletion of a different adenosine receptor. Regardless, the role of the A2aAR in brown adipose tissue appears to be clinically relevant and may lead to future therapeutic efforts.

Gnad et al. evaluated the effect of A2bAR stimulation or suppression in major energy dissipating tissues [86]. Their studies demonstrated a significant expression of A2bAR in skeletal muscle and brown adipose tissues. Upon genetic deletion of the A2bAR specifically in the skeletal muscle of mice, there was loss of muscle mass, exemplified by sarcopenia and reduced energy expenditure. A2bAR activation had opposite effects. Similarly, the deletion of A2bAR specifically in adipose tissue increased age-related reduction in brown adipose tissue, while stimulation of the A2bAR protected brown adipose tissue, and reduced white adipose tissue and obesity. This study demonstrates the beneficial effect of A2bAR stimulation on several cell types important in thermogenesis and in preventing obesity and age-related sarcopenia. Future studies are needed to understand how this receptor stimulation mediates this effect, especially in complex disease models consisting of other pathological processes accompanying obesity, such as diabetes.

5. Discussion: Implications for Therapeutic Benefit

Adenosine receptors play a role in regulating adipogenesis in peripheral adipose tissue and the bone marrow, and influence whole-body metabolism in terms of lipid storage and glucose homeostasis. Given the potential therapeutic benefits of adenosine receptors, this is an exciting time in adenosine biology. Given the complex biology and diverse, and, at times, contradictory effects of the adenosine receptors, continued development and study of drugs that target specific adenosine receptors will be key to the ultimate goal of creating therapeutic agents that lack potential adverse influences (thoroughly reviewed in [87]).

As shown in Table 2, several A1AR agonists, including SDZWAG994, ARA and RPR749, have been evaluated in humans as antilipolytic agents for hypertriglyceridemia [88–90]. RPR749 is an orally active A1AR agonist. In a double-blind, single increasing dose, placebo-controlled, parallel group randomized study, six parallel groups of eight men (six individuals in the active arm and two in the placebo arm in each group) were given RPR749 or control as a single dose. The average free fatty acid concentration in the serum decreased in all treatment groups [90]. Unfortunately, A1AR agonists have significant side effects, including in the heart and kidney, and also become desensitized after repeated exposure. Partial agonists like GS-9667 (previously known at CVT-3619), decreased plasma free fatty acids without desensitization or significant side effects in rodents [91,92]. In a single ascending dose study, healthy non-obese and obese individuals were given a single oral dose of GS-9667 and, in a multiple ascending dose study, healthy obese subjects received GS-9667 for 14 days. The studies found that GS-9667 resulted in decreased free fatty acid levels in a dose-dependent manner, which was maintained over the 14 days of treatment [93]. In addition to the above-cited side effects identified in these trials, the additional consequences of A1AR agonism in the long term may include effects on bone homeostasis with potential for greater risk of osteoporosis.

Table 2. A1AR Modulators Trialed in Humans.

Receptor	Activity	Drug Acronym	Therapeutic Target (Reference)
A1AR	Agonist	SDZWAG994	Hypertriglyceridemia [88]
A1AR	Agonist	ARA	Hypertriglyceridemia [89]
A1AR	Agonist	RPR749	Hypertriglyceridemia [90]
A1AR	Partial agonist	GS-9667	Hyperlipidemia [91]

Other exciting potential targets include the A2bAR, given its reported role in improving glucose and insulin homeostasis, and the A2aAR, given its reported ability to promote thermogenesis and browning of the white adipose tissue. However, given the potential for alterations in bone metabolism and other off-target effects, caution should be applied when evaluating adenosine receptor modulators. Much effort has been devoted to the synthesis of therapeutic drugs that target adenosine receptors, but several clinical trials have failed, and few such drugs are approved. Adenosine itself and regadenoson, an A2aAR agonist, are both approved for coronary stress imaging and for the treatment of paroxysmal supraventricular tachycardia in the case of adenosine. The first adenosine receptor antagonist to be FDA-approved is the A2aAR antagonist istradefylline (Nourianz), which is used with levodopa/carbidopa in Parkinson's disease [94]. While istradefylline had a good safety profile in over 4000 patients tested with this drug, a different A2aAR antagonist led to death from agranulocytosis in five patients in a phase III clinical trial [95]. An ongoing review of the approved A2aAR antagonist will be necessary to determine if there are long-term consequences, including impairment in bone or adipose tissue metabolism.

Author Contributions: Conceptualization, A.E., K.R.; writing—original outline and first draft preparation, A.E., K.R.; writing—review, supplementing and editing, A.E., S.V.C., K.R. All authors have read and agreed to the published version of the manuscript.

Funding: A.E. receives funding from the NIH/NIAMS 5T32AR007016-45 grant and from the Dermatology Foundation. K.R. is funded by NIH grant UL1TR001430.

Conflicts of Interest: The authors declare no conflict of interest.

Abbreviations

ATP	Adenosine triphosphate
ADP	Adenosine diphosphate
AMP	Adenosine monophosphate
A1AR	A1 adenosine receptor
A2aAR	A2a adenosine receptor
A2bAR	A2b adenosine receptor
A3AR	A3 adenosine receptor
cAMP	Cyclic adenosine monophosphate
DAG	Diacyglycerol
PKC	Protein kinase C
PLC	Phospholipase C
PI3K	Phosphoinositide 3-kinase
MAPK	Mitogen-activated protein kinase
MSC	Mesenchymal stem cell
PPARγ	Peroxisome proliferator-activated receptor gamma
CEBP	CCAAT enhancer binding protein
KLF4	Krüppel-like factor 4
Runx2	Runt-related transcription factor 2
TNFα	Tumor necrosis factor alpha
CPA	N6-Cyclopentyladenosine
PIA	Phenylisopropyladenosine
CCPA	2-Chloro-N6-Cyclopentyladenosine
DPCPX	8-Cyclopentyl-1,3-dipropylxanthine
CGS 21680	3-(4-(2-((6-amino-9-((2R,3R,4S,5S)-5-(ethylcarbomoyl)-3,4-dihydroxytetrahydrofuran-2-yl)-9H-purin-2-yl)amino)ethyl)phenyl)propanoic acid
SCH442416	2-(2-Furanyl)-7-[3-(4-methoxyphenyl)propyl]-7H-pyrazolo[4,3-e][1,2,4]triazolo[1,5-c]pyrimidin-5-amine
BAY 60-6583	2-[[6-Amino-3,5-dicyano-4-[4-(cyclopropylmethoxy)penyl]-2-pyridinyl]thio]-acetamide
MRS-1706	N-(4-Acetylphenyl)-2-[4-(2,3,6,7-tetrahydro-2,6-dioxo-1,3-dipropyl-1H-purin-8-yl)phenoxy]acetamide
MRS-1754	N-4(4-Cyanophenyl)-2-[4-(2,3,6,7-tetrahydro-2,6-dioxo-1,3-dipropyl-1H-purin-8-yl)phenoxy]acetamide
ATL-801	(N-[5-(1-cyclopropyl-2,6-dioxo-3-propul-2,3,6,7-tetrahydro-1H-purin-8-yl)-pyridin-2-yl]-N-ethyl-nicotinamide

IB-MECA	N^6-(3-Iodobenzyl)adenosine-5'-N-methyluronamide
C1-IB-MECA	2-Chloro-N^6-(3-Iodobenzyl)adenosine-5'-N-methyluronamide
MRS5698	(1S,2R,3S,4R,5S)-4-[6-[[(3-Chlorophenyl)methyl]amino]-2-[2-(3,4-difluorophenyl)ethynyl]-9H-purin-9-yl]-2,3-dihydroxy-N-methylbicyclo[3.1.0]hexane-1-carboxamide
MRS1523	3-Propyl-6-ethyl-5-[(ethylthio)carbonyl]-2 phenyl-4-propyl-3-pyridine carboxylate
MRS1220	N-[9-Chloro-2-(2-furanyl)[1,2,4]-triazol[1,5-c]quinazolin-5-yl]benzene acetamide
NECA	5'-N-Ethylcarboxamidoadenosine
SDZWAG994	N-Cyclohexyl-2'-O-methyladenosine
ARA	([1S,2R,3R,5R]-3-methoxymethyl-5-[6-(1-[5-trifluoromethyl-pyridin-2-yl]pyrrolidine-3-[S]-ylamino)-purin-9-yl]cyclopentane-1,2-diol)
RPR749	$C_{22}H_{26}F_3N_7O_3$
GS-9667	(2S,3S,4R,5R)-2-(((2-fluorophenyl)thio)methyl)-5-(6-((((1R,2R)-2-hydroxycyclopentyl)amino)-9H-purin-9-yl)tetrahydrofuran-3,4-diol

References

1. Fredholm, B.B. Release of Adenosine-like Material from Isolated Perfused Dog Adipose Tissue Following Sympathetic Nerve Stimulation and its Inhibition by Adrenergic α-Receptor Blockade. *Acta Physiol. Scand.* **1976**, *96*, 422–430. [CrossRef] [PubMed]
2. Schwabe, U.; Ebert, R.; Erbler, H.C. Adenosine release from isolated fat cells and its significance for the effects of hormones on cyclic 3',5'-AMP levels and lipolysis. *Naunyn Schmiedebergs Arch. Pharmacol.* **1973**, *276*, 133–148. [CrossRef] [PubMed]
3. Fredholm, B.B. Adenosine—a physiological or pathophysiological agent? *J. Mol. Med.* **2013**, *92*, 201–206. [CrossRef]
4. Borea, P.A.; Gessi, S.; Merighi, S.; Varani, K. Adenosine as a Multi-Signalling Guardian Angel in Human Diseases: When, Where and How Does it Exert its Protective Effects? *Trends Pharmacol. Sci.* **2016**, *37*, 419–434. [CrossRef] [PubMed]
5. Peleli, M.; Fredholm, B.B.; Sobrevia, L.; Carlstrom, M. Pharmacological targeting of adenosine receptor signaling. *Mol. Asp. Med.* **2017**, *55*, 4–8. [CrossRef]
6. Yang, D.; Zhang, Y.; Nguyen, H.G.; Koupenova, M.; Chauhan, A.K.; Makitalo, M.; Jones, M.R.; Hilaire, C.S.; Seldin, D.C.; Toselli, P.; et al. The A2B adenosine receptor protects against inflammation and excessive vascular adhesion. *J. Clin. Investig.* **2006**, *116*, 1913–1923. [CrossRef]
7. Dixon, A.K.; Gubitz, A.K.; Sirinathsinghji, D.J.; Richardson, P.J.; Freeman, T.C. Tissue distribution of adenosine receptor mRNAs in the rat. *Br. J. Pharmacol.* **1996**, *118*, 1461–1468. [CrossRef]
8. Fredholm, B.B.; Arslan, G.; Halldner, L.; Kull, B.; Schulte, G.; Wasserman, W. Structure and function of adenosine receptors and their genes. *Naunyn-Schmiedeberg's Arch. Pharmacol.* **2000**, *362*, 364–374. [CrossRef]
9. Poth, J.M.; Brodsky, K.; Ehrentraut, H.; Grenz, A.; Eltzschig, H.K. Transcriptional control of adenosine signaling by hypoxia-inducible transcription factors during ischemic or inflammatory disease. *J. Mol. Med.* **2012**, *91*, 183–193. [CrossRef]
10. Hilaire, C.S.; Carroll, S.H.; Chen, H.; Ravid, K. Mechanisms of induction of adenosine receptor genes and its functional significance. *J. Cell. Physiol.* **2009**, *218*, 35–44. [CrossRef]
11. Karmazyn, M.; A Cook, M. Adenosine A1 receptor activation attenuates cardiac injury produced by hydrogen peroxide. *Circ. Res.* **1992**, *71*, 1101–1110. [CrossRef] [PubMed]
12. Prabhakar, N.R.; Kumar, G.K. Oxidative stress in the systemic and cellular responses to intermittent hypoxia. *Biol. Chem.* **2004**, *385*, 217–221. [CrossRef] [PubMed]
13. Madi, L.; Cohen, S.; Ochayin, A.; Bar-Yehuda, S.; Barer, F.; Fishman, P. Overexpression of A3 adenosine receptor in peripheral blood mononuclear cells in rheumatoid arthritis: Involvement of nuclear factor-kappaB in mediating receptor level. *J. Rheumatol.* **2007**, *34*, 20–26.
14. Madi, L.; Ochaion, A.; Rath-Wolfson, L.; Bar-Yehuda, S.; Erlanger, A.; Ohana, G.; Harish, A.; Merimski, O.; Barer, F.; Fishman, P. The A3 Adenosine Receptor Is Highly Expressed in Tumor versus Normal Cells. *Clin. Cancer Res.* **2004**, *10*, 4472–4479. [CrossRef] [PubMed]

15. Ochaion, A.; Bar-Yehuda, S.; Cohn, S.; Del Valle, L.; Perez-Liz, G.; Madi, L.; Barer, F.; Farbstein, M.; Fishman-Furman, S.; Reitblat, T.; et al. Methotrexate enhances the anti-inflammatory effect of CF101 via up-regulation of the A3 adenosine receptor expression. *Arthritis Res. Ther.* **2006**, *8*, R169. [CrossRef] [PubMed]

16. Singh, A.; Misra, R.; Aggarwal, A. Baseline adenosine receptor mRNA expression in blood as predictor of response to methotrexate therapy in patients with rheumatoid arthritis. *Rheumatol. Int.* **2019**, *39*, 1431–1438. [CrossRef] [PubMed]

17. Morello, S.; Ito, K.; Yamamura, S.; Lee, K.-Y.; Jazrawi, E.; DeSouza, P.; Barnes, P.; Cicala, C.; Adcock, I.M. IL-1 beta and TNF-alpha regulation of the adenosine receptor (A2A) expression: Differential requirement for NF-kappa B binding to the proximal promoter. *J. Immunol.* **2006**, *177*, 7173–7183. [CrossRef]

18. Murphree, L.J.; Sullivan, G.W.; Marshall, M.A.; Linden, J. Lipopolysaccharide rapidly modifies adenosine receptor transcripts in murine and human macrophages: Role of NF-κB in A2Aadenosine receptor induction. *Biochem. J.* **2005**, *391*, 575–580. [CrossRef]

19. Ahmad, A.; Ahmad, S.; Glover, L.E.; Miller, S.M.; Shannon, J.M.; Guo, X.; Franklin, W.A.; Bridges, J.P.; Schaack, J.B.; Colgan, S.P.; et al. Adenosine A2A receptor is a unique angiogenic target of HIF-2 in pulmonary endothelial cells. *Proc. Natl. Acad. Sci. USA* **2009**, *106*, 10684–10689. [CrossRef]

20. Kobayashi, S.; Beitner-Johnson, D.; Conforti, L.; Millhorn, D.E. Chronic hypoxia reduces adenosine A2A receptor-mediated inhibition of calcium current in rat PC12 cells via downregulation of protein kinase A. *J. Physiol.* **1998**, *512*, 351–363. [CrossRef]

21. Di Bonaventura, M.V.E.; Pucci, M.; Giusepponi, M.E.; Romano, A.; Lambertucci, C.; Volpini, R.; Di Bonaventura, E.M.; Gaetani, S.; Maccarrone, M.; D'Addario, C.; et al. Regulation of adenosine A2A receptor gene expression in a model of binge eating in the amygdaloid complex of female rats. *J. Psychopharmacol.* **2019**, *33*, 1550–1561. [CrossRef]

22. Kong, T.; Westerman, K.A.; Faigle, M.; Eltzschig, H.K.; Colgan, S.P. HIF-dependent induction of adenosine A2B receptor in hypoxia. *FASEB J.* **2006**, *20*, 2242–2250. [CrossRef] [PubMed]

23. Lan, J.; Lu, H.; Samanta, D.; Salman, S.; Lu, Y.; Semenza, G.L. Hypoxia-inducible factor 1-dependent expression of adenosine receptor 2B promotes breast cancer stem cell enrichment. *Proc. Natl. Acad. Sci. USA* **2018**, *115*, E9640–E9648. [CrossRef] [PubMed]

24. Sitkovsky, M.V.; Ohta, A. The 'danger' sensors that STOP the immune response: The A2 adenosine receptors? *Trends Immunol.* **2005**, *26*, 299–304. [CrossRef]

25. Yang, M.; Ma, C.; Liu, S.; Shao, Q.; Gao, W.; Song, B.; Sun, J.; Xie, Q.; Zhang, Y.; Feng, A.; et al. HIF-dependent induction of adenosine receptor A2b skews human dendritic cells to a Th2-stimulating phenotype under hypoxia. *Immunol. Cell Biol.* **2009**, *88*, 165–171. [CrossRef] [PubMed]

26. Eckle, T.; Kewley, E.M.; Brodsky, K.S.; Tak, E.; Bonney, S.; Gobel, M.; Anderson, D.; Glover, L.E.; Riegel, A.K.; Colgan, S.P.; et al. Identification of Hypoxia-Inducible Factor HIF-1A as Transcriptional Regulator of the A2B Adenosine Receptor during Acute Lung Injury. *J. Immunol.* **2014**, *192*, 1249–1256. [CrossRef] [PubMed]

27. Kwon, J.H.; Lee, J.; Kim, J.; Jo, Y.H.; Kirchner, V.A.; Kim, N.; Kwak, B.J.; Hwang, S.; Song, G.; Lee, S.; et al. HIF-1α regulates A2B adenosine receptor expression in liver cancer cells. *Exp. Ther. Med.* **2019**, *18*, 4231–4240. [CrossRef] [PubMed]

28. Carroll, S.H.; Wigner, N.A.; Kulkarni, N.; Johnston-Cox, H.; Gerstenfeld, L.C.; Ravid, K. A2B adenosine receptor promotes mesenchymal stem cell differentiation to osteoblasts and bone formation in vivo. *J. Biol. Chem.* **2012**, *287*, 15718–15727. [CrossRef] [PubMed]

29. Børglum, J.; Vassaux, G.; Richelsen, B.; Gaillard, D.; Darimont, C.; Ailhaud, G.; Négrel, R.; Richelsen, D. Changes in adenosine A1- and A2-receptor expression during adipose cell differentiation. *Mol. Cell. Endocrinol.* **1996**, *117*, 17–25. [CrossRef]

30. Gharibi, B.; Abraham, A.A.; Ham, J.; Evans, B.A.J. Adenosine receptor subtype expression and activation influence the differentiation of mesenchymal stem cells to osteoblasts and adipocytes. *J. Bone Miner. Res.* **2011**, *26*, 2112–2124. [CrossRef]

31. Gnad, T.; Scheibler, S.; Von Kügelgen, I.; Scheele, C.; Kilić, A.; Glöde, A.; Hoffmann, L.S.; Reverte-Salisa, L.; Horn, P.; Mutlu, S.; et al. Adenosine activates brown adipose tissue and recruits beige adipocytes via A2A receptors. *Nat. Cell Biol.* **2014**, *516*, 395–399. [CrossRef] [PubMed]

32. Johansson, S.M.; Salehi, A.; Sandström, M.E.; Westerblad, H.; Lundquist, I.; Carlsson, P.-O.; Fredholm, B.B.; Katz, A. A1 receptor deficiency causes increased insulin and glucagon secretion in mice. *Biochem. Pharmacol.* **2007**, *74*, 1628–1635. [CrossRef] [PubMed]

33. Schoelch, C.; Kuhlmann, J.; Gossel, M.; Mueller, G.; Neumann-Haefelin, C.; Belz, U.; Kalisch, J.; Biemer-Daub, G.; Kramer, W.; Juretschke, H.-P.; et al. Characterization of adenosine-A1 receptor-mediated antilipolysis in rats by tissue microdialysis, 1H-spectroscopy, and glucose clamp studies. *Diabetes* **2004**, *53*, 1920–1926. [CrossRef] [PubMed]

34. Eisenstein, A.; Carroll, S.H.; Johnston-Cox, H.; Farb, M.; Gokce, N.; Ravid, K. An Adenosine Receptor-Krüppel-like Factor 4 Protein Axis Inhibits Adipogenesis. *J. Biol. Chem.* **2014**, *289*, 21071–21081. [CrossRef] [PubMed]

35. Ravid, K.; Lowenstein, J.M. Changes in adenosine receptors during differentiation of 3T3-F442A cells to adipocytes. *Biochem. J.* **1988**, *249*, 377–381. [CrossRef] [PubMed]

36. Vassaux, G.; Gaillard, D.; Mari, B.; Ailhaud, G.; Négrel, R. Differential Expression of Adenosine A1 and Adenosine A2 Receptors in Preadipocytes and Adipocytes. *Biochem. Biophys. Res. Commun.* **1993**, *193*, 1123–1130. [CrossRef]

37. Johnston-Cox, H.; Koupenova, M.; Yang, D.; Corkey, B.; Gokce, N.; Farb, M.G.; Lebrasseur, N.K.; Ravid, K. The A2b Adenosine Receptor Modulates Glucose Homeostasis and Obesity. *PLoS ONE* **2012**, *7*, e40584. [CrossRef]

38. Koupenova, M.; Johnston-Cox, H.; Vezeridis, A.; Gavras, H.; Yang, D.; Zannis, V.I.; Ravid, K. A2b adenosine receptor regulates hyperlipidemia and atherosclerosis. *Circulation* **2011**, *125*, 354–363. [CrossRef]

39. Kaartinen, J.M.; Hreniuk, S.P.; Martin, L.F.; Ranta, S.; LaNoue, K.F.; Ohisalo, J.J. Attenuated adenosine-sensitivity and decreased adenosine-receptor number in adipocyte plasma membranes in human obesity. *Biochem. J.* **1991**, *279*, 17–22. [CrossRef]

40. Hajjawi, M.O.R.; Patel, J.J.; Corcelli, M.; Arnett, T.; Orriss, I.R. Lack of effect of adenosine on the function of rodent osteoblasts and osteoclasts in vitro. *Purinergic Signal.* **2016**, *12*, 247–258. [CrossRef]

41. Barroso, I.; Gurnell, M.; Crowley, V.E.F.; Agostini, M.; Schwabe, J.W.; Soos, M.A.; Maslen, G.L.; Williams, T.D.M.; Lewis, H.; Schafer, A.J. Dominant negative mutations in human PPARgamma associated with severe insulin resistance, diabetes mellitus and hypertension. *Nature* **1999**, *402*, 880–883. [CrossRef] [PubMed]

42. Wang, F.; Mullican, S.E.; DiSpirito, J.R.; Peed, L.C.; Lazar, M.A. Lipoatrophy and severe metabolic disturbance in mice with fat-specific deletion of PPARgamma. *Proc. Natl. Acad. Sci. USA* **2013**, *110*, 18656–18661. [CrossRef] [PubMed]

43. Tontonoz, P.; Hu, E.; Spiegelman, B.M. Stimulation of adipogenesis in fibroblasts by PPAR gamma 2, a lipid-activated transcription factor. *Cell* **1994**, *79*, 1147–1156. [CrossRef]

44. Cao, Z.; Umek, R.M.; McKnight, S.L. Regulated expression of three C/EBP isoforms during adipose conversion of 3T3-L1 cells. *Genes Dev.* **1991**, *5*, 1538–1552. [CrossRef] [PubMed]

45. Steger, D.J.; Grant, G.R.; Schupp, M.; Tomaru, T.; Lefterova, M.I.; Schug, J.; Manduchi, E.; Stoeckert, C.J.; Lazar, M.A. Propagation of adipogenic signals through an epigenomic transition state. *Genes Dev.* **2010**, *24*, 1035–1044. [CrossRef] [PubMed]

46. Tanaka, T.; Yoshida, N.; Kishimoto, T.; Akira, S. Defective adipocyte differentiation in mice lacking the C/EBPbeta and/or C/EBPdelta gene. *EMBO J.* **1997**, *16*, 7432–7443. [CrossRef]

47. Wu, Z.; Bucher, N.L.; Farmer, S.R. Induction of peroxisome proliferator-activated receptor gamma during the conversion of 3T3 fibroblasts into adipocytes is mediated by C/EBPbeta, C/EBPdelta, and glucocorticoids. *Mol. Cell. Biol.* **1996**, *16*, 4128–4136. [CrossRef]

48. Gharibi, B.; Abraham, A.A.; Ham, J.; Evans, B.A.J. Contrasting effects of A1 and A2b adenosine receptors on adipogenesis. *Int. J. Obes.* **2011**, *36*, 397–406. [CrossRef]

49. Hinz, S.; Lacher, S.K.; Seibt, B.F.; Müller, C.E. BAY60-6583 Acts as a Partial Agonist at Adenosine A2B Receptors. *J. Pharmacol. Exp. Ther.* **2014**, *349*, 427–436. [CrossRef]

50. Katebi, M.; Soleimani, M.; Cronstein, B.N. Adenosine A2Areceptors play an active role in mouse bone marrow-derived mesenchymal stem cell development. *J. Leukoc. Biol.* **2008**, *85*, 438–444. [CrossRef]

51. Corciulo, C.; Wilder, T.; Cronstein, B.N. Adenosine A2B receptors play an important role in bone homeostasis. *Purinergic Signal.* **2016**, *12*, 537–547. [CrossRef]

52. Shih, Y.-R.V.; Liu, M.; Kwon, S.K.; Iida, M.; Gong, Y.; Sangaj, N.; Varghese, S. Dysregulation of ectonucleotidase-mediated extracellular adenosine during postmenopausal bone loss. *Sci. Adv.* **2019**, *5*, eaax1387. [CrossRef] [PubMed]
53. Kara, F.M.; Chitu, V.; Sloane, J.; Axelrod, M.; Fredholm, B.B.; Stanley, E.R.; Cronstein, B.N. Adenosine A 1 receptors (A 1 Rs) play a critical role in osteoclast formation and function. *FASEB J.* **2010**, *24*, 2325–2333. [CrossRef] [PubMed]
54. He, W.; Cronstein, B.N. Adenosine A1 receptor regulates osteoclast formation by altering TRAF6/TAK1 signaling. *Purinergic Signal* **2012**, *8*, 327–337. [CrossRef] [PubMed]
55. Kara, F.M.; Doty, S.B.; Boskey, A.L.; Goldring, S.; Zaidi, M.; Fredholm, B.B.; Cronstein, B.N. Adenosine A1receptors regulate bone resorption in mice: Adenosine A1receptor blockade or deletion increases bone density and prevents ovariectomy-induced bone loss in adenosine A1receptor-knockout mice. *Arthritis Rheum.* **2010**, *62*, 534–541. [CrossRef]
56. Mediero, A.; Kara, F.M.; Wilder, T.; Cronstein, B.N. Adenosine A2A Receptor Ligation Inhibits Osteoclast Formation. *Am. J. Pathol.* **2012**, *180*, 775–786. [CrossRef] [PubMed]
57. He, W.; Mazumder, A.; Wilder, T.; Cronstein, B.N. Adenosine regulates bone metabolism via A 1, A 2A, and A 2B receptors in bone marrow cells from normal humans and patients with multiple myeloma. *FASEB J.* **2013**, *27*, 3446–3454. [CrossRef]
58. Takedachi, M.; Oohara, H.; Smith, B.J.; Iyama, M.; Kobashi, M.; Maeda, K.; Long, C.L.; Humphrey, M.B.; Stoecker, B.J.; Toyosawa, S.; et al. CD73-generated adenosine promotes osteoblast differentiation. *J. Cell. Physiol.* **2012**, *227*, 2622–2631. [CrossRef]
59. Bradaschia-Correa, V.; Josephson, A.M.; Egol, A.J.; Mizrahi, M.M.; Leclerc, K.; Huo, J.; Cronstein, B.N.; Leucht, P. Ecto-5′-nucleotidase (CD73) regulates bone formation and remodeling during intramembranous bone repair in aging mice. *Tissue Cell* **2017**, *49*, 545–551. [CrossRef]
60. Wu, W.; Xiao, Z.; Chen, Y.; Deng, Y.; Zeng, D.; Liu, Y.; Huang, F.; Wang, J.; Liu, Y.; Bellanti, J.A. CD39 Produced from Human GMSCs Regulates the Balance of Osteoclasts and Osteoblasts through the Wnt/beta-Catenin Pathway in Osteoporosis. *Mol. Ther.* **2020**, *28*, 1518–1532. [CrossRef]
61. Gardinier, J.D.; Rostami, N.; Juliano, L.; Zhang, C. Bone adaptation in response to treadmill exercise in young and adult mice. *Bone Rep.* **2018**, *8*, 29–37. [CrossRef] [PubMed]
62. Wallace, J.; Rajachar, R.M.; Allen, M.R.; Bloomfield, S.A.; Robey, P.G.; Young, M.F.; Kohn, D.H. Exercise-induced changes in the cortical bone of growing mice are bone- and gender-specific. *Bone* **2007**, *40*, 1120–1127. [CrossRef] [PubMed]
63. Zarębska, E.; Kusy, K.; Słomińska, E.M.; Kruszyna, Ł.; Zieliński, J. Alterations in Exercise-Induced Plasma Adenosine Triphosphate Concentration in Highly Trained Athletes in a One-Year Training Cycle. *Metabolites* **2019**, *9*, 230. [CrossRef] [PubMed]
64. Honnor, R.C.; Dhillon, G.S.; Londos, C. cAMP-dependent protein kinase and lipolysis in rat adipocytes. II. Definition of steady-state relationship with lipolytic and antilipolytic modulators. *J. Biol. Chem.* **1985**, *260*, 15130–15138.
65. Fain, J.N.; Wieser, P.B. Effects of adenosine deaminase on cyclic adenosine monophosphate accumulation, lipolysis, and glucose metabolism of fat cells. *J. Biol. Chem.* **1975**, *250*, 1027–1034.
66. Szkudelski, T.; Szkudelska, K.; Nogowski, L. Effects of adenosine A1 receptor antagonism on lipogenesis and lipolysis in isolated rat adipocytes. *Physiol. Res.* **2008**, *58*, 863–871.
67. Vannucci, S.J.; Klim, C.M.; Martin, L.F.; LaNoue, K.F. A1-adenosine receptor-mediated inhibition of adipocyte adenylate cyclase and lipolysis in Zucker rats. *Am. J. Physiol. Metab.* **1989**, *257*, E871–E878. [CrossRef]
68. Hoffman, B.B.; Dall'Aglio, E.; Hollenbeck, C.; Chang, H.; Reaven, G.M. Suppression of free fatty acids and triglycerides in normal and hypertriglyceridemic rats by the adenosine receptor agonist phenylisopropyladenosine. *J. Pharmacol. Exp. Ther.* **1986**, *239*, 715–718.
69. Johansson, S.M.; Lindgren, E.; Yang, J.-N.; Herling, A.W.; Fredholm, B.B. Adenosine A1 receptors regulate lipolysis and lipogenesis in mouse adipose tissue—Interactions with insulin. *Eur. J. Pharmacol.* **2008**, *597*, 92–101. [CrossRef]
70. Yang, T.; Gao, X.; Sandberg, M.; Zollbrecht, C.; Zhang, X.-M.; Hezel, M.; Liu, M.; Peleli, M.; Lai, E.-Y.; A Harris, R.; et al. Abrogation of adenosine A1 receptor signalling improves metabolic regulation in mice by modulating oxidative stress and inflammatory responses. *Diabetologia* **2015**, *58*, 1610–1620. [CrossRef]

71. Faulhaber-Walter, R.; Jou, W.; Mizel, D.; Li, L.; Zhang, J.; Kim, S.M.; Huang, Y.; Chen, M.; Briggs, J.P.; Gavrilova, O.; et al. Impaired Glucose Tolerance in the Absence of Adenosine A1 Receptor Signaling. *Diabetes* **2011**, *60*, 2578–2587. [CrossRef] [PubMed]

72. Dong, Q.; Ginsberg, H.N.; Erlanger, B.F. Overexpression of the A1 adenosine receptor in adipose tissue protects mice from obesity-related insulin resistance. *Diabetes Obes. Metab.* **2001**, *3*, 360–366. [CrossRef] [PubMed]

73. Rice, A.M.; Fain, J.N.; Rivkees, S.A. A1 adenosine receptor activation increases adipocyte leptin secretion. *Endocrinology* **2000**, *141*, 1442–1445. [CrossRef]

74. Neméth, Z.H.; Bleich, D.; Csóka, B.; Pacher, P.; Mabley, J.G.; Himer, L.; Vizi, E.S.; Deitch, E.A.; Szabó, C.; Cronstein, B.N.; et al. Adenosine receptor activation ameliorates type 1 diabetes. *FASEB J.* **2007**, *21*, 2379–2388. [CrossRef] [PubMed]

75. Figler, R.A.; Wang, G.; Srinivasan, S.; Jung, D.Y.; Zhang, Z.; Pankow, J.; Ravid, K.; Fredholm, B.; Hedrick, C.C.; Rich, S.S.; et al. Links Between Insulin Resistance, Adenosine A2B Receptors, and Inflammatory Markers in Mice and Humans. *Diabetes* **2011**, *60*, 669–679. [CrossRef]

76. Johnston-Cox, H.; Eisenstein, A.S.; Koupenova, M.; Carroll, S.; Ravid, K. The Macrophage A2b Adenosine Receptor Regulates Tissue Insulin Sensitivity. *PLoS ONE* **2014**, *9*, e98775. [CrossRef]

77. Csóka, B.; Koscsó, B.; Törő, G.; Kókai, E.; Virág, L.; Németh, Z.H.; Pacher, P.; Bai, P.; Haskó, G. A2B Adenosine Receptors Prevent Insulin Resistance by Inhibiting Adipose Tissue Inflammation via Maintaining Alternative Macrophage Activation. *Diabetes* **2014**, *63*, 850–866. [CrossRef]

78. Merighi, S.; Borea, P.A.; Gessi, S. Adenosine receptors and diabetes: Focus on the A2B adenosine receptor subtype. *Pharmacol. Res.* **2015**, *99*, 229–236. [CrossRef]

79. Pei, Y.; Li, H.; Cai, Y.; Zhou, J.; Luo, X.; Ma, L.; McDaniel, K.; Zeng, T.; Chen, Y.; Qian, X.; et al. Regulation of adipose tissue inflammation by adenosine 2A receptor in obese mice. *J. Endocrinol.* **2018**, *239*, 365–376. [CrossRef]

80. Cai, Y.; Li, H.; Liu, M.; Pei, Y.; Zheng, J.; Zhou, J.; Luo, X.; Huang, W.; Ma, L.; Yang, Q.; et al. Disruption of adenosine 2A receptor exacerbates NAFLD through increasing inflammatory responses and SREBP1c activity. *Hepatology* **2018**, *68*, 48–61. [CrossRef]

81. Csóka, B.; Törő, G.; Vindeirinho, J.; Varga, Z.V.; Koscsó, B.; Németh, Z.H.; Kókai, E.; Antonioli, L.; Suleiman, M.; Marchetti, P.; et al. A 2A adenosine receptors control pancreatic dysfunction in high-fat-diet-induced obesity. *FASEB J.* **2017**, *31*, 4985–4997. [CrossRef] [PubMed]

82. DeOliveira, C.C.; Caria, C.; Gotardo, E.M.F.; Ribeiro, M.L.; Gambero, A. Role of A 1 and A 2A adenosine receptor agonists in adipose tissue inflammation induced by obesity in mice. *Eur. J. Pharmacol.* **2017**, *799*, 154–159. [CrossRef] [PubMed]

83. Ruan, C.; Gao, P. GW29-e1353 A2A Receptor Activation Attenuates Hypertensive Cardiac Remodeling via Promoting Brown Adipose Tissue-Derived FGF21. *J. Am. Coll. Cardiol.* **2018**, *72*, C41. [CrossRef]

84. Lahesmaa, M.; Oikonen, V.; Helin, S.; Luoto, P.; Din, M.U.; Pfeifer, A.; Nuutila, P.; Virtanen, K.A. Regulation of human brown adipose tissue by adenosine and A2A receptors—Studies with [15O]H$_2$O and [11C]TMSX PET/CT. *Eur. J. Nucl. Med. Mol. Imaging* **2018**, *46*, 743–750. [CrossRef]

85. Xiao, C.; Liu, N.; Jacobson, K.A.; Gavrilova, O.; Reitman, M.L. Physiology and effects of nucleosides in mice lacking all four adenosine receptors. *PLoS Biol.* **2019**, *17*, e3000161. [CrossRef]

86. Gnad, T.; Navarro, G.; Lahesmaa, M.; Reverte-Salisa, L.; Copperi, F.; Cordomi, A.; Naumann, J.; Hochhäuser, A.; Haufs-Brusberg, S.; Wenzel, D.; et al. Adenosine/A2B Receptor Signaling Ameliorates the Effects of Aging and Counteracts Obesity. *Cell Metab.* **2020**, *32*, 56–70.e7. [CrossRef]

87. Jacobson, K.A.; Tosh, D.K.; Jain, S.; Gao, Z.-G. Historical and Current Adenosine Receptor Agonists in Preclinical and Clinical Development. *Front. Cell. Neurosci.* **2019**, *13*, 124. [CrossRef]

88. Ishikawa, J.; Mitani, H.; Bandoh, T.; Kimura, M.; Totsuka, T.; Hayashi, S. Hypoglycemic and hypotensive effects of 6-cyclohexyl-2′-O-methyl-adenosine, an adenosine A1 receptor agonist, in spontaneously hypertensive rat complicated with hyperglycemia. *Diabetes Res. Clin. Pract.* **1998**, *39*, 3–9. [CrossRef]

89. Zannikos, P.N.; Rohatagi, S.; Jensen, B.K. Pharmacokinetic-pharmacodynamic modeling of the antilipolytic effects of an adenosine receptor agonist in healthy volunteers. *J. Clin. Pharmacol.* **2001**, *41*, 61–69. [CrossRef]

90. Shah, B.; Rohatagi, S.; Natarajan, C.; Kirkesseli, S.; Baybutt, R.; Jensen, B.K. Pharmacokinetics, Pharmacodynamics, and Safety of a Lipid-Lowering Adenosine A1 Agonist, RPR749, in Healthy Subjects. *Am. J. Ther.* **2004**, *11*, 175–189. [CrossRef]

91. Dhalla, A.K.; Santikul, M.; Smith, M.; Wong, M.-Y.; Shryock, J.C.; Belardinelli, L. Antilipolytic Activity of a Novel Partial A1 Adenosine Receptor Agonist Devoid of Cardiovascular Effects: Comparison with Nicotinic Acid. *J. Pharmacol. Exp. Ther.* **2007**, *321*, 327–333. [CrossRef] [PubMed]
92. Dhalla, A.K.; Wong, M.Y.; Voshol, P.J.; Belardinelli, L.; Reaven, G.M. A1 adenosine receptor partial agonist lowers plasma FFA and improves insulin resistance induced by high-fat diet in rodents. *Am. J. Physiol. Metab.* **2007**, *292*, E1358–E1363. [CrossRef] [PubMed]
93. Staehr, P.M.; Dhalla, A.K.; Zack, J.; Wang, X.; Ho, Y.L.; Bingham, J.; Belardinelli, L. Reduction of Free Fatty Acids, Safety, and Pharmacokinetics of Oral GS-9667, an A1Adenosine Receptor Partial Agonist. *J. Clin. Pharmacol.* **2013**, *53*, 385–392. [CrossRef] [PubMed]
94. Chen, J.-F.; Cunha, R.A. The belated US FDA approval of the adenosine A2A receptor antagonist istradefylline for treatment of Parkinson's disease. *Purinergic Signal.* **2020**, *16*, 1–8. [CrossRef] [PubMed]
95. Charvin, D.; Medori, R.; Hauser, R.A.; Rascol, O. Therapeutic strategies for Parkinson disease: Beyond dopaminergic drugs. *Nat. Rev. Drug Discov.* **2018**, *17*, 804–822. [CrossRef]

MDPI

St. Alban-Anlage 66

4052 Basel

Switzerland

Tel. +41 61 683 77 34

Fax +41 61 302 89 18

www.mdpi.com

International Journal of Molecular Sciences Editorial Office

E-mail: ijms@mdpi.com

www.mdpi.com/journal/ijms